Biotechnological Strategies in Agroprocessing

Biotechnological Strategies in Agroprocessing

Edited by:

S.S. Marwaha
J.K. Arora

API

Asiatech Publishers Inc.
New Delhi

First Edition, October 2003

ASIATECH PUBLISHERS INC.
New Delhi

ISBN: 81-87680-09-1

Published by N.K. Muraleedharan for Asiatech Publishers Inc.
E-20, Arya Nagar, CGHS, Plot No. 91, I.P. Extension, Patparganj,
Delhi 110 092. *Phone*: +91-11-22724601 • *Fax*: +91-11-22727254
e-mail: asiatech@bol.net.in

Typeset by Jayanti Graphics Pvt. Ltd., H-296, Uday Niketan,
Narain Vihar, New Delhi 110 028. *Phone*: +91-11-25791026
e-mail: jgpl@bol.net.in

Printed at Chaman Offset, Delhi.

Foreword

THE WTO REGIME has lead to the need to ensure quality and safety standards of agri produce/products for export as well as domestic markets. This warrants speedy adoption of biotechnological advancements that offer an edge over conventional cultivation as well as post harvest processing practices. During the last decade, the body of knowledge on this subject has grown substantially, hence, this volume has emerged from the sharing of experience of biotech experts from commercial and R&D establishments engaged in agriculture/agroprocessing, in a Conference on Biotechnological Strategies for Agroprocessing organized by PSCST.

The book is divided into two sections, the first section comprising chapters providing an overview of biotechnological developments for crop improvement through micropropagation; genetic transformation; and application of biofertilizers, biopesticides etc. for effective change over from chemical to organic farming as well as production of quality raw materials for processing. It also covers biotechnological strategies for post harvest processing of cereals, fruits, vegetables as well as dairy, meat products etc. This section also provides information on safety & quality aspects, initiatives being taken by government, financial institutions and technology providers as well as the importance of information technology back up for promotion of agri-biotech sector. The second section is a compilation of research findings on these aspects. This volume is expected to assist all those who grapple with these areas and I hope that it will be well received.

April, 2003

N.S. Tiwana
Executive Director
Punjab State Council for
Science & Technology (PSCST)

Preface

A STRONG AGRO BASED ECONOMY demands greater efficiency in production systems, the avoidance of losses through improvement in agro processing, enhanced nutritional quality and safety of processed products. In this context, there is much to be gained through biotechnological strategies in the coming years. The potential for this new coherence provided us the impetus to bring together eminent professionals from industries, institutions, government & non-government bodies to take stock of the technological advancements in this area during a Conference on Biotechnological Strategies for Agroprocessing organized by Punjab State Council for Science & Technology (PSCST) in the year 2000. Subsequently, it was pleasing to be able to persuade some of these experts to share their vision & experiences in the form of chapters in their respective areas of expertise and create a valuable collection of related, dependent and interwoven text that is expected to cater to the needs of the current and future workers in this upcoming area to help raise the competitive edge of agro-biotech sector.

Our effort was to present comprehensive account of the technical information on the various facets of biotechnology relevant to agroprocessing viz. crop improvement, crop production & protection aspects, agro industrial products based on cereals, fruits, vegetables, dairy, meat, etc. as well as financial & information technology back up for entrepreneurship development in Section-I of this volume in sufficient detail to enable the non-specialist reader also to understand. Hence, clearly we are responsible for any distortion or misrepresentations from these simplifications, although we hope that none has occurred. The second section of this compilation brings out the research findings in these areas in order to provide baseline data and research leads to the bench workers. The References/Further Reading sections also contain other sources that we used in general way, which might, if consulted, bring the reader closer to a particular subject. While all possible efforts were made to reach uniformity in presentation, the results and expressed ideas, however, remain the sole responsibility of respective authors. The delay in bringing out this volume is regretted, though, care has been taken to update the information covered, for which, the support of contributors is gratefully acknowledged.

We express our gratitude to Ministry of Food Processing Industries, Government of India for providing financial support to bring out this volume. The work could not have been completed without constant encouragement from Sh. D.S. Kalha, Secretary to Govt. of Punjab, Department of Science, Technology & Environment and Sh. N.S. Tiwana, Executive Director, PSCST.

April, 2003

S.S. Marwaha
J.K. Arora

Contents

Section I

Concept Papers

Biotechnological Strategies in Agro-Processing—An Overview

J.K. ARORA *and* S.S. MARWAHA

Punjab State Council for Science & Technology,
MGSIPA Complex, Sector 26,
Chandigarh 160 019.

THE LARGE AMOUNT OF FOOD grown in India including 84 million tonnes of rice, 72 million tonnes of wheat and 132 million tonnes of fruits and vegetables needs to be effectively utilized. In order to achieve this, there is need of a networking system for minimizing the spoilage as well as for value addition. Biotechnology has a major role to play in this chain of operations. The biotechnological interventions are required in augmenting yields of neutraceutically improved crops and livestock, processing the harvests, value addition, quality assurance, enhancement of shelf life as well as management of wastes of agro/food processing industry.

1. RAW MATERIAL UPGRADATION

1.1 Agro Resources

In spite of increasing availability of food, as per an estimate, nearly half of the world's micro-nutrient deficient people may be found in India with 5-7% (6.6-9.2 million) pre-school children suffering from xerophthalmia which causes atleast 60,000 children to go blind each year, the main cause being insufficient dietary intake of Vitamin A. Similarly, there is a worldwide concern about a major micro-nutrient deficiency of iron with 24% of global population suffering from iron deficiency anemia. The available data on India indicates that more than 70% of pregnant woman, approximately 50% of all women and 65-70% of adolescent girls may suffer from iron deficiency anemia. The deficiency of both the said micro-nutrients is especially severe in regions where rice is the predominant diet as its endosperm lacks provitamin A and contains very low amounts of iron. In view of this, in countries like India, superior nutritional quality of the crop produce and products assumes considerable significance. It is in this context that while 'first generation' transgenics were mainly devoted to traits like herbicide and insect resistance, the 'second generation' transgenics are focused on endowing the crops with traits such as desired protein & oil quantity and quality, improved vitamin and starch content etc. While the conventional breeding offered limited success in this context because of the tedious and time consuming selection procedure, biotechnological techniques have already provided various examples of foreign gene expression in plants. The transgenics of rice, brassica, moongbean, pigeonpea, cotton, potato, tomato and some vegetables like cabbage, cauliflower, etc, are likely to complete field assessment and some of them are expected to be ready for large scale seed production in India by 2005. Nutritionally enhanced potato, transgenic wheat with improved protein content and quality as well as with higher lysine content are the other important crops expected to be introduced in the country in near future. The edible vaccines, particularly for cholera, rabies and hepatitis B with an expression gene in tomato, cabbage and banana is another area under progress. The new avenues have also been opened up for reducing post harvest losses as well as improving the quality of horticultural crops, the delayed ripening of tomato being a salient example.

The consequences of indiscriminate use of chemical fertilizers and pesticides being realized world over in terms of emergence of pesticide resistant insect-pest races, more crops failures, groundwater pollution, narrowing down of genetic variability etc., a world wide cry is being heard from all quarters of people for organic farming. It is gaining popularity in both developed and developing countries with public ready to pay more for organic foods. In this context, process packages for vermiculture, production & use of biofertilizers & biopesticides have already been developed and are being increasingly adopted by the farming community.

Biotechnological applications such as micro-propagation are also being widely adopted especially in horticultural crops. It has resulted in increased yield and vigour in floriculture and has a big role to play in the rapid introduction of new varieties of both ornamental and vegetable crops in the market. It also holds a key position in conserving the germplasm and raising hi-tech nurseries. It can have tremendous applications in the upcoming sector of medicinal and aromatic plants. The Complementary and Alternative Medicine System with herbal drugs as its major component is emerging as a competitor of pharmaceuticals world over. India, though being the home of traditional herbal ayurvedic formulations, is having only around 0.1% share of global herbal market, which is largely controlled by China. Hence, biotechnological techniques can be of immense use in refining or standardizing herbal drug formulations e.g. clonal multiplication of uniform plants under controlled conditions could minimize manufacturer to manufacturer and batch to batch variation of the efficacy of herbal formulations. The production of high value compounds through *in vitro* culture could bring down the production costs significantly. The molecular markers could be of great value for proper identification of genuine herbal plant material & the adulterants and germplasm conservation using tissue culture could contribute immensely towards conservation of endangered herbal plant species.

1.2 Livestock Improvement

Improvements in efficiency of reproduction in farm livestock involving techniques such as super ovulation, embryo transfer and cloning are already having an impact in the livestock breeding industry. The recombinant DNA (r-DNA) derived bovine growth hormone (somatotropin) to increase milk production from dairy cow is another salient example of biotechnological application in livestock sector.

Further, as rumen microbial population plays a major role in the digestion of feeds, considerable work is being carried out for modifying rumen microflora with an objective of increasing production efficiency of cattle.

The livestock feeds largely comprising plant and vegetable material, the addition of enzymes to the feeds is being carried out to improve the feed conversion efficiency, thus reducing waste as well as digestive disorders in cattle. Also in case of young calves and pigs, enzyme addition facilitates continued growth during the weaning period when the young animal's natural enzyme levels are low. Over the last decade, the use of enzymes in broiler feeds has become universal and is thought to have added considerable profitability through improved efficiency of feed use. Also it has been reported that the use of enzymes reduces nitrogen excretion approximately by 20% besides reducing the amount of total waste produced. A number of neutraceuticals and microbial protein sources are being used increasingly by the feed industry. The *Methylophilus methylotrophus* (methanol used as substrate) based 'Pruteen' is a popular protein source being used by the feed industry in diets for broilers, piglets, pre-ruminant calves and fish.

The traditional silage making depended on the presence of microorganisms that could utilize the natural sugars of the plants and while doing so, they scavenge from the confined pile and, as this becomes anaerobic, lactic acid is produced that preserves the plant material for consumption. As an improvement of this process, sulphuric and formic acids were also used to preserve the plant material rather than relying on natural activity. Biotechnology has offered a substitute of these corrosive acids as bacterial inoculants can be added to achieve reliable and rapid acidification resulting in production of quality silage.

The stored grains used for feeding purposes could have mould infestations leading to presence of poisonous mycotoxins. Some of these could cause animal sterility while others like aflatoxins being carcinogenic, could find entry in human system through milk. The antibody based rapid analytical assays are available to test the presence of these chemicals. Such assays if used in routine could significantly reduce wastage of the grains as well as could have long term positive impact on livestock and human health.

2. PROCESSING AND VALUE ADDITION

The nearest term applications of biotechnology in agro/food processing for value addition, improving productivity, reducing time and processing costs include the production of speciality chemicals such as amino acids, vitamins, enzymes as well as polysaccharide biopolymers, sweetners, flavours, fragrances, biocolours, microbial proteins etc. (Table-1).

The enzymes in microbes have been traditionally used in the production of cheese, bread, beer and for food preservation. With the growth of microbial biotechnology, an increasing range of enzymes are commercially available now with a wide range of applications and are being used to add value or reduce costs in almost all sectors of agro/food industry (Table-2). They can serve as natural alternative to chemical treatments, increase operating efficiency of plant and have added advantages because of their high substrate and stereo specificity.

Table 1: Commercially Relevant Microbial Products

Product	*Example*	*Application*
Amino acids	Cysteine, lysine, glutamic acid, methionine, proline, threonine, phenylalanine, tryptophan	Food enrichments, flavorants
Polysaccharide biopolymers	Alginate gellan, curdlan, levan dextran, xanthan	Food additive for stabilizing liquid suspensions & gelling soft foods
Microbial cells	Yeast, lactobacilli, streptococci	Brewing, baking, wine and cheese making
Microbial proteins	*Methylophilus methylotrophus*, yeast, algae, fungi	Single cell proteins
Vitamins	B2, B12, D,C, nicotinic acid	Food
Peptides, proteins	Aspartame, thaumatin	Sweetener
Carotenoid	β-Carotene	Colorant
Organic acids, etc.	Geraniol, isobutylene, linalool, nerol, nucleotides, acetic acid, benzoic acid, citric acid, fumaric acid, lactic acid	Flavourants, preservation

Table 2: Enzyme Applications in Agro/Food Industry

Product	*Enzyme*	*Application*
Beer	Glycamylases	Hydrolysis of starch adjuncts
Cider	Pectinases improving filtration	Control of viscosity in juice concentrations;
Whey	Lactase	Hydrolysis of lactose to increase sweetness and stop crystallization in whey
Beer & cheese	Proteases	Flavour production
Cheese	Rennin	Production
Baked foods	Amylases	Hydrolysation of starch
Food, fruit juices & eggs	Glucose oxidases	Preservation of flavour and colour
Ethanol production	Cellulases	Conversion of cellulosics to glucose
Fruit juice	Pectinases	Increased extraction, improved filtration; clarification
Confectionery	Invertase	Inhibition of crystallization of sucrose
Starch syrups	Amylases Glucose Isomerase	Hydrolysis of starch to glucose, Conversion of glucose to fructose
Cereals	Hemicellulases Phytases	Increasing feed value of by-products
Meat	Papaine	Tenderisation
Fats and oils	Esterases	Modification of fatty acids of oils
Cheese	Lipases	Faster maturation; more intense flavour production

The animal/plant gene coding for a specific enzyme can also be inserted into micro-organisms commonly used for fermented food products, thereby producing the enzyme by fermentation in a more cost effective manner and at greater purity than would be possible from the original plant and animal. Further, with the help of molecular modelling, popularly known as protein engineering, the enzymes can be redesigned by making changes in their individual amino acids that alter their properties. This has led to the development of enzymes which can be switched on during processing and off when the required level of conversion has been obtained. The enzyme glucose isomerase that is used to convert sucrose to fructose and glucose (invert sugars), loses its activity over a few weeks and is required to be replaced at frequent intervals. By reconstructing the gene that codes for the enzyme (changing one amino acid i.e. lysine to arginine) and introducing it into a bacterium, the stability of the new enzyme has been increased five times thereby cutting down the costs, increasing converter capacity and allowing high temperatures to be used for production of high fructose syrups for application in confectionery and baking industry.

The use of immobilized biocatalysts (enzymes & cells) has further reduced the production costs, augmented the productivity and enhanced the product quality in agro/food industry.

Biotechnology could also add value to agro/food processing by providing biodegradable packaging materials or providing the enzymes or microbes that could be incorporated in the packaging to activate biodegradation on disposal. Odour control is another important environmental problem being faced by agro/food producers. The local communities demand that the release of odours from industrial sites be checked. Moreover, it has been established that volatile organic compounds emitted by industries processing food, animal by-products, etc. lead to increase in ground level ozone thereby reducing the air quality and causing an adverse effect on health and the environment. Therefore, currently, there is an international initiative from UNECE (United Nations Economic Commission for Europe) to reduce the emission of volatile organic compounds in Europe by 30% by the end of the decade. Manufacturing industry will be expected to play its part in this reduction through reduced emission of solvents and other volatile organics. For this, biofilters have been introduced recently in UK with applications in chemical industry, wastewater treatment plants, onion and garlic processing, coffee roasting, poultry processing, flavour and fragrance manufacturing. In biofilters, air is passed through a support material impregnated with microorganisms. The support materials being used include peat, heather, soil and compost.

Also there are increasing constraints on wastewater discharges by the industry. The waste emanating from industries can only be discharged after meeting the standards laid down by the regulatory agencies (Table 3 & 4) through suitable pollution control measures. Hence, the industrial organizations are always looking for effective and efficient management systems for achieving good environmental performance besides sustaining satisfactory economic capabilities. Therefore, in recent years, there has been an increased interest in biotechnology all over the world in agro/food processing industry to utilize their wastes and by-products for the production of food, fuel, fertilizer and some value added products with a view to increase the productivity of the process and reduce environment pollution.

Table 3: BOD Strengths of Some Common Industrial Effluents

Industry	*BOD (in mg/l)*
Dairy waste water	1800
Slaughter house waste water	3500
Distillery vinasse	60000
Brewery effluent	1400–1800

Table 4: Standards for Industrial Waste Water Disposal

Parameters	*Limits*
BOD (ppm)	
• Inland	30.0
• On land	100.0
• Sewage	350.0
pH	5.5
TDS (total dissolved solids)	9.0
Total soluble solids	210.0
Electrical conductivity	100.0
SO_4 (mg/l)	1000.0
Chloride (mg/l)	1000.0
COD (ppm)	2000.0

Further, agro/food industry wastes often contain suspended organic solids and fats that can foul surfaces and block drains. Many clearing protocols based on enzymes particularly lipases and proteases have been developed for use in plant operations for surface cleaning, thus, eliminating the need to use strong caustic agents or solvents for the purpose. The protein engineering is also playing an important part to bring structural changes in the enzyme proteins in order to evolve them to work at high alkalinity or acidity and keep them active and stable during the cleaning process.

3. QUALITY ASSURANCE

The drive for quality from production to consumption chain as well as growing global markets for food/agro products in view of liberalization (likely to grow further due to WTO regime coming to enforcement by 2005) have generated increasing requirements for assurance of safety and quality to meet the international regulations and standards. In this context, biotechnology offers new ways of assessing microbiological quality, contaminants, residues, authenticity, taste and flavour. These methods are based on highly specific recognition capability of components of biological system e.g. enzymes such as for distinguishing natural and synthetic isomers of the same molecule; antibodies such as for distinguishing hazardous food contaminant *Listeria monocytogenes* and other harmless species of *Listeria*; or DNA such as for detection of a particular toxin as well as surveillance and forensic work in food microbiology, authenticity and contamination assessment. The biotechnological quality assurance assays are more rapid, cost effective and require fewer skills and less investment in equipment than conventional methods. Though DNA techniques are presently being used by sophisticated laboratories, the methods based on antibodies and enzymes have been simplified and are being used by non-technical staff on the factory floor and in the field. Some food contaminant tests available as immunoassays are listed in Table-5.

Table 5 : Some Food Contaminant Tests Available as Immunoassays

Testing for	*Example*
Mycotoxins	Aflatoxin B1, M1; Zearalenone; Trichothecenes; Ochratoxin
Pesticide residues	Atrazine, Carbaryl, Methoprene
Veterinary residues	Clenbuterol, Neomycin, 19-nortestosterone
Authenticity/Adulteration	Meat and fish speciation, and durham wheat adulteration
Quality	Fish Freshness

In order to assess the microbiological safety of manufactured food/agro products and process equipments, the estimation of metabolic activities of microflora of the product and plant using commercially available hygiene monitoring kits based on ATP bioluminescence is quite helpful in real time decision making as compared to conventional plate counting. Further, in order to monitor the efficacy of HACCP (Hazard Analysis & Critical Control Point) procedures as well as to meet international quality assurance standards, the testing of specific pathogens in the product is required. The time taken for such investigations (e.g. identification of *Salmonella*) can be drastically cut down by use of antibodies for separating and concentrating the target bacteria from the product. Further, DNA fingerprinting is being employed by large brewing companies world over to distinguish between yeast strains as various yeasts are used to produce a range of products.

3. GOVERNMENT INITIATIVES

Agro/food processing sector plays a significant role in diversification of agricultural activities, improving value addition of agricultural produce, increasing employment opportunities and creating surplus for export of agro-food products. India offers a vast opportunity to this sector due to its natural advantage in terms of abundance of raw materials as also the mammoth size of its domestic market. This, however, needs to be backed up with appropriate policies and plans for improvement of processing infrastructure including upgradation of technology, enforcement of quality standards and activating domestic market with focus on exports.

The access to adequate and timely credit is recognized as one of the major constraints in the growth and development of this sector in the country, because, of reluctance of financial institutions to invest due to various factors such as uncertainties in availability of raw materials, seasonal processing, long gestation period and relatively smaller returns, huge storage costs involved due to perishable nature of raw materials etc. As a positive step in this direction, the Government has now included agro processing sector within the definition of "Priority Sector" for bank lending, thus bringing it at par with sectors like agriculture, small scale and information technology for the purposes of bank credit. As 40% of the net bank credit of all scheduled commercial banks is directed towards the priority sector, this step of government has ensured better availability of finances for agro processing.

With a view to develop adequate knowledge base as well as in-house techno-managerial expertise for better appreciation of agro/food processing projects at the time of appraisal, NABARD is examining the feasibility of setting up an exclusive cell for post harvest management and agro processing. The establishment of Rabo India Finance, a subsidiary of Rabo Bank of Netherlands, having expertise of food and agri-business is also very significant, as this will further strengthen the need for special treatment of the credit requirements of this sector. The British High Commission in India has also identified Agro-food as one of the priority areas, thus offering a wide range of potential business opportunities in this sector. The commercial department of the British High Commission is keen to assist Indian Agro-food related companies to do business with British Companies and the services being offered include identification of suppliers, manufacturers, consultants, joint venture partners & technical collaborators, supply of equipment or materials from Britain as well as opportunities to invest and participate in trade fairs & exhibitions in Britain.

Further, India being an agricultural nation and backed by a legacy of Organic farming, has a potential to make a mark in the International market. However, in view of the need for certification and maintaining strict quality standards in organic products, the national standards for organic products have been implemented under the National Organic programme. A trade mark of "Indian Organic", owned by Government of India will also be granted on the basis of compliance with the National Standards for Organic Production (NSOP). Only such exporters, manufacturers, and processors whose produce is duly certified by the accreditation and certifying agencies will be able to use the trademark. Government of India will register this Certification mark globally and its use will be governed by a set of regulations pertaining to Organic Products.

Also most of the agro/food processing industries barring a few have been exempted from the provisions of industrial licensing under Industries (Development & Regulation Act, 1951). As far as foreign investment is concerned, automatic approval for even 100% equity is available for majority of the processed food items. The new National Food Processing Policy emphasizes upon infrastructural development for application of biotechnology in the agro processing sector while in the document 'Biotechnology – A Vision : Ten Year Perspective' released by the Union Government, major thrust is on nutritional quality improvement of major crops, value addition to crops as neutraceuticals and therapeutics. Various other initiatives such as establishment of post-harvest infrastructure, cold chain facilities, food processing industrial estates/food parks, mobile fruit & vegetable processing units, mobile food testing & certification laboratories and the facilities for R&D and quality upgradation & control are also underway to ensure that India emerges as a strong player in agro processing sector in the world.

FURTHER READING

Anderson, A. Biotechnology Application in Pharmaceutical and Food Processing—A Study. Publ. FICCI (2001).

Biotechnology—A Vision : Ten Year Perspective. Publ. DBT, Ministry of Science & Technology, Govt. of India (2001).

Biotechnology Means Business Initiative—Status Report: Food. Publ. Department of Trade & Industry, London (1995).

Biotechnology Means Business Initiative—Fact File. Publ. Department of Trade & Industry, London (1995).

Marwaha, S.S. and Arora, J.K. Food Processing: Biotechnological Applications. Publ. AsiaTech Publishers Inc. New Delhi (2000).

Pandey, A. Advances in Biotechnology. Publ. Educational Publishers and Distributors, New Delhi (1998).

Rice Improvement Strategies to Combat Food Security Risks

R.K. SINGH

Representative, IRRI Outpost India
IARI, Pusa Campus
New Delhi

RICE IS THE PRINCIPAL STAPLE food crop of humid and sub-humid Asia. More than 90% of the rice is produced and also consumed in this region. Rice accounts for 30-50% of agricultural income and provides 50-80% of the daily calories consumed by the people. Despite multiple growth in rice production, more and more éfforts are required to meet the growing demand for rice. The world population is likely to reach 8 billion by 2025; about 60% of increases will be , from Asia, excluding Japan. International Rice Research Institute (RRI) has projected that by 2025, the world must produce 880 million tons of rice, 57% more than the current level to meet the demand created by increased population and rising incomes. The problem is compounded by the fact that increase in production must come from less land, with less water, less labour and less pesticides. Increasing yield potential of rice varieties is, therefore, considered the most important strategy for meeting this challenge.

The year 1966 marked the beginning of green revolution with the release of the first high yielding dwarf variety — IR8 — which shifted the yield potential from 4 t to 10 t/ha allowing farmers to significantly increase production. During 70s and 80s, rice scientists developed varieties which had resistance to major insects and diseases. This helped farmers to reduce dependence on harmful agrochemicals and decrease costs. Varieties with shorter duration allowed farmers to take more than one crop. These and other technological advances changed the face of rice cultivation in Asia over the past 40 years. Rice production in Asia grew from 240 million tons in 1966 to about 530 million tons in 1999, much faster than the regional population, which has almost doubled over the past 35 years.

1. RICE IMPROVEMENT

1.1 Current Status

Stabilizing rice production was the major concern of the scientists during 70s and 80s. The objective was well achieved and many countries in South and Southeast Asia became self-sufficient in rice supply. A few of them — Thailand, India. Vietnam—are now exporters. However, the unabated growth in population and declining resources forced rice scientists to reorient their strategies for a second yield jump. About a decade ago, two programs were launched to enhance yields of irrigated rice by 20-25%. First, a new plant type (NPT) was conceived with increased biomass (22 t/ha), high harvest index (0,6), fewer, but, all productive tillers and long filled panicles etc (Table 1). The crosses were made in 1989 and now the prototypes are being field-tested (IRRI, 1999-2000). In another 2-3 years time, these will be available to the farmers for large scale cultivation. The second program was to exploit hybrid vigour. The hybrid rice technology has been successfully utilized in China. Today, about 50% of rice area in China is covered with rice hybrids. Chinese success in rice hybrid technology encouraged some other countries also to initiate work on hybrid rice technology. About 14 hybrid rice varieties have al-

Table 1: Attributes of New Plant Type

- Lower tillering capacity (3-4 tillers when direct-seeded, 8-10 tillers when transplanted)
- No unproductive tillers
- 200-250 grain panicle–1
- Harvest index 0.55/0.6
- 90-100 cm tall
- Very sturdy culms
- Dark green, thick and erect leaves
- Vigorous root system
- 100-130 d growth duration
- Multiple disease and insect resistance
- Acceptable grain quality

Table 2: Salient features of hybrids released in India

Hybrids/year	*Parentage*	*Duration (d)*	*Yield (t/ ha)*		*Yield advantage over check*	
			Hybrid	*Check*	*t/ha*	*%*
APHR-1 (1994)	IR58025A/ Vajram	130-135	7.14	5.27(Chaitanya)	1.87	35.4
APHR-2(1994)	IR62829A/ MTU 9992	120-125	7.52	5.21 (Chaitanya)	2.31	44.2
MGR-1 (1994)	IR62829A/ IR10198	110-115	6.08	5.23(IR50)	0.85	16.2
KRH-1 (1994)	IR58025A/ IR9761	120-125	6.02	4.58 (Mangla)	1.44	31.4
CNRH-3(1995)	IR62829A/Ajaya	125-130	7.49	5.45 (Khitish)	2.04	37.4
DRRH-1 (1996)	IR58025A/ IR40750	125-130	7.30	5.50(Tellahamsa)	1.80	32.7
KRH-2(1996)	IR58025A/ KMR3	130-135	7.40	6.10(Jaya)	1.30	21.3

ready been released in India, a few of them are doing excellently well (Table 2). Vietnam is the other country in line to exploit hybrid technology. Efforts are on to further improve hybrid vigour, develop two-line hybrids and use apomixis to avoid seed replacement every year.

For the rain fed ecosystems, emphasis is on improved germplasm with enhanced resistance to biotic and abiotic stresses. Grain quality and nutritional values are the other considerations in breeding objectives. A number of germplasms with higher Fe, and Zn content have been identified (Table 3). An IRRI line, IR 68144, is being tested for its effectiveness in increasing serum ferritin levels in the blood. A trial on this line is going on with 27 young sisters of a Roman Catholic Convent in Manila (IRRI, 1999-2000).

Table 3: Fe and Zn content in some rice varieties grown in a greenhouse with optimum fertility and full protection from insects and diseases, IRRI, 1998

Variety	*Fe (mg/kg)*	*Zn (mg/kg)*
Ganjay Roozy	26.4	58.9
Zuchem	23.4	51.0
YR4194	23.2	54.0
Banjaiman	22.7	53.0
Xua Buc Nuo	22.5	46.6
IR64446	22.2	53.5
IR74	13.0	36.4
IR72	11.7	32.5
IR36	10.1	31.4

A dispersed breeding program based on physiological undertaking of plant adaptation and environmental characterization has been established at key representative sites in different countries. It operates through a shuttle breeding partnership in collaboration with NARS. This program has led to identification/release of some very promising materials for rainfed ecosystem which suffer from floods and droughts.

The third development relates to the use of molecular tools to address the dominant problems of these ecosystems. Using molecular mapping and marker-aided selection (MAS) for traits such as root depths, presentation ability, osmotic adjustment and tolerance to submergence, flooding and salinity, improved types are being developed and tested on farmers fields. For instance, seven QTLs affecting plant elongation ability and submergence tolerance are identified and their use for MAS is in progress. Similarly, QTLs for potassium (K) and sodium (Na) absorption and for the Na-K ratio are identified to use them for MAS for salinity tolerance in rice varieties (Table 4). The status of MAS in rice variety improvement is described in Fig. 1. As evident, identification and use of marker-aided technique has reached the application level for traits like resistance to tungro, bacterial blight,

Table 4: QTLs identified by interval analysis for K and Na absorption and Na-K ratio in the shoot IRRI, 1998

Index	*Marker interval*	*Chromosome no.*	*LOD*
	K absorption	1	17.2
195–209	P3/M9-8 - Saltol		
206–200	RG375 - P4/M3-2	4	5.3
39–3	PI/M1-3-P2/M1-3	12	3.4
	Na absorption		
195–209	P3/M9-8 - Saltol	1	14.5
39–188	P1/M5-3-P3/M9-1	3	3.2
88–67	P1/M10-6-P1/M7-10	3	3.0
27–25	P1/M3-10-P1/M3-8	10	4.0
	Na-K ratio		
195–209	P3/M9-8 - Saltol	1	14.5
207–65	G29I-P1/M7-8	10	3.6
3–93	P1/M1-3-P2/MI-3	12	3.1

blast, brown plant hopper, white brown plant hopper and gall midge. From among abiotic traits, marker-aided selection is at the advance stage of testing for salinity, drought and submergence tolerance, and some other traits like aroma, CMs, TGMS and iron content. Similarly, the transgenics for BB and stem borer are also at the advance stage of testing.

1.2. New Developments

Several other new developments are in progress which show great promise. A project that offers dramatic improvement in human nutrition and has attracted attraction is the development of so called "golden rice", which has been genetically engineered to produce high levels of beta-carotene, the precursor of Vitamin A. Three genes, two from daffodil and one from a bacterium, have been inserted in a japonica variety, T309. IRRI scientists are trying to insert this complex into the commercial indica varieties.

The new field of genomics that means deaphaning DNA sequences (structural genomics) and their biological functions (functional genomics) is expected to drive the future advances in plant breeding (IRRI, 1999-2000). The international sequencing effort is being led by the Japanese Rice Genome Program and involves several national laboratories including India. However, the challenge lies beyond the sequencing of genes. Discovering the functions of genes is the next most vital step. IRRI is leading the efforts on functional genomics and aims, among other things, to create a public platform, involving NARS, to ensure that the information generated from genomics remains broadly available.

The quest for a perennial rice plant for cultivation in uplands began at IRRI in 1995. There has been a slow, but, a steady progress. The success may be as far as six to eight years away. A perennial variety would allow upland farmers to harvest

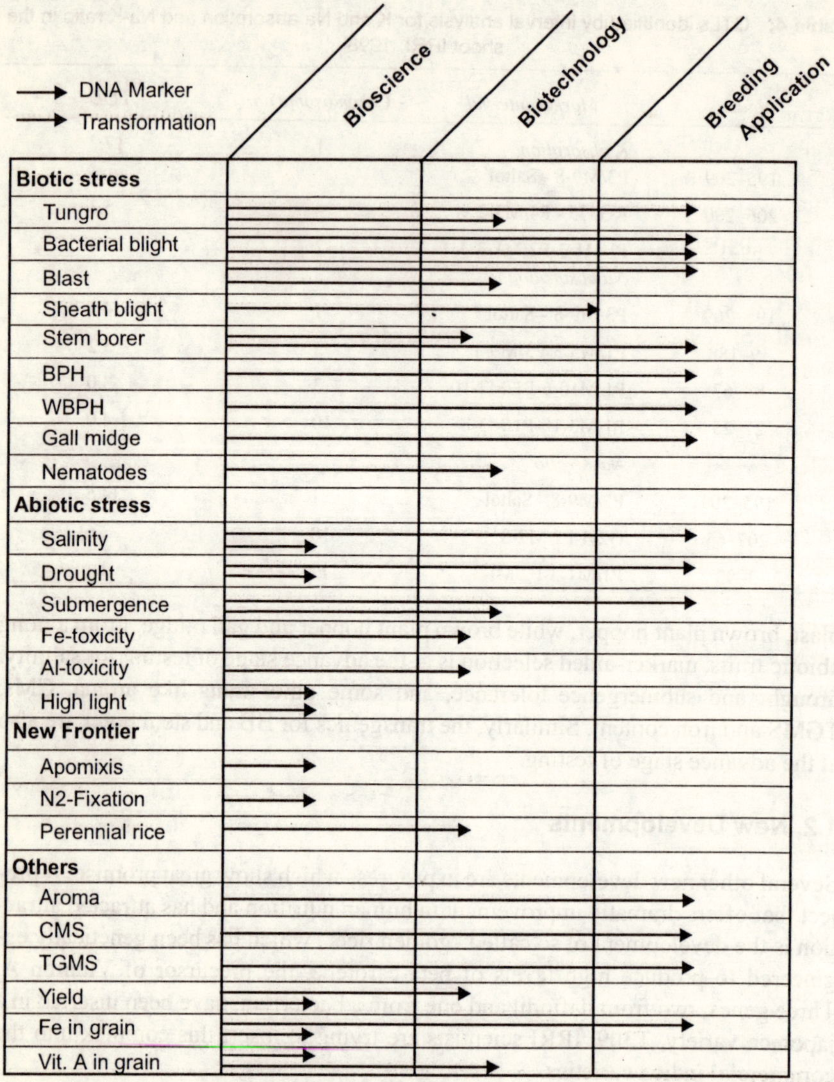

Fig. 1: *Status of Biotech Activities at IRRI*

rice year after year. Thus, food production, and soil stabilization would go hand in hand.

Another most challenging effort being made is to convert rice, a C_3 plant, into a C_4 plant like maize and sugarcane. The C_4 type plants are photo-synthetically more efficient. All the relevant genes for a C_4 plant type are present in rice, but it lacks triggering in of the system to function like a C_4 plant. Once achieved, raising yield potentials beyond 10 t/ha will be much easier.

2. FUTURE CONCERNS

Free exchange of germplasm in the past has helped all. However, the restriction on germplasm exchange is fast increasing largely due to misuse of the intellectual property right (IPR) provision. Conflict between small and commercial farmers interest is growing. Private versus public good research is another area which needs to be carefully planned and coordinated so that the farmers in developing countries can also benefits from new developments.

Loss of biodiversity is another cause of concern Appropriate measures are required to be taken to collect conserve and utilize the valuable genetic resources.

A lot of development is taking place in rice science today. The question, however, is if the developing NARS are prepared to absorb and utilize these modern technologies towards their 'benefits'. In this respect, launching of the international molecular breeding program and the forum for functional genomics by IRRI are well thought ideas to involve developing countries NARS so that they become partners in the outcome of these programs. Their prepared-ness includes human resource development, bio-safety measures and responsiveness to WTO regulations.

Challenges to grow more rice in future will be more complex and, hence, will require more sophisticated research. Unfortunately, however, instead of increase, international support for research is declining. Rice growing NARSs have, therefore, to come forward to share the research cost to keep the efforts going.

FURTHER READING

Bhasin, V.K. India and emerging global rice trade. In: Aromatic Rices: (Eds. R.K. Singh, U.S. Singh and G.S. Khush). Oxford and IBH Publishing Co. Pvt. Ltd., New Delhi: (2000) 257.

IRRI (International Rice Research Institute). The rewards of rice research. IRRI, Los Banos, Philippines: (1999-2000) p. 72.

Singh, S., Singh, O.N., Singh, R.K., and Sarkarung, S. A shuttle breeding approach to rice improvement for rainfed lowland ecosystem in eastern India. James and James (Science Publishers) Ltd., U.K.: (1998) 105.

Singh, R.K., Gautam, P.L., Saxena, S., and Singh, S. Scented rice germplasm : conservation, evaluation and utilization. In: Aromatic Rices: (Eds. R.K. Singh, U.S. Singh and G.S. Khush). Oxford and IBH Publishing Co. Pvt. Ltd., New Delhi: (2000) 107.

Virmani, S.S. and Zaman, F.U. Improving grain quality of hybrid rice: challenges, strategies, and achievements. In: Advances in hybrid rice technology: (Eds. S.S. Virmani, E.A. Siddiq and K. Muralidharan). Proc. 3rd Int'l Symp. on Hybrid Rice. Nov. 14-16, 1996, Hyderabad, India: (1998) 177.

Virmani, S.S., Jachuck, P.J., Chatterjee, S.D., and Ahmed, M.I. Opportunities and challenges of developing hybrid rice technology for rainfed lowland and boro rice ecosystem. In: Rainfed rice for sustainable food security (Ed. S.K. Mohanty), Cuttack, India: (1998) 53.

Genetic Transformation and Crop Improvement

D.K. SRIVASTAVA
Department of Biotechnology
Dr. YS Parmar University of Horticulture & Forestry
Solan 173 230, Himachal Pradesh

THE GENETIC MANIPULATION of crop plants has been carried out by man for thousands of years, albeit in early times by a process of unconscious hybridization and selection of wild species. This evolution of more useful varieties has been speeded up greatly by virtue of a recognition, during the early part of this century that traits segregate in an ordered and predictable fashion and can be recombined in a directed way to generate a progeny possessing novel genetic characteristics. The transmission of traits in plants is complex, in that, unlike animals, plants have three genomes, located in the nucleus,, mitochondria and chloroplasts. Since the male gametophyte contributes insignificant quantities of cytoplasm to the zygote, genes which are encoded in the mitochondria and chloroplast are inherited in a non-Mendelian manner through the female line.

Plant breeders and biologists interested in crop improvement have been attempting all along to achieve the ability to direct the course of event that can lead to creation of desired gene assemblies in more efficient and predictable ways. Advances in molecular biology have made it possible for identifying, isolating and transferring desired genes from one species to another for improving agronomic traits. It is now possible to transform several of the major crop plants. Foreign DNA can be introduced into plant cells by vector-mediated transfer (*Agrobacterium*-mediated) or direct transfer (electroporation, microinjection particle gun bombardment, liposome-mediated, PEG-mediated), both of which essentially involve precise tissue culture methods. Such genetically manipulated crops are referred to as "transgenic plants" and to-date, transgenic plants are reported in about 70 plant species such as rice, wheat, maize, oats, potato, cotton, tomato, *Brassica* and soybean. Agronomically important genes for biotic and abiotic stresses and quality attributes have been the major focus of gene transformation research and several gene constructs have been made available (Table 1) that are being used in production of transgenic plants with insect resistance, herbicide resistance, virus resistance, fungal resistance, male sterility, delayed fruit ripening, osmotolerance and cold tolerance.

Table 1: Expression of foreign genes of agronomic interest in transgenic plants

Trait (s)	*Gene(s)*
Insect resistance	Bt toxin, protease inhibitor
Virus resistance	Tobacco mosaic virus coat protein. Alfalfa mosaic virus coat protein, potato virus coat protein, cucumber mosaic virus antisense coat protein, rice stripe virus coat protein, papaya ring spot virus coat protein, maize dwarf mosaic virus coat protein, plumpox virus coat protein, satellite RNA
Fungal resistance	Chitinase, ribosome inactivating protein, phytoallexin
Delayed fruit ripening	ACC oxidase, ACC synthase, Polygalacturonase
Male sterility	Ribonuclease (*barnase*)
Fertility restoration	Ribonuclease inhibitor (barstar)
Seed storage protein	*Phaseolin, Ama I*
Osmotolerance	Pyroline-5-carboxylase
Cold tolerance	Glycerol-3 phosphate acetyl transferase
Herbicide tolerance	*aroA*, (glyphosate), *bar* (phosphinotricin), *bxn* (bromoxynil), *ALS* (Sulfoxylurea)

1. PLANT GENETIC TRANSFORMATION

There are currently a number of techniques available by which genes may be introduced non-sexually into plants. Broadly these can be described as 'Genetic Transformations' in which usually one or two defined genes are transferred and become integrated into the host genome. This approach results in a novel combination of genetic material and it requires a mean of selecting or screening for the desired recombinants, which may represent, a small proportion of the total cell population. Genetic transformation techniques are now in widespread use, not only in the genetic modification of crops, but, also as a tool to study the relationship between gene structure and function. For example by deleting or mutating putative regulatory sequences in gene promoters and transforming plants with either the mutant gene or a chimeric gene in which the modified regulatory sequencers are fused to a reporter gene (the translation product of which is readily detectable, such as an enzyme), the consequent effects on expression can be defined. Such analytical experiments are commonly carried out in 'model' species such as tobacco, which is easy to transform by a variety of techniques. Indeed, the relative difficulty and lack of reproducibility in transforming crop rather than 'model' species has led to the development of the variety of approaches to genetic transformation which now exist. These approaches can be described either as exploiting vector systems or as being vector free.

2. VECTOR SYSTEM—*AGROBACTERIUM*-MEDIATED TRANSFORMATION

The soil bacterium *Agrobacterium tumefaciens* causes tumors called crown galls on dicot plants when these bacteria infect, a wound site usually on the stem close to the ground. The infection disturbs the normal wound healing process and instead cells proliferate into cancerous growth. Cell(s) from this tumour can be grown in tissue culture in the absence of normal plant growth regulators, in the medium. It has been found that, the bacteria on infection transfer some of their genes to the plant cell which carry the information necessary to make auxin and cytokinin at abnormally high levels. Thus the genes transferred from single A*grobacterium* to a single plant cell get integrated into the nuclear genome of the plant cell. The cell is now transformed with the agrobacterial genes, which will be present in every cell derived from the transformed cells.

Molecular biologists realized that the plants cells can be transformed with any desired bacterial genes. These transformed cells can then be regenerated to get a transformed plant (transgenic plant.) in which every cell carried the foreign gene. Thus combining molecular biology and plant tissue culture and by relying on the natural gene transfer system of the soil bacterium, it has become possible to transform plants with a variety of desirable important genes (Table 1).

The property of crowngall formation is due to the presence of a mega plasmid called Ti (tumor inducing) plasmid in the *Agrobacterium* strains. *Agrobacterium tumefaciens* introduces a part of the Ti-plasmid called T-DNA into plant cells during infection. This T-DNA carry the genes for the opines as well as for the plant growth regulators (auxin and cytokinin).

The production of transgenic plants (genetic transformation of plants) can be divided into four major steps i.e.

- Introduction of foreign genes into modified *Agrobacterium* strain
- Co-cultivation of A*grobacterium* strains with protoplasts, plant cell or tissues
- Selection and regeneration of transformants
- Analysis and verification of gene expression in transformed plants

The first step involves the removal of genes in the Ti-plasmid that cause tumorous growth and replace them with the genes one want to transfer to the plant. These genes on the Ti-plasmid occur in one continuous DNA segment i.e. T-DNA which is flanked by a right border (RB) and left border (LB). Other *Agrobacterium* genes which are required in trans for T-DNA transfer are *vir* genes (A-G). The two types of vectors developed for the genetic transformation of plants are:

(a) Cointegrate vector: These vectors include a region of homology between the vector plasmid and Ti-plasmid. There are two types of cointegrating systems. The first utilize the disarmed *Agrobacterium* Ti-plasmid pGV3850. In this plasmid the plant growth regulator genes (auxin and cytokinin) of C5B plasmid have been excised and replaced by pBR 322 sequence. Any plasmid containing the pBR322 sequence homology can be cointegrated into this disarmed Ti-plasmid. The border sequences as well as nopaline synthase gene are part of the Ti-plasmid and the cointegration places the new sequences between the T-DNA borders. In the second system a left border and a small part of the original T-DNA referred to as limited homology (LIH) remain intact and the right border and all of the phytohormone genes are removed. The vector to be introduced into *Agrobacterium* (also called intermediate vectors) contain the LIH region for homologous recombination as well as a right border. The reintegrated DNA reconstructs a functional T-DNA with a right border and left border. This system has been used extensively for introduction of many genes into plants. The cointegrate system, while more difficult to use do offer advantages. Once the cointegrate vector has been formed, plasmid is stable in *Agrobacterium* and is virtually impossible to remove.

(b) Binary vector: They contain broad host range replication origin to facilitate their replication in *E.coli* as well as *Agrobacterium*. They also contain the border sequences and the foreign gene is cloned in between the border sequences under suitable promoters and the binary vectors are then used for transforming *E.coli* host. The binary vectors are introduced into *Agrobacterium* host strain which carry a disarmed Ti-plasmid or Ri-plasmid by triparental mating. The resident disarmed Ti-plasmid will provide the *vir* function for transfer of DNA to plants.

Binary vectors, are not completely stable in *Agrobacterium* in the absence of drug selections as in cointegrate vectors. But, the advantage of binary vectors is their lack of dependence on a specific Ti-plasmid. The vector may be introduced into virtually any *Agrobacterium* host containing any Ti or Ri plasmid as long as their *vir* helper functions are provided.

3. DIRECT GENE TRANSFER TECHNIQUES

Agrobacterium has been found to be a very-effective gene transfer system for genetic transformations. However, monocot species except *Asparagus,* rice and wheat do not respond well to *Agrobacterium*-mediated transformation. Thus, alternative direct DNA transfer techniques have been developed which require highly plant regeneration systems. Some of the methods are discussed below:

3.1 Microinjection

Microinjection technique involves the direct injection of DNA into recipient cell, protoplasts, pollen, ovules, microspores or embryos. One of the ways is to inject DNA into an *in vitro* germinated pollen followed by transfer of the pollen to a receptive stigma. The DNA if maintained may become incorporated into the genome during natural fertilisation eliminating the need for artificial regeneration techniques. The difficulty of this technique lies in the physical coordination between the system to inject DNA and the means to immobilise the recipient cell. Immobilization of protoplasts is achieved by using various embedding media like agarose, agar, polylysine treated glass surfaces and by using suction holding pipette. Following injection, the injected cell is cultured to ensure its continued growth and development. The first graminereous transformed plant has been obtained by mechanical injection of naked DNA into immautre (premeotic) inflorescence of rye. Thereafter, a number of transgenic plants and transformed calli are obtained by using this method of gene transfer.

3.2 Electroporation

Electroporation involves a sudden electrical discharge in a very small container that holds plant protoplasts and DNA molecules in a liquid suspension. The discharge creates pore in plant plasma membrane simultaneously at different places, allowing the DNA molecules to go into the cell. The protoplasts after electroporation are cultured to get regenerated plants. This is the most difficult step, because many plant species cannot be readily regenerated from single protoplast. Electroporation has been successfully used to introduce foreign genes and produce transgenic plants in rice, maize, wheat.

3.3 Direct DNA Uptake

The chemicals like polyethylene glycol (PEG) and polyvinyl alcohol together with Ca^{+} and high pH induce protoplast to take up naked DNA. The high transformation frequencies are obtained using chemically mediated DNA. Uptake have been low when compared with A*grobacterium*-mediated transformation. Using the PEG method, foreign genes in the form of plasmid DNA were inserted into plant protoplasts, resulting in the recovery of fertile regenerated tobacco. Eversince, several instances of PEG mediated gene transfer have been reported.

3.4 Biolistic (microprojectile bombardment) Mediated Transformation

This technology was developed by John Sanford in 1987 and allows direct delivery of naked DNA into intact plant cell by shooting it through the cell wall and cell membrane. This method involves the acceleration of dense DNA-coated micro-particles (tungston or gold) by means of an explosion or burst of gas, upto a sufficient velocity to penetrate the cell wall. The tremendous advantage of particle gun bombardment over other technologies is the wide range of cellular compartments, cell types and plant species to which foreign DNA can be transferred. Using this technique foreign DNA has been transferred to nuclear genome of most monocotyledonous plant species as well as into plant chloroplast and yeast genomes. Initially, the use of this method was limited to some extent by the cost and the complexity of the particle guns. However, the development of particle guns using a direct acceleration of micro-particles in an helium flow has facilitated the use of this technology. A large number of transgenic plants have been obtained using this technique of gene transfer.

3.5 DNA Transfer Using Silicon Carbide Fibres

DNA delivery into plant cells using silicon carbide fibres was investigated by Kaeppler and co-workers in 1990. The method is very simple and involves vortexing a mixture of plasmid DNA encoding scorable and selectable marker genes, silicon carbide fibres and plant cells to be transformed. Expression of GUS was observed in corn and tobacco, suspension culture cells following vortexing with silicon carbide fibres and DNA. Electron microscopic examination of cells following fibre treatment reveal that fibres appear to penetrate the cell walls suggesting that DNA adhered to the fibre surface which is propelled through the cell wall via collisions between cell aggregates during vortexing. Corn plant transformation has been accomplished using silicon carbide fibres. However, it appears that routinely achieving high frequency DNA delivery in tissue culture that retain plant regeneration capacity will remain challenging. This is likely due to differences encountered in cell wall rigidity and thickness, cell colony size in regenerable versus non-regenerable suspension culture systems. It is likely that improvements in DNA delivery via silicon carbide fibres and plant tissue culture methods will increase the number of species that are amenable to this transformation procedure.

4. SELECTABLE AND NON-SELECTABLE MARKERS FOR PLANT GENETIC TRANSFORMATION

Several selectable and non-selectable marker genes viz. neomycin phosphotransferase-II (NPT-II), luciferase (LUX), streptomycin phosphotransferase (SPT), hygromycin phosphotransferase (HPT), chloramphenical acetyl transferase (CAT), dihydrofolate reductase (DHFR), nopoline synthase (NOS), octopine synthase (DCS), β-glucuronidase (GUS), green fluorescent protein (GFP) are widely available today for plant transformation (Table 2) and have also been transferred in large number of plant species. Several requirements are considered in the development of a truly useful selectable marker system. It is most critical that the selective agent be inhibitory to plant cells. However, not all compounds toxic to plant cells are necessarily useful as selective agents. Cells that are not transformed can be killed in such a manner that they become toxic to adjacent transformed cells. This presumably happens because of leakage of toxic compounds such as phenols, from the dying cells. If this occurs, even high level expression of a resistance gene in the transformed cells is insufficient to rescue these cells. The best selective agents are compounds that arrest growth of nontransformed cells or slowly kill them.

The most widely used selectable marker has been the neomycin phosphotransferase-II (NPT-II) enzyme, which was originally isolated from the prokarytic transposan Tn5. This enzyme detoxifies aminoglycoside compounds such as kanamycin and G418 by phosphorylation. This gene, fused to constitute plant transcriptional promoters has been used successfully to transform a large number of plant species. The most widely used reporter/marker genes have been the bacterial genes chloroamhenicol acetyl transferase (CAT), neomycin phosphotransferase (NPT-II) and β-glucuronidase, which encode enzymes with specificities not normally found in plant. These genes fused to constitute plant transcriptional promoter, used successfully to transform a large number of plant species (Table 3) and have been incorporated into numerous plant transformation vectors. In addition, NPT-II can tolerate aminoterminal fusions and remain enzymatically active, making it useful for studying organelle transport in plants. However, CAT and NPT-II are relatively difficult, tedious and expensive to assay. Competing reactions catalyzed by endogenous esterases, phosphatases, transferases and other enzymes also limit sensitivity and make quantitation of CAT or NPT-II by enzyme kinetics very

Table 2: Selectable and non-selectable marker genes for plant genetic transformation

Selectable/non-selectable marker gene	*Enzymes encoded by the gene*	*Selected for resistance against*
Nopaline synthase gene (NOS)	Nopaline synthase	—
Octopine synthase gene (OCS)	Octopine synthase	—
Neomycin phospho-transferase II gene (NPT-II)	Neomycin phospho-transferase	Neomycin Kanamycin
Luciferase gene (LUX)	Luciferase	—
Streptomycin phospho-transferase gene (SPT)	Streptomycin phospho-transferase	Streptomycin
Hygromycin phospho-transferase gene (HPT)	Hygromycin phospho-transferase	Hygromycin
Chloramphenicol acetyl transferase gene (CAT)	Chloramphenicol acetyl transferase	Chloramphenical
Dihydrofolate reductase gene (DHFR)	Dihydrofolate reductase	Methotrexate
β-glucuronidase gene (GUS)	β-glucuronidase	—
Green fluorescent protein gene (GFP)	Light is produced by the bio-luminescent jelly fish, (*Aequorea victoria*) when calcium binds to photoprotein aequorin, the green fluorescent protein. It functions both as a sensitive reporter of *in vivo* gene expression and an efficient selectable marker. Activation of "Aequorin" *in vitro* or in heterologous cells produce blue light.	

difficult. β-glucuronidase (GUS) encoded by the *uidA* locus is a hydrolase that catalyses the cleavage of a wide variety of β-glucuronides many of which are available commercially as spectrophotometric, fluorometric and histochemical substrates. This gene has been used successfully to transform large number of plant species (Tab les 3, 4 and 5).

5. GENE, TRANSFORMATION CONSTRUCTS AND GENE EXPRESSION

To achieve effective transfer of genetic material into a host, appropriate genetic constructs need to be made to facilitate integration and expression of the foreign gene(s). It is preferable to include a selectable or screenable marker to allow effective identification of those few cells that have been transformed. Selectable markers are preferred, as they allow, under ideal conditions, establishment and proliferation only of transformed cells. These cells may regenerate into transgenic plants following transfer to appropriate media containing plant growth regulators. In a number of situations, the use of selectable marker is not possible and alternative procedure had to be developed to allow identification of transformation events.

Table 3: Transgenic field crops

Plants	*Method and gene transferred*
Tobacco	At, NPT-II
Alfalfa	At, NPT-II
Cotton	At, NPT-II
Sunflower	At, NPT-II
Moth Bean	At, NPT-II
Soybean	At, GUS
Safflower	At, GUS
Sugarcane	MB* GUS
Green Bean	At, GUS
Cowpea	EL* GUS
Chickpea	At, NPT-II

At = *Agrobacterium tumefaciens*
EL = Electroporation
MB = Microprojectile Bombardment

Table 4: Transgenic vegetable crops

Plants	*Method and gene transferred*
Carrot	Ar, NPT-II
Tomato	At, NPT-II
Cucumber	Ar, NPT-II At, Ar, NPT-II
Potato	At, NPT-II
Oil seed rape	At, NPT-II
Lettuce	At, NPT-II
Cauliflower	Ar, Mannopine, At, NPT-II
Celery	At, NPT-II
Brinjal (Egg plant)	At, NPT-II
Watermelon	At, NPT-II
Sugar beet	At, NPT-II
Muskmelon	At, NPT-II
Bell Pepper	At, GUS
Peas	At, NPT-II
Sweet Potato	MB , GUS
Cabbage	At, GUS
Broccoli	At, Bt gene
Hot Pepper	At, GUS
Amaranthus hypochondriachus	At, GUS

At = *Agrobacterium tumefaciens* Ar = *Agrobacterium rhizogenes*
MB = Microprojectile Bombardment

Table 5: Transgenic fruit crops

Plants	*Method and gene transferred*
Walnut	AT, NPT-II
Apple	Ar, NPT-II
Peach	At, Octopine
Grapevine	At, NPT-II
Papaya	MB, NPT-II At, cp gene
Kiwifruit	At, GUS
Strawberry	At, GUS
Citrus spp.	At, NPT-II
Apricot	At, cp gene
Pecan	At, GUS
Pear	At, GUS, NPT-II
Lime	At, GUS, NPT-II
Sour orange	At, GUS

At = *Agrobacterium tumefaciens*
Ar = *Agrobacterium rhizogenes*
MB = Microprojectile Bombardment

Table 6: Transgenic monocots

Plants	*Method and gene transferred*
Asparagus	At, NPT-II
Rice	At, NPT-II At, GUS
Maize	EL, NPT-II At, BUS
Rye	MI, NPT-II
Tulip spp.	MB, GUS
Wheat	EL, GUS, bar
Barley	MB, GUS
Sugarcane	MB, GUS
Sorghum	MB, GUS

At = *Agrobacterium tumefaciens*
MI - Microinjection
EL = Electroporation
MB - Microprojectile Bombardment

A typical gene construct will include a promoter, a coding sequence (transgene) and a terminating signal. Transgenes may be introduced into plant tissue on separate plasmids or on cointegrate vectors. In the latter case, multiple genes, including selectable marker or screenable markers are present on the same plasmid. Commonly used promoters for plant cell transformation include the cauliflower mosaic virus (CaMV35S) promoter, which is a constitutive promoter suitable for driving the expression of foreign gene in dicotyledons and the maize ubiquitin promoter, suitable for driving the expression of foreign genes in monocotyledons. Organ/tissue specific promoters are also available to derive expression of transgenes in particular part of the plant and specific examples include the vicilin and phytohemaglutinin promoters derived from pea and bean, respectively, suitable for seed specific expressions, the high molecular weight glutenin promoter from wheat, also suitable for seed specific expression in the aleurone of cereal grains. The most common selectable marker genes encode protein that detoxify metabolic inhibitors such as antibiotics or herbicides.

Efficient methods of plant regeneration to obtain transgenic plants from the transformed cells is an integral part of plant genetic transformation. A wide spectrum of plant species have been successfully regenerated from the transformed cells and protoplasts (Tables 3, 4, 5, 6).

6. PRACTICAL APPLICATION OF PLANT GENETIC TRANSFORMATION TECHNIQUES

The first stage in the useful manipulation of crop by gene transformation technology is the development of a suitable gene transfer system. Since the first report on the production of transgenic plant, much effort has been put into this and many crop species (field crops, vegetable crops, fruit crops, monocots and cereal crops— Tables 3, 4, 5, 6) have been stably transformed with marker/reporter gene(s) either by *Agrobacterium*-mediated or direct gene transfer techniques.

7. CROP IMPROVEMENT THROUGH GENETIC TRANSFORMATION

Many agronomically and economically useful traits have been incorporated into crop plants through genetic transformation. The traits which have been incorporated into plants include disease resistance (viral, bacterial and fungal diseases), insect pest resistance, herbicide resistance, improved quality traits and engineering of male sterility for hybrid seed production. Some of the applications of genetic transformation in the area of crop improvement are discussed below:

7.1 Insect Resistance

The production of insect resistant plant is an application of plant biotechnology with important implications for crop improvement and for both the seed and agrochemical industries. Many different insects, damage crop plants, reducing the yield or quality of the harvested food. Several species of bacteria produce proteins that kill insect larvae that ingest these bacteria with their food. The most studied of these bacteria is *Bacillus thuringiensis* or *Bt*, a species that is present all over the world and have a large number of sub-species and strains. When *B.thuringiensis* form spores, and large crystals like structure (made up of proteins and one of them is called the *Bt*-protoxin) in its cytoplasm are produced. When insect larvae eat the bacterial cells, the spores, and crystal like structure containing the protoxin are released in the larval gut, where the digestive enzymes cleave the protoxin to produce the active toxin. The toxin binds to the cell membrane that line the gut, inserts itself into that membrane and creates a channel through which other molecules can freely pass. Punctured by many pores, the gut cells cannot survive long. So the insect larvae starve for lack of nutrition. Hence the introduction of *Bt* gene i.e. *cry* genes into a wide spectrum of plants have been attempted (Table 7). Since there are a large number of *Bt* strains occurring in nature, their action towards different insects may also differ. Another difficulty which has been observed is their expression, since their origin is prokaryotic. Efforts to enhance *Bt* gene expression in plants became a priority for many groups involved in insect control programmes. Scientists from Monsanto Company St. Louis, USA introduced the gene from Kurtsaki strain of *Bacillus thuringensis* into tomato plants and checked the resistance of these plants to three lepidopteran insects, tomato hornworm, tomato fruit worm and tomato pin worm. The results showed that. tobacco hornworm is particularly sensitive to *Bt* kurtsaki toxin, while tomato fruit worm and tomato pinworm are less sensitive. Scientists from CIBA research centre have succeeded in transforming corn plants with a Bt toxin gene that makes the corn resistant to the European corn borer. Transgenic cotton developed by Monsanto Company (USA) carrying *Bt* gene to protect them against the devastating boll weevil has been commercially released.

Table 7: The cry gene classification

cry gene class	*Insect order against which the encoded cry protein has activity*
cry I	Lepidoptera
cry II	Lepidoptera and Diptera
cry III	Coleoptera
cry IV	Diptera
cry V	Lepidoptera and Diptera
cry V and VI	Nematocera

Another approach taken to protect plants against insects is to introduce genes that encode lectins or inhibitors of digestive enzymes (amylase and protease). The gene which codes for a wheat germ agglutinin, chitin binding lectin that inhibits the growth of insect larvae is being introduced into corn. Trypsin inhibitor gene from an insect resistant strain of cowpea under the control of '35S' CaMV promoter when transferred to tobacco conferred protection against some of the insect group. Cloning of lectin gene has been reported to impart resistance against aphids in many crops. Transgenic tobacco and potato plants carrying protease inhibitor genes have been found to have resistance to tobacco horn worms. A number of transgenic plants have been developed in which *Bt* gene has been transferred (Table 8).

Table 8: Some important Bt transgenic plants developed

Plant	*Toxin gene*	*Target insects*
Tomato	Synthetic *cry* I Ab, *cry* I Ac	*Manduca sexta* (Horn worm)
Cotton	Modified *cry* IAb, *cry* IAc	*Heliothis* zea (cotton boll worm) *Spadoptera exigua* (indigo caterpillar)
Cotton	Synthetic *cry* IAb	*Pectinophora gossiepella* (Pink boll worm) *Spodoptera exigua* (Indigo caterpillar)
Potato	Synthetic *cry* IIIA	*Leptinotarsa decemlineata* (Colorado potato beetle)
Populus	*cry* IAa	*Lymantria dispar* (Gypsy moth)
Tobacco	Synthetic *cry* IIIA	*Leptinotarsa decemlineata* (Colorado Potato Beetle)
Corn	Synthetic *cry* IAb	*Ostrinia nubilalis* (European maize borer)
Rice	Synthetic *cry* IAb	*Chilo suppressalis* (Rice stem borer)
Cabbage	*cry* IAc	*Plutella xylostella* (Diamondback Moth)
Broccoli	*cry* IAc	*Plutella xylostella* (Diamondback Moth)

7.2 Herbicide Resistance

Engineering herbicide resistance into crops represents a new alternative for conferring selectivity and enhancing crop safety. Research has already concentrated on those herbicides with properties such as high unit activity, low toxicity, low soil mobility, rapid biodegradation and broad spectrum activity against various weeds. The development of crop plants that are resistant to such herbicides would provide more effective and less costly weed control,

Two successful approaches for creating herbicide tolerant plants are known. Glyphosate is a herbicide that acts by inhibiting one of the enzymes that is necessary for the synthesis of amino acids in the chloroplast. Researchers of Monsanto Company, St. Luis, USA and Institute of Molecular Genetics, Moscow, Russia have isolated a gene for an enzyme involved in amino acid biosynthesis (EPSP-synthase, 5-enolpyruvyl shikimate-3-phosphate synthase) from resistant *E.coli*, modified the gene in such a way that it could express in plants and then used to generate transgenic tobacco, tomato and soybean plants. The bacterial gene before insertion was modified to get it expressed in cytoplasm and transported to chloroplasts.

Phosphinotricin is a herbicide that acts by inhibiting enzyme glutamine synthase, necessary for amino acid biosynthesis and nitrogen assimilation by plants. Inhibiting the activity of this enzyme leads to accumulation of ammonia within the cell. Because ammonia is toxic except at very low levels, the plants are killed. A gene called *bar* gene which encodes an enzyme that converts phosphinotricin to a non-herbicidal derivative have been isolated from the fungus *Streptomyces hygroscopicus*. A number of transgenic plants have been developed in which these herbicide resistant genes are transferred (Table 9).

Table 9: Herbicide resistance gene transferred in different crops

Transferred Gene	*Plants*
'*aroA*' gene confers the resistance to Glyphosate	Tomato, Poplar, Tobacco
'*bar*' gene confers the resistance to Bialaphos/Phosphinothricin	Tobacco, tomato, and potato, Oil seed rape cauliflower, Rice, Wheat
'*bxn*' gene confers the resistance to Bromoxynil	Tobacco

7.3 Virus Resistance

Expression of virus resistance in crop species has been approached through cross protection. Infection with a mild strain of a virus can provide resistance to more virulent strains. Cross protection has spurred the use of chimeric genes specifying *viral coat protein gene* as a way of protecting plants from viral diseases. Potato 'Virus X' is an important disease in potato. The gene encoding the coat protein of potato 'Virus X' has been introduced into potato varieties which are highly susceptible to this virus attack and transgenic plants with resistance to this virus have been obtained. Transgenic tomatoes carrying the TMV coat protein gene have been evaluated in green house and field tested and shown to be highly resistant to viral infection. Likewise resistance has been created against various groups of viruses in number of crop species (Table 10).

Table 10: Disease resistance genes transferred in different crops

Transferred Gene	*Plants*
Viral coat protein gene	Tobacco, Tomato, Potato, Apricot, Cucumber, Rice, Papaya, Green bean, Soybean, Sour orange and Hot Pepper
Chitinase gene	Tobacco, Canola, Tomato and Cucumber
Phytoalexin gene	Tobacco

7.4 Resistance to Fungal and Bacterial Diseases

The approach used to make plants resistant to fungal and bacterial pathogens involves introducing into the plants the genes that encode enzyme that degrade cell walls. Fungal cell wall consists of two polymers glucan and chitin, that are degraded by glucanases and chitinases, respectively. Since chitin of one species differs in the polymer from other-species, a particular chitinase may be effective against one species of fungus, but, not against other. Thus effective use of chitinase genes for genetic engineering depends on exact matching. Chitinase and β–1-3 glucanases belong to the so called pathogenesis related proteins (PR proteins) because their synthesis increases on pathogen attack. The chitinase gene from *Serratia marcescens* has been successfully introduced into tobacco, potato and sugarbeet and transgenic plants have shown resistance to *Rhizobium solani.* Phytoallexins depending on their concentration have also shown to exhibit antifungal properties. A substantial resistance probably only can be achieved by using a combination of several genes also called gene pyramiding. Chitinase and phytoallexin genes have been transferred to some crops which show resistance for the specific pathogen (Table 10).

8. REGULATION OF FRUIT RIPENING AND INCREASING SHELF-LIFE OF VEGETABLE AND FRUIT CROPS

Genetic engineering to alter ripening of vegetable crop (tomatoes) has been attempted by several researchers and the approach involves the suppression of ethylene synthesis by the tomatoes. The final two steps in the biosynthesis of ethylene involve first the synthesis of the key chemical amino cyclopropane carboxylic acid (ACC) by the enzyme ACC-synthase and its subsequent conversion to ethylene by the enzyme ACC-oxidase. Ethylene synthesis can be suppressed by antisense cloning (of any of these two enzymes coding genes in reverse orientation) and cloned to a control region (promoters) that directs gene expression in the entire plant. Transgenic tomato plants have been developed by applying this approach and ethylene synthesis in transformed tomatoes was down to 5% of control fruits and ripening took four weeks instead of one week. Researchers at Monsanto Company have used a different approach to get the similar result. They introduced into tomatoes a bacterial gene that encodes an enzyme that degrades ACC (Amino cyclopropane carboxylic acid) as soon as it is synthesized which led to the suppression of ethylene synthesis and delay in ripening (10% reduction).

Genetic engineering technique has also been used to alter the softening of tomatoes which is associated with ripening. Softening occurs when the cell walls are degraded by cell wall degrading enzymes. In tomatoes, two such enzymes namely pectin methyl esterase and polygalacturonase have been identified. Transgenic tomato plants produced tomatoes that can be left on vine longer that normal tomatoes. The tomatoes do not get spoiled easily and can be harvested later. Calgene markets these tomatoes as Flavar Savr (R) tomatoes.

9. ENGINEERING OF MALE STERILITY

As many crop species are self-fertile, a desirable and increasingly necessary part of programmes for producing hybrid seeds is to make male sterile lines of these crops. Cytoplasmic male sterile lines of maize and other plants have been used in the past, but, were limited to particular cultivars, thus leaving them susceptible to rapid spread of disease.

A biotechnological solution to this problem was developed in 1990. Male sterile line of tobacco and oil seed rape were created by transforming them with chimeric RNase T1 and the *Bacillus amyloliquefaciens* RNase "*barnase*" genes driven by the anther tapetal layer specific promoter TA29. In transgenic plants containing these chimeric genes, the cells of the tapetal layer surrounding developing pollen were selectively destroyed by the RNase activity, which in turn caused failure of the pollen to develop. To restore fertility, TA29 :: *barnase*, rape plants, were crossed with plants containing a TA29 :: *barstar* construction. *Barstar* is a specific RNase inhibitor that forms a complex with *barnase* and inactivates it.

Table 11: Transgenic crops approved in USA for commercial use

Plants	*Genetically altered traits*	*Company*	*Year of approval & product if any*
Tomato	Delayed ripening	Calgene	1994, Flavr Savr
	Delayed ripening	DNA plant Tech.	1995, Endless Summer
	Delayed ripening	Monsanto	1995
	Thicker skin & altered pectin content	Zeneca/Petaseed	1995
Cotton	*Bt* gene incorporated Plants (Ballworm and Budworm resistant)	Monsanto	1995, Bollgard
		Calgene	1995 BXN Cotton
	Resistant to Bromoxynil	Monsanto	1996, Round Up Ready 1996
	Resistant to Glyphosate	Dupont	
	Resistant to Sulphoxylurea urea		
Soybean	Resistant to Glyphosate	Monsanto	1995, Round Up Ready
Potato	*Bt* Gene incorporated (Colorado Potato Beetle Resistant)	Monsanto	1995, New Leaf Ready
	Insect resistant (*Bt* gene incorporated)	Monsanto	1996
Maize	*Bt* gene incorporated (resistant to corn borer)	Ciba Geigy	1995, Maximizer
	Bt gene incorporated (resistant to corn borer)	Monsanto	1996, Yield Gard
	Bt gene incorporated (resistant, to corn borer)	Northrup King	1996
	Resistant to Glufosinate	Devalb	1996
	Resistant to Glufosinate	Agro Evo	1996, Liberty Link
	Male sterility	Plant genetic system	1996
Rape-seed/ Canola	Altered oil composition (high lauric acid)	Calgene	1995, Laurical
Squash	Resistant to viruses	Asgrow	1995, Freedom II

Thus the pollen in the anthers of F_1 progeny which contained both *barstar* and *barnase* had normal tapetal development and restored fertility. It has been observed that the TA29 promoter also functions in lettuce, chicory, cauliflower, tomato, cotton and maize making it generally useful promoter for engineering male sterility in these crops.

A second approach to the bioengineering of male sterility is based on the observation that certain naturally occurring male sterile lines of petunia are caused by the precocious expressing of callose wall degrading enzymes before pollen tetrad formation. Using this information, β-1,3 glucan hydrolase was placed downstream of the tapetal-specific *Arabidopsis thaliana* A9 promoter causing premature degradation of the callose cell wall surrounding microspores and abnormal microspore development in transgenic plants.

10. GLOBAL STATUS OF GENETIC TRANSFORMATION OF PLANTS

Progress in plant, genetic transformation has been spectacular since the recovery of the first transformed plants in the early 1980s. Molecular techniques have been applied to an array of species, resulting in the generation of numerous transgenic plants. During the last 15 years (1985-2000), about 25,000 field trials have been conducted with transgenic crops covering 60 species and 10 traits in 45 countries. The most frequent transgenic crops engineered herbicide tolerance, insect resistance, product quality and virus resistance under field trials are maize, tomato, soybean, canola, potato and cotton. By the end of 1999, about 50 transgenic crop products involving 12 crops and 6 traits were approved for commercialisation. A number of transgenic crops have been approved in USA for commercial use and some broad patents for plants, traits and technology or processes have also been granted (Tables 11, 12).

Table 12: Some broad patents for plants, traits and technology or processes

Patent	*Subject of Patent*
Plant patented	
W.R. Brace & ed.	All transgenic cotton All transgenic soybean
Calgene Inc.	All transgenic plants of *Brassica family*(Agrobacterium-mediated only)
Escagenetics Corp.	All transgenic coffee (*Coffea arabica*)
Biosem, Subsidiary of Limagrain, France.	All transgenic Cucumis *melo* modified using *Agrobactrium tumefaciens*
DNA plant technology	All transgenic pepper (Capsicum spp.)
Fresh World	All transgenic pea (*Pisum sativum*)
Plants with specific traits patented	
Dekalb Genetic Corpn.	All transgeenic plants with increased Lysine content
Lubrisol Corpn.	All sunflower plants/seeds with increased Oleic fatty acid (80%)
Pioneer Hi-Bred	* Sunflower products with low level of saturated fatty acids
Lucky Biotech Corp. & Univ. of California	Fruits/seeds/vegetables of transgenic plants with supersweet Thaumatin/Monellin genes.
Plant genetic systems	All transgenic plants with Bt gene (using *Agrobacterium*)
Upjohn Company	All transgenic plants from Cucurbitaceae and Solanaceae. Carrying "Cucumber Mosaic: Virus Coat Protein Gene"
Technology/Processes	
Enzo Biochem Inc.	Antisense Technology
Mycogen Corpn.	Process used to synthesize Bt genes

These are only few examples to the long list of successful genetic engineering. Crop improvement via genetic engineering is limited only by the ability to transform certain plants and to isolate genes that will give these plants new useful properties. Projects on regulation of fruit ripening in tomatoes, engineering plants resistant to insects and viruses, creating herbicide tolerance in plants and improving plant breeding techniques using genetically engineered male sterile lines are going on at various laboratories. It will also benefit industry by engineering plants to enable them to produce industrially useful chemicals and pharmaceuticals. By incorporating one additional gene into already established crop plant, it will be possible to develop plants that make valuable oils, starches and even biodegradable plastics. But there is always a caution needed in this work. Whatever may be the pros and cons, genetic engineering has proved a powerful technology and is certain to make impact on agriculture in coming decades.

FURTHER READING

Barnason, A.R. and Fry, J.E. Genetically engineering plants for crop improvement. *Science.* **224** (1989) 1295.

Copping, L.G. Crop protection agents from nature. In: Natural Products and Analogues. Royal Society of Chemistry, Cambridge, UK. (1996).

Day, A.G. and Lichtenstein, C.P. In: Plant genetic transformation. In: Plant Biotechnology (Eds. M.W. Fowler and G.S. Worren) Pergamon Press, Oxford (1992) 151-182.

Ghosh, P.K. Transgenic plants: Some current issues. *Current Science.* **70** (1996) 654.

Gupta, P.K. Transgenic plants and biosafety concerns in India. *Current Science.* **72** (1997) 172.

Horsch, R.B.; Fry, J.E.; Holfman, N.L.; Wallroth, M.; Eichholtz, D.; Roggers, S.G. and Fraley, R.T. A simple and general method for transferring genes into plant. *Science.* **227** (1983) 1229.

Lichtenstein, C. and Draper, J. Genetic engineering of plants. In : DNA cloning, volume II (Ed. D.M. Glover) IRL Press, Oxford (1985) 67-119.

Sanford, J.C. The biolistic process. *Trends in Biotechnology.* **6** (1988) 299.

Srivastava, D.K. Plant genetic engineering in crop improvement. In: Biotechnology in Agriculture and Environment (Eds. J.K. Arora, S.S. Marwaha and R. Grover) Asiatech Publishers, India (1998) 39-56.

Zambryski, P.; Joos, H.; Genetello, C.; Leeman, J.; Van-Montagu, M. and Schell, J. Ti-plasmid vector for the introduction of DNA into plant cells without alteration of their normal regeneration capacity. *Eur. Molecular Biology Organization J.* **2**: (1983) 2143.

Altering Biological Software in Crop Plants: Quantum and Quality of Produce

C.P. MALIK, A. SHARMA*, N. DILBAGHI
and
M. PAUL

Department of Biotechnology
** Department of Food Technology*
Guru Jambheshwar University
Hisar 125 001 (Haryána)

IN THE COMING FEW DECADES, global demand for farm products is likely to increase several-fold owing to increased population and enhanced purchasing power. The past two decades have seen adoption of modern techniques of agricultural production to increase yield. Several thinkers have estimated that India shall be required to produce 250 million tonnes of foodgrains in the next few years to meet the dietary needs of the people. Since 1965, various programmes were launched to boost the yield of various crop species and these include expanding irrigation facilities, promotion of fertilizers and pesticide industries, invoking of 'High Yielding Varieties Programme' as multiple strategies to increase production. Both geneticists and plant breeders were called upon to reconstruct crop varieties which suited new systems of farming and matched improved agronomic management. The intent was to evolve varieties with a better balance of source and sink capacity and such accomplishments were clearly visible in wheat and rice and in them, dwarf genes were increasingly exploited to redistribute dry matter in order to harness enhanced harvest index. Consequently, several wheat and rice varieties were evolved and released. Over the years, the environmental, economic and social sustainability issues have been consistently raised and called for afresh look on the prevailing system(s) of agriculture. It seems that 'Gene Revolution' over the years has fatigued and now one has to exploit biotechnology which integrates the principles of ecology, economics and equity'. It seems now that a biological software industry for sustainable agriculture can be promoted in various ways and thus desired archetypes in cereals, oilseeds, pulses and forest trees could be evolved. In this discussion two biological softwares viz. bioregulators and genetic engineering are shown to alter plant archetypes to achieve high yield.

Till recently, the crop productivity was increased through the use of fertilizers, pesticides, better water management and exploitation of high yielding varieties. Over the years, the yields of several major food crops appears to have plateaued and even with increased inputs, the total productivity has not gone up. In the last two decades, some economists in USA suggested that crop productivity could be altered favourably by increasing photosynthesis; stimulating biological nitrogen fixation; through micropropagation; genetic engineering and through the use of plant growth regulators. It is well accepted that normal plant growth and development is controlled by chemicals produced endogenously in a plant. The biosynthesis of these chemicals is under genetic control and these compounds are called plant hormones or phytohormones. There are large number of synthetic chemicals which are becoming increasingly valuable in the regulation of crop growth, both in agriculture and horticulture. One view is that synthetic growth regulating substances cause their effects through altering the endogenous levels of the naturally-occurring hormones-thereby modifying growth and development in the desired direction to the intended extent.

The use of plant growth regulators has been practised since times immemorial and induction of rooting of cuttings is the best example of their use. The best known PGR is Indole Butyric acid (IBA) which has recently been shown to be naturally occurring. It is interesting that Benfey, P. of New York University in a publication has reported identification of a gene (SHR) that governs as to how plant cell organizes into root systems.

1. PLANT GROWTH REGULATORS AND CROP PRODUCTIVITY

The suppression of sprouting of onions and potatoes was affected by treating with maleic hydrazide and this compound was extensively used till 80's. It was also used to suppress the growth of lawn grasses in golf courses etc. The next few years experienced the wide use of a regulator CCC (chlormequat; cycocel) which dwarfed the height of wheat and barley without affecting the size and quality of the ears. This reduced height prevented lodging after rain or storm. Another process which reduces yield is abscission of fruits and leaves. Naphthaleneacetic acid was used to thin apple blossoms. Thus, lesser fruits resulted in heavier fruits and hence added yields. Imposed abscission was used to harvest/pick oranges, walnuts, cherries, olives and other fruits. Thus, hormones could help in hand picking, mechanical picking or by shaking.

Indeed plant growth regulators have an impact on the economic production of crops by saving energy and also saving in cash outlay by increasing profitability. The yield of a crop if increased for a given area of land resulted in substantial gains. Thus no extra land needs to be purchased and no extra soil was to be tilled and no extra fertilizers or weed control practices were to be taken into account. Thus growth regulators could be exploited to synchronize flowering and uniform fruiting and seeding and therefore eliminating the need for repeated harvests. The economic products such as oil, protein or sugar could be increased in a given crop by the proper use of pertinent plant growth regulators. In Indian scenario what can plant growth regulators achieve is best illustrated by some of the examples like groundnut, cotton, mustard, rice, etc. To mention, the control of rooting, delayed abscission, early flowering, regulation of fruit set, development of more flowers, control of plant organ size and regulation of abscission, regulation of metabolic processes, control of maturity of fruit and overcoming environmental stresses could be caused. Additionally, early and uniform seed germination; induction or breaking of dormancy, causing plant resistance to disease(s), maintenance of fruit firmness in storage condition, prevention of fruit harvest, spoilage and regulation of colour of the fruits as well as their taste could also be achieved by controlled spraying with specific plant growth regulators. The plant growth regulators that have been profitably used in some of the major crop plants have demonstrated that applications of some of the growth regulators like cycocel, ethephon, paclobutrazol in rice, wheat and maize could cause retardation of height and preparation of the plant for better harnessing of solar radiation, water and nutrients resulting in enhanced crop productivity. Additionally, compounds like humacil, aliphatic alcohols, cytozyme, hico-emulsion were successfully tested and shown to increase number of tillers and increase yield. Some of the steroidal regulators that have been demonstrated to be potent biological activity promoters, result in boosting of yields and also altering the quality of lipids as well as carbohydrates. In pulses, groundnut, chickpea, lentil, pigeon pea and moongbean, long chain aliphatic alcohols, phenolic compounds and PCB cause retention of more flowers and enhanced fruiting, delayed senescence of fruits, prolonging of flowers, increased photosynthesis resulting in boosting of their yields. Similar studies have been reported in case of cotton, sugarcane, oilseeds and some of the horticultural crops like apples, mango and vegetables. In oilseeds e.g. groundnut, poor yields are generally attributed to several physiological constraints. Apart from diseases, pests and unpredictable rains also reduce yields. One of the major bottlenecks for high yields is indeterminate growth habit of plants resulting in interorgan competition for photosynthates. Some of the growth regulators like PCB have been effectively used to suppress excessive vegetative growth concomitant with enhanced production of flowers at short time intervals and enhanced pod filling. There is also stimulation of photosynthetic enzymes and increased oil accumulation.

In mustard, there are several constraints which limit yield and these are indeterminate growth habit; lodging of plants; shattering of fruits, etc. Some of the growth regulators like PCB and phenolic compounds have been used to demonstrate that indeterminate growth could be checked; plants could be dwarfed to prevent lodging and improve the plant archetype and canopy; induce uniformity of flowering and fruit set; induce synchrony of flowering and fruit set by reducing the flowering period. Seed quality is also improved by altering fatty acid composition and lowering glucosinolate contents by stimulating the mobilization of stored assimilate from stem to developing fruits and seeds. Studies have demonstrated that PGRs could be used to stimulate metabolic and cell division activities in shoot meristem and hence altering the vegetative and reproductive structures resulting in increased biomass and mobilization of dry matter to the sinks.

Genetic studies in *Arabidopsis* have led to the identification of many loci concerned with ABA responses. The ABA response mutant phenotypes were of various types e.g. defects in seed storage reserve accumulation, maturation, dormancy and changed sensitivity to ABA, or control of germination inhibition, stomatal regulation, root growth and expression of several stress-induced genes. Interestingly most of the mutants were stage-specific defects and possibly these loci acted in multiple overlapping response pathways. So far only five mutationally identified ABA response loci have been cloned and these represent three classes of proteins. These are : two orthologous transcriptional regulators (Vpl) in maize, ABA-insensitive 3 (Abl3) of *Arabidopsis,* two homologous members of phosphatases 2C family (Ab11 and AB12) of *Arabidopsis* and farnesyl transferase (ERA1) of *Arabidopsis.* In a recent study, positional cloning is used to isolate AB14 gene and its identity was confirmed using complementation analysis. Interestingly, the protein product showed homology to APETALA2 domain. Expression analyses indicated that AB14 expression was not seed specific. Further AB14 and AB13 appeared to be similar and both appeared to be transcriptional regulators. But expression analyses suggested that AB14 expresses in vegetative and seed tissues. Some research groups used dwarf mutant (saxl) of *Arabidopsis thaliana* which exhibited changed sensitivity of growth responses to ABA, auxin, gibberellins and ethylene and partially overcame by exogenously applied brassinosteroid. The studies provided direct evidence for the existence of interactions between brassinosteroid and auxin, ABA, GA and ethylene signaling pathways.

Two classes *of Arabidopsis thaliana* ABA mutants have been characterized and these are (1) biosynthetic mutants which have extremely reduced amounts of ABA and (2) those which are insensitive (abl) or hypersensitive *(era)* to ABA. All the three mutants (aba-1, aba-2, aba-3) do not synthesize ABA *de novo.* Several putative intermediates concerned with ABA responses are reported. The genetic analyses have been carried out to show that the *adel* mutant is a monogenic recessive trait and a role has been defined for negative regulator function in ABA signaling. New methods have been developed to understand the biology of cytokinins. These include: activating tagging—to identify genes whose overexpression mimics cytokinin activity e.g. *CK11, CK12* which code for proteins with homology to histidine kinases of the bacterial two-component signaling system. Database analyses—to demonstrate the existence of G-protein-coupled seven transmembrane receptor kinases. Studies have been done in *Arabidopsis* plants wirth reduced expression of one of the corresponding coding genes, *CCR1* having reduced cytokinin sensitivity.

From these studies it has emerged that *CKII* and *CCRI* proteins play a role in perception and/or transducation of the cytokinin signal. Recently primary cytokinin response gene has been identified and these genes code homologs of bacterial response regulators *(ARR/IBC).* These rapid response genes will provide the clue to establish tests for fast cytokinin action. Recent years have also seen added information on the study of cytokinin mutants and cytokinin over producing transgenic plants.

A pivotal development in the investigation of plant hormones over the last one decade has enhanced use of single-gene mutations and the studies pertaining to their molecular actions. In cytokinins, the genetic approach was tricky since it caused pleiotropic effects in auxtrophic mutants.

There are large number of studies which implicate the role of salicylic acid (SA) in plant disease resistance. SA is shown to be a key component in triggering resistance via a signal transduction system. In fact when the plants are challenged with an avirulent pathogen, the interaction of the products of the pathogen's avirulence and host's resistance genes trigger resistance and this signal transduction system comes into play. Tobacco cultivars having the N resistance gene, when inoculated with TMV induces the HR leading to host cell death at the infection site. The necrotic lesions produced restrict the spread of TMV. The triggering of the HR is accompanied by enhancement of SA levels. Concomitantly there is systemic acquired resistance (SAR) accompanied by an increase in SA and the plant parts acquire an increased degree of resistance to secondary infection. Several studies have demonstrated that SAR is effective against TMV and a broad spectrum of other pathogens. The rise in levels of SA and the induction of SAR are accompanied by the activation of range of genes including those which encode pathogenesis-related proteins. Some workers have used PR proteins as molecular markers for SAR induction and several of these proteins are antibacterial or antifungal. Several evidences have shown that it is SA which confers resistance to TMV and SA was required to establish both local and systemic acquired resistance to TMV. The alternate oxidase in the signal transduction pathway leading to SAR against viruses has also been implicated. It has also been mediated via localization early in the HR to TMV.

Ethylene also plays an important role in ripening of kiwifruit though it does not produce ethylene for a long time without exposure to exogenous C_2H_4. Recent studies establish that there are two systems of C_2H_4 production involved in the ripening process. Suggestively system 1 is the low level of ethylene present and exists before the ripening process while system II represents autocatalytic burst of gas production and accompanies ripening. Fruit ripening represents a biological process unique to plants where developmental and hormonal signaling systems orchestrate several biochemical and physiological changes which finally result in 'ripe' stage of fruit maturation. Most studies have focussed on the genetic basis of ethylene biosynthesis and resulting terminal. Genes coding in the regulated steps in ethylene production have been isolated, and genes responding to ethylene during ripening have been identified. Promoter sequences from several of these ethylene-inducible genes have been isolated and characterized. Gibberellin (GA) is involved in regulating a number of processes during growth and development, including cereal grain germination, elongation, flowering, and fruit development. Such regulations are dependent on the presence of active GAs and the ability of cells to perceive and respond to the hormone. Enough studies exist on the regulation of GA levels, elucidation of its biosynthesis and more recently by cloning GA biosynthesis enzymes. GA response mutants have been used to undertake molecular genetic approaches and two kinds of mutants have been used: GA-insensitive dwarfs and constitutive GA response slenders. Most of the mutants have been obtained from barley or *Arabidopsis*. More recent study has demonstrated regulation of gibberellin biosynthesis genes during flowering and early fruit development of tomato. Several genes for GA biosynthetic enzymes have been isolated and characterized. This is a simplified GA-biosynthetic pathway. The available studies have shown that GA 20-oxidase may have an important role in regulating levels of active GAs and hence in controlling plant development. That GA 20-oxidase is encoded by more than one gene in several plants suggesting that the members of this gene family may be differentially regulated. The cloning of tomato GA 20-oxidase genes would enable to investigate the involvement of GAs in flower bud development at the molecular level. Very recently the cloning of three GA 20-oxidase cDNAs from tomato has been done. It has been demonstrated that transcript levels of GA biosynthesis genes are highly regulated during flower bud development. Thus GA biosynthesis could be stimulated by over expressing GA 20-oxidase, a key enzyme responsible for the production of biologically active GAs from inactive 20-carbon precursor gibberellins. In fact in some of the trees where such a cloning was done, xylem fibre increased by 71% and fibre length enhanced by 8%.

Despite great advances, information regarding the site of synthesis of bioactive GAs and their specific mode of action in precise cells and tissues is far from clear. Again it has been shown that GA_1, the bioactive GA, is not transported unlike its precursor GA. Further it is now well accepted that GAs are mainly present in actively growing and elongating tissues such as shoot apices, young leaves and flowers. One study has also demonstrated that GA action in various organs may depend upon the amount of endogenous GA present which in turn depends on the regulation of expression of GA biosynthesis enzymes, rather than on translocation of bioactive GA to the site of its action. It has also been demonstrated that Nty (*Nicotiana tabacum* GA 3B-hydroxylase) expression is localized in regions that are known to have GA activity. The possibility that GA action depends on modulation of the level of bioactive GA and controlled by the expression of GA 3B-hydroxylase has been suggested. Some workers have demonstrated that synthesis of EPB, a cysteine proteinase is responsible for the degradation of seed endosperm storage proteins in *Hordeum vulgare* and is induced by GA and repressed by ABA. This gene consists of two very similar members, EPB-1 and EPB-2 with the former being highly induced by GA. This investigation has also demonstrated that ABA was effective in blocking the GA induction of EPB-1, but, did not affect constitutive expression of a transcription factor GAM $_{yb}$-mediated expression of EPB-1. It has been concluded that ABA acts upstream of the formation of functional GAM $_{yb}$ which co-ordinates the hormonal regulation of a diverse group of genes in cereal's aleurone layers, including those encoding EPB and α-amylase.

Briefly, genetic engineering techniques will be useful chiefly in the initial step of plant breeding; enhancing the gene pool available to the breeders, providing better selection criteria for some traits and development of new stocks. Plant growth regulators (PGRs) have demonstrated substantial upsurge in synthesis and utilization in the last decade or so and have emerged as magic chemicals which could be profitably exploited to overcome physiological constraints leading to increased crop productivity, and alter the produce qualitatively. The performance of these chemicals is unprecedented and some of them are demonstrated to circumvent many of the barriers, which are otherwise imposed by environment or even genetics. This is certainly an important area in bioengineering. Recent developments in molecular biology of seed development and plant biotechnology have sparked a scientific revolution which, in the 1980's resulted in the first generation of biotechnology derived products based on rDNA and cell fusion techniques. Contemporary food industry biotechnology is still at the stage of testing novel products. The general public needs to be educated on safety, health and economic benefits and risks of bioengineered products. Plant sciences are also filled with excitement, intrigue fascination and fulfillment in some of the oil-seed crops. Their role has been clearly demonstrated to cause quantum jumps in productivity and change the quality of oil.

The classical breeding approach involved the mixing of genetic information from the two parents through sexual reproduction. Then repeated backcrossing to the parental commercial variety eliminates trait of the wild parent mostly. The newly developed variety had sweetness trait, but, lowered fertility, since the breeder did not have absolute control over exactly what information was retained in the new tomatoes, they tried to minimize the unwanted information from the wild species. Despite this effort, they were unable to eliminate everything, but the sweetness characteristic. They ended up with a new variety that still had a substantial input from the wild species and in that information was the sweetness characteristic and something, that caused lower fertility.

Through genetic engineering approach, a single piece of genetic information was located which when removed, slowed the breakdown of sugar. Through genetic engineering, that characteristic was introduced into the commercial variety, which stops sugar degradation in the commercial tomato. This is done by specific change in a single gene. In contrast to traditional, new technology helped in securing information on breeding, in fact it would have been possible to take that new genetic information from another plant species, a microbe, or any other organism. It is possible to understand precisely when and where this newly acquired information will be applied using characterized regulatory regions that function when and where they are required.

2. IMPACT OF BIOTECHNOLOGY ON AGRICULTURE

Through the rDNA technology, there are several engineered crop species whose products are available in the field and the market. The genetic engineering can result in more environment friendly plant pest-protection, foods with increased nutrition, more accurate and sensitive diagnostics, foods with improved processing and marketing traits, better and more efficient medical delivery, new methods for removing contaminants from soils and waters, and prepare products that are presently being made from nonrenewable resources. Many of these alterations could be accomplished through plant growth regulators.

The products of this genetic engineering (GE) are no more confined to the laboratory. These and other new crops represent a substantial percentage of actual production acreage in USA and 11 more countries. In 1998, 50% of the cotton acreage, 30% of the soybean, and 20% of the maize acreage in the USA were genetically engineered. In USA, some 60 million acres of GE crops were planted in 1998. Between 1996-2000, twelve countries (eight industrialized and four developing) have contributed to more than 20-fold increase in global area of transgenic crops. During 1999, global market for transgenic crop was US $ 2.2 billion. In Table 1, a rough estimate of land under transgenics is given.

Transgenics are being made both in food and cash crops. In USA, several crop plants have been released for commercial cultivation and these are herbicide tolerant, and insect and disease resistant. Some transgenics have nutritional and

Table 1: Estimate of land under transgenics

Country	*Land under transgenic (mha)*
USA	28.7
Canada	4
China	0.3
Argentina	6.7

post-harvest quality improvement traits. The transgenic tomatoes (*Flavr savr*—a Calgene product), having delayed fruit ripening genes were first introduced in USA on commercial scale. The major transgenic crops are soybean, maize, cotton, rapeseed, and potato, in descending order of acreage under cultivation. Most of these crops had high yields and hence added profits. One estimate has shown that farmers who cultivated transgenic cotton (for *Bt*) economically gained $128 million and $133 million during 1996 and 1997, respectively, in USA. China has commercialised *Bt* cotton; sweet pepper, delayed ripening tomatoes and have also accepted GM food. In India, Review Committee on Genetic Manipulation (RCGM), Department of Biotechnology (DBT) have permitted limited trials of *Bt* cotton. MAHYCO received permission to conduct large-scale trials and seed production of GM *Bt* cotton. In fact MAHYCO has been conducting trials for the past four years. RCGM has recommended large-scale trials of Bt cotton in different parts of the country. The transgenics are least dependent on insecticides, herbicides, give high yields, produce clean and high quality grains/products. There is an increased acceptance by the consumer and also transgenics give high economic returns to the farmers.

One report mentions that India's RCGM has approved ProAgro-PGS plans to conduct limited trials on multiple locations of its transgenic mustard. The trials shall be conducted at 19 locations in India. The company has inserted the *barnase barstar* and *bar* genes into *Brassica juncea* plants for hybrid seed production.

During 2000, the total world hectarage of genetically improved crops increased by 11% i.e. 39.9 million ha (mha) in 1999 to 44.2 mha in all six continents as reported by the International Service for the Acquition of Agri-biotech Applications (ISAAA), an independent agency monitoring the use of biotech crops globally. The number of countries planting these crops has gone up to 13-eight of them are industrialized countries and five are developing nations. The figure of 44.2 million hectares of genetically improved crops represents a 25-fold increase in five years, from the 1.7 million ha planted in 1996. The adoption rate of biotech crops is the highest for any new technology by agricultural industry standards. According to ISAAA, global acreage of GM crops planted from 1996-2000 is given below:

Year	*Million ha*
1996	1.7
1997	11.0
1998	27.8
1999	39.9
2000	44.2

The above figures clearly show growing acceptance of transgenic crops by farmers using the technology in both industrial and developing countries. This acceptance speaks volumes of the confidence and trust farmers have placed in transgenic crops. One estimate indicates that 10% of China's cotton area is now under transgenic cultivation.

Globally, herbicide tolerant crops were more in demand, with insect resistant crops among the next. During 2000, herbicide tolerance deployed in soybean, corn and cotton occupied 74% of the 44.2 mha with *Bt* insect resistance crops taking up 19% and stacked genes for herbicide tolerance and insect resistance in cotton and corn accounting for 7%. The world figures for the total hectarage planted is as follow:

- 36% of the 72 mha of soybean
- 16% of the 34 mha of cotton
- 11 % of the 25 mha of canola and
- 7% of the 140 mha of corn are genetically improved.

In all, 44.2 mha of the total global hactarage of 271 mha under these four crops is now transgenic. Of the 271 mha, two-thirds are in the developing countries where yields are lower, natural resource constraints are high and require improved food, feed and fibre production. One additional study has demonstrated that the US grew genetically improved crops on 30.3 mha followed by Argentina with 10 mha, Canada with 3 mha and China with 0.5 mha. On percentage basis, US grew 68% of the global transgenic crops followed by Argentina (23%), Canada (7%) and China (1%).

2.1 Herbicide Tolerance

The identification of selective herbicides that could eliminate weeds without affecting the crop plant was a pivotal advancement in agriculture. But selectivity has been problematic in several crops. One alternative was to identify new selective herbicides by genetically engineering the crop to withstand deleterious effects of herbicides. The mechanisms included modifications in the site of action of the herbicide, detoxification or in the uptake and translocation of the herbicide to the target site. By the end of 1998, USDA Animal and Plant Health Inspection Service (APHIS) issued 5900 permits and 1569 pertained to herbicide-tolerant crops. Currently six herbicides: bromoxynil, glyphosate, imadazolinones, phosphinothricin, sulfonylurea, and 2,4-D are being investigated. The herbicide tolerance in most instances is achieved by modifying site of action through mutants or altered forms of the target protein. It is desirable that the altered enzyme must be properly compartmentalized (chloroplast, cytoplasm, nucleus etc.) and should be assembled in a metabolically active form at its target site, as in glyphosate where the herbicide blocks a pathway that causes synthesis of several amino acids; blockage of this pathway leads to non-availability of amino acids. Genetic engineering techniques have endeavoured on using mutant target enzymes resistant to glyphosate. Similar strategy is used with the imadazolinones and sulfonylureas, which inhibit a pivotal step in the biosynthesis of different amino acids.

Herbicide resistance can also be engineered by introducing genes for enzymes that degrade and detoxify the herbicides e.g. bromoxynil where resistance was engineered by introducing a bacterial gene that codes for an enzyme that converts bromoxynil to an inactive metabolite. Glufosinate is another example of this category where resistance is engineered by using a bacterial gene that acetylates and inactivates the herbicide. Hence, a bacterial gene that encodes an enzyme that degrades 2,4-D is engineered in cotton, which is resistant to this herbicide to prevent crop losses due to spray drifts.

In engineering herbicide-resistant crops, some problems are encountered. There is need for finding an effective strategy for engineering resistance and identifying the target site of the herbicide or a degradative or inactivating factor. Further, the resistance factor must express in the target place and in sufficient levels.

The herbicides for which tolerance is being biotechnologically engineered are generally used at low rate and have low-toxicity. Even then some environmental issues must be taken into account and these are:

- **Origin of herbicide-tolerant weeds following monoherbicide application.** Frequent and long-term applications of a single herbicide may lead to specific herbicide-tolerant varieties. Thus engineering a large number of crop species tolerant to a single herbicide might create some problems.
- **Transfer of resistance genes to related weeds.**
- **Shifting of weeds not affected by the herbicide,** i.e. weed shift.

2.2 Insect Resistance

Insect resistance in crop species is engineered through the use of genes from *Bacillus thuringiensis* e.g. in bollgard cotton and the commercially available maize which is resistant to European corn borer. These toxins are activated by cleavage in the midgut of insects; however, making engineered toxins that are effective against certain insects has been complicated. For example *Spodoptera* armyworm insect pest of alfalfa required extensive *in vitro* testing of synthetic toxins led to identification of specific synthetic delta-endotoxin gene, which when expressed optimally in alfalfa, ultimately provided protection against the Egyptian cotton leaf worm (*Spodoptera littorals*) and the beet armyworm (*Spodoptera exigua*). In some cases, new types of *Bt's* and also some insecticide compounds from other microbial species, spiders, scorpions are being explored.

2.3 Viral Resistance

Viral infection in many cases is a problem in several crop species; however, the sources of natural resistance are limited and in many cases, chemical protection is not effective. Viral resistance is engineered by the use of the coat protein gene. This approach has been used successfully for many viruses of different plant species and the Freedom II squash was released in last year. In the coming years, genetically engineered papaya will be released protected against papaya ringspot virus, which once devastated certain production areas in Hawaii. Due to limitations to the coat protein approach for uniform applicability and broad protection to related viruses, it is imperative to develop strategies that complement or replace the coat protein strategy. These include the use of viral polymerase and movement protein genes; such strategies require an in-depth understanding of the architecture of the virus. Use of all these viral genes has led to a reduction in the damage incurred by the engineered plant following viral attack.

Natural defence genes have been isolated for resistance to various pathogens in different plant species. Of the first three such genes that were isolated, it was discovered that they share common domains. Once these domains and their function are understood, it will be possible to devise plant protection strategies based on these genes that confer broad-spectrum resistance against various pathogens.

2.4 Nematode Resistance

To protect the crop against damages from nematodes several approaches have been made. Hence, regulatory regions have been identified which are concerned with the nematode burrowing into the root; thus, physical damage of the root is not enough to turn on the promoter. This regulatory region is then hooked to nematicidal proteins, which are delivered to the nematode at its precise site of attack. The nematicidal proteins are several different proteinase inhibitors and lectins, which act in some instances synergistically to interfere with the sexual development of the nematode.

Recently a natural resistance gene has been isolated from tomato, which provides protection in engineered plants against nematodes and aphids. A comparison of the gene to various natural resistance genes (fungal, viral and bacterial) from other species indicate that the nematode resistance gene shares common domains with these other genes suggesting, universality of these mechanisms and raising hopes to induce some engineered proteins with widespread efforts.

2.5 Harvest and Post-Harvest Characteristics

Another area of active research relates to harvest and post-harvest characteristics. The ripening of fruits is commercialized. The approach of the *Flavr Savr* tomato utilized a technique allowing growers and producers extra time to delayed picking and delivering in two ways in tomatoes by interfering with polygalacturonase synthesis and hence ripening is delayed by several days, but, was not stopped. The approach pertains to the stopping of the ripening process by interfering with the production of ethylene, which is essential for the ripening process. In the second approach, it is possible to leave a tomato fruit on the plant for up to 3 months and then picked. The ripening is induced in tomato when exposed to ethylene/ethaphon. The endless summer tomato was marketed in 1997. This approach is also being applied in melons, raspberries, and other spoilage sensitive fruits. The approach is being applied in the cut flowers industry and the gene involved is called the "anti-aging gene." This gene stops the production of ethylene and thus cut flowers senescing.

Uniform ripening is another post-harvest characteristic where genetic engineering technologies are now being applied. This was successfully done in cherry tomato and the basic approach is also being used with the processing tomatoes to lessen waste and help in harvesting. In this technique, genes involved in slowing the degradation of ethylene are manipulated.

Fruits and vegetables contain high percentage of water and thus show high metabolic activity and during post harvest stage are vulnerable to perishability. For successful transportation and marketing of fresh fruits, it is essential to prolong the shelf life of major commercial crops. In several fruit crops, refrigeration is not a successful venture and hence solutions to post harvest problems of fruit are sought in biotechnology. Hence molecular biology techniques are increasingly exploited to regulate ethylene synthesis, pigment production and starch degradation. Such manipulations are restricted to climacteric fruits and tomato is one good example where fruit softening has been influenced by the down regulation of the polygalacturonas gene. Similarly, ethylene synthesis has also been manipulated through genetic modification of fruits.

2.6 Male Sterility

The hybrid varieties of crops show increased yields, but, the hybrid production is a complicated process and in many instances, manual emasculation of the male parent needs to be done to preclude self-fertilization. If large-scale production of hybrid seed is intended, it is imperative to use efficient mechanisms to suppress pollen production and/ or function. For several years, geneticists identified naturally occurring male-sterile lines of some crops and used them for commercial hybrid seed production. Such strategies also required the identification of genetic "restorer lines" where counteracting of the male sterile genes was possible to restore male fertility. This was easy in some crops, but, was tedious in others. Through genetic engineering, it has been possible to develop male sterile lines. Thus, regulatory sequences were identified which had very specific tissue expression. In one instance, a promoter was identified that caused expression in the tapetum. A bacterial enzyme, which degrades RNA was hooked to this promoter and hence development of pollen grains was disturbed. A restorer of fertility was also identified; this protein is a chemical coupler of the RNA-degrading enzyme; the tight coupling of these two proteins inactivated the RNA-degrading enzyme, and therefore restored fertility. This technique was employed in canola initially and is now being used in several different crop and vegetable species.

2.7 Technology Protection System or Terminator Technology

This approach is a relatively new one patented by a small cotton company, 'Delta and Pine Land, and the USDA; the Monsanto Corporation has made a bid to buy DP&L, but that deal is still awaiting approval by the Department of Justice. The fundamental concept behind this technology is to preclude reproduction by interfering with seed development, thus making sure that the user of the seed cannot replant, necessitating purchase of the seed afresh. In this way, the company's investment in developing the variety is protected.

The system is based on participation of three introduced genes. One of these genes produces a toxin that kills the seed in the final stage of development. This gene is initially interrupted by a stretch of DNA between the toxin gene itself and the regulatory region or promoter that is responsible for its synthesis. This gene configuration can then be transplanted or crossed with another genetically engineered plant that produces a scissors-like enzyme, capable of removing the segment in the toxin gene, thereby activating it and causing the seed to abort. The third part of the genetic system is a represser gene that prevents the synthesis of the scissor-like enzyme. If all three components are present, the terminator technology is dormant and the seed is fertile. If the seeds are then treated with a stimulator, in one case an antibiotic called tetracycline, this compound can then block represser of the scissor-like protein and this gene product can then snip out the DNA that prevents the synthesis of the toxin, thereby turning on synthesis of the toxin gene. Plants grown from such seeds will then produce sterile seeds incapable of reproducing. DP&L is currently pursuing this technique in cotton and wheat and expect these products to be on the market by 2005. With the entry of Monsanto into the equation application of the technology could extend to many more crops and its availability could be much more rapid. There have been many debates about this technology since it is thought to interfere with farmer's rights and ensures that farmers return every year to purchase a new lot of seeds.

While certainly there is a negative aspect to the technology related to farmer's rights, There is also a counter perspective on the use of such technology. The concern held by environmentalists and certain agriculturists is that certain traits, like herbicide tolerance and certain other traits, might escape from the crop plant and enter into wild species around, thereby making it difficult to control the wild species. This issue has been raised with, for instance, sorghum and Johnson grass, oat and wild oat, rice and red rice, canola and its wild *Brassica* relatives.

2.8 Value-Added Products

The another area that involves plant growth regulators and is being vigorously pursued by the public and private sectors, pertains to the improvement of quality products, over the available ones. This aspect also calls for new or improved functions which could be identified and help enhance the value of the product. Thus speciality oils from rape or soybean, which are designed to have improved heat stability are evolved. Another instance would be the genetic engineering of sugarbeet to produce fructans, a sweet, but no-calorie sugar substitute. Recent years are experiencing upsurge interest in the area of nutraceuticals or functional foods. Natural products in foods are identified that can either cause adverse human condition or prevent disease(s). Examples of those products focused on factors in foods that cause disease or undesirable pharmacological effects include ameliorating the allergenic potential of foods, such as wheat, milk, and peanut or removing caffeine from coffee. Transgenic potatoes are being raised which produce high quantities of GAD, an enzyme believed to be involved in the prevention of type II juvenile diabetes. Researchers have identified an element in grapes that increases blood flow in blood vessels; grapes could be engineered to increase the level of such an element. Several groups are corning up with margarine and oils, based on natural plant sterols and stanols, that are "low in cholesterol". Another approach is designing/engineering food to vaccinate the consumer against disease, an idea that was once confined to the research laboratory, but, is a reality now.

FURTHER READING

1. Cercos, M., Gomez-Cadenas, A. and David Ho, T.H. *The Plant J.* **19** (1999) 107.
2. Ephritikhine, G., Fellner, M., Vannini, C., Lapous, D. and Barbier-Brygoo, H. *The Plant J.* **18** (1999) 303.
3. Finkelstein, R.R., Wang, M.L., Lynch, T.J., Rao, S. and Goodman, H.M. *The Plant Cell* **70** (1998) 1043.
4. Foster, A. and Chua Nam-Hai, *The Plant J.* **7** (*1*999) 363.
5. Giraudat, J., Parcy, F., Bertauche, N., Gosti, F., Leung J., Morris, P.C., Bouvier-Deerand, M. and Vaxtamian, N. *Plant Mol. Biol.* **26** (1999) 363.
6. Hedden, P. and Kamiya, Y. *Annu Rev. Plant Physiol. Plant Mol. Biol.* **48** (1997) 431.
7. Itoh, H., Tanaka-Ueguch, M., Kawaide, H., Chen, X., Kamiya, Y. and Matsuoka, M. *The Plant J.* **20** (1999) 15.
8. Malik, C.P. Advances in Plant Hormones Research: Indian Scenario, Agro-Botanical Publishers, Jodhpur, 1999.
9. Malik, C.P., Thind, S.K. and Bhatia, D.S., *Ann. Rev. PI. Physiol.,* **2** (1995) 13.
10. Malik, C.P., Thind, S.K. and Bhatia, D.S., *Current Researches in Plant Sciences* . (1994) 271.
11. Robers, M., Kaneta, T., Kawaide, H., Yamaguchi, S., Yang Young-Yell, Imai, R., Sekimoto, H. and Kamiya, Y. *The Plant J.* **17** (1999) 241.
12. Schmalling, T., Rupp, H.M, Frank, M and Schaffer, S. In: Advances in regulation of plant growth and development. (Eds. M. Strand and P.P. Beck) Peres Publ., Prag (2000) 85.
13. Swaminathan, M.S. *Natl. Acad. Ag. Sci.,* IARI (1993) 1.

Biofertilizers for Sustainable Rice Production

S. KANNAIYAN

Tamil Nadu Agricultural University
Coimbatore (Tamil Nadu)

RICE CONTRIBUTES 20 PER CENT of total world food grain production and it is the staple food for major population in Asia. The increase in the human population continues to reduce the per capita rice land in most of the rice growing countries in the world. Nitrogen is an important growth limiting factor for rice, which largely determines the yield. N_2 fixation plays an important role in the nitrogen economy of the rice field ecosystem.

The role of biofertilizers in sustainable agriculture assumes special significance particularly in the present context of very high cost of chemical fertilizers. Flooded rice fields are an ideal ecosystem for N_2 fixation by non symbiotic, anaerobic and aerobic microflora. The main agents of biological nitrogen fixation (BNF) in rice fields are free living BGA, symbiotic BGA with *Azolla* and heterotrophic bacteria in the rhizosphere and soil. *Sesbania rostrata*—a leguminous stem nodulating green manure also plays a major role in sustainable agriculture through N contribution.

1. *AZOLLA* BIOFERTILIZER

Azolla is a free-floating waterfern which in symbiotic association with the cyanobacterium — *Anabaena azollae* contributes substantial amount of biologically fixed nitrogen to the rice crop. The nitrogen fixing waterfern, *Azolla* has been used as a biofertilizer for rice crop in India, China, Vietnam, Thailand and Philippines. The *Azolla-Anabaena* association is a "living floating nitrogen manufacturing factory" which is utilizing the energy from photosynthesis to fix atmospheric nitrogen amounting to 150 to 200 kg per hectare per year. *Azolla* is used as biofertilizer for rice crop and it could contribute 40 - 60 kg N ha^{-1} per crop The important factor in using *Azolla* as a biofertilizer for rice crop is its quick decomposition in soil and efficient availability of its nitrogen to rice plants.

In tropical rice soil, the applied *Azolla* mineralizes rapidly and its nitrogen is made available to the rice crop in a very short period. The growth rate of *Azolla* is very rapid so much so that it doubles its weight in mere 2-3d. Due to fast relative growth rate, the biomass production is heavy in a given time and hence the biomass is exploited as biofertilizer for rice.

Sporulation in different *Azolla* species is not well understood and the factors responsible for sporulation and spore germination are not well defined. *Azolla microphylla* SK-TNAU, a sporulating culture was selected for spore inoculum development. Fresh young fronds of *A. microphylla* SK-TNAU were inoculated in puddled and well prepared field under flooded condition. A basal nutrition of 10 kg fresh cattle dung was mixed in 20 litres of water and applied in 40 m^2. A week after inoculation of *A. microphylla* SK-TNAU, the growth regulator GA_3 at 100 ppm was applied as foliar spray. Foliar spray with GA_3 induced the growth and stimulated the production of sporocarps. These sporulated fronds were harvested after four weeks and formed a heap with flat surface. One hundred kilogram of sporulated fresh fronds of *A. microphylla* SK-TNAU were used to form a heap. The heap was then covered with a thin layer of clay soil slurry (20%) and allowed to undergo partial decomposition for a period of 21d. Subsequently the heaps were removed and dried well under shade for a week. The decomposed and well dried frond materials are called "Frond based spore inoculum" of *Azolla*. The viability of the spore inoculum was found to be good for 10-15 months.

2. BLUE GREEN ALGAL BIOFERTILIZER

Diazotrophic Cyanobacteria have been found to play a vital role in sustaining the cultivation of wet land rice without added nitrogen fertilizer due to their inherent ability to add fixed nitrogen to such habitats under aerobic photosynthetic conditions. In recent years, the practice of utilizing Cyanobacteria as an efficient source of biofertilizer for rice has been advocated and adopted in India. The rice field ecosystem provides an environment favourable for the growth of blue green algae with respect to their requirements for light, water, high temperature and nutrient availability. However, cyanobacteria in rice fields are subjected to various stresses like salinity, acidity *etc.* which reduced the population and nitrogen fixation. Many agricultural ecosystems are increasingly becoming salt affected, which ultimately have an adverse effect on crop production. Salt affected soils collectively called as 'usar' soil comprising saline, saline sodic and sodic soils cover more than 7 million ha in India. High pH, low organic matter content and high calcium carbonate in soil strongly limit the availability of micronutrients to crops grown in sodic soil. The salt tolerant cyanobacteria immobilize Na^+ in an organic form and thereby play a major role in saline soil condition.

2.1 Acid Tolerant Cyanobacteria

Many cyanobacterial cultures were isolated from acid soils of Kerala having a pH of 3.8. The survival of cyanobacteria under acid soil rice field condition was studied.

Recently, several efficient cyanobacterial cultures were isolated from the acid soils of Tamil Nadu. The predominant genera are *Anabaena* (25%). *Westiellopsis* (17.5%), *Nostoc*(15.0%) *Oscillatoria* and *Phormidium* (25.0%) and the rest were unicellular algae *Microcystis* (5%). The acid tolerant cyanobacterial culture *Westiellopsis* AT-TGK-4A7 performed better both under normal and acid pH levels such as at pH 4,5 and 6. The effect of inoculation of acid tolerant cyanobacterial cultures on rice variety ASD 20 was studied. The study revealed that the inoculation of composite cultures of the acid tolerant cyanobacteria comprising *Anabaena*-AT-TGK-5CW, *Nostoc*-AT-TGK-5A4, *Oseillatoria* AT-TGK-5C9, *Westiellopsis* AT-TGK-4AT and *Westiellopsis*-AT-TGK-5A9 was found to be better in increasing the growth and yield of ASD 20 besides excreting maximum ammonia in the flood water.

2.2 Saline Tolerant Cyanobacteria

Cyanobacteria react to salt stress either through osmotic adjustment or by secretion of excess polysaccharide. Salt tolerance in cyanobacteria is mediated by an osmo regulatory mechanism which involved the accumulation of carbohydrates, amino acids and quaternary ammonium compounds. Salt adaptation strategy of cyanobacteria includes two main processes viz., (i) Maintenance of low internal contents of inorganic ions by active export mechanism and (ii) Synthesis and accumulation of osmoprotective compounds corresponding to the osmotic potential of the surrounding medium.

3. AMMONIA EXCRETION

Ammonia excretion by *A. azollae* strains isolated from *A. filiculoides* and *A. microphylla* under free-living condition in N-free medium has been demonstrated *A. azollae* and *A variabilis* when immobilized in polyurethane and polyvinyl foams and calcium alginate beads have been shown to release ammonia extracellularly. The algal symbiont *A. azollae* is known to colonize and immobilize well in polyvinyl foam. Significantly higher quantities of ammonia were excreted by *Anabaena* sp. immobilized in polyurethane foam in the presence of the glutamine analogue MSX (L-methionine DL-Sulphoximine). It has been shown that the systemic fungicide Benlate (methyl-1-butyl-carbamoyl-2 benzimidazole carbonate) stimulated ammonia production by *A. azollae* in the immobilized state in a photobioreactor system.

Further, it has been found that the systemic fungicide Benlate due to its systemic action is likely to diffuse into the cyanobacterial cells where it might like MSX inhibits ammonia assimilating enzyme, glutamine synthetase and thereby increasing ammonia production. The inhibitory action of the systemic fungicides Bavistin and Vitavax on GS enzyme in the strains *of A. azollae* resulted in higher amounts of ammonia excretion under immobilized state in PU foam.

It has also been noticed that the inoculated immobilized algal symbionts have excreted significantly higher amounts of ammonia into the flood water and this was reflected on the higher ammonia level in the flood water during the growth of rice seedlings. The inoculation of immobilized algal symbiont would increase the growth of rice seedlings by virtue of their continuous photoproduction of ammonia in flood water system.

The effect of inoculation of immobilized cyanobacteria in the transplanted rice was also significant The inoculation of polyurethane foam and sugarcane waste immobilized strains of A *azollae*-AS-DS-SK and *A. variabilis*-SAo in transplanted rice fields significantly increased the flood water ammonia content. Inoculation of immobilized cyanobacteria in solid matrix excreted significantly higher amounts of ammonia into the flood water of rice. The inoculation of immobilized cyanobacteria have positively influenced the plant growth, yield components, biochemical constituents, grain and straw yield of rice.

4. STEM NODULATING *SESBANIA*

Biological nitrogen fixation by leguminous green manure crops in symbiotic association with *Rhizobium* is a low cost input for rice crop. The cultivation of nitrogen fixing legume green manure crop would be an efficient way of improving soil fertility status and sustainability Many *Sesbania* species are widely used in Asia, of which *S. acideata* and *S. speciosa* are considered tc be the most predominant species grown in India. These two are more commonly grown as N_2 fixing green manures because of their vigorous growth habits, higher rates of N_2 fixation and their general ability to withstand a wide range of soil conditions *viz.,* waterlogging, salinity and alkalinity The use of stem nodulating *S. rostrata* and root nodulating *S. aculeata* as green manure have not only increased the yield of rice but also improved the soil microbial population and soil enzymes.

5. INDUCTION OF ROOT NODULE STRUCTURES IN CEREAL CROPS FOR N_2 FIXATION

Non-legumes such as cereal crops, grasses etc. do not form the nodules in their root system. Recently it has been investigated at TNAU and found that the symbiotic nitrogen fixing bacterium, *Azorhizobium caulinodans* isolated from the stem nodules of *Sesbania rostrata* could colonise and produce nodular structures in rice root. Among the cereals tested, maize crop responded well for the inoculation of *A. caulinodans* and prominent nodular structures have been observed.

The bacterium, *A. caulinodans* is a unique one which fixes nitrogen under free-living condition and also tolerant to combined nitrogen at low concentration. The bacterium could also tolerate flooded condition and *A. caulinodans* also produces the pectic and cellulolytic enzymes with good penetration capacity and in general it fixes nitrogen in rice root system at micro-aerophilic condition (2-3%, O_2 level). The nitrogenase enzyme which is responsible for nitrogen fixation was detected from the root nodule structures of rice, maize, sorghum, cumbu and ragi.

The nitrogen fixing rate in the nodular structures of cereal crops is low but the colonization of the bacterium could be induced by biotechnological manipulation of rice plant. Flavonoids are a class of phenolics secreted by plant roots as part of the regulatory system for colonization *of A. caulinodans.* Four families of flavonoids such as flavones, flavonols, flavonones and iso-flavonones have been shown to have nod-gene inducing activity. Naringenin-a specific flavonoid is known to stimulate the colonization of *A. caulinodans* in rice roots. It is a signal molecule for attracting the population of *A. caulinodans.* Recently it has been established that the flavonoid 'Naringenin' could induce the *A. caulinodans* to colonize the xylem of rice.

The induction of nodulation in cereal crops has excellent scope in terms of nitrogen transfer to rice crop for higher productivity and this finding has a potential practical significance in cereal crops nitrogen fixation and it is a strong signal to intensify the research on this direction of "Endophytic Bacterial Nitrogen Fixation" inside the root of the rice plant which would reduce the fertilizer consumption in wetland rice in sustainable rice fanning. Rice root associated nitrogen fixation through nodular structures in rice is like "mini fertilizer factory" in rice root itself. The new approach on nodule induction in rice and other cereal crops would be a break through for intensive colonization and N_2 fixation by 'diazotrophic bacteria' in rice root. To move further in this direction, it is possible to introduce new genes into rice through transformation by 'Plant genetic manipulation' which would create an excellent opportunity for incorporating N_2 fixation capability to form nodules in rice.

6. FUTURE OUTLOOK

The management of native soil N which is the principal source for rice, and biological N_2 fixation may play a role in sustaining the soil N under intensive rice cropping. Biological nitrogen fixation is an important process in rice farming systems because it is an inexpensive source of nitrogen for increasing the productivity of crops. The nitrogen fixing bio systems such as *Azolla,* cyanobacteria and legume green manures were able to adopt well under wetland rice field ecological conditions and fix considerable amounts of nitrogen. The biomass of these bio systems decomposes rapidly in rice soil and supply the nitrogen to rice crop. Besides nitrogen addition, these biofertilizers or biomanures also contribute significant amounts of phosphorus, potassium, sulphur, zinc, iron, molybdenum and other micronutrients. The organic acids released during the mineralization process of the biomass/biomanures would accelerate the phosphorus availability in rice soil. These biological fertilizers or biomanures were known to influence the total soil microbial population such as bacteria, fungi, actinomycetes and also substantially activate the population dynamics of nitrogen fixing free-living bacteria like *Azotobacter, Azospirillum* and also the free-living symbiotic bacterium such as *Rhizobium.* It is evidently clear now that the application of biological fertilizer were greatly involved in the accumulation of soil enzymes which directly reflects on soil fertility index. The use of biofertilizers would reduce the cost of chemical fertilizers involved in crop production. The effective utilization of biofertilizers for crops will not only provide economic benefits to the farmers, but, also improves and maintain the soil fertility and sustainability in natural soil ecosystem. Since these biosources represent a great diversity in chemical, physical and biological characteristics, their efficient use depends upon the particular agro-ecological environment and local availability. Therefore, it is very much essential to develop a strong, workable and compatible package of nutrient management through organic sources for various crops based on scientific facts, local conditions and economic viability.

FURTHER READING

Kannaiyan, S. Biotechnology of Biofertilizer for rice crop. Tamil Nadu Agric. Univ., Coimbatore, Tamil Nadu, India, (1990) 212.

Kannaiyan, S. Biofertilizers for rice. Tamil Nadu Agric. Univ., Coimbatore, Tamil Nadu, India, (1993) 60.

Kannaiyan, S. Immobilization of *Anabaena azollae* in solid matrix on ammonia excretion and its effect on rice crop. Paper presented in the National Seminar on *Azolla* and Algal Biofertilizers for rice, Tamil Nadu Agric. Univ., Coimbatore, Tamil Nadu, India, (1994) 25-26.

Kannaiyan, S. Summary report on EEC Project on imobilized cyanobacteria for continuous ammonia fertilization in reice production. Kings College, Univ. London, London. (1997) 49.

Kannaiyan, S., Rao, K.K. and Hall, D.O. Immobilization of *Anabaena azollae* from *Azolla filiculoides* in polyvinyl foam for ammonia production in photobioreactor system. *World J. Microiol. Biotechnol.*, **10**: (1994) 55.

Mahesh, G. and Kannaiyan, S. Immobilization of cyanobacteria on their effect on rice seedling growth. In: Rice Management Biotechnology (Ed. S. Kannaiyan) Assoc. Pub. Co., New Delhi, India. (1994) 229–311.

Reed, R.H. and Stewart, W.D.P. The responses of cyanobactria to salt stress. In: Biochemistry of the Algae and Cyanobacteria. (Eds. L.J. and J. R. Gallon) Clarendon Press, Oxford, (1988) 217–237.

Roger, P.A. and Ladha, J.K. Biological N2 fixation in wetland rice fields: Estimation and contribution to nitrogen balance. *Plant soil*, **141**: (1992) 41.

Samal, K.C. and Kannaiyan, S. Ammonia excretion by *Anabaena azollae* immobilized in alginate and its effect on the growth of rice seedlings. *Int. Rice Res. Newsletter*, **17**: (1992) 21.

Emerging Bio-Technologies in Agriculture: Production & Promotion of Biofertiliser as a Nutrient Source to Crops

M.R. MOTSARA

Department of Agriculture & Cooperation
Indian Council of Agricultural Research, (ICAR)
New Delhi 110 001.

THE SUCCESS STORY of progress in Indian Agriculture and attainment of self-sufficiency in the production of foodgrains is associated with the deployment of scientific technologies relating to breeding of varieties with high yield potentials coupled with improved management of agricultural inputs like water, plant nutrients and plant protection measures supported by credit and improvement in crop management practices. These efforts have resulted in doubling of food production twice since beginning of planning era in the year 1950-51 when the foodgrains production was 50.8 million tonnes in the country, which increased to 99.5 million tonnes during 1969-70. This period, however, was the period of growth mostly through traditional system of farming particularly upto 1966-67 when the production was 74.2 million tonnes. The period of green revolution started from mid-seventies onwards. Production in the year 1970-71 was 108.4 million tonnes which progressively kept on increasing and had further reached to a level of about 200 million tonnes in the year 1996-97. The foodgrain production is required to be further by next decade, which is necessary keeping in view the continuous growth in population resulting into enhanced requirement of food. Notwithstanding attainment of self-sufficiency in food production, the quality of nutrition and the availability of food to every Indian yet remain to be improved.

An analysis of factors of production reveals that contribution of the plant nutrients and their management is most significant. It is believed that about 50% of increase comes through the addition of plant nutrients in combination with other factors of crop production. A response of 1:7.5 in the production of foodgrains is taken as a national level average. However, the response is not so fixed and uniform and varies from soil to soil, crop to crop, region to region, depending upon a number of factors including nutrient management efficiency. The response to the nutrient application is declining in intensively cultivated and irrigated areas where a platue seems to be attaining. Breaking of this barrier would require greater efforts and improvement in management of factors of crop production particularly the water and plant nutrients. The responses in marginal farming areas including rainfed areas are generally lower due to unfavourable moisture regime in the soils due to lack of irrigation facilities coupled with scanty rainfall. The responses are reported to be much higher than the national average in irrigated and better managed areas which have not yet been fully exploited and intensification of agriculture is still not of a very high order. Real task is to improve the crop response in all such areas of crop production.

Further an analysis of plant nutrient use in the country shows that chemical forms of nutrients are a major source. A consumption level of about 18 million tonnes of nutrients in the country represents an average of 90 kg of nutrient per ha during 1998-99. The organic manures and biofertilisers, which are secondary/supplementary sources of nutrients though, have a great potential, but, their actual consumption is not significant. However, this situation has to be improved.

It is due to tropical and sub-tropical climate, organic carbon status, which is an index of nitrogen availability in soils, is low in the country. Its level is going down further. Phosphate status of Indian soil ranges between medium to low and high only in pockets. Thus the requirement of phosphate application is growing fast. Potash, a nutrient which was considered adequate during pre-green revolution period, is found to be deficient, in vast areas in recent years. Often talked, average NPK consumption ratio of 4 : 2 : 1 as an ideal balance is found to be actually 1 : 0 : 0.5 : 1.0 in Kerala, 3.2 : 1.7 : 1.7 in Karnataka and 2.1 : 0.8 : 1.0 in Tamil Nadu, which means higher amount of potash use. This ratio being 45.2 : 12.9 : 1.0 in Punjab and 171.0 : 47.8 : 1 in Haryana, shows that potash consumption in the two loading agriculture states is negligible in comparison to N and P while the yields of crops in these states are highest/ha. The proportions of the nutrients to be used are, therefore, required to be decided on the basis of inherent fertility status of the soils as determined by soil and the crops to be raised on them and not on the basis of pre-determined nutrient ratio.

In terms of the level of nutrient consumption in different States, as influenced by the type of soil, irrigation management practices and economic condition of the farmers and the crop variety. Average consumption of plant nutrients (NPK) in the West Zone is 59.5 kg/ha, in East zone it is 67 kg in South zone 114 kg/ha while in the North zone it is 128 kg/ha. State wise variation can be appreciated when observed that Punjab consumes over 170 kg NPK/ha, while the consumption in Haryana is 140 kg, in Tamil Nadu 134 kg, in West Bengal 112 kg, in Madhya Pradesh 49 kg and in Rajasthan 39 kg/ha. Arunachal Pradesh consumes about 2 kg/ha, and Sikkim 6.5 kg NPK/ha.

Apart from disparity in consumption in different states, the low efficiency and the declining crop response to nutrient application is another area of major concern. Fertilizer use has to be promoted in low consuming potential areas including rainfed areas, which are about, 117 million ha or 63% of the gross cropped area of 188 million ha and are crucial for agriculture, where special methods of nutrient application, dose and timings etc. need to be standardized. The application of nutrients according to the soil test based fertility assessment ensures a better use efficiency and improved profitability.

The average nutrient consumption at 90 kg/ha in India is much lower than the consumption in developed countries like UK (362 kg/ha), Netherlands (548 kg/ha), and France (260 kg/ha). It is even lower than the neighbouring countries with broadly similar agroclimate and soil conditions as in Pakistan (111.5 kg/ha) and Bangladesh (142 kg/ha). This may be an indication for further increasing the nutrient consumption.

A district wise examination of fertilizer consumption in India would show that 19 districts in the country consume more than 200 kg nutrients/ha while 110 consume between 100-200 kg nutrient/ha. A total of 138 districts consume between 50-100 kg nutrient/ha and about 175 districts consume less than 50 kg NPK/ha. The post green revolution era has witnessed a multiple nutrient deficiency because of higher crop harvest in the intensively cultivated areas where use of organic manures had declined while chemical fertilisers like urea, DAP and MOP became the major source of plant nutrients. Nutrient consumption can be taken as an index of the level of growth in agriculture and, therefore, fertilizer consumption has to be improved in low consuming districts to accelerate the crop production while in high consuming districts, an impact study on the quality of underground water in terms of nutrient content and deterioration of the physical and biological properties of the soil will be timely. Environmental hazards of excessive use of chemicals have to be kept in view particularly in the districts consuming more than 200 kg nutrients/ha. The integrated use of chemical nutrients, organic manures and biofertilisers is a possible solution to the problem of imbalance in nutrient use, lower use efficiency and other associated problems to ensure sustained fertility and enhanced crop yields.

1. INTEGRATED PLANT NUTRIENT MANAGEMENT AND ROLE OF BIOFERTILISER

The sustainability of soil fertility and a gradual and steady increase in the crop productivity would have to be the priority in future for ensuring food security of the nation. Exhausted soils with poor or low amounts of nutrient content, impeded drainage causing salinity and alkalinity, erosion due to both, water and wind which take away the nutrients of the soil and increase soil acidity in highly leached condition and continuous fragmentation of the farm holdings making the mechanical operation difficult and all such factors combined together make the yield increase difficult. Poor soils are becoming limiting factor more particularly due to the deficiencies of micronutrients. Agriculture, devoid of recycled organic manures/crop residues is facing disorder in physico-chemical and biological properties of the soil.

To achieve the proposed doubling of food production, nutrient addition in the soils has to be improved. In order to minimize the use of chemical fertilizers, the use of other sources of the nutrients have to be promoted. A two-fold strategy may be useful in nutrient management. Firstly to continue to encourage the use of fertilizer for a faster growth in agriculture and secondly to popularise the use of recycled crop residues, green manures, use of compost (both urban and rural) and biofertilisers as a supplementary source of nutrients in conjunction with chemical forms of plant nutrients to ensure sustainability in crop production system. Soil amendment/reclamation measures and use of micronutrients have to be simultaneously popularised. A check on further deterioration of soil fertility is critical and improvement of physical and biological properties of the soils coupled with a balance in the nutrient content in the soil alone can ensure sustained crop production. Integratecd nutrient supply system (IPNS) and integrated pest management (IPM) system have become an accepted strategy to bring about improvement in the declining soil fertility and protecting environment. A high proportion of organic sources of nutrients such as organic manures and biofertilisers, as a renewable source and supplement to chemical fertilizers is considered environment friendly in terms of protecting the quality of underground waters, soil property and the environment in general.

2. BIOFERTILISERS—SCOPE, POTENTIAL AND USE

Biofertilisers, the products of biological origin, can be advantageously blended in IPNS to replace a part of energy intensive non-renewable and costly fertilizers. The benefits of biofertilisers, in terms of biological N-fixation are known for nearly a century, but, the need for commercial exploitation of these natural endowments for improving agriculture productivity was hardly felt in pre-independence era. The commercial production of biofertilisers was started for the first time in 1956 at IARI and agriculture college (now TNAU), Coimbatore, which was followed by production of different types of biofertiliser in smaller quantities mainly by State Agricultural Universities (SAUs). Concerted R&D efforts for promotion of biofertilisers were started with the initiation of an All India coordinated research project on BNF (ICAR in 1978) and launching of a national project on Development and Use of Biofertilisers (GOI) in 1983.

Among biofertilisers for increasing N-supply, *Rhizobium* for legumes and *Azoatobactor*, *Azosperillium*, *Azolla* and Blue green Algae for cereals are important. Those associated with increased P-availability to the crops are P solubilising microorganism and Mycorrhizae.

In a situation where higher and higher amounts of nutrient use is called for and at the same time excessive use of chemical fertilizers is known to be unfavourable for sustaining soil fertility and crop productivity, organic sources of nutrients including biofertilisers can play a significant role. Biofertilisers are renewable and environment friendly source of crops nutrients.

3. POTENTIAL DEMAND OF BIOFERTILISERS

Since biofertilisers can successfully be used in all the crops grown under different agro-ecologies, their potential requirement is quite large and far exceeds the present production levels. It is also assumed that the biofertiliser can substitute about 25% of the nutrient requirement of the crops. The nitrogen supply through biofertilisers may be in the range of 15-20 kg/ha while P solubilisation can be effective in the range of 10-20 kg/ha.

4. BIOFERTILISER PRODUCTION SCENARIO

During 1997-98, 10,000 tonnes of biofertilisers were produced in the country, utilising 61% of the installed capacity (Table 1). However, there has been constant increase in biofertiliser production over years, as the production in 1997-98 was 5 times higher than that in 1992-93. Since biofertiliser production is demand driven, the capacity utilisation of different production units depends largely on creation of demand of biofertilisers.

Table 1: Biofertilizer production scenario in the country

Year	*Capacity (000 t)*	*Production (000 t)*	*Capacity Utilization (%)*
1992-93	5.40	2.00	37.1
1993-94	6.12	3.08	50.3
1994-95	8.11	5.18	63.8
1995-96	10.68	6.69	62.6
1996-97	12.65	6.86	54.2
1997-98	16.36	10.00	61.1

Depending on the source of funding, production units located in different states are categorized as

- those financed by Government of India (GOI) and
- those financed by other sources.

So far, 60 biofertiliser production units are established with the GOI support. These units share above 9 thousand tonnes of the total installed capacity and 6 thousand tonnes of the total biofertiliser production. Besides, 28 biofertiliser production units other than GOI financed ones, with a total installed capacity of 8 thousand tonnes produce nearly 4 thousand tonnes of biofertiliser annually. These units are established by private entrepreneurs, NGOs and some of the agricultural universities. The installed biofertiliser (*Rhizobium*, *Azotobacter*, *Azospirillum* and P solubilisers) production capacity of about 16500 tonnes is much lower than the potential demand. This underlines the need for expansion of biofertiliser production net work in the country.

5. GOVERNMENTAL EFFORTS FOR PROMOTION OF BIOFERTILISERS

Recognising the potential necessity of biofertilisers in sustaining agricultural production, Govt. of India had launched in 1983, a central sector scheme "National Project on Development and Use of "Biofertilisers", with the establishment of National Biofertiliser Development Centres one each at Ghaziabad and six Regional Centres at Hissar, Nagpur, Bangalore, Jabalpur, Bhubaneswar and imphal with the broad objectives.

- Production and distribution of biofertilisers.
- Acquisition and maintenance of mother cultures of different biofertilisers and development of quality standards.

- Providing financial assistance for establishment of biofertilizer production units in different parts of the country.
- Human resource development in the areas of production, marketing and use of biofertilisers through training and orientation programmes.
- Field demonstrations and promotion by publishing literature on the benefits of biofertilizers.

6. BIOFERTILISER PRODUCTION AND QUALITY CONTROL

At present biofertilisers are prepared using manual mixing technique using charcoal/lignite as the carrier material. However, media sterilization and its inoculation under aseptic conditions ensure a check on contamination by unwanted microbes. The final product is packaged in LDPE bags and marketed directly as well as through state governments and by appointing dealers.

The national project has a key role in developing standards for different biofertilisers in association with the experts and BIS authorities. The quality standards developed recently as also circulated to different production units are given in Table 2, though these are yet to be notified by the BIS. In order to promote the

Table 2: Indian Standard specifications for different biofertilisers

Parameters	*Rhizobium*	*Azotobacter*	*Azospirillum*	*Phosphate Solubilising Micro-organism (PSM)*
Base	Carrier based	Carrier based	Carrier based	Carrier based
Cell number at the time of manufacture	10^7	10^6	10^7	10^8
Contamination	Nil at 10^5 dilution	Nil at 10^5 dilution	Nil at 10^5 dilution	Nil at 10^6 dilution
Expiry period from date of manufacture	6 months	6 months	6 months	6 months
pH	6.5–7.5	6.5–7.5	6.5–7.5	6.5–7.5
Strain	Should be checked serologically	Nothing specific, but, chroococcum species is mentioned	Nothing specific	Nothing specific
Carrier should pass through	150-212 micron IS sieve	106 micron IS sieve	106 micron IS sieve	106 micron IS sieve

production and use of quality products. NBDC/RBDCs and agricultural universities informally analyse the biofertilisers samples produced by different units for assuring the quality. The quality checks carried out by NBDC/RBDCs and SAUs help the production units in improving their produce and also help the states in monitoring the quality of biofertiliser produced by different units.

7. FINANCIAL SUPPORT TO PRODUCTION UNITS

The national project on development and use of biofertilisers provides financial assistance as one time grant-in-aid up to Rs. 20 lacs per unit of 150 tonnes annual production capacity. Till 8th five year plan, the grant was provided only to Institutional agencies and the state governments. A more liberalised approach is, however, adopted in the 9th five year plan by extending the grant-in-aid to private entrepreneurs and NGOs also. The GOI grant is offered mainly for purchase of equipment and partly for the procurement of glass wares, chemicals and raw materials. A grant of Rs. 20.0 lacs constitutes approximately I/3rd of the total cost for a project of 150 tonnes/y capacity.

8. CROP RESPONSE TO BIOFERTILISERS

8.1 N-fixing microorganism

The terrestrial flux of nitrogen from biological nitrogen fixation has been estimated to be 139 to 170 million tonnes as N/y. While the accuracy of these estimates might be questioned as they are indicative of the importance of biofertiliser in global N cycle. Only a few procaryotic organisms are able to fix atmospheric N directly through a biological process, among which symbiotic bacteria (*Rhizobium*), non-symbiotic free living (*Azotobacter* and *Azospirillum*) and cyno bacteria (*Azolla* and Blue green algae) are important.

8.1.1 Rhizobium

Rhizobium has been most widely studied and extensively used biofertiliser because of its proven ability to fix atmospheric nitrogen in symbiosis with leguminous host plants. The N-fixation by different legumes in association with specific species of *Rhizobium* under ideal and controlled conditions is estimated to be 100-300 kg/ha in alfalfa, 102-150 kg/ha in clover, 37–196 kg/ha in cluster bean, 112-152 kg/ha in groundnut, 68-200 kg/ha in pigeon pea, 49-130 kg/ha in soyabean, 53-85 kg/ha in cow pea.

8.1.2 Azotobacter

Azotobacters are aerobic fee living, heterotrophic N-fixers. Increase in yield due *Azotobacter* innoculation could be obtained in crops like rice, wheat, shorghum, maize, pearlmillet, cotton, vegetables and sesamum. In irrigated wheat, significant yield increase due to *Azotobacter* innoculants is reported in 342 out of 411 trials conducted on farmers fields and the response ranged from 34 to 247 kg/ha.

8.1.3 Azospirillum

Azospirillum group of bacteria is an associative microaerophillic N-fixer which colonises the root mass of cereals, millets and grasses and fixes N in loose association with plants. Field demonstrations conducted on the use of *Azospirillum* biofertilisers on diverse soils and climatic conditions revealed sizeable gains in productivity of different crops. *Azospirillum* has shown positive interaction with applied N in several cereal crops with an average response equivalent to 15-30 kg/ha of applied N. Dipping the roots of rice seedlings in 2% solutions of *Azospirillum* culture resulted in the yield increase by 100 kg/ha over un-inoculated control. Findings of All India Coordinated Project of Cropping System Research, recorded that out of 108 trials conducted in Monsoon rice on cultivators fields (Tamil Nadu). In 48 trials, it was possible to reduce the application of N-fertiliser by 25-50%. In winter rice, such reduction in N fertilizer was possible in 31 out of 55 trials.

8.1.4 Blue Green Algae

Blue green algae (BGA) or cyanobacteria are free living photosynthetic N-fixing microorganism. These are distributed world wide and considered ideal biofertilisers for rice based cropping systems. Estimates indicate that judicial use of these algae could provide to the country's entire rice areas as much N as obtained from 15-17 lakh tonnes of urea. The predominant genera of BGA in Indian soil are *Anabaena*, *Nostoc*, *Aulosira*, *Calothrix* and *Tolypothrix*. Various studies conducted in the country have shown that BGA applied @10 kg/ha after one week of rice transplantion contributes 20-30 kg N/ha per season and grain yield increase by 5-14%. These benefits are ascribed to supply of combined N and growth promoting substances of BGA. Nitrogen from BGA becomes available to rice. Only a part of BGA nitrogen is absorbed by existing rice crop and rest remains in soil as residual N, and becomes available to the ensuing crop repeated use of BGA produces a cumulative residual effect.

8.2 Phosohate Solubilising Microorganism

Most of the Indian soils are low to medium in P fertility and the efficiency of P fertilizers is also low due to fixation of a large fraction of applied P into sparingly available or insoluble forms of P in the soil. Introduction of efficient P solubiliser in the rhizosphere increases the availability of insoluble forms as also the use efficiency of P fertilizers. Early work on crop response to phosphate solubilising microorganisms (PSM) as reviewed critically (revealed that, of the total inoculation experiments conducted,) only 25% studies exhibited an yield increase to the extent of 5-10%. The crop responses to phospo-bacteria applied along with rock phosphate were, however, quite striking in the rice-wheat system on trails conducted under All India, Coordinated Project on Cropping System Research.

9. CONSTRAINTS

Since biofertilisers are the living microscopoic organisms utmost care is needed at all the steps beginning from the production till the end use to maintain the organisms in required number and vigour. Despite the governmental efforts and R&D support, large scale acceptability of biofertilisers is hindered owing to several physical, technical and infrastructural constraints. In India, biofertilisers are generally manufactured under septic conditions using unsterilised carrier and manual mixing of broth with the carrier, which encourages contamination. Because of low price structure of biofertilisers, hi- tech instrumentation required for producing biofertilisers under aseptic conditions is not getting acceptance. However, the area needs greater attention with a view to eventually take up the production under total aseptic condition.

10. SELECTION OF SUITABLE STRAIN

The effectiveness of microbial strains is decided by complex interaction among host, native population and the environment. Thus the strains possess specificity to soil and agro climatic conditions. Use of non-specific strain often results in poor quality biofertilisers, producing practically negligible crop responses. Since biofertilisers are transported over long distances and stored at uncontrolled temperatures, there is also a need for continued R&D on isolation and evaluation of strains that can withstand abberant weather particularly temperature variations. Localized or decentralised production may also need to be encouraged.

11. SHELF LIFE

Short shelf life of biofertilisers is one of the major constraint, limiting the quality and large scale accesbility of biofertilisers. BIS specification of six months shelf life of aseptically prepared biofertilisers are generally not adhered to and in most circumstances, it is reduced to 3-4 months only. Since shelf life is a culmination of several factors like production technology, carrier and packaging material used, mode and distance of transport and storage facilities with the dealer, an improvement at all these levels is desired to enhance the shelf life. The LDPE bags used at present for packaging of biofertilisers are not ideal to maintain the viability of strains for a longer period. As an alternative, HDPE material has also been tried, but, was not found a viable proposition due to low gas exchange and relatively higher cost. Alternative packing material and packaging system needs to be explored.

12. QUALITY CONTROL

Bureau of Indian Standard (BIS) specifications for *Rhizobium* and *Azotobacter* have been notified for use by biofertiliser producers. Recently, an expert committee constituted by the Government reviewed these specifications and suggested some modifications. The Committee has also suggested specifications for phosphate solubilising micro-organisms and *Azospirillum*. While BIS may take some time to finally notify these new specifications, the same have been circulated to the State Departments of Agriculture and the agencies engaged in biofertiliser production are following so as to ensure the production and availability of quality biofertilisers. For ensuring a greater control on the quality of biofertiliser, state governments are required to ensure regular checking of the quality of biofertilisers being produced/marketed.

13. INCONSISTENT CROP RESPONSE

Unpredictability in field performance of biofertilisers is not an uncommon phenomenon. The success level with any biofertiliser in terms of impact on yield increase ranges between 10 and 30%. Though this inconsistency in response to biofertilisers is not always due to substandard material, but, may be because of unfavourable environment and improper storage/handling of biofertilisers. Site and crop specific use needs to be ensured.

14. MARKETING CONSTRAINTS

Marketing constraints include weaknesses in marketing network, retail outlets and storage facilities. Unfortunately there is no standardization, in packaging, labelling and pricing of biofertilisers. The cost of culture required per ha varies from Rs 20 to 100 depending on the crop and type of biofertiliser. The size of packing varies from 75-500 g. per packet. Some manufacturers engage private dealers with lucrative trade discounts. Otherwise, most of the sale is through state governments or institutional agencies.

15. FUTURE AREA OF RESEARCH AND PROMOTION OF BIOFERTILISERS

Potential utility biofertilisers is well established through various experiments as reported above. For exploitation of available potential, good quality biofertilisers with proven efficiency have to be made available to the farmers. This is to be preceded by large scale field demonstrations on various crops under different agro-climatic and soil conditions as these factors are responsible in determining the response.

Research has to be continued on strain identification and even on their further improvement though genetic engineering. The quality to be incorporated in the newer strains has to provide for greater shelf life, higher capacity to mobilise nutrients like N and P. Standardization of carrier material keeping in view of their easy availability and cost may be useful. Charcoal, though is considered as a good carrier, but, it is wood dependent source hence needs substitution. Reports about liquid biofertilisers are appearing, though sporadic. This area needs examination with possible advantages, if any. Packing material including the packing size needs to be standardized.

The efforts may be made to look for some strains which may be capable of helping in increasing the potash availability in the soils.

FURTHER READING

1. Chhonkar, P.K. Crop response to phosphatic biofertilisers. *Fertiliser News*, **39**, (1994) 39.
2. Hedge, D.M. and Dwivedi, B.S. Nutrient management in rice-wheat cropping system in India, *Fertiliser News*, **37**, (1992) 11.

3. Hedge, D.M. and Dwivedi, B.S. Crop response to biofertilisers in irrigated areas. *Fertilisers News,* **39**, (1994) 19.
4. Kundu, S. and Barman, K.K. Biological Nitrogen Fixation. In: 50 years of Natural Resource Management Research (Eds. G.B. Singh and B.R. Sharma), ICAR, New Delhi. (1998) 199–213.
5. Lee, K.K. and Wani, S.P. Significance of biological nitrogen fixation and organic manures in soil fertility management. In: Soil Fertility and Fertiliser Management-Semi-Arid Tropical India (Ed. Christianson), IFDC, Muscle Shoals, Albama, USA (1989) 89–108.
6. Mahapatra, B.S. and Sharma, G.L. Effect of Azospirillum, Cyanobacteria and Azolla biofertilisers on productivity of low-land rice. *Indian J. of Agronomy* **33**: (1988) 368.
7. Misha, S. and Kaushik, B.D. *Proc. Indian Natl. Sci. Acad.* (1989). 56(B).
8. Subba Rao, N.S. In: Proc. Mycorrhiza Round Table, IDRC-CRDI-CIID, New Delhi, (1988) 211–225.
9. Venkataraman, G.S. and Shanmugasundaram, S. Biofertiliser Technology for Rice. DBT Centre for BGA Biofertiliser, Madhuraj Kamraj University, Madurai, (1992) 1–24.
10. Venkataraman, G.S. and Tilak, K.V.B.R. Biofertilisers in Sustainable Agriculture. In: Soil Fertility and Fertilisers vol. -IV, IFFCO, New Delhi. (1990) 137–148.

Biofertilizers: Their Development

R.P. GUPTA[1]; M.S. PANDHER[1]; S.K. GOSAL[1]; R.K. HUNJAN[1]; A.S. BRAR[2]; N. SINGH[3] *and* G.K. SEKHON[1]

[1]Department of Microbiology, [2]Plant Breeding, [3]Agronomy
Punjab Agricultural University
Ludhiana-141004 (India)

AGRICULTURAL PRODUCTION RATE has been the major emphasis in Punjab, hence the State though occupying only 1.5% of total geographical area of country and 2.98% of cultivated area, consumes 10% of the total fertilizer of the country that comes to 1.01 million tonnes of nitrogen and 256 thousand tonnes of phosphorus nutrients. The high nutritional requirement of paddy and wheat, the major crops in rotation, in the state, have exhausted the soil of nutrients. Application of higher doses of major nutrients as chemical fertilizers has not only resulted in increase in the cost of production of crops, but, also in imbalance of soil reserve like microbial ecology, nutrient deficiency & environmental pollution and deterioration of soil structure. To maintain long-term soil productivity and ecological sustainability, there is need to fully exploit the biopotential of biological resources including the beneficial microbial system.

Agricultural prosperity of any nation can be easily traced to the inherent fertility status of the soil of that region. In this regard, the use of biofertilizers will be an environmentally benign approach to nutrient management and ecosystem function. Biofertilizers are cost effective, ecofriendly and renewable source of plant nutrients to supplement chemical fertilizers in sustainable agricultural system. The gap between nutrient removal and replenishment is significantly high in India, hence, biofertilizers assume special significance in the present context of very high cost of chemical fertilizers.

1. BIOFERTILIZERS

These are the preparations containing living cells of efficient nitrogen fixing, phosphorus-solubilising/mobilizing or cellulose decomposing microorganisms, which when applied to seed or inoculated into the soil enhance availability of nutrients to plants either working symbiotically/asymbiotically or through solubilization of soil nutrients such as phosphorus or decomposition of complex materials. Biofertilizers have been classified into two major categories.

- **Nitrogen fixing microorganisms** — *Rhizobium* (for legumes), *Azotobacter*, *Azospirillum* and blue green algae for cereals.
- **Phosphorus mobilizing microorganisms**

1.1 Nitrogen-Fixing Microorganisms

1.1.1 Rhizobium

Rhizobium in association with specific leguminous crop forms root nodules which act as mini factories for the fixation of atmospheric nitrogen into ammonia. Because of their specific symbiotic association, nitrogen fixation can be as high as 360 kg N/ha/season (Table 1). This association depends upon the effectiveness and infectiveness of the introduced rhizobia, soil characteristics, status of soil nutrients, climatic conditions and above all the intensity of native rhizobia. As per nodulation survey of different legume crops conducted from time to time, Punjab soils have been found deficient in both effective and infective rhizobia. This fact also comes from the studies that the repeated applications of *Rhizobium* in the same field give response to inoculation. These observations indicate the need of introducing effective and infective rhizobia for different crops. The rhizobial biofertilizers for seven legume crops namely, moong, soybean, lentil, gram, berseem, leucerne and pea have been developed and now regularly practiced in Punjab, India. The use of these biofertilizers has not only enhanced the legume crop grain yield, but, it has also been found to reduce the initial input of N fertilizers (Tables 2, 3, 4) besides enhancing soil-fertility, by showing a residual effect of 30 kg N/ha for the subsequent crop (Table 5). However, it is still felt that there is a gap between the yield of inoculated and fertilized crops which needs to be bridged by studying ecological factors and manipulation of genetic potential of inoculated strains through mutagenesis/genetic engineering. In this regard, important genes have been incorporated to improve upon their efficiency.

Table 1: Major N-fixing microorganisms and beneficiary plants

Microorganism	*Association*	*Beneficiary crops*	*N-fixation (Kg N/ha/ crop)*
Rhizobium	Symbiotic with root nodules	Legumes	Upto 360
Azotobacter	Free-living in soil	Cereals, Vegetable, Cotton	20-30
Azospirillum	Associative symbiosis	Maize, Millets, Sugarcane	20-30
Blue-green algae	Free-living in water	Rice	10-80
Anabaena azollae	Symbiotic with Azolla	Rice	20-100

1.1.2 Azotobacter/Azospirillum

These are non-symbiotic N-fixing bacteria that contribute to the nitrogen content of the soil in a substantial way, when used as an inoculant. These bacteria have been found to produce vitamins, growth promoting substances, antibacterial and antifungal substances, besides helping in expansion of root system, better seed germination and thus play important role in integrated nutrient management. The beneficial effects of these bacteria on crops like wheat, maize and cotton have been

Table 2: Effect of *Rhizobium* strains on N-economy and grain yield of moong

Rhizobium strain	*Grain yield (q/ha)*		
	N_0	$N_{12.5}$	*C.D (5%)*
1995-96			
Control	6.20	8.00	
IARI	8.80	7.80	
M-1	8.90	7.90	
MA-1	9.50	10.20	
FDR-1	10.10	9.10	
ML-45	10.40	10.00	
ML-55	9.20	8.30	
Mean*	9.48	8.92	3.18
1996-97			
Control	7.40	8.90	
ML-45	9.70	10.20	
KL-Kaoni	10.00	9.70	
MJ-57	10.00	9.60	
MS-32	9.80	9.80	
MA-1	10.60	10.20	
Mean*	10.02	9.90	14.9
1997-98			
Control	7.60	9.90	
MA-1	9.30	8.90	
J-57	9.50	9.80	
KI	8.50	8.70	
IARI	8.70	8.90	
MS-32	9.10	9.60	
Mean*	9.02	9.18	

* Inoculated treatments only
N_0 - No nitrogen, $N_{12.5}$ - 12.5 kg N/ha

demonstrated. The technology for the production has been standardized and government institutions and private companies are supplying these biofertilizers to the Punjab farmers . In Punjab, *Azotobacter* inoculants have no consistent results for correct probability due to low organic content of soils and extreme climatic conditions. Moreover, low priority has been given for research on the development of these biofertilizers . As there is large area of more than seven million hectare under cereal cultivation in Punjab, there is urgent need to develop biofertilizers for cereals, suitable to geoclimatic conditions of the State.

Table 3: Effect of *Rhizobium* strains on N-economy and grain yield of Soybean

Rhizobium strain	*Grain yield (q/ha)*		
	N_{16}	N_{32}	*C.D (5%)*
1995-96			
Control	8.63	12.93	
IARI	12.29	13.86	
C-42	13.62	12.02	
PAU	12.65	12.51	
J-57	13.74	12.77	
J-63	11.45	12.07	
Mean*	12.75	12.65	3.07
1996-97			
Control	17.75	21.60	
dctlparGup 3	19.80	20.20	
SB-117	19.90	21.90	
L-43	23.60	20.10	
BTA	23.00	20.60	
L-42	21.00	21.80	
Mean*	21.46	20.92	4.25
1997-98			
Control	16.90	19.60	
L-43	19.90	20.50	
BTA	19.30	19.60	
Gup 3	18.30	19.30	
SB-120	20.10	18.90	
USDA 110	19.90	19.40	
Mean*	19.50	19.54	

* Inoculated treatments only

1.1.3 Blue Green Algae (BGA)

BGA are ideal biological system which represent a self supporting biological agent because they can photosythetically provide energy for nitrogen fixation .The rice field ecosystem provides an ideal and favourable environment for the proper growth of cyanobacteria with respect to light, water, high temperature and nutrient availability. Blue green algae besides a good source of nutrients, contribute 25-30 kg N/ha, add organic matter to the soil, synthesize and excrete growth promoting substances like IAA, vitamins, sugar and free amino acids, solublize insoluble phosphate and improves soil health. Algal inoculation to rice crop has been found to be effective in different agroclimatic conditions and soil types. The technology

Table 4: Effect of *Rhizobium* strains on N-economy and grain yield of Lentil

Rhizobium strain	*Grain yield (q/ha)*		
	N_0	$N_{12.5}$	*C.D (5%)*
1994-95			
Control	16.20	18.20	
IARI	18.30	19.40	
L-Ropar	19.50	20.40	
PAU	19.33	18.20	
L-LDH	19.10	18.10	
L-Phillaur	18.63	19.90	
Mean*	18.97	19.21	2.80
1995-96			
Control	8.42	13.45	
IARI	14.22	11.22	
L-LDH	14.11	11.72	
PAU	15.76	11.30	
L-Sangrur	13.86	13.78	
L-1-93HSP	13.86	13.04	
Mean*	14.36	12.21	4.55

* Inoculated treatments only

Table 5: Effect of *Rhizobium* inoculation to moong on subsequent crop-wheat (1995-)

Strain inoculation to moong crop	*Kg N/ha*	*Grain yield (q/ha) of wheat*
C0	120	40.8
C1	90	36.7
C2	90	37.6
C3	90	46.6
C4	90	45.5
C5	90	42.2
C6	90	41.0

CD (5%) N.S.

for large scale production of composite culture of BGA under the rice field condition was developed by Indian Agricultural Research Institute(IARI), New Delhi and Tamil Nadu Agricultural University (TNAU), Coimbatore.

BGA potentials could not be harnessed in Punjab due to two main reasons:

- High input of inorganic N-fertilizers by Punjab farmers.
- High temperature prevailing during rice cultivation.

Table 6: Effect of *Azotobacter* inoculation on N-economy and grain yield of maize

Treatment	Grain yield (q/ha)		
	N_{60}	N_{90}	N_{120}
Control	1426	1852	1852
Azotobacter	1500	2028	1963

C.D (%) N.S

Table 7: Effect of *Azotobacter* inoculation on N-economy and grain yield of wheat and gobi sarson

Treatment	Yield (q/ha)	
	Wheat	*Gobi sarson*
Control (No fertilizer)	3320	916
Control + Biotech *Azotobacter*	3700	1034
Control + Local strain	3760	1068
NPK(125:62.5:5)	5404	1504
50 % NPK	4900	1364
50% NPK + Biotech strain	4868	1360
NPK + Local strain	5596	1548
NPK+ Biotech strain	5620	1540

* Results of demonstration trials conducted at village Galib Kalan in collaboration with IFFCO.

In the last couple of years, Punjab Agricultural University (PAU), Ludhiana, Punjab has identified certain BGA strains capable of saving 30 kg N/ha and suitable for Punjab conditions for basmati rice (Table 8). Similarly efforts were also made to study the effect of BGA applications at various growth stages of rice growth (when temperature becomes less) and using split doses of N-fertilizer. The results have shown a saving of up to 40 kg N/ha when BGA applied at tillering /panicle stage. Keeping in view the huge area of 2.5 million hectares under rice cultivation in Punjab, there is need to fast develop this culture and popularise this cheap technology for rice ecosystem which will not only reduce cost of production but will improve soil health and reduce environmental pollution problems.

1.2 Phosphorus-Solubilizing/Mobilizing Microorganisms

Phosphorus is a major plant nutrient involved in the energy transformation in plants and is essential for many metabolic processes involved in plant growth and development to achieve optimum crop yields. The fertilizers use efficiency of P fertilizers is about 15-20% only. About 98% of Indian soils contain insufficient

Table 8: Effect of BGA application on grain yield of Basmati rice at different N-levels and its residual effect on subsequent crop-wheat

Treatment (N kg/ha)		*Grain yield (q/ha)*	
Basmati	*Wheat*	*Basmati*	*Wheat*
BGA (IARI) +30	90	36.35	39.40
BGA(IARI) +60	90	39.55	40.30
Nostoc +30	90	37.89	39.40
Nostoc +60	90	42.59	38.80
34A2 +30	90	37.50	42.50
34A2 +60	90	39.68	39.50
28 B +30	90	39.68	39.40
28B +60	90	37.17	40.60
Mixture +30	90	37.74	43.30
Mixture +60	90	40.51	41.60
No BGA +30	120	31.33	42.60
No BGA +60	120	39.76	44.20
C.D 5%	7.30	NS	

amount of available P. The manufacture of phospahatic fertilizers require non-renewable source like high grade rock phosphate which are getting depleted and becoming costly. This necessitates to find a cheaper alternative to meet the P-requirements of the crops. Phosphorus solublizing microorganisms hold the potential to mineralize organic forms of phosphorus in soil and solublization of insoluble inorganic phosphates to make it available to plants. Soil microorganisms such as *Bacillus*, *Pseudomonas* and fungi like *Aspergillus*, *Penicillin*, *Trichoderma*, *Alternaria*, *Fusarium* play a significant role in mobilizing P for the use of plants from the native soil pool as well as from added insoluble phosphate. However, the rate of P-solublization vary with source of inorganic P and the microorganisms involved. P-solublizing microorganisms produce organic acids including citric, fumaric, oxalic, succinic, α-ketoglutaric, tartaric acids etc. which help in solublizing phosphorus. Besides organic acids, phosphorus solubilizing microorganism also produce chelating substances, humic substances, H_2S, CO_2, mineral acids and siderophores. The proton extrusion mechanism involving release of protons accompanying respiration or ammonia assimilation without release of organic acids in environment is also observed. Phosphobacterium application has been found to reduce 25% of phosphatic fertilizer input in various legumes, cereals and other useful crops. Inoculation of various crops with PSM has been found to increase P uptake, plant growth and grain yield in soybean, maize and gram. The potential of seed or soil inoculation with P-solublizing organism has been very well documented and commercial production of PSM has been taken by many private companies, but, the future of this technology to meet P requirement of crop will depend greatly on their impact under practical farming conditions. There is a great scope and potential of this technology because of large acreage under cereals and vast reservoir of insoluble form of phosphorus in soils.

1.3 Vesicular Arbuscular Mycorrhiza (VAM)

The use of VAM as inoculant in agriculture would prove highly beneficial as most plants in ecosystem have mycorrhizal associations. VAM enhances plant growth by improved mineral nutrition ,helps in increased P uptake, improves growth, nodulation and N fixation through legume rhizobium symbiosis and better utilization of other nutrients like N, Cu, Zn, S, etc. Studies on the effectiveness of VAM inoculation conducted at PAU have shown the potential saving of 10-20 kg P_2O_5/ha in chick pea and soybean and enhancing biomass, uptake of N and P in Leucaena and Acacia trees. The potential efficacy of VAM inoculants have certain constraints because of variation in soil nutrients status, phosphorus inhibition of mycorrhizal development ,slow growth of mycorrhizal fungi and incompatibilities with established cultural practices. Above all, being obligate symbiont, it has not been possible to bulk multiply VA mycorrhiza on synthetic medium.

1.4 Biofertilizer Use

1.4.1 Constraints

Being a biological entity, biofertilizers are subjected to many biological, technical and marketing constraints.

1.4.1.1 Biological/technical constraints

- Inconsistent performance in fields mainly due to poor stability, competitiveness and stress due to ecological factors
- Non-availability of region/soil/ crop specific strains
- Short shelf life period
- Substandard quality of inoculants
- Absence of well equipped quality testing laboratories

1.4.1.2 Marketing constraints

- Lack of perception of concerned officials about the importance of biofertilizers from sustainability point of view
- Limiting demand because of farmer's unawareness about biofertilizers
- Lack of qualified technical staff
- Lack of storage and transport facilities
- Seasonal demand

1.4.2 Measures

- Use efficient, region and crop specific coded strains
- Use good quality sterilized carrier material

- Employ technically competent human resources for production, storage and marketing
- Ensure good quality biofertilizers having ISI standards/mark
- Establish nodal agency for quality monitoring during production, transport, storage and use.

1.4.3 Promotional strategies

- Realization on the part of administrators about the usefulness of the technology in large perspectives of sustainability
- Include the use of the biofertilizers in the package and practices of the State Agriculture Universities/(SAUs)
  - — By carrying demonstration trials in collaborations with SAU, Directorate of Agriculture and the concerned firm
  - — By evaluating the product from SAU
- Educate/train agriculture staff of various agriculture oriented agencies regarding the use and importance of biofertilizers for their further dissemination to farmers
- Lay demonstration trials and organize field rallies
- Highlighting usefulness of biofertilizers in the offices of all Agriculture and block development officers
- Distribute pamphlets during field days and kisan melas to the farmers
- Give wide publicity through radio, television etc.
- Provide biofertilizers along with the seed mini kits being supplied by the state govt. under different programmes
- Sell biofertilizers through companies engaged in production and distribution of seed and fertilizers

1.4.4 Research Gaps

To promote the application of biofertilizer, the thrust of research needs to be directed to:

- Evolve competitive and crop/region specific strains through selection/ genetic manipulation
- Carry out research regarding integrated use of various biofertilizers

Biofertilizers being cheap and eco-friendly, have a large scope to reduce the dependence of crop productivity entirely on chemical fertilizers and their application can be a step towards sustainable agriculture. However, the viability of biofertilizer production technology depends upon development of good quality inoculants suitable for a particular situation with effective delivery system. The technology is very healthy with no harmful effects and need to be popularized by the administrators and practiced by the farmers. Even though being biological input one always does not get visible glaring effects, but, the slow build-up of the desirable microflora itself is likely to prove useful.

FURTHER READING

Gaur, A.C. Phosphate solublizing microorganism as Biofertilizer. Omega Scientific Publishers, New Delhi, (1990) 283.

Kaur, R., Interaction studies on VAM, Rhizobium and P source in relation to symbiotic nitrogen fixation. Ph.D. Thesis, PAU, Ludhiana, (1996).

Ladha, J.K., Pareek, R.P., So, R. and Becker, M. Stem nodule symbiosis and its unusual properties. In: 'Nitrogen Fixation: Achievements and Objectives. (Eds. P.M. Gresshoff, L.E. Roth, G. Stacey and W.L. Wewton) Chapman and Hall, New York, London. (1990) pp. 633–640.

Kannaiyan, S. Biofertilizers for rice. TNAU Press, Coimbatore, Tamil Nadu, India (1993).

Kannaiyan, S., Sundaravarathan, S. and Palanippan. Nitrogen fixing green manures for sustainable rice production. TNAU Press, Coimbatore, Tamil Nadu, India. (1998).

Kannaiyan, S. Biofertilizer technology and quality control. TNAU Press, Coimbatore, India, (2000) 4–234.

Current Scenario and Constraints in the Use of Biofertilizers in India

A.L. KHURANA *and* S.S. DUDEJA

Department of Microbiology
CCS Haryana Agricultural University
Hisar 125 004. India

SUPPORTING EVER INCREASING population levels on a near stabilized agricultural land places a heavy burden on the soil resource-particularly its nutrient supply power. During last few decades the number of persons for each hectare of land has doubled, while the availability of land has decreased due to industrialisation and urbanisation. Estimated removal of four major nutrients viz. NPK and S is expected to reach to 32 million tonnes by 2000 A.D. Looking at the heavy demand for food grains, it appears impossible that the trend of increased nutrient depletion will ever be stopped or decreased. The increased use of different nutrients in terms of increased cost will put poor farmers under great stress. The only possible alternate route is effective and optimal utilization of plant nutrients and other resources. Nutrient depletion can be reduced and comparatively more favourable base can be built for sustainable and moderately high crop yield, in order to meet the needs of an expanding population, its growing aspirations and urgencies of stepping up farm based exports. The use of mineral rich fertilizers is the quickest and surest way of boosting crop production. Their high cost and other constraints frequently deter farmers from using them in recommended quantities. No doubt the use of inorganic fertilizers increases the yields dramatically but without a good management, there is rapid deterioration of soil. The excessive use of N and P fertilizer has caused a great concern among environmentalists and health specialists. The use of monoculture crops which require very high inputs has resulted in disappearance of traditional practices like crop rotation, ultimately resulting in loss of physical and chemical properties of soils.

Biological routes of improving the soil fertility for optimum crop production are vital components of integrated nutrient supply system. These routes are operated by microorganisms who either synthesise plant usable forms of nutrients or increase the availability of nutrients already present in the soil as in case of phosphorous. Such organisms are called Biofertilizers.

1. CURRENT SCENARIO

Biofertilizers or microbial inoculants can be defined as biologically active products containing active strains of specific bacteria, fungi, algae or combinations which may help in increasing crop productivity by way of helping in the biological nitrogen fixation, solubilization of insoluble nutrients, stimulating plant growth or in decomposition of plant residues. Biofertilizers are ecofriendly and ensure constant supply of specific nutrient throughout the growth period unlike chemical fertilizers. Use of one tonne of *Rhizobium* inoculant is equal to 100 tonnes of nitrogen considering minimum fixation of 50 kg N/ha from 0.5 kg/ha application dose. While use of one tonne of *Azotobacter* is equivalent to 40 tonnes of nitrogen considering minimum fixation of 20 kg N/ha from 0.5 kg/ha application dose. One tonne of phosphate solubilizers are equivalent to 24 tonnes of P_2O_5 considering 30% solubilization at minimum dose of 40 kg P_2O_5/ha with 0.5 kg PSM/ha application dose.

Estimated requirement of biofertilizers (excluding blue green algae) is about 85000 tonnes per year in India and the production levels are much less than the requirement. It is an important task to ensure the supply of such a large quantity of biofertilizers to farmers in India as availability of good quality inoculants has been identified as one of the major constraints responsible for limited use of biological nitrogen fixation (BNF) technology. For success of biofertilizers in India, concerted efforts right from production of quality biofertilizers, demonstration to distribution are needed. Performance of biofertilizers is highly unpredictable because of their biological nature and their susceptibility to biotic and abiotic factors. Establishment of any introduced microbe in rhizosphere through biofertilizer application depends on the competitiveness of the strain. Efficient and competitive strains with selective traits perform even under adverse environmental conditions such as extremes of temperature, water stress, pH variability and nitrogen tolerance. Recent biotechnological methods can be used to tailor the strains with desirable characteristics. The quality of biofertilizer can be judged based on standards laid down by Indian Standard Institute (Bureau of Indian Standards).

The major biofertilizer producers in India are Government (Central and State—54%), Institutional agencies (28%), State Agricultural Universities (2%) and private companies (16%). The Rhizobial biofertilizer occupies the bulk share of the production (50%) followed by *Azotobacter* + *Azospirillum* (27%) and subsequently PSM (21%) and others (2%). Rough estimates indicate nearly 150 million hectare of cultivated land covers crops such as legumes, rice, wheat, other cereals, sugarcane and cotton. Nearly 12000-15000 tonnes of rhizobial biofertilizers (assuming 400 g/ha) are required for legumes alone whereas for all other crops, the total requirement would be about 85000 tonnes including BGA. The total installed capacity of India is about 13000 tonnes. At present National Project on Development and use of biofertilizers, the Govt. of India and DBT are engaged in the promotion of biofertilizers.

A profile of different biofertilizers which fix nitrogen is given in Table 1.

2. CONSTRAINTS IN THE USE OF BIOFERTILIZER

Inspite of low cost and ecofriendly input in production technology of pulses, cereals, oilseed and vegetable crops, the biofertilizer is not as popular as other chemical inputs used in crop production. The major constraints faced by

Table 1: Profile of various biofertilizers

Biofertilizer	*Contribution*	*Limitation*	*Used for crops*
Rhizobium (Symbiotic)	Fixation of 20-100 kg N/ha, 10-20% increase in grain yield, leaves residual nitrogen in soil	Visible effects not reflected in areas where chemicals have been used for years.	Pulses, Oil seeds, Soybean, groundnut, Forage, Barseem, Guar, Leucaena
Azotobacte and Azospirillum (Non-symbiotic)	Fixation of 10-20 kg N/ha, 10-15% increase in grain yield, production of growth promoting substances.	Demands high organic matter	Wheat, maize, cotton, sorghum, mustard, sugar-cane, pearl-millet, rice, vegetable and other crops
Blue green Algae	Fixation of 20-25% N/ha, 10-15% increase in yield	Effective only in submerged rice. Demand bright sunlight and mild temperature	Flooded rice

development and extension agencies in popularizing the use of biofertilizers are as under:

- With the introduction of intensive agriculture, the farmer has accepted the use of chemical fertilizers, pesticides and insecticides on the basis of visual differences demonstrated by development and extension agencies at farmers fields and in turn the farmers became tuned to the perception that only these practices are useful where the visual difference is exhibited between the control and treated plots. Biofertilizer being a biological material sometimes lacks in this criterion of showing marked visible differences among treated and untreated plots which is a major stumbling block in popularization and adoption of this technology by the Indian farmers. Although the increase in total biomass and grain yield (10-20%) has been reported in trials conducted in All India Corrdinated Research Schemes of different crops at research farms and front line demonstration trials conducted by scientists at farmer's fields, however, the farmer's response is still not encouraging.
- Awareness regarding the use of biofertilizer among the farmers is not upto the required level and the technology has not percolated upto grass root level. The development and extension agencies have to play an important role regarding this aspect by organising field demonstrations and field days at village level and inviting the experts for explaining the usefulness of this input in supplementing the nutrient demand of the crops.
- Availability of quality inoculant in the market is matter of great concern for extension agencies and farmers. A large number of production units have started biofertilizer production without any technical back up and started supplying substandard product in the market. Although the Bureau of Indian Standards has laid down I.S.I. standard for different biofertilizers but it has

not been made mandatory for producers to apply for I.S.I. mark. Hence to check the proliferation of spurious production units, appropriate legal provisions are required to make the I.S.I. mark mandatory for such units so that proper monitoring may be carried out to weed out the substandard suppliers from the market. Otherwise the fake producers will kill the product in infancy and it will become very difficult for field staff to regain the confidence of farmers.

- To produce quality inoculant, the efficient strains must be used by producers after exhaustive screening under laboratory, greenhouse and field conditions. The response under field conditions may be major selective criterion for selection of strain to be used for mass production. The selection of appropriate carrier is very important for ultimate quality of product and it must be capable of maintaining high viable count under room temperature and hence have longer shelf life.
- Distribution network is not extensive and small farmers are handicapped for purchasing biofertilizer. Efforts should be made to make biofertilizer available at seed distribution agencies, Krishi Gyan Kendras (KGK) and Krishi Vigyan Kendras (KVK). Unless the easy availability of biofertilizer is not assured, the percolation of this technology to grass root level is not feasible.
- Small producers lacking in technical resources need to coordinate with experts in the Universities and Government agencies to produce high quality inoculants. These agencies in turn must formulates policies to cooperate with private sector for transfer of technology to the small producers at very nominal charges.
- The responsibility of mass production may now be shifted from the universities to different agro processing units/state agricultural departments so that the product may reach to large number of farmers through their vast distribution net work.

The biofertilizers have immense scope for rapid penetration into cropping system due to high input : output ratio as compared to other inputs used in crop production technology. However, concerted efforts are required by the universities, development and extension agencies, media and production agencies for recommending the use of biofertilizers for supplementing the nutrient requirement of the different crops under Indian situation in near future.

FURTHER READING

1. Chanda, T.K., Dubey, A.C., Sati, K. and Robertson, C. Biofertilizer Statistic. The Fertilizer Association of India. (1998) 11.
2. Dogra, R.C. and Khurana, A.L. Microbiology and Industry (Eds. R.S. Kahlon, R.K. Sedha and J.K. Arora). PAU Press Ludhiana, (1999) 53.
3. Khurana, A.L. and Dudeja, S.S. Biological Nitrogen Fixation Technology for Pulses Production in India. IIPR Kanpur. (1998) 15.

Mycorrhizal Biotechnology for Improved Crop Productivity

D.J. BAGYARAJ

Department of Agricultural Microbiology
University of Agricultural Sciences, GKVK Campus
Bangalore 560 065

THE TERM 'MYCORRHIZA' LITERALLY means 'fungus root' associations. The common mycorrhizal association of most of the crop plants is the vesicular arbuscular type. Pot experiments conducted earlier in sterilized soil showed that vesicular arbuscular mycorrhizal (VAM) inoculation can drastically improve plant growth. Later investigations indicated that even in unsterile soils, plants do respond to inoculation with efficient strains of VAM fungi. The increased growth of mycorrhizal plants is favoured in soils with low to moderate fertility. Greater soil exploration by mycorrhizal roots as a means of increasing P, Zn, Cu, S, etc. is well established. The other beneficial effects are their role in the biological control of root pathogens, biological nitrogen fixation, hormone production and greater ability to withstand water stress. If the beneficial effects of VAM fungi observed under pot culture studies could be emulated under field conditions, it would have a large economic impact on agriculture.

The effects of VAM fungi on crop plants include a saving in application of phosphatic fertilizer, increased survival rate and growth, early flowering and fruiting of plants and enhanced yield.

1. MYCORRHIZAL DEPENDENCY OF PLANTS

Plants differ greatly in their mycorrhizal dependence. Relative mycorrhizal dependency (RMD) has been as the degree to which a plant is dependent on the mycorrhizal condition to produce its maximum growth or yield at a given level of soil fertility. This was to know the extent of growth increase attributed to mycorrhizal condition. The literature suggests a formula to calculate relative mycorrhizal dependency (RMD) of crop plants under field conditions comparing plants in fumigated and unfumigated soils. This measures the extent of growth increase due to native endophytes and the calculated values lie between 0 to 100%.

2. HOST PREFERENCE

It is well known that VAM fungi are not host specific. Any plant species can be infected by any VAM fungal species, but the extent of VAM colonization and their effects can differ with different host-endophyte combinations. Since the early endophyte screening trials it has become more and more obvious that VAM fungi differ greatly in their symbiotic effectiveness. Though a particular VAM fungus can infect and colonize many host plants, it has a preferred host which exhibits maximum symbiotic response when colonized by that particular VAM fungus. This lead to the concept of 'host preference' in VAM fungi. Several recent studies uphold the existence of host preference by VAM fungi.

3. SELECTION OF EFFICIENT FUNGI

There have been some attempts to select efficient VAM fungi on a rational basis, such as ability to stimulate plant growth. Literature suggests that several fungi can be screened for symbiotic response using test host through pot culture, followed by microplot and then field trials. Seedlings were first raised in unsterile nursery soils containing indigenous VAM fungi to which different test VAM fungi had been added. It is important to add the same number of infective propagules of different fungi based on most probable number counts. Once the infection had developed, the seedlings were transplanted to experimental plots and plant growth monitored. This technique followed precisely the procedure used by farmers in India and many developed countries. The procedure is extremely good to select fungi for the pre-inoculation of transplanted crops raised in unsterile soils. The fungi thus selected are ones which will compete well with indigenous VAM fungi. In this selection process, it is essential to include VAM fungi isolated from the root zone of the test host. This procedure for selecting efficient fungi has lead to the selection of inoculant VAM fungi for many crops important in agriculture, horticulture and forestry. Inoculation of citrus and mango root stocks with efficient VAM fungi made them ready for budding/grafting 2-6 months early, which has a great impact in citrus and mango nurseries.

4. PRESPONSE CURVES

Using the most efficient fungus for inoculating a particular crop, the amount of P fertilizer that can be saved may be worked out. It should then be possible to select the fertilizer level at which the response to inoculation can be optimized. These kinds of experiments suggest the amount of P fertilizer that could be saved with mycorrhizal inoculation to produce the same amount of growth and yield. Mycorrhizal inoculation will only be useful in agriculture if it enables the farmer to reach his optimum yield level with a significantly lower level of P fertilizer. The saving from the reduced fertilizer requirement will be dependent on the shape of the P response curves, and then can be compared with the costs of supply and introduction of the mycorrhizal inoculant.

5. INOCULUM PRODUCTION

VAM fungi are obligate symbionts. Attempt to culture VAM fungi on artificial media has met with little or no success. At the present time, the only method to produce these fungi is in association with the host plant root. Novel techniques to produce VAM inoculum in almost sterile environment through nutrient film technique or circulatory hydroponic culture system, root organ culture and tissue culture are available. But for large scale field trials, the only convenient method to produce large quantities of inoculum is by the traditional pot culture technique. Recently the technology has been standardized for mass producing VAM fungi using Rhodes grass as the host and Soilrite (TC grade) : Perlite (1 : 1 by volume) mix. The inoculum thus produced can be air dried and stored in polythene bags for up to two years at room temperature. Because of the difficulty in inoculum production technology it is currently recommended only for nursery raised transplanted crops.

6. APPROPRIATE TECHNOLOGY FOR NURSERY RAISED CROPS

VAM inoculum of suitably selected strains can be used for inoculation in the nursery bed. Growers only need to incorporate inoculum in the nursery beds or seedling trays at the appropriate rate by hand. Seedlings thus raised will be colonized by the introduced fungus and then can be planted out in the field. In developed countries, seedlings are usually raised in fumigated soil or potting mix. Inoculation of such substrate, with selected VAM fungi for the crop, is a easy and appropriate technology. Commercial use of VAM inoculation by this method, in the nursery production of citrus is done in USA. Inoculation of the nursery with efficient VAM fungi has been successful in the large scale production of many horticultural plants in France.

In many developing countries and sometimes in developed countries, seedlings are raised in unsterile soil or substrate. Inoculation with appropriate fungi in seedling trays containing unsterilized potting mix found to give vigorously growing seedlings of asparagus which performed well when planted out in the field. It was found that inoculation of unsterile nursery beds with efficient VAM fungi enhances the growth of root stocks of citrus and mango. Further it has been observed that inoculation increased yield of chillies, tomato, capsicum and other vegetables. Increased growth of the seedlings of plantation crops like cardamom and coffee was observed, when inoculated with VAM. It was also observed that in VAM inoculated plants, there is increased number of flowers and vase life of cut flowers of chrysanthemum and China asters.

7. USE OF VAM FUNGI IN ROOTING OF CUTTINGS, AIR LAYERING AND IN OVERCOMING TRANSPLANT STOCK

Some horticultural plants are propagated through cuttings. In such cases, rooting of cuttings is important. Enhanced rooting of cuttings through inoculation with VAM fungi has been reported. More studies are needed in this aspect. More rooting has been observed in tamarind and cashew plants propagated through air layering. Such plants withstood transplant stock and established better when planted in polybags. VAM inculated plants withstanding transplant stock have also been reported in avocado. The recent studies showed higher percentage of grafting success in cashew.

8. USE OF VAM FUNGI IN HIGH VALUE HORTICULTURAL PLANTS RAISED IN FUMIGATED SOIL OR SOILLESS MEDIA

There is a great potential of utilizing VAM fungi in horticultural crops raised in fumigated nursery and/or field soil. Fumigation not only destroys VAM fungi, but, also other microorganisms. Inoculation with VAM fungi in such situations results in increased plant growth and yield. Such growth increase in fumigated nursery has been demonstrated in citrus and avocado etc. Similarly large acreages of strawberry, tomato and rose are produced in fumigated soil to control pathogens in several countries. Inoculation of the field or transplanting inoculated seedlings to the field markedly increased yield of these crops.

Soilless media like vermiculite, perlite, potting mixes etc. are used for raising several horticultural crop plants. Such soilless media are usually fumigated or heat sterilized. Inoculation of such soilless media enhancing seedling growth and finally yield has been reported in woody ornamentals and asparagus.

Significant increases in plant growth and yield of several crop plants because of VAM inoculation in unsterile soils containing less or inefficient indigenous endophytes have been reported by several workers in crops like easter lily, poinsettia, tomato, capsicum, chillies, papaya, mulberry, cardamom. etc.

9. USE OF VAM FUNGI IN THE MICROPROPAGATED PLANTLETS

In recent years, many high value crops are raised through tissue culture. Micropropagated plantlets cultivated on agar rooting medium are usually delicate and hence hardened before planting in the field. During the process of hardening, micropropagated plantlets get acclimatized to the outside environment. During the process usually 20 to 40% mortality is observed. Inoculation with mycorrhizal fungi at the time of hardening is known to considerably reduce the mortality rate. Inoculation of micropropagated plantlets with VAM fungi after hardening also improved plantlet vigour and growth in coffee, grapevine, apple, avocado, pineapple, kiwifruit, strawberry, raspberry, asparagus and banana.

The information discussed above has emphasized the potential benefits of the use of VAM fungi in crop production. Many high value crops are grown in nursery beds and then planted in the main field. Some of them are raised in fumigated soils or soilless media. More recently micropropagation is widely and successfully used for mass production of several crop plants. Inoculation of micropropagated plantlets with appropriate VAM fungi appears to be critical for their survival and growth. Studies conducted so far brought out that VAM inoculation can increase plant growth in the nursery, and result in better survival and long term growth when transplanted to the field. The method of inoculation is very simple and can easily be followed by farmers. Inoculum production must be given top priority and should be undertaken by private entrepreneurs, to make the inoculum available to growers. The growers should be properly educated through demonstrations and other extension methods so that the inoculation with VAM fungi can become a standard nursery practice. In the current era of organic farming emphasis is given to reduce the use of chemicals and to promote the use of organic manures and pesticides. Thus use of VAM fungi can be considered as an alternate strategy to more rational and sustainable agriculture.

FURTHER READING

Bagyaraj, D.J. Mycorrhizal fungi in crop production. In: Biotechnological approaches in soil microorganisms for sustainable crop production. (Ed. K.R. Dadawal), Scientific Publishers, Jodhpur, (1997) 309–319.

Bagyaraj, D.J. and Rashmi, A., VA mycorrhizal inoculum production, *Biofertilizer Newsletter,* **8**: (2000) 3.

Jeffries, P. Use of mycorrhizae in agriculture. *Critical Rev. Biotechnol.,* **5**: (1987) 319.

Lakshmipathy, R., Balakrishna, A.N. and Bagyaraj, D.J. Symbiotic response of cashew root stocks to different VA mycorrhizal fungi. *The Cashew,* **14**: (2000) 20.

Shashikala, B.N., Reddy, B.J.D. and Bagyaraj, D.J. Response of micropropagated banana plantlets to *Glomus mosseae* at varied levels of fertilizer phosphorus. *Indian. J. Exptl. Biol,* **37:** (1999) 499.

Varma, A. and Schuepp, H. Mycorrhization of the commercially important micropropagated plants. *Crit. Rev. Biotech.,* **15**: (1995) 313.

Transgenic Bt Crops as a Component of Integrated Pest Management

H. KUMAR[a] *and* A. KAPUR[b]

[a]*Proagro Research Station, Proagro Seed Company Ltd.*
[b]*Nunhems Seeds, Gurgaon 122 001, India.*

INSECTS, MITES AND NEMATODES reduce crop yields worldwide each year by more than 20% on average. However, individual fields may sustain losses of 50-100% from one or more pests. A single pest like Diamondback moth, *Plutella xylostella* has been reported to destroy the entire cauliflower or cabbage crop. Similarly, cotton bollworm, *Helicoverpa armigera* is also capable of destroying the entire cotton, pigeon pea or tomato crops. In spite of extensive application of pesticides on these crops, the percentage of crops lost to these pests has either remained stagnant or it has increased over the passage of time. Pesticides probably continue to be the most effective to control insect/pests that threaten to destroy crops. The detrimental effects of some pesticides on wildlife, the environment and human health have been well documented and brought to the public attention, first in the classic book *Silent Spring*, written in 1962 by the biologist Rachel Carson. As a result of increasing public awareness about the adverse effects of the pesticides, a number of pesticides have been banned from use in agriculture in a number of countries including India. From broad -spectrum pesticides like DDT to species — specific organo-phosphoros and carbamates to synthetic pyrethroids, killed not only the target pests but the natural enemies of the pests as well. Synthetic pyrethroids are being replaced with safer, more selective microbial toxins, new synthetic pesticides and transgenic pesticides-producing crops (Table 1).

Pesticides are one of the elements in the overall pollution of the earth's biosphere. Like many other human activities, pesticides are threatening to harm our valuable ecosystems, fauna elements, and, for that matter man himself. Why do we need non-contaminating, predominantly biological approach? Three basic reasons are:

1. Although pesticides do not rank first as environmental contaminants, specific pesticides may offer a wide variety of problems:
   (a) Disturbance of the natural ecosystem by killing natural enemies and thus inducing fauna elements to reach the pest status,
   (b) Accumulation of appreciable quantities of pesticides and their residues in the soil, natural waters and in food chains
   (c) Endangering the safety of the grower and consumer by direct hazards and residues in food products and
   (d) Development of strains of pests that are resistant to pesticides.

2. The above listed risks make the pesticides legislation very cumbersome, thereby increasing the costs involved in pesticides registration. Thus, this factor considerably limits the development of new pesticides thus reducing the availability of adequate technical control methods.
3. The application of biological control agents serves as the dynamic, self-sustaining mortality factors, which are able to regulate the pest numbers and are not subjected to resistance development.

Table 1: Comparative features of insecticides and transgenic B.t. toxin that contribute to resistance development

Feature	*Insecticide*	*Bt toxin*
Persistence	Very variable	Potentially continuous, can be controlled through tissue and time-specific promoters
Species selectivity	Low to high, including variable safety to natural enemies	High, including safety to natural enemies
Stage specificity	Variable, may be toxic to several stages of insects	Toxic to larvae but not adults
Distribution on plants	Vary variable, including inability to reach various tissues	Potentially vary and present throughout the plant including parts very difficult to reach with conventional pesticides
Distribution in crop ecosystem	Widely distributed on target and non target crops	Confined to the tissue of crop
Distribution beyond crop ecosystem	Drift to non-target areas common	Initial distribution restricted on the modified plants; in long term, distribution possible due to cross ability
Dose	Low to high at application but depends on biotic and abiotic factors	Highly consistent
Genetic basis	Insect resistance is monogenic, may be complex	Single gene with potential for deployment of multiple genes
Compatibility with IPM	Low to high depending on the selectivity	High

The recognition of widespread pest resistance to pesticides and the unacceptable consequences of excessive pesticide use led to the development of integrated pest management (IPM) approach. Integrated pest management of crop pests involves a complex integration of biological economic data, requiring a thorough understanding of the pests, beneficial organisms, crop phenology, cropping system, ecology and pest control strategies. It is an ecology based philosophy that emphasizes the need for pest management strategies to keep the pest population below economic threshold level with minimum use of pesticides. The IPM approach does not mean the complete eradication of the pests. The use of pesticides is strictly need-based when the other pest management tactics fail to prevent crop losses by the pests.

A reduction in the use of pesticide to allow sufficient scope to use other pest control tactics like biological control has been designed. The concept of Integrated Control probably emerged in California, whereas, the term pest management arose in Canada and in Australia.

The proponents of IPM see the following goals:

1. Maximization of economic profit.
2. Protecting the environment by minimal use of pesticides, thereby conserving the natural enemies, biodiversity and genetic diversity.
3. Substituting the local skill and inputs for imported inputs.
4. Increasing the cooperation among the farmers.

The basic considerations for the development of an integrated pest management (IPM) program are the following:

1. The ecosystem
2. Pest population dynamics
3. Compatibility of the tactic with other control methods.

1. STRATEGIES USED IN INTEGRATED PEST MANAGEMENT

1.1 Biological Control

Biological control is the use or encouragement of beneficial living organisms for the reduction of pest populations. It is the most ecologically compatible, inexpensive, sustainable, effective, and long-term method available to the farmers to control certain (not all) pests. This strategy includes the:

- Introduction and permanent establishment of natural enemies of the pest (classical biological control).
- Repeated mass release of natural enemies (predators and parasites).
- Conservation or enhancement of natural enemies
- Application of mass-cultured parasitic nematodes, bacteria, fungi and viruses of pests (biopesticides).

1.2 Cultural control (Modification of Agronomic Practices)

Such type of control strategies require timely modifications of following everyday agronomic practices:

- Planting, harvesting and irrigation timing
- Cultivation and tillage practices
- Crop rotation (most widely used IPM tactic) and disposal of crop residues
- Trap crops
- Pest-resistant varieties (This is the foundation of IPM program). Pest-resistant varieties can also improve the effectiveness of natural enemies of the pests.
- Intercropping

1.3 Mechanical and Physical Control

- Traps, screens, barriers to attract, repel or kill pests.

1.4 Reproductive and Genetic Control

- Chemical to alter pest reproduction
- Mass release of sterile individuals
- Introduction of harmful genes into the pest populations.

1.5 Chemical Control

- Pesticides
- Behavior-disrupting pheromones that attract or repel
- Pest hormones that disrupt growth, survival or reproduction
- Pesticides resistance management

The strategies described above can be woven together in an integrated pest management (IPM). The IPM technology offered serves as the Technology basket or menu of technologies from which the farmers can make choices. No general recipe for pest control works for every pest.

If so many environmental - friendly pest control strategies are available, then what is the need to have Bt crops? Why should we invest so much in developing another strategy inter-woven with fears and controversies?

2. CONVENTIONAL VERSUS TRANSGENIC PEST-RESISTANT CROP VARIETIES—DUPLICATION OR COMPLENTARY

Host plant resistance in crops against insects through conventional breeding approach has been considered as the most ecological, economical and sustainable pest management tool. Breeders the world over have developed very useful germplasm for resistance against insects. In some cases, the germplasm has been used effectively to breed commercially useful varieties for the farmers. However, not all the insect problems could be solved with the successes achieved through conventional plant breeding. For example, in cotton, boll weevils have been partially controlled by incorporation of resistance genes in commercial varieties in the United States as the biggest problematic cotton bollworm could not be controlled with conventional resistance breeding or through the pesticides. The pest continued to cause heavy damage to cotton every year. Even in India, the pest outbreaks caused extensive losses to the cotton producers of Punjab and Andhra Pradesh. Bt genes in cotton were very effective to control *Heliothis* species in almost every part of the world. Besides, *Heliothis* sp, Bt genes could also control pink bollworm and armyworm, which otherwise cannot be controlled with pesticides. Bt technology can thus complement the conventional breeding efforts to control a wide array of insects with different feeding habits and infestation patterns on the cotton crop.

Similarly, in rice, conventional breeding has been very useful to develop brown plant hopper (BPH) (*Nilaparvata lugens*) resistant varieties. However, such resistant varieties could not be bred against certain other pests, which are capable of causing crop losses of much bigger magnitudes to the farmers. Yellow Stem Borer (YSB), *Scirpophaga incertulas* is considered to be a very serious pest of rice not only in India but also in other parts of Asia. The pest cannot be controlled either by pest-resistant varieties or by the pesticides. Certain Bt genes are very effective to cause almost 100% mortality of the pest. Besides controlling YSB, the Bt genes can also control certain other pests like leaf-folders, *Cnaphalocrocis medinalis,* which is a serious pest of rice. Thus, breeding rice through conventional and biotechnological routes can be very effective to contain almost all the serious pests of rice. Use of conventional and biotechnological techniques in isolation will allow certain pests to make heydays.

More often, the crop germplasm developed through conventional breeding for resistance against insects suppresses the target pest but because of agronomic penalties, can not be used effectively to generate commercial varieties. For example, certain corn inbred lines developed for resistance against stem borers and armyworm could not be used effectively for producing commercial hybrids because the lines failed to combine well to yield agronomic advantage. On the other hand, no such agronomic practices have been observed in the commercial varieties produced via biotechnology. In view of the above, the germplasm developed for resistance against insects via biotechnological techniques should be used effectively in the conventional breeding programs for producing agronomically desirable crop varieties.

3. INTEGRATED PEST MANAGEMENT (IPM) IN INDIA

For more than 50 years, the pest control specialists in India have been developing IPM programs that carefully describe strategies for controlling agricultural pests. However, the adoption of IPM programs by the farmers have not been very successful. Village level experiments demonstrate a reduction of pesticide usage by non-chemical method but large-scale adoption of such IPM strategies is yet to be established. In Punjab, IPM modules are largely chemical-intensive which are causing unwarranted disruption of agro-ecosystem. Fourteen constraints for not using IPM at community level have been listed. Over-dependence of farmers on dealers for credit purchase of inputs for crop production has been reported to be the major cause for non-adoption of IPM programs. In South India, the farmers are sustaining high yields by applying frequent applications of insecticides at high rates.

In view of the above, the farmer always compares the new pest control strategy with the one he has been using i.e., pesticides to suppress the pests. The farmer always looks for convenience and results-oriented pest control methods. Therefore, the strategy, which goes with the seed, is highly acceptable to the farmer because of the simple mode of delivery.

In spite of its complementarities to conventional breeding, the agricultural biotechnology is the focus of controversy regarding the food and environmental safety. The principal food safety issues for the new crop varieties are:

- Toxicity of the newly introduced proteins,

- Potential changes in allergenicity,
- Changes in nutrient composition,
- Safety of antibiotic resistance marker encoded proteins included with the transgene.

After extensive safety testing, and five years of experience with transgenic crops in the market place, there is not a single report that would lead an expert food scientist to question the safety of such transgenic crops being grown the world over. Crops produced through biotechnology have proven to be as safe as or safer than crops produced by conventional breeding. However, it does not mean that the food produced through biotechnology present zero risks to the consumers. More often, because of lack of understanding for the biotechnologically-produced crops, the issues regarding the bio-safety are greatly exaggerated by critics and are not supported by science.

The one sided projection of discussion on biotechnology in the media creates undue doubts in the minds of public. Hence, balanced public awareness campaigns on this aspect are the urgent need of hour.

4. ROLE OF FARMER IN IPM

Knowledge about the pest population density is the most important parameter for the success of the integrated pest management for any crop. A key goal of IPM is to provide the farmer with sufficient information to decide whether he should spend money for treating the crop with a biological or chemical control agent. This is quite a complex decision, which is influenced by various factors. The farmer needs to determine:

- Number of pest individuals present on the crop at the time of applying control measures,
- Number of individuals of natural enemies on the crop
- Age of the crop
- Weather,
- Crop variety, etc.

A very few elite farmers may have access to the knowledge necessary to monitor and understand the interactions of their pest crop agro-ecosystem and implement a successful IPM program. In developing countries, the lack of knowledge results in using pesticides as insurance against crop loss, so farmers make applications of pesticides which may not be required at all and economically may not be justified. In the developing countries, the international agencies in collaboration with various national agricultural universities are making all round efforts to sponsor pest control educational programs to help improve and implement IPM practices by the farmers. However, though the programs are highly successful in the developed countries, the impact of such initiatives is far from realization in the developing countries. A country like India has almost 500-600 million farmers and of these, a less than 5% farmers possess necessary know-how about the real benefits of IPM.) The constraints for not using IPM at the community level in the Punjab state of India are listed below:

- IPM programs are quite complex in comparison to the currently available control tactic (pesticide).
- Farming community is highly influenced by the dealer network, which lack the basic training about the pest-crop-environment interaction.
- A majority of the farmers are completely dependent on the dealers for getting seeds/pesticides on credit.
- Farmers have a suspicion in their mind regarding the new technology
- Pest is a dynamic entity and changes year after year in the field
- Farmers are mostly ignorant about the new pest technologies.
- Farmers lack the basic education required for understanding the complex IPM strategies.
- Lack of ready cash for hiring private pest management practitioners.

5. TRANSGENIC BT CROPS AND IPM

Pest-resistant crops in IPM are useful because of the following advantages:

- Provide season long pest control
- Less expensive than the pesticides
- Less expensive and lesser labour oriented
- Provide higher returns to the farmers than pesticides
- Resistant varieties are compatible with other control tactics like biological control and pesticides (Table 2).
- Environmentally safe and harmless to people
- Provide cumulative benefits, more widely a pest-resistant variety is used, the more effective will be the pest suppression.

Table 2. Interaction of *Bacillus thuringiensis* (HD1 strain) with parasitoids and predators

Target insect	*Host plant*	*Parasitoid/predator*	*Interaction*
Boarmia selenaria	Avocado	*Trichogramma platneri*	No effect
Choristoneura occidentalis	Forest	Apanteles fumiferanac	No effect
C. pinus	Forest	*Cotesia melanoscela*	No effect
Lymantria dispar	Forest	*Cotesia melanoscela*	Synergistic
Lymantria dispar	Forest	*Cotesia melanoscela*	Synergistic
Pieris rapae	Cabbage	*Cotesia rubecula*	Antagonistic

Pest resistant crops developed through genetic engineering are capable of achieving a higher level of target pest control than that of conventional pest-resistant varieties. In fact, this control is also greater than the control achieved through the use of pesticides.

The first genes available for genetic engineering of crops for pest resistance were the Cry genes of *Bacillus thuringiensis*, also called *Bt genes*. "Bt" means *Ba-*

cillus thuringiensis, which is a rod shaped bacterium commonly, found in the soil, plants, air, etc. The bacterium was first isolated from silkworm in 1901. In 1911, Berliner named it as *Bacillus thuringiensis*. The bacterium is harmless to human beings and to birds and other animals.

The selectivity of the Bt toxins to insects is due to the specific mode of action that the toxins possess. The bacterium has a unique life cycle, reproduces by asexual method and produces millions of cells. On encountering the adverse conditions like low temperature, etc, the single celled bacteria form heat and radiation resistant spores. It is during the sporulation stage, that the bacterium produces a protein called as Cry (Crystal) protein. Detailed nomenclature of Cry proteins has been dealt. This protein is toxic to insects, which are serious pests of field and vegetable crops.

For commercial use of the bacterium, *B. thuringiensis* for insect control, the bacterium is multiplied in three stages: Fermentation, recovery and formulations. The formulations containing Bt spores are sprayed over the crops for pest control. The first commercial product, 'Sporein' was released in France. In 1957, Thuricide was commercialized in U.S.A. By mid-eighties, this biological pesticide was widely used all over the world for pest management. In India, these formulations are also available.

However, Bt sprays had the following limitations:

- Narrow spectrum of insecticidal activity regardless of environmental implication.
- Efficacy is more variable than pesticides because :
  - (a) it has to be ingested,
  - (b) it gets washed out by rains and
  - (c) it gets inactivated by the sunlight.
- Transmission from the diseased to healthy insects is non- existent.

The cry protein is produced in the bacterium by plasmid DNA. These genes are isolated from the DNA by first cutting the desired genes (smaller fragments) by molecular scissors, "restriction enzymes". The genes are then separated by "gel electrophoresis", on a gel plate. Various negatively charged DNA fragments can be separated by applying the electric current.

5.1 Brief History of *Bacillus Thuringiensis* (Berliner)

1901	Japanese Bacteriologist isolated *Bacillus* from diseased silkworm
1911	Berliner isolated an organism from flour moth and named it. *Bacillus thuringiensis.*
1938	First commercial product ' Sporeine' was released in France
1957	Thuricide was commercialized in U.S.A
1957–1977	Research work led to understand the mechanisms of action of Bt spores

Early 1980	Market was dominated by the products based on HD–1 isolate of the kurstaki subspecies and that of israelensis species.
Mid 1980's	Wide spread applications for pest management particularly in the forestry
Late 1980's	The molecular biologists found new pest-resistance genes of *B. thuringiensis* (*Bt genes*) for effective pest control.

Significant developments of molecular genetics are:

- Location of protein genes on plasmids
- Cloning of protein genes in *E. coli.*
- Availability of cloned genes and their sequencing
- Use of Crown gall bacterium, *Agrobacterium tumefaciens* for inserting foreign genes into plant cells or the use of high velocity microprojectile for monocots.

Out of more than 50 stably transformed crops, tobacco was the first plant species to be transformed.

Initially, *Agrobacterium tumefaciens* — mediated transformation was used to introduce full-length ICP (Insecticidal Crystal Protein) genes in tobacco and tomato with low level of expressed protein and insecticidal activity. Subsequently, the gene fragments encoding the N-terminal half of the crystal proteins were used for transformations. Toxic protein levels were increased significantly when the Tobacco plants were engineered with truncated *cry1Aa* and *cry1Ab* genes. Tomato plants expressing truncated form of *cry1Ab* showed higher mortality or growth inhibition by the *Heliothis virescens* and *Heliothis zea*.

Bt genes have significant higher A+T content than plant genes, which are rich in G+C contents. A+T-rich regions could cause splicing of the transcript or premature 3' end formation resulting in messengers which do not encode a functional Bt protein. Removal of potential plant polyadenylation signals and ATTTA sequences from a *cry1Ab* and *cry1Ac* coding sequence resulted in a higher frequency of insecticidal plants in tobacco and tomato transformations. Besides, increasing overall G+C content and introduction of plant-preferred codons increased expression levels up to 0.3% of the total plant protein. The resulting partial or fully-synthetic gene encoded, exactly the same protein as the native gene (Table 3). In 1990, the performance of transgenic cotton plants (var. Coker 312) containing cry1Ab or cry1Ac gene, which were effective in controlling cotton bollworm has been reported. Field tests were conducted with transgenic cotton in 1992 and the two very important pests, cotton bollworm and pink bollworm were effectively controlled. Similarly, potato plants were transformed with delta-endotoxin gene, *cry 3A*, which gave very good level of protection against Colorado potato beetle, *Leptinotarsa decemlineata* under high levels of natural field infestations. Similarly, having enhanced the G+C contents from 37 to 65% in a synthetic *cry1Ab gene*, the transgenic maize plants provided excellent control of European corn borer, *Ostrinia nubilalis* under insect pressure several hundred fold higher than the natural infestations. Transgenic cabbage and cauliflower expressing the modified *cry9c* gene resulted in complete protection from *Plutella xylostella* larvae.

Table 3. Expression levels of endotoxins in plants

Crop	*Gene structure*	*% endotoxin per total soluble protein*	*Insecticidal activity*
Tobacco	*cry1A(b)*: full length	0.0001–0.0005	–
	cry1A(b): truncated	0.003–0.012	+
	cry1A(b): neofusion	0.003–0.013	+
Tomato	*cry1A(b)*: truncated	0.0001	+
Cotton *cry1A(b)*:			
wild type		< 0.002	–
	cry1A(b): modified	0.05–0.1	+
Tomato	*cry1A(b)*:truncated (modified)	=0.3	+

The transgenic cotton, maize and potato were commercialized in mid-1990s. In 2001, commercial varieties of insect pest resistant Bt-maize, Bt-cotton and Bt-potato were planted on almost 12 million acres in the United States, Argentina, Canada, China, Australia, South Africa, Mexico, Spain, France, Romania and other countries. With India joining the elite group of transgenic crop-growing countries, the area under such crops is likely to increase in 2002. Farmers have used these insect resistant varieties to control insect/pests that have become resistant to conventional pesticides such as budworms or insects such as stalk borers of maize, which are difficult to control because they feed inside the stalks and cannot be targeted easily with pesticides. For China and its subsistence farmers, Bt cotton is an incredibly beneficial and valuable IPM technology. The Chinese farmers who grew Bt cotton in 1999 observed yield increases averaging 25% over conventional varieties with less than one insecticide spray on the Bt cotton compared to 14-15 on conventional cotton. Similarly, the farmers observed a significant increase in natural enemies of cotton pests on Bt cotton. In India, Bt cotton has also been reported to show significant improvement of cotton yields and according to the Indian government report, Bt cotton has been very effective for controlling insecticide-resistant bollworm (*Heliothis armigera*) in the farmer's field.

6. ADVANTAGES OF GROWING THE BT CROP

Following are the advantages of cultivating Bt crops:

- Bt crop provides season long protection from the pests
- Insect control shifts from reactive, in-season decisions of insecticide use to pre-season, planned, proactive decision of seed purchase.
- No elaborate scouting is required except for the resistance management.
- Insecticide-related concerns like timing, health risks, environmental impacts are eliminated by growing the Bt crop.
- No direct impact of Bt crop on beneficial insects or wild life has been ob-

served. The negative side effects of GM plants on natural enemies have been confined to small-scale tests in the laboratory but field studies have not shown any adverse effect on the natural enemies
- The production benefits include reduced yield and harvest loss, and lower incidence of fruit and stalk rot.
- The environment of the farm becomes safer to live and work in.
- The risk of accidental poisoning is reduced.

Many species of lepidopteran insect pests damage maize, some by boring into the stalks and ears where they cannot be controlled with the pesticides. The larvae tunneling and feeding inside the stalks of maize also increase the incidence of bacterial and fungal infection like mycotoxins (e.g., aflatoxin) produced by *Aspergillus flavus*. The spores of this fungus affect maize kernels and it is very poisonous. Reducing crop pest injury to crops such as maize and rice by using pest-resistant Bt varieties, has the potential to significantly reduce the mycotoxin levels in food.

Actual cost-benefit data is not yet available but benefits will vary from time to time at different fields. Transgenic crops can effectively control pests that feed on parts of plants which are difficult to access through conventional means e.g., fruit borers, stalk borers that feed inside the stalks, second generation stalk borers which can not be controlled through the application of pesticides.

A 10%, penetration of global insecticide market by Bt crops could lead to an 8% reduction in pesticide usage, which is equivalent to almost 50,000 fewer tons of insecticides introduced into the environment. This would cause a significant reduction of problems related with insecticides.

The transgenic Bt crops exhibit a tremendous efficacy in controlling the important pests like *Plutella xylostella, Helicoverpa armigera*, and *Leucinodes orbonalis*. Scientifically, this is a highly feasible pest management tactic because of its simple mode of delivery with the seeds. The Bt crops provide an effective alternative to chemical pesticides for preventing the crop losses. On Bt crops, the larvae die after taking the first bite on the leaf. The larval bite is so tiny that the feeding is not manifested in the form of visible damage on the plant.

Bt crops on their own justify all the considerations of IPM as explained below:

- *Environmental implications:* The Bt is specific and attacks only a target group of insects, which are dangerous to the crop. It is not toxic to the other vertebrates and non-vertebrates. The experiments conducted by Nunhems Proagro Seeds, India and elsewhere show that Bt toxins have no detrimental effect on the beneficial organisms like earthworms, honeybees and green lacewing. The toxins kill only the pest but have no effect on the predatory or predaceous species, thus making it environmentally sound tactic of pest management.
- *Pest population dynamics:* Of the four stages of insect i.e., egg, larva, pupa and adult, Bt is toxic to larval stage, which actually causes damage. The adult moths are not affected by Bt toxin, which can fly about and lay eggs on Bt and non-Bt plants alike. The larvae emerging from the eggs deposited on the Bt plants will die while those hatching on the non-Bt plants will continue

to survive, thus keeping the pest population below the damaging levels. This tactic is thus in conformity with the concept of IPM, which entails containment rather than eradication of the pest.

- *Compatibility with the other tactics:* Bt crops can be sprayed with pesticides or can be treated with other biocontrol agents to further strengthen the insect control.

7. RESISTANCE MANAGEMENT STRATEGIES

For sustainable agriculture, the methods to delay or avoid development of pests that are resistant to control are necessary. It is already known that genes for resistance to Cry proteins exist in wild populations of some target pests of new Bt crops. It has been therefore predicted with some confidence that resistance will develop. Resistance management strategies try to prevent or diminish the selection of the rare individuals carrying resistance genes and hence to keep the frequency of resistance genes below levels resulting in inefficient insect control. Following strategies would delay the development of pest resistance to the specific Cry protein insecticides in transgenic Bt crop varieties (Table 4).

Table 4. Operational Procedures for the Management of Resistance to Bt in Transgenic Crops

Retention of susceptible insects for dilution of selection for resistance	
a. Refugia	— Nontransgenic (susceptible) crop cultivars within the Bt crop
	— Other nontransgenic host plants of crop pests within and aroun Bt crop;
	— Use of tissue— and time-specific promoters to leave non essential crop tissues unprotected with Bt toxin
b. Immigration	
	— Distribution of non-transformed crops and alternative host plants of the crop pests in relation to the distribution of Bt crops
	— Release of susceptible insects.
Dose and its persistence and distribution	
a. Dose control through Bt gene expression	
	—Low to high dose option
b. Duration and distribution of dose	
	—Tissue-specific promoters e.g., buds, fruits, seeds etc.
	—Time-specific promoters e.g., fruit ripening, flowering etc.
	—Wound-specific promoters
Gene Management	
a. Gene options:	
	—Single vs multiple Bt genes
	—Different Bt genes in different crops in relation to target pest
	—Bt genes combined with other genes.
b. Rotations or alternations	
Use in integrated pest management programmes	

7.1 Reduction of selection pressure and maintaining a susceptible population via refugia

- Preservation of susceptible genes by reducing the selection pressure by:
- Patchwork of completely transgenic and completely non transgenic plots in the same field i.e., spatial mixture and
- Season-to-season sequence that alternates between transgenic and non transgenic plantings i.e., temporal refugia.

The susceptible pests are expected to maintain overall susceptibility in the population by mating with partially or fully resistant pests. For such strategies to use, the biology of the insect pest and the environment in which it is prevalent is very important. For example, for cotton bollworm, the strategy deployed for preventing the development of resistance in insects will be different in developed and developing countries. In developed countries, where Plant Variety Protection Act is well placed and the farmers' understanding of the overall pest-crop relationship is very rich, the spatial or temporal mixtures will delay or prevent the development of resistance in insects against the Bt toxins. There is sufficient evidence in support of this argument as no insect resistance has been reported so far against any Bt crop in the USA even 6-7 year after commercialization of Bt crops. In the developing countries, like India, where farmers are not fully aware of the insect-crop relationships and the Plant Variety Protection laws are still being implemented, the above-mentioned strategies perhaps may prove to be of less use for delaying the development of resistance against Bt toxins. This argument is based on the fact that for a mobile insect such as Cotton bollworm, there will be sufficient refuge for the insect available naturally. The resource-poor farmer may find it very difficult to afford the cost of Bt crop seeds. Their fields will offer a natural refuge for the insect. Besides, alternative host plants of the polyphagous insects will provide the insect with sufficient refuge to dilute the resistance genes of insect population. In the developing countries, the planting of the crop is always staggered and the farmers hardly plant their crop simultaneously. The most sensitive phonological stage of the crop is always available to the insect for infestation and survival.

In addition, the current experience with the Indian market shows clearly that farmers tend to buy uncertified seeds, which are not strictly of F_1 hybrids. Farmers tend to grow F_2 or even F_3 seeds of the Bt crop. Such kind of anomalies, though prevent the farmer to have a rich harvest, provide a fair amount of refuge for the insect to develop and mate with those carrying resistance alleles. Under such conditions, the selection pressure will never build up on the insect population to evolve adaptation.

7.2 Mixtures of Cry Proteins with different Receptor Binding Sites in the Insect Gut

In addition to resistance genes from *Bacillus* species, other potential resistance genes available (Table 5) for insect control should be deployed to delay or prevent the development of resistance against insects.

7.3 High dose expression of crystal protein to control pest heterozygous for the resistance alleles.

A potential resistance management strategy for Bt plants is the expression of a dose high enough to kill homozygous resistant insects. Such an ultra high dose seems difficult to achieve with a sprayable product, it may be possible with Bt plants. A diamondback moth population less susceptible to a Cry1Ab spray could not survive on cabbage expressing the same protein. Similarly, a Colorado potato beetle population resistant to a Cry3A Bt spray could not survive on potato plants expressing the same protein.

Table 5. Resistance genes from Bacteria, Fungi and higher plants transferred to crop plants for protection against insect pests

Insecticidal protein	*Common name*	*Source of protein*	*Species targeted*
Bt endotoxins	Cry proteins (Crystal secreted at sporulation phase of Bacteria)	*Bacillus thuringiensis subsp, kurstaki*	Specific insect pest
Vegetative insecticidal proteins	Vip protein (secreted at the vegetative phase of bacteria)	*Bacillus thuringiensis* and *B. cereus*	Armyworm, Corn root worm, Cutworms
Cholesterol oxidases	Oxidases or CO	*Streptomyces*	Cotton boll weevil
Proteinase inhibitors	Proteinase inhibitors	Higher plants such as cow pea, beans	Specific insect pests
Alpha amylase inhibitors	Amylase inhibitor	Common bean	Specific insect pests
Chitinase	Chitinase	*Plants*	Specific insect pests
Lectins	Lectins	*Plants*	Specific insect pests

7.4 Rotation of Varieties Expressing Different Bt Genes

For this, different toxins are used in rotation. In the first year, one toxin producing variety is grown and in the second year, another toxin-producing variety is cultivated to prevent the selection of resistant pest on the same Cry protein.

7.5 Diversification of Eradication Measures in an Integrated Pest Management Program

The other pest management approaches available to the farmers should be used judiciously with the Bt crops to reduce the dependency on any one-control tactic to suppress the insect pests. Use of pesticides is limited to kill the pests to release the pressure on insect due to Bt crop only. The natural biological control is encouraged to release pressure on insect due to Bt crops. Similarly, classical biological control should also be practiced to suppress the target insect pest on the crops. All the tactics for pest management will eventually establish a new population balance at a lower density and the pest no longer would cause outbreaks and economic losses.

The laboratory selection experiments have documented the development of resistance in several insects against Bt toxins. After 15 or more generations (selections) on the Bt toxins in the laboratory, a significant shift in the susceptibility of insect toward resistance has been observed in these studies. Under controlled laboratory experiments, the frequency of resistant genes has been estimated to be 1×10^{-3} or less than this. However such estimations in the field are difficult to make simply because of several factors influencing the insect population, as explained above. The experiments demonstrate the extreme genetic capacity of insects to adapt to the novel proteins but such studies failed to correlate the data with the field selection even 6 years after the commercialization of the Bt crops.

The pattern of arthropod invasion of a crop is also likely to influence the performance of Bt crops for host plant resistance through conventional breeding. For this, the implementation of an IPM program in which various pest control tactics are used in conjunction with transgenic crops is also very important. For example, it has been observed that *Helicoverpa armigera* emigrating from other crops can infest tomatoes in large numbers and most of the fruits suffer extensive damage by the pest. When Bt tomatoes are grown, the plants are heavily oviposited by the invading *Helicoverpa armigera* moths. The newly emerged larvae invade the tomato tissue for feeding. Even though, the Bt crop would eventually kill the neonate larvae, that has caused a considerable amount of damage just by initial penetration and probing of the tissue prior to suffering mortality. The crop loss suffered by the farmers can be very high if the larvae attack tomato at the flowering stage. The larvae penetrating the reproductive parts of the plant can cause huge degree of flower loss and thereby affecting the tomato fruit harvest. Therefore in order to implement an IPM program with Bt tomato or cotton, the following tactics should be used:

- Bt tomato/cotton as the main component with resistance against *Helicoverpa armigera*,
- An effective pest monitoring system e.g., light trap / pheromone trap to monitor the magnitude of invasion by the moths,
- A selective pesticide along with the natural and/ or inundative release of bioagents at the time of pest invasion.

The implementation of such an IPM program will also be very useful for the management of yellow stem borer, *Scirpophaga incertulas* of rice in the farmer's field. Like *Helicoverpa armigera*, this pest also invades rice field in large numbers but it does not attack flowers. Extensive oviposition will occur on the Bt rice crop by the invading moths. Little probing of the stalk tissue for feeding by the thousands of neonate larvae can adversely affect the rice plants.

Each pest-crop complex will require a specific implementation of certain resistance management strategies. A thorough understanding of the agronomic practices employed by the growers is very essential to define what is required in terms of resistance management. The resistance management strategies should be practical and acceptable to all involved in the decision-making of the Bt-crop commercialization.

8. RESISTANCE MANAGEMENT STRATEGIES IN RELATION TO PEST BIOLOGY

Prior to using the appropriate resistance management strategy, it is important to know the detailed pest biology. A thorough knowledge of this will enable the workers to develop the practical and effective strategy. The following biological factors could be important for influencing the resistance management strategies:

- It is important to know the biotic and abiotic factors, which cause mortality of the insect pest. It is known that the effectiveness of most of the natural bio-agents is density-dependent. If the activity of these bio control agents is high on the corresponding high insect pest population present on the susceptible host used as refuge, the population of the susceptible insect pests destined to mate with those on the Bt crop will decrease and the effectiveness of the refuge in reducing the development of resistance in the target pest. Knowledge of such factors could enhance the effectiveness of the refuge.
- Certain pests have greater mobility and are capable of flying long distances, while others may be restricted within the infested field only. Therefore, before resistance management strategy is implemented, it is important to know how far the given insect pest or its natural bio control agent can fly. Such factors can be extremely critical in decision making of refuge placement in the field where Bt crop is grown.
- The insect pest developmental rate on the Bt as well as susceptible host used as refuge should be known. If the insect pest develops faster on the susceptible refuge than the Bt-expressing plants, the adults emerging on the former will not find a mate from the latter so that the effectiveness of the refuge will be reduced tremendously. Thus success or failure of the refuge will be determined by synchronous development on the Bt- expressing plants and the susceptible host.
- Certain insect pests such as diamondback moth have a number of generations per cropping season, while others like cotton bollworm have relatively lesser number of generations per cropping season. The placement of refugiea will be determined by the number of generations a pest can have in the given agricultural background. Whether all farmers use the Bt crop or a few of them use Bt crop while other grow the non-Bt cultivars will be crucial along with the number of generations and other pest biological attributes in deciding the placement and size of refuge.
- The success or failure of the refuge in reducing the adaptation of the insect pest on a given Bt crop will also be determined by the polyphagous or monophagous behavior of the insect pest and whether two or more hosts are grown in the same cropping season. For example, for controlling *Helicoverpa armigera*, the Bt cotton is grown during the Kharif season when the alternate and preferred non Bt host like pigeon pea is also grown by the farmers. Under such conditions, the pigeon pea will by all probability serve as the natural refuge for cotton bollworm resistance development. Unless all the crops become transgenic, the probability for which is extremely remote, the

natural refuge will be adequate to provide the susceptible insects for diluting the resistance genes. However, other factors need to be considered as described above to get a normal synchrony in the sex ratio of the individuals developing on the Bt and non-Bt crop.

9. MONITORING TARGET INSECT POPULATIONS FOR SUSCEPTIBILITY TO CRY PROTEINS

As more and more Bt crops are being commercialized the world over, it is very important that the pest population is monitored carefully for its mortality on the Bt crops and any change from the current status is observed and immediately communicated to the agency responsible for their release and sale. For this, all the players involved with the crop production need to play an active role in determining the shift in the crop resistance against the target pest. The NGO's, agricultural universities and the farmers need to play a key role in the resistance-monitoring program. The role of farmers in particular is very important for resistance-monitoring program. The farmers need to be educated comprehensively for accomplishing the job of resistance-monitoring in the Bt crops. Special training program should be conducted by the agency responsible for commercialization of the Bt crops so that farmers can give them adequate and desired feedback about the status of the pest in their fields. Any damage noticed should be communicated to the desired agency.

Genetically modified (GM) crops are becoming a part of agriculture throughout the world. The appropriateness and safety of this technology is still being questioned by various group of people. As such no technology is risk free. The public because of various myths and doubts created by certain activists is not accepting the technology, which has a potential benefit for the farmer and the consumer alike. As people come to understand the real risks and benefits of the technology, and as they become familiar and educated about it, then they will be in a position to judge and accept it according to its merits.

GM plants have the potential to play an important role in developing an agricultural system that can support an ever-growing food needs.

The following tactics will have to be used in an IPM program:

- Reduced and selective usage of currently used tactic of pesticides along with the Bt crop to reduce the selection pressure on the pest by either of these tactics.
- Use of bio-control agents like *Chrysopa* and *Trichogramma* to further reduce the selection pressure.
- Supply of susceptible individuals through temporal and spatial placement of refugia, immigration, rotation of Bt toxins and gene stacking/pyramiding, etc.
- In the case of DBM, which can infest plants of mustard family equally i.e., cabbage, cauliflower and mustard, it is often seen that the farmers grow them in the same field. However, because of economic considerations, farmers protect only cabbage or cauliflower with pesticides but leave mustard without spray. The unsprayed mustard serves as the reservoir of DBM infes-

tations. Farmers go on spraying cabbage/cauliflower, but, pest does not go away though temporarily, the insecticides do control the pest on the target area. In fact, Bt crop will fit very well in such situations because mustard crop will serve as a kind of refugia for the Bt crop and immigration of susceptible population from the mustard reservoir will help to delay the development of resistance in insects on Bt crop.

For any program to succeed, the indigenous forces need to function and flourish. It is required to consolidate the knowledge base for various components of integrated pest management at various levels in the country. Once it is done, then serious efforts can be made to identify and develop a practical approach to implement IPM for various crops and agro ecological zones. The constraints (technical aspects, organizational aspects and social aspects) for IPM implementation in the country need to be analysed and addressed cohesively. The decision-makers in the government, industry, NGO's and farmers will have to decide whether it wants at this moment an agricultural production at lower costs (unsustainable) or a system, which is slightly more expensive but is sustainable for long run. There is need to make deliberate efforts to develop sustained-yield agro-ecosystems having IPM modules as necessary component. The public has to be informed of the possible consequences of both the alternatives in the right perspective. It is probably the time to pay now in order to assure the establishment of a durable system of integrated pest management for future generations. Else, we continue to watch helplessly the flow of pesticides residue into our biological systems, the degradation of our environment, the depletion of biodiversity and pushing our successors into a potential hell.

10. BT CROPS, IPM AND COMMERCIAL SEED COMPANIES

As discussed above, Bt crops have a lot of potential benefits and are very important and basic component of an integrated pest management program. The investments in developing Bt crops are very high and the financial returns to the seed companies may not be on the expected target. The possibility of financial returns are always there when investments are made on transgenic Bt crop—insect pest projects like, Cotton-*Helicoverpa*, Cabbage/Cauliflower-Diamondback moth or egg-plant—shoot and fruit borer combination. In a tropical country like India, where pest damage on the crop is tremendous and the losses suffered by the farmers are high, projects with Bt crops can give considerable financial returns to the seed companies. However, the efforts and investments in such projects should be rewarded with equal support and recognition by the other agencies. Others should not view the efforts of the seed companies with apprehensions and profit generating tactic. After all, any new technology introduced in a country will help its agricultural community to enhance the crop production. No technology is introduced in a country without examining all the potential risks and benefits. The introduction of the technology gets underway once the highest educated agricultural experts in the country study all the facets of its introduction (ecological, food safety in terms of health and environment, potential to cause allergenicity, invasiveness, effects on non-target organisms, potential benefits to farmers with a systematic and proven testing procedure, etc.). The private seed companies feel let-down, defeated in their goal and suffer heavy economic loss if the project gets blocked in the way in spite of satisfying all the parameters for deregulation. Host plant resistance through conventional breeding is an ecologically safe and economical method of pest management, but, little has been done to exploit it commercially because of complex nature of polygenes involved, time consuming, not very effective in suppressing the target pest, and perhaps the necessary yield gains are seldom realized. Monogenic resistance of Bt crop is providing all the yield advantages and can be the tool for next Green Revolution in India provided the resistance is managed judiciously as has been done so far in the United States.

ACKNOWLEDGEMENTS

The author wishes to thank Proagro Seed Company for providing the facilities. The opinion expressed in this article is strictly that of author and the views expressed do not necessarily reflect that of the seed company or the Indian government.

FURTHER READING

Adang, M.J., Firoozabady,E., Klein, J., Deboer, D., Sekar, V., Kemp, J.D., Murray, E, Rocheleau, T.A., Rashka, K., Staffield, G.,Stock, C., Sutton, D. and Merl, D.J. Application of a *Bacillus thuringiensis* crystal protein for insect control. In: *Molecular Strategies for crop protection*. (Eds. C.J. Arntzen and C. Ryan). Alan R. Liss, New York. (1987)

Beirne, B.P. Pest management. Leonard Hill. London. (1967) 123.

Chassy, B.M. Food safety evaluation of crops produced through biotechnology. *J. Amer. College of Nutrition*. **21**: (2002) 1665.

Entwistle P.F., Cory, J.S., Bailey, M.J. and Higgs, S. *Bacillus thuringiensis*, An environmental biopesticide: Theory and Practice. John Wiley & Sons (1993).

Geier, P.W. Management of insect pests. *Ann.Rev. Entomol.* **11**: (1966) 471.

Gould, F. Sustainability of transgenic insecticidal cultivars: integration of pest genetics and ecology. *Ann. Rev. Entomol.* **43**: (1998) 701.

Gould, F.A., Anderson, A., Reynolds, A., Bumgarner, L. and, Moar, W. Selection and gentic analysis of a *Heliothis virescens* (Lepidoptera: Noctuida) strain with high levels of resistance to *Bacillus thuringiensis* toxins. *J. Econ. Entomol.* **88**: (1995) 1545.

Jumin Tu, Zhang, G., Datta, Karabi, Caiguo Xu, Yuqing He, Zhang, Q., Khush, G.S. and Datta, S.K. Field performance of transgenic elite commercial hybrid rice expressing *Bacillus thuringiensis. Nature Biotechnology.* **18**: (2000) 1101.

Kalkat, G.S. Key-note address. National level group discussion on "Critical issues of IPM in the changing agricultural scenario in India", Nov. 28-29, 1998. Punjab Agricultural University, Ludhiana, Punjab (1998).

Kennedy, G.G., Gould, F., DePonti, O.M.B. and Stinner, R.E. Ecological, agricultural, genetic, and commercial considerations in the deployment of insect resistant germplasm. *Environ. Entomol.* **16**: (1987) 327.

Matthews, G.A. Implementing cotton integrated pest management. *Expl. Agric.* **33**: (1997) 1.

Maxwell, F.G., Jenkins, J.N. and Parrott, W.L. Resistance of plants to insects. *Adv. Agron.* **24**: (1972) 187.

McGaughey, W.H. and Whalon, M.E. Managing insect resistance to *Bacillus thuringiensis* toxins. *Science* **258**: (1992) 1451.

Pathak, M.D. and Khush, G.S. Studies of varietal resistance in rice to the brown planthopper at the International Rice Research Institute. In: *Brown planthopper: Threat to rice production in Asia*, Philippines: IRRI, (1979) 285.

Roush, R.T. Managing pests and their resistance to Bacillus thuringiensis: Can transgenic crops be better than sprays? *Biocontrol. Sci. Technol.* **4**: (1994) 501.

Schnepf, E., Crickmore, N., Van Rie, J., Lereclus, D., Baum, J, Feitelson, J., Zeigler, D.R., and Dean, D.H. *Bacillus thuringiensis* and its pesticidal proteins. *Microbiol. Mole. Biol. Rev.* (1998) 775.

Schuler, T.H., Poppy, G.M.,Kerry, B.R., and Denholm, I. Insect-resistant transgenic plants. *TIBTECH.* **16**: (1998) 168.

Stewart, C.N. (Jr) and Wheaton, S.K. Urban myths and scientific facts about the biosafety of genetically modified (GM) crops. In: *Plants, genes and crop biotechnology.* (Eds. M.J. Chrispeels, and D.E. Sadava). Jones and Bartlett publishers. Canada. (2003) 529.

Tabashnik, B.E., Liu, Y.B., Malvar, T; Heckel, D.G., Masson, L., Ballester, V., Granero, F., Mensua, J. L., and Ferre, J. Global variation in the genetic and biochemical basis of diamondback moth resistance to *Bacillus thuringiensis*. *Proc. Natl. Acad. Sci.* USA. **94**: (1997). 12780.

Wearing, C.H. and Hokkanen, H..M. T. Pest resistance to *Bacillus thuringiensis*: ecological crop assessment for Bt gene incorporation and strategies of management. In: *Biological Control: benefits and risks.* (Eds. H.M.T Hokkanen and James M. Lynch). Cambridge University Press. (1995).

Biopesticides: Potential for Indian Agriculture

M.C. SHARMA

Biotech International Limited
"VIPPS Centre", 2, L.S.C, Block E.F.G.H
Masjid Moth, Greater Kailash-II, New Delhi-110048.

THE ADOPTION OF MODERN AGRICULTURAL technologies in India, while leading to the so called "Green Revolution", has culminated into serious problems of crop losses, pesticide resistance in major pests, resurgence in minor pests, elimination of populations of natural parasites, predators & pollinators, environmental pollution, pesticide residues in food chain and health hazards. Recent estimates of annual crop losses in India due to pests and diseases amount to around Rs. 2,90,000 million and one single insect Helicoverpa is responsible for losses over Rs. 35,000 million crores per annum because of very high pesticide residue contents.

World Health Organization (WHO) estimates 1 million pesticides poisoning cases and 20000 deaths every year globally This is due to high pesticide residues in food chain including cereals, pulses, vegetables, fruits, milk and milk products (including mother's milk), fishes, poultry, meat products and water. Twenty five per cent of Indian food products contain pesticide residues above tolerance level compared to only 1.2% globally. No detectable residues are found in only 2.5% Indian food compared to 80% globally. Chemical pesticides cause significant health hazards to man including convulsions, diarrhoea, dermatitis, allergies, anaemia, organ disfunctions, blindness, immunosupression, neurotoxicity, vasodilation, impairing of reproductive functions, leukaemia, carcinogenecity, tumorogenecity and paralysis, etc. Apart from these, they also cause harm to non target beneficials, soil microflora and crops. Suicides by farmers have been frequent in Andhra Pradesh (A.P.), Maharashtra, Karnataka, Punjab on account of total and sole dependence on chemical pesticides.

Most of the chemical pesticides presently used in India fall under extremely hazardous and highly hazardous categories. Use of globally banned/restricted pesticides to the tune of 55% in Agriculture sector and 95% in Public Health Sector still continues in India. The pesticides banned in India continue to flow into the market inspite of the notifications. DDT which is permitted for use in health sector only gets pilfered for use on crops and being a dustable powder, it is specially preferred by marginal farmers. Having carcinogenic potential, this is extremely hazardous because of its direct dermal contact and needs strict checks.

This adverse impact has led to Governments to re-orient protection strategies towards ecologically sound "Integrated Pest Management" (IPM).

1. IPM APPROACHES

Though initiated as back as 1962, Insect Pest Management (IPM) has made very slow progress in the country inspite of being funded by International Organizations like FAO, UNDP, ADB-CABI and Central and State Governments. Only 1% of 143 million hactare crop area and only 2500 villages out of over 6 lac villages in the country have been covered so far. Various Indian Council of Agricultural Research (ICAR) Institutes, State Agricultural Universities (SAUs) and Directorate of Plant Protection (Government of India) have developed comprehensive IPM packages for various crops.

The IPM concept involves integration of cultural, mechanical & behavioural methods, use of biocontrol insects, biopesticides, botanical pesticides and need based use of chemical pesticides. The biological control/ biorational approaches include use of behavioural methods, biocontrol insects, biopesticides and botanical pesticides.

The biopesticides based on pathogens (microbial biopesticides) could be an organism, part of an organism, a mixture of organisms, applied periodically to control crop pests. The most commonly used microbial pesticides are based on *Bacillus thuringiensis* (Bt), Nuclear Polyhedrosis Viruses (NPV) and *Trichoderma*. The Bts are highly effective on lepidopterous caterpillars, generally polyphagous and resistant to traditional chemical pesticides. Viruses are also very effective products for different borers. Fungal agents like *Trichoderma* are highly potential seed treatment agents to control wilts and rots etc, while *Baeuveria and Metarrhizium* are effective for control of sucking pests.

The botanical pesticides based on extracts of leaves/kernels/seeds of neem/karanja/chillies/garlic are equally efficacious insecticidal agents. Of these, the neem is the most promising and largely used one with broad spectrum action in insects, particularly for control of sucking pests.

2. ADVANTAGES OF BIOPESTICIDES

Biopesticides are advantageous due to their eco-safety, target specificity, no development of resistance, reduced number of applications, yields & quality improvement, higher acceptability and value of produce for exports and suitability for rural areas.

- Biopesticides are products based on living biologicals which are prone to decay/degradation in soil and water after 20-30 days. The chemical pesticides, however, persist in air, soil and water for very prolonged periods causing environmental pollution and their residues cause significant health hazards.
- Biopesticides kill the target pests only and do not harm other insects. For example, *Bacillus thuringiensis* var. *Kurstaki* kills only lepidopteran caterpil-

lars and has no impact on housefly, mosquito or adult moths. The real advantage of this comes to farmer in the form of conservation of natural parasites and predators like *Trichogramma, Chrysoperla,* coccinellids, ants and spiders which prey/parasitise the pests if left uncovered by the biopesticide target range. Additionally the beneficial pollinators like honey bees, etc. are also conserved, resulting in increased fruit setting in crops. Chemical pesticides however, kill the natural parasites and predators, thus eliminating their contribution in plant protection. The case of mango in Lucknow (Malihabad) area is an eye opener where the use of chemical pesticides results in damage to pollinators causing immense "Flower Drop" and very little fruit setting.

- The injudicious and indiscriminate use of chemical pesticides has led to the development of resistance in the pests (adaptation to chemical pesticides) and therefore, the farmers have to resort to repeated applications of variety of pesticides and their mixtures, even at times as "Prophylactic Sprays", inspite of no such recommendations in PACKAGE OF PRACTICES issued by SAUs. Contrary to this, only 2-4 applications of biopesticides may be required as they are highly efficacious with no development of resistance reported in India so far. This cuts down the cost of application of pesticide very significantly. It is to remember that pesticide spray operation cost constitutes about 60% of the total pesticide application costs.
- Another advantage of biopesticides is that they do not cause any changes in the biology of minor pests, whereas the chemical pesticides are known to have resulted in the resurgence of many minor pests of yester years. The best example of this is the whiteflies which were not considered as pests about 20 years ago, but, cause severe loss of cotton and vegetable crops currently. *Heliothis armigera* was initially known as "gram pod borer", but, currently it causes damages to almost all crops, contributing almost 15% of the total crop losses in the country.
- The use of biopesticides results in economic gains such as increased yields due to effective disease/pest control, improvement of quality, extra shine, better taste, immediate sale of their products (no waiting time after harvest) and higher price for their produce.
- The exportable agriculture products like cotton, fruits, vegetables, tobacco, tea etc. with certification of origin from biopesticide/organic farming areas find enhanced acceptability and fetch higher price of the produce. It is well known that organic tea fetches Rs. 1000/- per Kg more than the normal tea consignments in European markets. So is the case with grapes, bananas and mangoes. Therefore, the biopesticides are essential for food processing industry particularly for exports.
- Not surprising is the use of organic cotton by U.K. garment designers who sell their "organic Jeans" at a price of $ 30 over conventional denims. M/s Maikaal Fibres in M.P. have been purchasing entire organic cotton produce from 570 small growers at 20% premium price.
- The biopesticides being eco-safe and harmless to human beings are the most

ideal for use in rural areas where the labour/marginal farmers are not only ignorant about the precautions to be taken during chemical pesticide application operations, but, at times they do not have the capability of purchasing such protective gears. It is worthwhile noting that even F.A.O and W.H.O. refuse the grant "Financial Support" to the proposals involving the use of hazardous chemical pesticides for Locust control/Mosquito control programmes.

In view of the advantages of biopesticides, there has been wide acceptance of these globally, amounting to around 10% share of agrochemical market expected by 2000 A.D. with the growth rate of 10-15% per annum. There has been a total ban on use of chemicals in green houses in Europe and there is a niche market for "Green Label" (no chemicals used) vegetables and fruits, apart from clothes from organic cotton. The use of Bts in neighbouring countries are high. Malaysia-250 T, Thailand–160T and Phillipines–300T in comparison to India's Bt market approximately at 80T.

3. CONSTRAINTS IN ADOPTION OF IPM APPROACHES

A number of constraints plague the widespread adoption of IPM and biopesticides by farmers. Some of these are:

- Farmer suspicious of advise from Government Officials/Industry Scientists.
- Loans for plant protection inputs as well as other agricultural inputs are often provided by dealers (with vested interests).
- Non adoption of IPM practices in large areas mainly because of "Farmer Individualism".
- Linking of "Progressive Farmer" status with expenses on chemical inputs.
- Language barriers (state to state) and lack of appropriate conversion of scientific language to popular language compatible to the level of understanding of farmers.
- Lack of simple illustrative materials.
- Degradation of information during knowledge dissemination from scientists to extension workers and further to farmers.
- More emphasis on "Routine IPM" restricted to seminars and workshops rather than on really well executed IPM demonstrations at farmers fields.
- Timely availability of non-chemical inputs.

4. PROPOSED STRATEGY

The key to success of biological control lies in good pest surveillance, robust ETLs and precise applications at the egg/young/larval stages late in the evening to provide temperature and humidity congenial for microbial multiplication and actions coinciding with the most active phase of the pests. The precise timing only can save the crops from damage. There is no use to spray them once the damage has been caused and the larvae are at an advanced age ready to change to pupae. The proper and thorough coverage through appropriate sprayers and nozzles are desirable for effective results. At times skip row/alternate row/alternate pair row coverages are also advantageous enabling more numbers of parasites/predators in the field. In case of transplantation crops like paddy and vegetables, a spray of the microbial pesticides at the nursery stage about a week before the transplantations is a must to eradicate the pests carry over from nurseries. The emphasis needs to be laid on controlling the damage rather than on the instant kill of the pests.

A variety of corrective measures including no prophylaxis, dissemination of knowledge, demonstrations, recommendations from State Agriculture Universities, dealer training, dealer qualifications, prescription sales, proper application technology and increased availability at acceptable price are important steps to increase the adoption of the biopesticides.

4.1. Dissemination of Knowledge

The officials from Directorate of Agriculture, SAU Scientists and Scientists of Industry should educate the farmers/dealers/ extension officials regarding the beneficial & harmful insects, list of biopesticides & biocontrol agents & manufacturers, recent list of banned pesticides, disadvantages of chemical pesticides, no mixing of pesticides, sprays on E.T.L, at reduced/desired frequency & precise timings, standardization of sprayer design, nozzle design, volume of spray, norms of tractor speed, etc., special attention to off-season hosts like sunflower, berseem, vegetables to disrupt the continuity of pest populations and sales of pesticides only on the basis of prescriptions from SAU scientists/Directorate officials/industry scientists. Universities should come up with publications on biopesticides.

The efficacy of biocontrol agent may also go down if proper spray equipments are not used, resulting in the death of IPM movement. Only in few states proper nozzles as recommended by SAUs are used. Dealer is the sole advisor to the farmer. Due to the credit oriented market, the farmer has to depend for seed, fertilizers and pesticides on him. In view of this, dealer's qualifications assume great importance and only Agriculture/ Science graduates should be allowed to obtain "Sale Licences". This can be done only through "legislation". Director Agriculture, Punjab has already started action in this direction and other State Directorates may like to follow the same.

4.2. Reorientation

The training of officials and Farmer Field Schools do provide impetus to IPM, but, the successful implementation and acceptance of IPM at large scale can be only through timely availability of high quality inputs with sufficient shelf life. In fact all the energy and money spent in IPM seminars, workshops and farmers field schools shall go waste, if quality biocontrol inputs are not available or are not applied properly. This requires a total retuning of trainers and farmers, through close interactions between plant protection officials, the suppliers of biocontrol agents and the farmers.

Unfortunately, today, the decision of the pesticide application lies with the dealers, who in their interest, advise farmers to use a large number of applications of a variety of chemical pesticides and their mixtures, whether required or not or whether efficacious or not. Prescriptions from plant protection officials or from entomologists of the manufacturers can form a better base for purchase of plant protection chemical, just like a prescription from a doctor for purchase of medicines. This would also help remove the social stigma of adherence to past practices of continuing with hazardous and not so efficacious chemical pesticides. Dealer training and imposition of qualifications for dealers shall definitely create a positive change towards IPM.

The knowledge of biology of pests, damage caused by pests, beneficial insects, new ecosafe products and proper application technology need be transmitted to the farmers for early detection of infestations (egg & young larvae stages) and timely applications of appropriate biocontrol agents. It becomes imperative that farmer changes from philosophy of calendar based spray programmes (5 days interval/week or fortnight) to economic threshold level (ETL) based sprays. The 'ETL' and 'Plant Protection Modules' have mostly been designed by Universities/ States/ ICAR Institutes considering the individual pests/diseases and these need to be revalidated keeping in view total pest/disease complex of the particular crop. Even the mobility, fecundity of parasites, predators and pests and feeding pattern and propensity to gain insecticide resistance of pests/diseases should be considered for designing the "Modules".

5. ACCEPTANCE OF BIOLOGICAL CONTROL

The biological control has proved to be highly efficacious, wherever tried, but, the issue of "as cheap as conventional" puts a hurdle to its widespread use and social acceptability. Though cheaper to develop and register, the biocontrol agents and biopesticides have high production costs because of highly specific environmental conditions for optimum productivity and "purity" of the agents. The consumer has to decide between "the environmental friendliness and effective control more important" or "the lowest possible price with meager control as important" and this decision divides the globe into "Developed countries–more eco-conscious" and "Under developed/Developing Countries not so eco-conscious but more cost conscious".

The significance of biological control is evidenced by the interest in use of "Organic Cotton" by garment designers of U.K. and U.S. Companies who sell their "Organic Jeans" at a premium of $ 30 over conventional denims and European markets selling vegetables and spices labeled "Organic (Chemical pesticide free)" at premium prices. Even Indian Government has founded organic cotton pilot production programme in 1994. M/s Maikaal Fibres, Indian Textile Company has offered to buy all organic cotton produce from 570 small growers at about 20% premium price.

6. TRANSGENIC CROPS

A number of transgenic crops are being introduced world over and are under trials in India also. Cautious approach is a must for these, in view of some inherent problems. Transgenic crops having target specific genes incorporated in them shall be able to take care of specific pest/disease problems and shall require pesticidal sprays for the remaining pest complex. For example, transgenic cotton with Bt. gene may be able to withstand the bollworm infestations, but, shall require pesticide applications for aphids, jassids, whiteflies and diseases. The presence of the "terminator gene" shall not only eliminate the viable cotton seed production in the transgenic crop fields, but, might result in loss of viable seed production in adjacent fields of normal varieties as well.

Biological control and integrated pest management has become priority and thrust area of Central and State Governments in India, resulting in initial positive impact and adoption by few farmers. The availability and quality of biocontrol inputs, however, deserves more attention from industry as well as government. Several immediate steps, including abolition of import duty, excise duty, sales tax and octroi, strict adherence to quality and specifications, nodal agency for quality evaluations, increasing plant alterations and procurement of biocontrol inputs, widespread publicity and training of trainers, dealers and farmers, are essential for adoption of biopesticides at large scale by farmers.

Apart from this, strengthening of state level Biocontrol Laboratories, State/ Industry partnership for their functionalisation, utilization of global funds for procurement of biocontrol inputs and for technology/quality upgradation, quantification of bad impact of pesticides on human life and expenditure incurred on the treatment of affected human beings and loss of beneficial insects, microflora, animals and plants could help in faster implementation of biopesticides technology.

There is no doubt that the adoption of eco-friendly biological control approaches through co-ordinated and planned efforts from government, industry and NGOs with sincerity and accountability shall play vital role in preventing the damage to economy, environment and mankind by reducing the consumption of hazardous chemical pesticides, conserving natural beneficial biodiversity and increasing the confidence in farmer and extension workers in managing crops and handling potential pest problems in the new millennium.

FURTHER READING

Copping, L.G. and Menn, J.J. Biopesticides: A review of their action, application and efficacy. *Pest Manag. Sci.* **56** (2000) 651.

Entwistle, J.S. Cary, Bailey, M.J. and Higgs, S. *Bacillus Thuringensis*—An environmental biopesticides: Theory & Practice. Wiley Publishers, John Wiley & Sons, (1993).

Manjunath, T.M. Biopesticides: Importance of their commercial production, constraints and strategies. In: Changing Scenarios in Indian Agriculture (Eds. S. Kannaiyan and S. Uthamasamy), TNAU Press, Coimbatore, Tamil Nadu, (2001). 23–28.

Milner, R. and Chandler, C. Proceedings: Workshop on *Bacillus thuringiensis*. Abbott Chemical and Agricultural Products Division, CSIRO Institute of Plant Protection and Processing. Rhone-Poulene Rural Division, Sandoz Australia Agro Division, Australia, (1991).

Navarajan Paul, A.V. Commercial production of biocontrol agents—Problems and Prospects. AIBA Newsletter, July–August, (1998) 33–35.

Singhal, V. Status of adoption of biopesticides and biocontrol agents in Indian Agriculture. Published by All India Biotech Association (AIBA), New Delhi. (2000) 1–20.

Sticher, I., Mauch-Mani, B. and Metraux, J.P. Systemic acquired resistance. *Ann. Rev. Phytopathol*. **35** (1997) 271.

Enzymes in Agro-Processing: Trends and Future Prospects

S.K. SONI *and* J.K. GUPTA

Department of Microbiology
Panjab University
Chandigarh 160 014 (India)

MAN'S USE OF ENZYME reactions dates back to antiquity. The making of cheese, vinegar and wine, the leavening of bread, and the brewing of beer are all enzymatic processes which have their origins in prehistory. In the past few decades industrial uses for enzymes have been developed, and enzymes are now widely employed in many sectors of industry. They often have the advantage that they can replace harsh chemicals and reduce high temperatures and pressure and can thus have a beneficial impact on the environment. Enzymes are totally biodegradable and have no harmful effects in waste water treatment plants or in the aquatic environment at large. A summarised history of enzymes is listed in Table 1.

Alcohol fermentation is probably the oldest known enzymatic reaction, but this and other similar phenomena such as souring of milk were considered to take place only through action of living organism. In 1833 Payen and Persoz partially isolated the active agent breaking down the starch to sugar from malt and named it diastase, now known as amylase. These and other active preparations were given the general term 'ferments'. Liebig recognized that these ferments could be non-living material obtained from living cells, but, Pasteur and others believed that ferments must contain living material. While this dispute continued, the term 'ferments' was gradually replaced by the term enzyme. The term 'enzyme' was first proposed in 1867 by German physiologist Wilhelm Kuhne and comes from Greek meaning 'in yeast'. In 1897 German chemist Edward Buchner discovered that a cell-free extract of yeast can cause alcoholic fermentation. The earliest research to use and produce enzymes industrially was performed by Otto Rohm in 1906. Rohm used pancreatic tissue from offal to produce trypsin, which was used in tanning of hides. After Buchner's discovery it was assumed that enzymes in general caused fermentation and vital reactions.

In 1926 the American biochemist James B. Summer finally succeeded in isolating and crystallizing urease from Jack-bean extracts. Four years later, pepsin and trypsin were isolated and crystallized by American biochemist John H. Northop. All these three enzymes were found to be protein. This was the beginning of enzyme biotechnology and subsequently hundreds of enzymes were purified and characterized. Years of investigations in biochemistry and biotechnology have

Table 1: A summarized history of industrial enzymes

Pre-history	Cheese production. Carbohydrate fermentation for beverages.
Pre-biological sciences	Oriental fermentations—beverages, sauces, pickles, wines, beers, malting of cereals, milk and cheese products.
1874	Chr. Hansen—commercial rennet.
1878	W. Kuhne—term 'Enzyme' proposed.
1890	Takamine—takadiastase (fungal amylase).
1900-1930	Malt extract for textile desizing. Start of animal enzyme extraction on commercial scale.
1930	Pectinases for food processing.
1954	Submerged fermentation—proteinases for detergents; amylases for textiles and starch.
1960s	Diversification of fermented enzyme types—wider range for starch and detergent industries; development of fungal milk coagulants.
1970s	Further expansion of fermented enzymes—immobilized enzymes; thermostable and thermolabile varieties produced.
1980s	New, more refined enzymes; multi-activity specifically formulated products; energy saving systems, bio-fuels production; chemical bio-transformations and diagnostics; environmentally significant developments; practical genetic manipulations.
1990s	Rapid development of large numbers of speciality enzymes for R&D, diagnostics and industrial processing; enzymes from GMO; enzymes modified by protein engineering; large scale expansion of bio-transformations by enzymes.

boosted knowledge of enzymes for industries as well as researchers. Many new techniques have been established to modify enzymes or increase their yields. New techniques for purification of enzymes are constantly developing and so are being discovered new application of enzymes in medicine, research and industries.

During the early part of last century, enzyme technology was developing slowly but surely outside Europe. In the Far East, an old aged tradition prevailed where moulds called koji were (and still are) used in the production of certain foodstuffs and flavour additives based on soya protein (*shoyu*, *miso, tempeh*) and fermented beverages (sake, alcohol). Koji is prepared by inoculating the steamed rice with fungi. The composition of the mixture passes down from generation to generation. This formed the basis on which the Japanese scientist Takamine developed a fermentation process of growing a culture of *Aspergillus oryzae* on moist rice or wheat bran for industrial production of fungal amylase. The product was called 'Takadiastase' and is still used as a digestive aid.

A very important field in which enzymes have proved to be of great value over the last 15-20 years is the starch industry. As early as the 1950, fungal amylase was used in the manufacture of specific types of syrup, i.e. those containing a range of sugars which could not be produced by conventional acid hydrolysis. The real turning point was reached in the early 1960s with the launch of the first enzyme preparation, amyloglucosidase, that could completely break down starch into glucose. Within a few years, almost all glucose production was reorganized and enzyme hydrolysis was used instead of acid hydrolysis. The process was further improved by the introduction of a new technique used for the enzymatic pre-treatment (liquefaction). A highly successful and fully enzymatic starch hydrolysis was achieved when the bacterial amylase Termamyl® which was stable at high temperature, was introduced by Novo in 1973. Since then, the interest of the starch industry has focussed mainly on an enzyme of a different nature, glucose isomerase by which glucose containing an aldehyde group can be converted into the corresponding ketone sugar, fructose. Hence starting from pure glucose it is now possible to use glucose isomerase to produce a mixture of glucose and fructose with the same calorific content and sweetening effect as ordinary cane or beet sugar.

1. COMMERCIAL APPLICATIONS OF ENZYMES IN AGRO-PROCESSING

Enzymes find applications in many fields, including organic synthesis, clinical analysis, pharmaceuticals, food production, agro-processing and fermentation. The discovery of new microbial enzymes through extensive and persistent screening is opening new, simple routes for synthetic processes and consequently new ways to solve environmental problems. When designing a new synthetic process, a suitable catalyst for the reaction must be found and enzymes are ideal candidates for this role. However, in many cases, the substrates in industrial processes are artificial compounds and enzymes known to catalyse suitable reactions for such processes are still unknown. Therefore, screening for novel enzymes that are capable of catalysing new reactions is constantly needed. In addition, the discovery of new enzymes will provide clues for designing new enzymatic processes. One of the most efficient and successful means of finding new enzymes is to screen a large number of micro-organisms, because of their characteristic diversity and versatility. Such enzymes are useful catalysts for the synthesis of many biologically and chemically useful compounds (Table 2).

Enzymes have become an efficient means for processing agricultural products and bringing better value to food and beverage products. These are used in a wide range of foods, to improve gains, processability, shelf-life, taste and other properties in flour, starches, pop drinks, fruit juices, oils, beer, wine, cheese and meat. In the oil palm industry, enzymes are used to fully extract oil from palm kernels. The new applications of enzymes in agro-processing are emerging day by day and it is expected that these will be used in every catalysed factory process and in every home in the years to come. The upcoming applications of industrial enzymes reveal that they can even be used to:

- peel the fresh fruits which are sold in some supermarkets as ready-peeled fresh fruit. These appealing products, which are sold chilled in see-through plastic trays and the shelf-life in a refrigerator is about three weeks, have become possible due to enzymatic peeling for which Novo Nordisk Ferment in Switzerland has developed a range of preparations called 'Peelzym'.

Table 2: Novel microbial enzymes and the final products of their processes

Product	*Enzyme*	*Origin*
D-Amino acids	D-Hydantoinase D-Decarbamoylase	*Pseudomonas putida, Bacillus* sp. *Blastobacter* sp., *Agrobacterium* sp.
L-Amino acids	L-Hydantoinase L-Decarbamoylase	*Pseudomonas putida* *Alcaligenes xylosoxidans*
L-3,4-Dihydroxyphenylalanine	β-Tyrosinase	*Erwinia herbicola*
Trans-4-Hydroxy-L-proline	Dioxygenase	*Dactylosporangium* sp.
Cis-3-Hydroxy-L-proline	Dioxygenase	*Streptomyces* sp.
L-Serine	Serine hydroxy-methyltransferase	*Methylobacterium* sp.
Ethyl R-4-chloro-3-hydroxybutanoate	Aldehyde reductase	*Sporobolomyces salmonicolor*
Ethyl S-4-chloro-3-hydroxybutanoate	Carbonyl reductase	*Candida magnoliae*
Acrylamide	Nitrile hydratase	*Rhodococcus rhodochorus*
Nicotinamide	Nitrile hydratase	*Rhodococcus rhodochorus*
Acrylic acid	Nitrilase	*Rhodococcus rhodochorus*
Nicitinic acid	Nitrilase	*Rhodococcus rhodochorus*
5-Hydroxypyrazine-2-carboxylic acid	Nitrilase and dehydrogenase	*Agrobacterium* sp.
2S,3R-3-(4-methoxyphenyl)glycidic acid methyl ester	Lipase	*Serratia marcescens*
Carbacephem	o-Phthalyl amidase	*Xanthobacter agilis*
Chiral epoxide	Alkene monooxygenase	*Nocardia corallina*
R-2-(4-Hydroxyphenoxy)propionic acid	Hydroxylase	*Beauveria bassiana*
S-p-Chlorophenylethanol	Alcohol dehydrogenase	*Rhodococcus erythropolis*
R-p-chlorophenylethanol	Alcohol dehydrogenase	*Lactobacillus kefir*
Chiral2,3-dichloro-1-propanol	Halohydrin hydrogen-halide lyase	*Alcaligenes* sp., *Pseudomonas* sp.

Product	*Enzyme*	*Origin*
Chiral 3-Chloro-1,2-propanediol	Halohydrin hydrogen-halide lyase	*Alcaligenes* sp., *Pseudomonas* sp.
S-1,2-pentanediol	Alcohol dehydrogenase and reductase	*Candida parapsilosis*
Treharose	Maltooligosyl treharose synthase Maltooligosyl treharose trehalohydrolase	*Arthrobacter* sp. *Sulfolobus acidocaldarius*
D-Pantoyl lactone	Carbonyl reductase	*Candida parapsilosis*
D-Pantoic acid	Lactonase	*Fusarium oxysporum*
Theobromine	Oxygenase	*Pseudomonas putida*
Adenosylmethionine	Adenosylmethionine synthetase	*Saccharomyces sake*
Adenosylhomocysteine	Adenosylhomocysteine hydrolase	*Alcaligenes faecalis*
Adenine arabinoside	Nucleoside phosphorylase	*Enteronacter aerogenes*
Ribavirine	Nucleoside phosphorylase	*Erwinia carotovora*
5-Methyluridine	Nucleoside phosphorylase	*Erwinia carotovora*
Dideoxyadenine	Nucleoside phosphorylase	*Escherichia coli*
Dihomo-γ-linolenic acid	Multistep conversion	*Mortierella alpina*
Arachidonic acid	Multistep conversion	*Mortierella alpina*
Eicosapentaenoic acid	Multistep conversion	*Mortierella alpina*
5,8,11-cis-Eicosatrienoic acid	Multistep conversion	*Mortierella alpina*

- make the impossible 'unmalted barley beer'. To a traditional brewer, these words are a contradiction. How can a beer be made with only unmalted barley when malt has always been one of the essential ingredients in beer making? The answer is the use of auxiliary enzymes.
- transform an apple into juice by coating the slices of apple in an enzyme mixture of pectinase, cellulase and hemicellulase and leaving overnight. The enzymes rapidly disintegrate the fruit and release the juice thus improving yields. The quality of the juice also improves due to better colour and aroma extraction.
- make the bread stay fresh longer. Bread is generally only saleable until it goes stale by mould spoilage which is currently controlled with an antistaling enzyme preparation, Novamyl®.

2. EXTREMOZYMES

Most enzymes work in mild conditions, not too hot, not too cold; not too acidic and not too alkaline, but, there are notable exceptions. Many chemical processes used in industry require high temperatures and high pressure or both to drive reactions. In contrast, one of the advantages of enzymes is that they work in mild conditions. In general, they operate best between 20°C (68°F) and 60°C (140°F). In recent years the emphasis has been on the thermostable enzymes having better stabilities under extreme conditions, as many of the industrial enzymatic processes operate at temperatures above 50°C (Table 3). The properties which allow thermostable enzymes to withstand high temperatures also confer resistance to denaturing agents, solvents and chaotropic agents. Some of the advantages of using thermostable enzymes are listed in Table 4.

Table 3: Enzymatic agro-processes at high temperatures

Enzyme	*Operating temperature (°C)*	*Major applications*
α-amylase (bacterial)	90-100	Starch hydrolysis, brewing, baking.
α-amylase (fungal)	50-60	Maltose production
β-amylase	50-60	Production of maltose syrups
Glucoamylase	50-60	Maltodextrin hydrolysis
Pullulanase		
Acid protease	30-40	Food processing
Fungal protease (neutral protease)	40-60	Baking, brewing and food processing
Lipase	30-70	Food processing
Xylose isomerase	45-55	High temperature syrups
Pectinase	20-50	Clarification of juices/ wines
Cellulase	45-55	Cellulose hydrolysis, food processing
Lactose	30-50	Lactose hydrolysis, Food processing

Table 4: Advantages of thermostable enzymes

Property of thermostable enzyme	*Advantage in process*
Thermostability	Tolerate high temperature, lasts longer
High optimum temperature	Little activity at low temperature, long shelf life
Resistance to denaturing agents	Tolerate organic solvents, high and low pH
General robustness	Tolerate harsh purification, gives better yield
Genes can be cloned in *E. coli*	A heating step makes purification easier
Chemical reaction rates	Diffusion and other chemical processes are accelerated
Solubility	Higher concentrations of poorly soluble compounds
Viscosity	Decreases, mixing and pumping can be accelerated, mass transfer rate increases
Microbial contamination	Growth of all pathogens and most environmental mesophiles prevented
Biological activity in raw materials	Heating kills most interfering enzyme or microbial activities

The predominant reasons for selecting thermostable enzymes are major cost savings resulting from longer storage stability and the higher activity at higher temperature. The driving force behind the isolation and characterization of thermostable enzymes is the promise of being able to use enzymes effectively at high temperatures in industrial processing. The efficient use of enzymes in industrial processes requires that the overall amount of substrate converted per unit quantity of enzyme is maximised. Reaction temperature is probably the most important physical factor which can be optimised. For every 10°C increase in temperature, reaction rate approximately doubles. Therefore, assuming the enzyme is more stable at higher temperature, the amount of enzyme needed can be reduced or the conversion time shortened.

Thermostable enzymes can be categorised in two groups according to their source, those derived by natural selection and those created by physical/chemical modification of thermolabile enzymes. In general, the enzymes obtained from thermophiles are thermostable and the availability of these will continue to expand as more new novel thermophiles are isolated. High temperature environments are generally found only in a few restricted locations. these include geothermal areas like hot water springs. The potential advantages of using enzymes from thermophiles as catalysts in industrial processes lie in their specificity and efficacy in giving fewer side products, less toxic wastes and reduced handling problems. In general the higher the growth temperature of the organism the more stable are its enzymes. Following are some of the common enzymes available in the market which are known to thrive in extreme conditions.

2.1. Heat-stable Termamyl®

Termamyl, first discovered in the early 1970s at Novo Nordisk, Denmark, is one of most thermostable enzymes. It is a bacterial α-amylase that functions perfectly well at 105°C (221°F). In fact, Termamyl can bring about a significant reduction in the viscosity of a starch mass under high pressure at temperatures above 150°C (300°F). As this amylase can survive temperatures above boiling point, it revolutionized the degradation of starch because it could be used in a jet cooker.

2.2. Acid pectinase

The industrial enzymes with the best tolerance to low pH are pectinase preparations called Pectinex® BE 3-L and Citrozym® Ultra. The first is used for making blackcurrant juice and the second for lemon juice, both of which are extremely acidic. The first step in making blackcurrant juice is to crush the blackcurrants and heat them to 50°C (122°F) in a holding tank. Soluble pectin makes the resulting mash very viscous in a matter of hours. It is therefore important to depectinize the berries before pressing. This is the job of pectin-splitting enzymes. The problem is that blackcurrants are high in acidity. For standard pectinases, activity falls away rapidly at low pH values, but, Pectinex BE 3-L is more stable in acidic conditions, even at pH as low as 2.3.

3. IMPORTANT INDUSTRIAL ENZYMES

Industrial enzymes are generally obtained from micro-organisms, usually bacteria, fungi or yeasts. One micro-organism contains over 1000 different enzymes. A long period of trial and effort in the laboratory is needed to isolate the best micro-organism for producing a particular type of enzyme. When the right micro-organism has been found, it is to be modified to make it capable of producing high yields of the desired enzyme. Then the micro-organism is 'grown' in trays or fermentation tanks where it produces the desired enzyme. With the latest technological advancements, it is possible to produce enzymes economically and in virtually unlimited quantities. The end product of fermentation is a broth from which the enzymes are extracted.

Industrial enzymes have seen a spectacular rise in their production in the last three decades. The growth of industrial enzyme market has expanded with the discovery of a variety of new and more active enzymes. At least 75% of all industrial enzymes are hydrolytic in action and are used to depolymerize natural substrates. Global sale of the enzymes is estimated at over 1.1 billion US dollars and the major industrial applications are in detergents (40%), textile (14%), starch processing (13%), food industry (23%) and other (10%). The enzyme market has been forecasted to go up to 1.7-2.0 billion by 2005. The conventional distribution of current world sales of enzymes is to assess them by their application sectors (dairy, detergent, starch, textiles and 'other'; Figure 1). This has been the practice for many years, but, it is important to consider the composition of the 'other' areas in order to develop future markets. There are currently 12 main sectors under the 'others' category including alcohol, animal feed, baking , chemical biotransformations, diagnostics, fats and oils, flavour, fruit and wine, leather, protein (other than for milk coagulation, flavour and detergents), pulp and paper and waste. The detergent industry has been the largest market for industrial enzymes for over 25 years. In 1998, it accounted for 37% of the total world sales of enzymes. The growth of

Table 5: A summary of era of enzyme economics

Before 1965	The size of enzyme market was only a few million dollors.
1972-1974	With the introduction of detergent proteases sales grew upto a hundred million dollors. Enzymes for making high fructose cane syrup were introduced. This application has become almost as large as the use of enzymes for detergents. Microbial rennet was also introduced.
1974-1986	No major new enzymatic application emerged. Still the market grew at an average rate of 10–15%.
1983	The estimated sales of all industrial enzymes was US $ 400 million (approx.).
1989	The estimated sales grew upto $ 500 million (approx.). Detergent, starch and dairy industry were the main consumers with total consumption in terms of sales was 40–45%, 20–25% and 12–15% respectively. In only starch processing industry, Bacillus amylase acquired 10–12% global industrial market.
1995	The estimated sales touched $ 1 billion mark of which some $ 100 million was assessed for the 'in-house' processing.
By 2005	The whole world market for industrial enzyme is expected to go upto $ 1.7 to 2.0 billion.

these sectors in the coming years is expected to be such that the distribution of sales for 2005 will be markedly different from 1994 (Figure 2). It is expected that these 'other' sectors will collectively become the single largest part of enzyme sales at approximately 47% of total sales. The forecast of sales growth for these sectors is estimated to exceed $500 million by 2005 and will be approximately 70% of all the growth in industrial enzyme sales which are expected to be atleast $800 million increase over 1994 sales. There are 12 major producers of enzymes and 400 other minor companies. Almost three quarters of the production by weight comes from Europe. The two large producers of enzymes are Denmark and Holland (Table 6). Novo Nordisk, the Danish market leader, makes about $500 million per year with industrial enzymes. The gains in the industry by using these enzymes are even bigger (billions $).

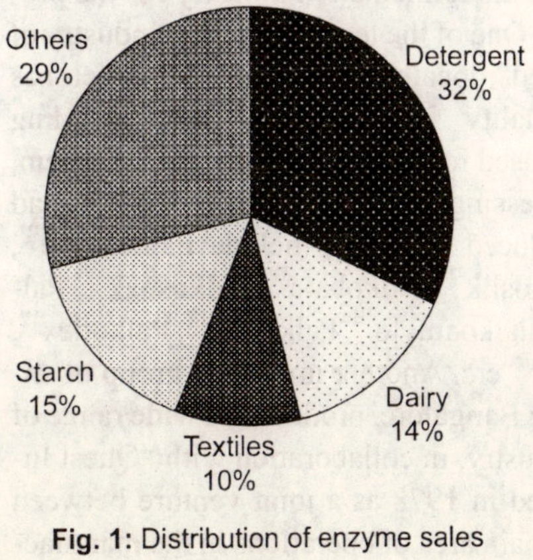

Fig. 1: Distribution of enzyme sales (1994)

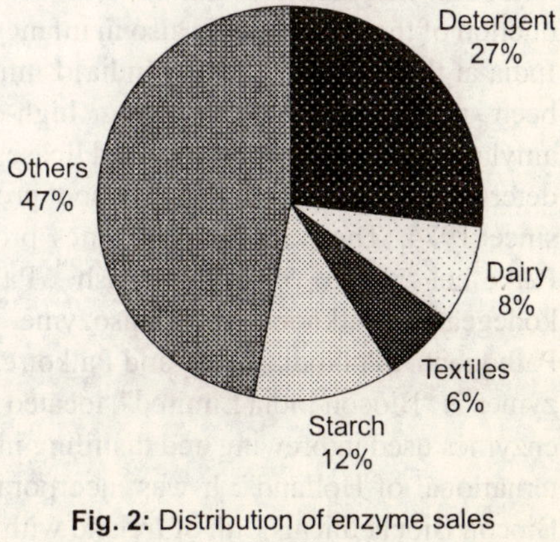

Fig. 2: Distribution of enzyme sales (2005)

Table 6: Production of industrial enzymes in Western world

Nation	*Tonnage*	*%*
USA	6360	12
Japan	4240	8
Denmark	24910	47
France	1590	3
Germany	3180	6
Holland	10070	19
UK	1060	2
EEC	40810	77
Switzerland	1060	2
Others	530	1

In contrast to the extensive uses of industrial enzymes in the world, the status of this in India is in its infancy. The use of microbial enzymes in food industry is confined to few cases in India and the enzymes are mainly imported from the developed countries. The details of these applications are not known precisely due to the trade secrets. Bacterial α-amylase and glucoamylases are used in Indian starch processing industry for the production of dextrose, liquid glucose and potable alcohol. All of the dextrose marketed, is either made by acid-enzyme or enzyme-enzyme conversion process. The latter involves pre-hydrolysis of starch with bacterial α-amylase to about 15-20 dextrose equivalent (DE) followed by saccharification to about 95% glucose with fungal glucoamylase. Although liquid glucose is traditionally manufactured in India by acid process, the use of enzyme-enzyme process technology, involving bacterial α-amylase and fungal glucoamylase, is fast emerging in few units. These two enzymes have also been used by certain distilleries in the fermentation of potable alcohol from tapioca flour, sorghum, corn, wheat and rice etc. by employing enzymatic liquefaction and saccharification of pre-gelatinized slurry. Similar to commercial exploitation of microbial enzymes in Indian food industry, the fermentation industry for the production of these enzymes is also in infancy. One of the leading enzyme industry of India at the moment, "Maps (India) Limited", located in Ahmedabad, which has been producing and supplying high-quality industrial enzymes including amylases, proteases, cellulases and lipases used for textile, leather, starch, protein, detergent, pharmaceutical and food processing industry in India and abroad since 1974. The important enzymes produced by Maps include Palkozyme™, Palkofeel™, Palkosoft™, Palkowash™, Palkosilk™, Palkobate™, Palkosoak™, Palkodegease™, Palkodehair™, Mapsozyme™, Palkoamylo™, Palkolase™, Palkodex™, Palkogent™, Palkodiastase™ and Palkotreat™ etc. Another important group of enzymes is "Biocon India Limited", located in Bangalore, producing a wide range of enzymes used in brewing and distilling industry, in collaboration with "Quest International of Holland". It was incorporated in 1978 as a joint venture between Biocon Biochemicals Ltd. of Ireland with main area of operations in the manufacture of papain, a plant enzyme. Today, it is an undisputed market leader in industrial enzymes, particularly those used in brewing and textiles. It is the first enzyme company in the world to achieve ISO 9001 accreditation. The company's research and development base has developed a number of innovative technologies. In fact, Biocon India filed its first US patent in November 1998 for a novel Bioreactor based on solid substrate technology. Unilever acquired Biocon Biochemicals Ltd. in Ireland and merged it with its subsidiary, Quest International. Consequently, Biocon India became an affiliate of Unilever in 1989. The enzymes produced by BIOCON are marketed under the trade names of Bioglucanase™, Amylo™, Promalt™, Profix™, Whirlfloc™, Biofoam™ and Biase™ etc. The production of thermostable enzymes has been thoroughly studied in India, but on laboratory scale, by several workers in submerged fermentation (smF) as well as solid state fermentation (SSF) systems. The Central Food Technological Research Institute (CFTRI), Mysore has carried out research and development on SSF systems for more than 2 decades and Central Research Laboratories, Trivandrum have contributed significantly towards the fundamental understanding of these systems. They have developed a number of industrially viable technologies. In addition to a laboratory scale process for the production of an enzyme complex by *Aspergillus terreus* for use in Jute industry for upgrading jute fibres and for sizing of jute fibres, developed by The Indian Jute Industries Research Association, Calcutta, has been scaled up.

Enzymes are generally classified according to their source of origin and application. The enzymes from the plants include the well known proteases, papain, bromelain, ficin and the amylolytic enzymes of the cereals, soyabean lipoxygenase. Most plant enzymes are available as comparatively unpurified powder extracts. The influence of cultivation conditions, growth cycle and climatic requirements make plant enzyme production a comparatively long term project. The adverse agricultural economics and unavailable land for cultivation severely affects the plant enzymes production. The enzymes obtained from animals include pancreatic lipases and proteases, pregastric esterases and rennets. They are produced in both ultra-pure industrial bulk quantities. Enzymes produced by microorganisms represent approximately 90% of all enzyme materials produced for industrial processing. The participation of many, largely unspecified, enzymes in the agro-processes, such as fermentation of foods is a powerful stimulus for the development and use of isolated, specified, food-grade enzymes for the development of food systems. Investigations on the metabolic activities of microorganisms in the past half century have brought a number of enzymes into commercial scale supply (Table 7) that derive from inside the cell. Following are some of the important enzymes being exploited worldwide for agro-processing applications.

4. STARCH HYDROLYZING ENZYMES

The conventional method of starch hydrolysis using acid has been replaced by processes using starch saccharifying enzymes, accounting for approximately 15% share in world enzyme market. Starch is not hydrolyzed completely by a single enzyme but is attacked by a variety of enzymes, generally known as amylolytic en-

Table 7 : Enzymes for commercial applications from microbial sources

Enzyme	*Source Genus*	*E.C. No.*
α-Acetolactate decarboxylase	*Thermoanaerobium*	4.1.1.2
Amylases (diastases)	*Aspergillus*	3.2.1.1
	Bacillus	3.2.1.1/3.2.1.2
	Endomyces	3.2.1.1
	Rhizopus	3.2.1.1/3.2.2.2
Amyloglucosidase	*Aspergillus*	3.2.1.3
(glucoamylase)	*Rhizopus*	3.2.1.3
Anthocyanase (α-glucosidase)	*Aspergillus*	3.2.1.21
	Penicillium	3.2.1.21
Catalase	*Aspergillus*	1.11.1.6
	Micrococcus	1.11.1.6
	Penicillium	1.11.1.6
Cellulase	*Aspergillus*	3.2.1.4
	Humicola	3.2.1.4
	Penicillium	3.2.1.4
	Trichoderma	3.2.1.4
Galactomannase	*Aspergillus*	3.2.1.78
	Bacillus	3.2.1.78
	Humicola	3.2.1.78
	Trametes	3.2.1.78
	Trichoderma	3.2.1.78
α-Galactosidase (melibiase)	*Aspergillus*	3.2.1.22
	Saccharomyces	3.2.1.22
β-Galactosidase (lactase)	*Aspergillus*	3.2.1.23
	Bacillus	3.2.1.23
	Escherichia	3.2.1.23
	Saccharomyces	3.2.1.23
	Kluyveromyces	3.2.1.23
β-Glucanase	*Aspergillus*	3.2.1.6
	Bacillus	3.2.1.6
	Humicola	3.2.1.6
	Penicillium	3.2.1.6
	Saccharomyces	3.2.1.6
	Trichoderma	3.2.1.6
Glucose isomerase	*Aspergillus*	5.3.1.5
	Bacillus	5.3.1.5
	Mycobacterium	5.3.1.5
	Streptomyces	5.3.1.5
Glucose oxidase (notatin)	*Aspergillus*	1.1..3.4
	Penicillium	1.1..3.4
Hemicellulase (Pentosanase)	*Aspergillus*	3.2.1.8/3.2.1.32
	Humicola	3.2.1.8/3.2.1.32
	Trichoderma	3.2.1.8/3.2.1.32

Enzyme	*Source Genus*	*E.C. No.*
Lipase (including esterase)	*Achromobacter*	3.1.1.3
	Alcaligenes	3.1.1.3
	Aspergillus	3.1.1.3
	Bacillus	3.1.1.3
	Candida	3.1.1.3
	Mucor	3.1.1.3
	Penicillium	3.1.1.3
	Rhizopus	3.1.1.3
Lipoxygenase	*Rhizopus*	1.13.11.12
Microbial Milk Coagulants	*Aspergillus* (GMO)	3.4.23.6
	Cryphonectria	3.4.23.6
	Endothia	3.4.23.6
	Escherichia (GMO)	3.4.23.6
	Kluyveromyces	3.4.23.6
	Mucor	3.4.23.6
	Rhizomucor	3.4.23.6
Pectinase	*Aspergillus*	3.2.1.15
	Humicola	3.2.1.15
	Penicillium	3.2.1.15
	Trichoderma	3.2.1.15
Peptidase (amino)	*Aspergillus*	3.4.11.11
	Rhizopus	3.4.11.11
Peptidase (carboxy)	*Aspergillus*	3.4.17.?
	Rhizopus	3.4.17.?
Peptidase (leucine amino)	*Lactobacillus*	3.4.11.1
	Rhizopus	3.4.11.1
Peroxidase	*Aspergillus*	1.11.1.7
	Penicillium	1.11.1.7
Proteinase (serine)	*Bacillus*	3.4.21.4
Proteinase (alkaline)	*Aspergillus*	Not allocated
Proteinase (carboxyl-acid)	*Aspergillus*	3.4.23.6
	Mucor	3.4.23.6
Proteinase (metallo)	*Aspergillus*	3.4.24.4
	Bacillus	3.4.24.4
	Streptomyces	3.4.24.4
Pullulanase	*Bacillus*	3.2.1.41
	Klebsiella	3.2.1.41
Tannase	*Aspergillus*	3.1.1.20
Xylanase (hemicellulase/pentosanase)	*Aspergillus*	3.2.1.32
	Humicola	3.2.1.32
	Trichoderma	3.2.1.32

zymes, produced by a number of microorganisms including bacteria, fungi, yeasts, algae and actinomycetes which have the ability of utilizing starch as energy and carbon source. Table 8 describes the various enzymes involved in starch degradation and the microorganisms capable of producing them.

Table 8: Starch degrading enzymes produced by amylolytic microorganisms

Amylolytic Enzyme	*Enzyme Code*	*Specific substrate and end products of amylolysis*	*Microorganisms producing the enzyme*
α-amylase (1,4-α-D-glucan glucanohydrolase	E.C. 3.2.1.1	An extracellular enzyme that catalyze the hydrolysis of α-1,4-glucosidic linkages and is capable of bypassing the α-1,6 linkages, producing poly and oligosaccharide chains of varying length	*Bacillus amyloliquefaciens* *B. licheniformis* *B. stearothermophilus* *B. subtilis* *Aspergillus oryzae* *A. niger* *Rhizopus oryzae*
β-amylase (1,4-α-D-glucan maltohydrolase)	E.C. 3.2.1.2	An extracellular enzyme that hydrolyzes alternate α-1,4-glucosidic linkages from the non-reducing end and incapable of bypassingα-1,6-glucosidic linkages to produce dextrin and theβ-anomeric from the maltose.	*Bacillus polymyxa* *B. subtilis* *B. cereus* *B. megaterium* *Clostridium thermosulfurogenes*
γ-amylase/ glucoamylase Amyloglucosidase	E.C. 3.2.1.3	An extracellular exoacting enzyme that splits α-1,6-glucosidic linkages and also some α-1,3 glucosidic linkages from the non-reducing ends of α-glucans to yieldβ-D-glucose	*Aspergillus awamori* *A. phoenicis* *A. niger* *A. oryzae* *Rhizopus delemar* *R. oryzae* *Saccharomyces cerevisiae*var *diastaticus*
Pullulanase (α-dextrin-6-glucanohydrolase)	E.C.3.2.1.41	An extracellular debranching enzyme that hydrolyzes α-1,6-linkages of pullulan and other branched oligosaccharides to form maltotriose and dextrins respectively	*Bacillus stearothermophilus* *B. acidopullulans* *Klebsiella pnumoniae*

Amylolytic Enzyme	*Enzyme Code*	*Specific substrate and end products of amylolysis*	*Microorganisms producing the enzyme*
Isoamylase (glycogen-6-glucanohydrolase)	E.C.3.2.1.68	An extracellular debranching enzyme that hydrolyzes α-1,6-glucosidic linkages of amylopectin, glycogen, various branched dextrins and oligosaccharides with no activity on pullulan	*Pseudomonas amylodera* *Lipomyces kononenkoae*
Cyclodextrin glucosyl transferase (1,4-α-D-glucan 4-α-D-[1,4-α-glucano]-transferase)	E.C.2.4.1.19	An extracellular enzyme that produces a series of nonreducing cyclodextrins from starch and other polysaccharides	*Bacillus macerans* *Klebsiella pneumoniae*
α-glucosidase (α-D-glucoside glucohydrolase)	E.C.3.2.1.20	An extracellular or intracellular enzyme that appears to hydrolyze short chain α-1,4 or α-1,6 linked saccharides arising from the action of other enzymes on starch	*Bacillus licheniformis* *Schizosaccharomyces pombe*

4.1. Industrial uses of amylases

There are a wide range of industrial applications of amylases some of which are discussed below:

4.1.1. Manufacture of dextrins

The average molecular size of dextrin to be manufactured is obtained by regulating the degree of enzymatic hydrolysis of starch. Dextrins are widely used as viscosity improvers, filler or ingredients in food, production of glucose or maltose.

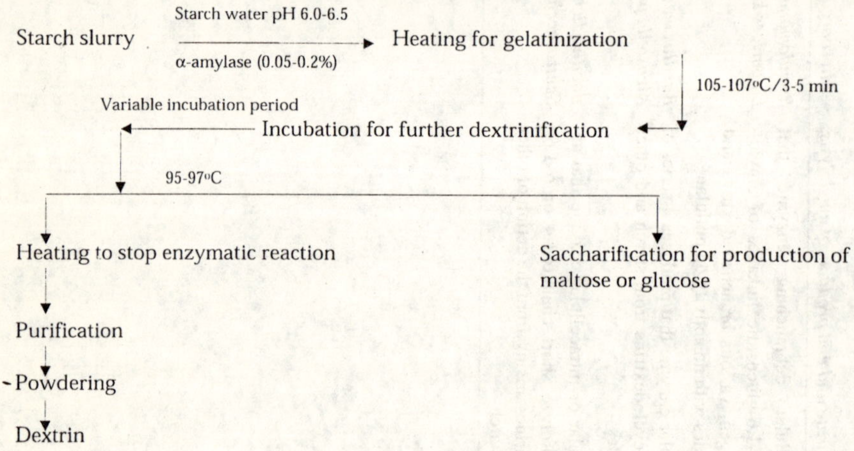

Dextrin Production Process

4.1.2. Manufacture of glucose

Sucrose is the sugar, most of us are familiar with in our homes as white or brown sugar. It comes from cane or beet sugar and is composed of equal parts of glucose and fructose. However, a syrup with almost exactly the same sweetness and properties can be made from corn using enzymes. Corn is hardly known for its sweet taste but it is rich in starch. Using enzymes, starch can be broken down into simple sugars.

Conversion of glucose by acid hydrolysis of starch had disadvantages of low yields, formation of undesirable bitter sugars (Gentibiose), formation of sugar complexes with salt and large amount of colouring matter. Enzymatic hydrolysis of starch has several advantages. Firstly, the starch is liquefied into a stable dextrin substrate using an α-amylase. Then saccharifying enzymes, a combination of pullulanases and glucoamylases, are applied to break down the dextrin almost completely to glucose.

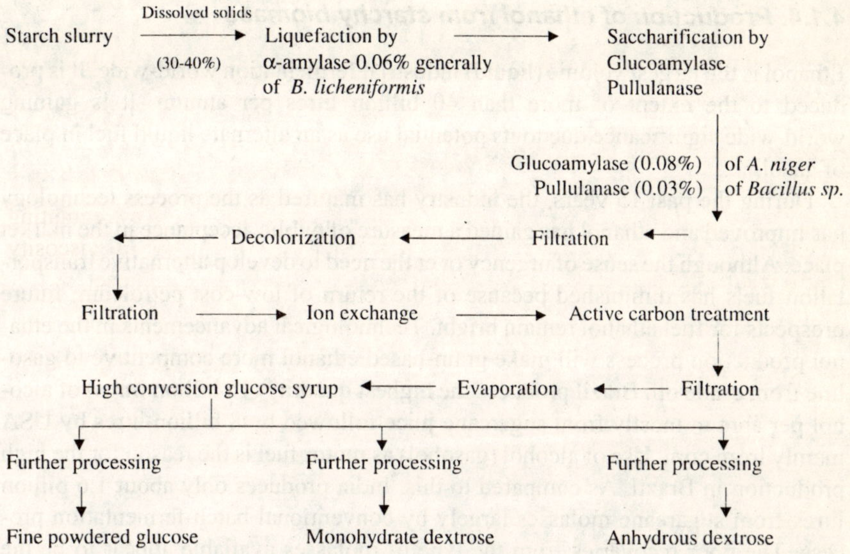

Glucose Production at Industrial Scale

4.1.3. Manufacture of anhydrous crystalline fructose

Fructose exists in five different structures and the most commonly marketed form is α-D-fructopyranose. In addition to liquefying and saccharifying amylases, glucose isomerase also plays a very important role. Firstly, the starch is liquefied into a stable dextrin substrate using an α-amylase. Then saccharifying enzymes, a combination of pullulanases and glucoamylases, are applied to break down the dextrin almost completely to glucose. The final step is isomerization where the purified glucose is converted into fructose by an immobilized glucose isomerase. The final product is called high fructose corn syrup (HFCS). In the USA, where corn is plentiful but sugar has to be imported, HFCS is cheaper than sugar. As a result, it has largely replaced sugar in food processing wherever technically feasible. In USA, the soft drinks industry uses about 2/3rd of all the HFCS consumed.

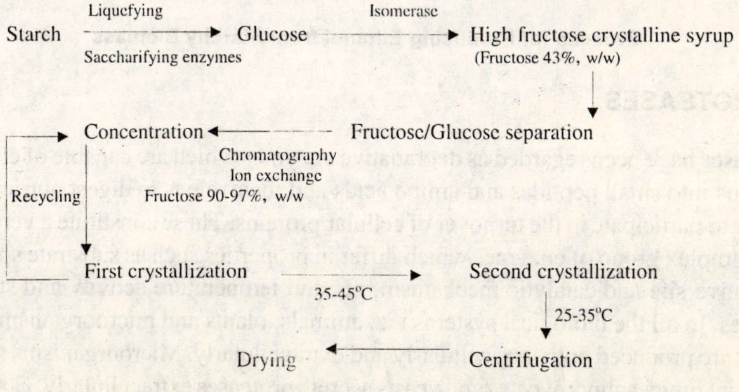

Manufacture of Anhydrous Crystalline Fructose

4.1.4. Production of ethanol from starchy biomass

Ethanol is the largest volume (liquid) industrial fermentation world-wide. It is produced to the extent of more than 40 billion litres per annum. It is gaining world-wide significance due to its potential use as an alternate liquid fuel in place of gasoline.

During the past 15 years, the industry has matured as the process technology has improved and ethanol has gained a measure of public acceptance in the market place. Although the sense of urgency over the need to develop alternative transportation fuels has diminished because of the return of low-cost petroleum, future prospects for fuel ethanol remain bright. Technological advancements in the ethanol production process will make grain-based ethanol more competitive to gasoline from crude oil. Brazil produces the highest quantity (16 billion litres) of alcohol per annum mostly from sugarcane juice followed by 8 billion litres by USA mainly from corn. Use of alcohol (gasohol) as motor fuel is the reason for the high production in Brazil. As compared to this, India produces only about 1.6 billion litres from sugarcane molasses largely by conventional batch fermentation process. The poor recoveries from the type of molasses available appear to be the main reason for low production level in India. Some alternative substrates are under trials to improve the commercial production of alcohol. Some of the modern distilleries have already adopted the standardized technologies for using starchy substrates including sorghum, wheat, rice and corn, and this has generated a tremendous requirement for the starch degrading enzymes.

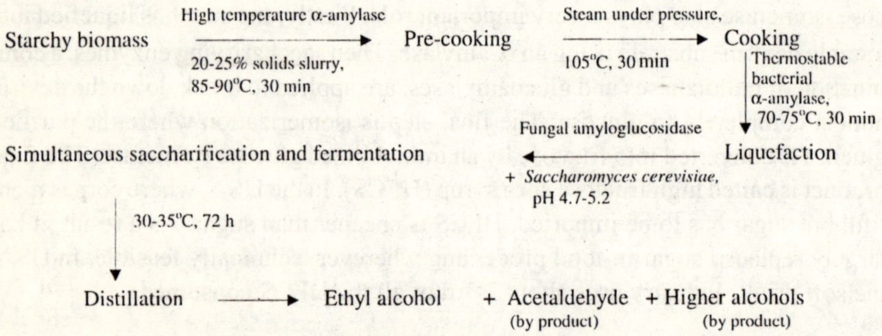

Process for Producing Ethanol from Starchy Biomass

5. PROTEASES

Proteases have been regarded as degradative enzymes which are capable of cleaving proteins into small peptides and amino acids and their role is to digest nutrient protein or to participate in the turnover of cellular proteins. These constitute a very large and complex group of enzymes which differ in properties such as substrate specificity, active site and catalytic mechanism, pH and temperature activity and stability profiles. In all the biological systems viz. animals, plants and microorganisms, proteases are produced both intracellularly and extracellularly. Microorganisms such as bacteria, fungi, actinomycetes and yeasts secrete proteases extracellularly. Commercially all proteases are produced by selected hyperproducing microorganisms although enzymes from other sources are continually being researched. Most commercially used enzymes are produced by *Bacillus* sp. which are harmless and can grow on simple media. *Aspergillus oryzae* and *A. niger* are also the important producers of protease. Two commercial proteases, subtilisin Carlsberg, and subtilisin Novo or Bacterial Protease Nagarse (BPN') are synthesized by *B. licheniformis* and *B. amyloliquefaciens* respectively. Proteases have a wide range of functions in nature and industrial transformations/processes. In all cell systems there is a balance between metabolic processes involving protein synthesis and breakdown and intracellular proteinases play a vital role in these protein turnover processes. Protein turnover is essential for the adaptation of cells to new environmental condition, particularly in environments lacking in amino acids, as it generates an amino acid pool for synthesis of newly required enzymes and other proteins.

Proteases find their application in a wide range of industrial processes viz. detergent, baking, pharmaceuticals, leather tanning, meat tenderization and peptide synthesis. Over the years, the use of proteases in various industrial processes has expanded enormously with concomitant increase in the isolation and characterization of many new proteases. The search for novel proteases is still continuing around the world.

5.1. Applications and industrial importance of proteases

The proteases are the most extensively industrially used enzymes. Almost 2/3rd of the world enzyme market is represented by microbial proteases which reflects their relative commercial importance as compared to other enzymes. Proteases for detergents are the dominant use with sales turnover of 89% of the total microbial proteinase use. In a very distant second place are the microbial rennets which account for 8% of the microbial protease market.

5.1.1. Proteases in Dairy Industry

Cheese making is the most important process employing proteases in dairy industry. There are currently over 400 varieties of cheese, although only about 18 are of distinct types. Selected species of lactic acid bacteria, such as propionibacteria and lactobacilli are added as starters for cheese production. The increasing demands for cheese and higher cost of calf rennet, resulted in the development of microbial rennet substitutes from *Mucor miehei*, *M. pusillus* and *Endothica parasitica*. A milk clotting enzyme resembling rennet chymosn has been reported from a starin of *M. miehei* and recommended for dairy processes. Multiple enzyme systems (proteases and lipases) are also used to produce strongly flavoured cheese. It has led to the development of enzyme modified cheese (EMC) which are used in formulation such as processed cheese, cheese spreads and cheese dips, Neutrase, from *B. subtilis* is added to cheese to enhance flavour. Encapsulated proteinase from *B. subtilis* and *Cardoon cyprosins* has been reported to stimulate the release of liposomes for cheese ripening and to enhance cheese flavour intensity.

5.1.2. Proteases in Soy-Product formation

Proteases play an important role in the formation of soy sauce and other traditional fermented foods.

5.1.2.1. Production of soy sauce

Soy sauce is a dark brown liquid of distinct fragrances and with a salty taste which is used as a seasoning agent in the preparation of food and as table condiment. Japnese type soysauce use *A. oryzae* and *A. sojae* strain as starter. The neutral and alkaline proteases from *Aspergillus* sp. play significant role in soyabean protein digestion during soy sauce processing. In Thailand, addition of *A. flavus* var *columnaris* as a controlled inoculum resulted in production of soy-sauce of superior quality. Neutral and alkaline proteases play a significant role in the improvement of quality of soysauce.

5.1.2.2. Other products

Soyabeans have been used as a raw material for traditional foods such as tofu (bean curd), sufu, (softcream cheese type product) and miso (bean paste). Efficient curd formation from soymilk can be achieved with plant protease, bromelain.

5.1.3. Protease in Baking

The unique and variable rheological properties of wheat flour doughs are largely due to the insoluble gluten proteins they contain. Wheat gluten is not susceptible to attack by flour proteinases and protease supplementation is usually essential. Addition of proteases to doughs improves their handling properties and increases loaf volume, thereby producing a more symmetrical loaf with good crumb and crust texture. Proteases from fungus *A. oryzae* are normally used in bread production due to their lower temperature stability and also their amylase content, which also contributes to the baking process. Both fungal and bacterial proteases have application in the manufacture of cracker and biscuit dough. These enzymes are also used in high protein flour doughs for cookie, pizza and biscuit production. A proteolytic system from *Lactobacillus sanfrancisco* CB1, a key sourdough lactic acid bacterium having a particular capacity to degrade proteins on peptides during sour dough fermentation has been reported .

5.1.4. Proteases in brewing

Proteases have two major applications in brewing. They are used during the cereal mashing process to increase the yield of extract and level of α-amino nitrogen of the wort produced. Microbial proteases from *B. subtilis*, *A. oryzae* and plant proteases can be used effectively to enhance yield of extract and α-amino nitrogen in such cases. Proteases are also used during the finishing stages of beer production to remove chill haze (chill proofing) from beer. The plant protease papain is used to counteract the problem of chill haze formation in beer. A procedure whereby protease from *Streptomyces fradiae* is used successfully for beer clarification has also been reported.

6. LIPASES

Lipases (triacylglycerol acylhydrolases) are ubiquitous enzymes of considerable physiological significance as well as industrial potential. According to a classical definition lipases are the enzymes which, in contrast to esterases, become active only when adsorbed on to an oil-water interface and do not hydrolyse soluble substrates. The unique property of lipase is interfacial activation. These catalyze the hydrolysis of triacyl glycerols to glycerol and free fatty acids. A true lipase is one which splits emulsified esters of glycerine and fatty acids with long chains such as triolein and tripalmitin. Lipases are serine hydrolases and unlike esterases, which show a normal Michaelis-Menten activity. Lipases display little activity in aqueous solutions with soluble substrates. In eukaryotes, lipases are involved in various stages of lipid metabolism including fat digestion, absorption, reconstitution and lipoprotein metabolism. Lipases with industrial potential are usually obtained from microorganisms which produce a wide spectrum of extracellular lipases. Many are active in organic solvents where they catalyze a number of useful reactions including trans esterification, regioselective acylation of glycols and menthols and synthesis of peptides. The expectation is that lipases will be as important industrially in the future as the proteases and carbohydrases. Lipases have a number of biotechnological applications in the areas of detergent industry, oleochemical industry, dairy industry, paper manufacturing industry and a number of pharmaceutical industries.

Microorganisms producing lipases are widespread and include fungi, yeast, bacteria and actinomycetes. The important molds belonging to the genera *Rhizopus, Aspergillus & Mucor*, yeasts belonging to *Candida, Rhodotorula & Saccharomyces* and bacteria belonging to *Bacillus & Pseudomonas* are reported to produce significant levels of extracellular lipases.

6.1. Industrial applications of lipases

The potential applications of lipases in agro-processing have been the subject of recent interest. The most important applications of lipases in agro-processing are discussed below:

6.1.1. Lipases in food industry

The position, chain length and degree of unsaturation greatly influence not only the physical properties, but, also the nutritional and sensory value of a given triglyceride. Cocoa butter contains palmitic and stearic acids and has a melting point of approximately 37°C, leading to its melting in the mouth which results in a perceived cooling sensation. In 1976, Unilever filed a patent describing a mixed hydrolysis and synthesis reaction to produce a cocoa butter substitute. The technology is now commercialized by Quest-Loders Croklaan, based on immobilized *Rhizopus miehei* lipase which carries out a transesterification reaction replacing palmitic acid with stearic acid. The modification of butter fat by inter-esterification of butter fat that resulted in a considerable decrease in long chain saturated fatty acids with a concomitant increase in C 18:0 and C 18: 1 at sn-2 position of selected triacyl glycerol has been reported. Polyunsaturated fatty acids play an increasingly important role as biomedical and so called 'neutraceutical' agents. Many of them belong to the essential fatty acids, the uptake of which is required for membrane-lipid and prostaglandin synthesis, a fact that is often ignored when recommending a fat free diet. Microbial lipases are used to enrich PUFA's from animal and plant lipids such as menhaden, tuna or borage oil. Free PUFA's and their mono and diglycerides are subsequently used to produce a variety of pharmaceuticals including anticholesterolemics, antiinflammatories and thrombolytics.

In addition to the above, lipases have been used as flavour agents in hydrolysis of fat, cheese ripening, bakery products, improvement in aroma of beverages and fat removal in meat and fish products.

6.1.2. Lipases in pulp and paper industry

Another application of increasing importance is the use of lipases in removing the pitch from pulp produced in paper industry. 'Pitch' is a term used to describe collectively the hydrophobic components of wood, namely triglycerides and waxes which cause severe problems in pulp and paper manufacture. Nippon Paper Industries in Japan developed a pitch control method that uses a fungal lipase from *Candida rugosa* to hydrolyze up to 90% of the triglycerides.

6.1.3. Lipases in organic synthesis

Over the past decade, non-aqueous enzymology has emerged as a major area of biotechnology research and development. Although the reaction of enzymatic processes in organic solvents was initially met with scepticism by scientific community, but now non-aqueous enzymology is widely studied and exploited world wide. The reason why non-aqueous enzymology has generated so much excitement is that the enzymes exhibit striking new properties in organic solvents. Lipases have been employed by organic chemists to catalyze a wide variety of chemo-, regio- and stereo-selective transformations. These enzymes work at hydrophilic lipophilic interfaces and thus tolerate organic solvents in reaction mixtures. They catalyze the hydrolysis of water immiscible triglycerides at water-liquid interface. Under given conditions, the amount of water in the reaction will determine the direction of lipase catalyzed reaction. In the absence or presence of trace quantities of water, only esterification and transesterification are favoured, whereas with excess of water hydrolysis is favoured.

The lipase-acyl transferase from *Candida parapsilosis* had been shown to catalyze fatty hydroxamic acid biosynthesis in a biphasic liquid/aqueous medium. The substrates of the reaction were acyl donor (fatty acid or fatty acid methyl ester) and hydroxylamine. The transfer of acyl group from a donor ester to hydroxylamine (aminolysis) was catalyzed preferentially to the reaction of free fatty acids. This feature made it an enzyme of choice for the direct bioconversion of oils in aqueous medium. Stereoselectivity of lipases for resolution of various racemic acid mixtures in immiscible biphasic system has been well demonstrated. Racemic alcohols can also be resolved into their enantiomerically pure forms by transesterification. Monoacylated sugars had been produced in anhydrous pyridine from triethyl carboxylates and various monosaccharides.

Lipases had been successfully used as catalyst for synthesis of esters both on small scale and on industrial scale. The esters produced from short chain fatty acids had applications as flavour constituents in food industry, while methyl or ethyl esters of long chain acids were used to enrich diesel fuels. Within the field of organic synthesis an increasing interest in application of lipases is apparent from the literature. In this field the lipases are now utilized experimentally for synthesis of simple achiral esters, synthesis and production of chiral building blocks, for synthesis of chiral esters and for regio selective esterification of polyfunctional alcohols. The quality of esters synthesized in this fashion is often superior to that of esters prepared using conventional method. Probably this is the chronical area where lipase will gain first industrial importance.

Lipases can be applied for regioselective esterification of poly functional alcohol. The process is carried out simply by mixing the reactants in the presence of the enzyme and removing water from the reaction mixture in the course of the reaction. The glycoside esters shown are excellent surfactants and can be prepared at a low cost using the lipase based process. There is a great interest for low molecular weight esters. Indeed, they are valorized in food, cosmetic and pharmaceutical industries as flavour and fragrance compounds. Moreover, when prepared by enzymatic process, the molecules may obtain the "natural label".

One of the most attractive field of lipase application is the production of monoacyl glycerols (MAG), which are most widely used as emulsifiers in food, pharmaceuticals and cosmetic industries. Chemically monoglycerides can be synthesized at high temperature (200-250°C) leading to unwanted side products. Different approaches for synthesis of MAG by non specific *Pseudomonas capacia* lipase are followed. It has been found that pure MAG synthesis occur as (a) solid-phase glycerolysis in trilaurin; and (b) transestirification of protected glycerol, 1,2-0-iso propylidene glycerol by Vinyl laurate, in the presence of a solvent.

7. CONCLUSION AND FUTURE OUTLOOK

Greater awareness of the environment has brought about a more intensive search for alternative cleaner technologies. Enzyme technology has provided some of the alternatives and has replaced many chemical industrial processes. This trend is already clear in starch processing where enzymes have largely replaced the use of strong acids and high temperatures once used to breakdown starch. The commercial success of amylases is linked to the emergence of starch as an industrial raw material extensively used in the production of baby foods, beverages, chemicals, confectionery, desserts, fermentations, paper, soups and sugars. Most of these agro-products have become indispensible for mankind, hence their supply has to be increased to meet the demands of ever increasing population. This would lead to an increase in consumption of starch. The wide spread use of starch for many of these requirements has been made possible by the availability of efficient enzymatic hydrolysis process which has advantages of specificity, versatility and efficient hydrolysis over the traditional acid hydrolysis method. The high fructose corn syrups (HFCS), the products of starchy hydrolysis process, have already emerged as cheaper alternatives of sucrose. By 1985 these HFCS earned 60% (USA) and 30% (Japan) of industrial sweeteners market and this is expected to increase many fold in the near future. Starch also has the potential to emerge as a renewable energy source of the future in the form of "Gasohol" with its annual global production to compete out cellulose as well as fossil fuels in the form of mixtures with other components. Starchy biomass too appears to be the cheapest future raw material of Indian alcohol industry in view of the rising cost and deteriorating quality of fermentable molasses due to the Govt's decontrol policy and technological advancement of the sugar recovery processes. Thus, a significant interest exists in research on the amylolytic enzymes as the conventional production of ethanol from starch requires an initial pretreatment, whereby, the action of amylolytic enzymes on the starch molecules produces low molecular weight sugars. The main problem in using starchy substrates as alternate substrates for alcohol production is that the thermostable microbial enzymes for the hydrolysis of starch are not produced in abundance in India and are generally imported from Novo Industries Denmark. A lot of R&D inputs are expected in the near future for developing indigenously cost-effective processes for production of thermostable amylases for the Indian alcohol industry to save a lot of foreign exchange. One of such efforts aimed at producing very high titres of thermostable α-amylase from a *Bacillus* sp., by mutagenesis and genetic recombination technology, has been undertaken at shake flask and laboratory scale fermenter level and the enzyme has been evaluated by various user industries. Another effort developing an indigenous in-house technology using solid state fermentation for Indian alcohol industry is in the advance stage of development at the Department of Microbiology, Panjab University, Chandigarh.

Proteases have been used since long in food industry and beverages. Besides these, proteases are now widely used in peptide synthesis that is carried out in non-aqueous solvents. Peptides are important class of biologically active molecules that find their use in pharmaceuticals, agro-chemicals, nutritional dipeptides and tripeptides, as harmones and polymeric materials. Many useful peptides have been synthesized using protease enzymes including aspartame (artificial sweetner), kytorphin, captopril, dermorphin (drugs), enkephalins, precursors, dipeptide of glutamine and cystein (as additive in intravenous solutions). Some of these compounds have already been commercialised while others are expected to do so in the near future. Aspartame, an artificial sweetner (200 fold as sweet as sucrose), is being used in place of sugar in some of the variants of soft drinks, marketed by multinational houses, including Pepsi-diet and Coke-diet, which are claimed to have the calorific value less than 1. Sucralose (trichlorotrideoxygalactosurose), another high intensity sweetner, produced from glucose by microbial transformation, is undergoing final trials and is 600 times as sweet as sucrose.

Lipases find promising future in fat & oil industry and organic synthesis. Development of lipase technologies for the synthesis of novel compounds will probably also result in their expansion into new areas and the major impact on the range of industry. However, as the applications increase, the availability of lipases possessing satisfactory operating characteristics is a limiting factor. Some of the processes involving lipases have been developed/commercialized while efforts are being made in various laboratories worldwide to make some others possible at large scale. Some of the major products of future importance in fat and oil industry include: γ-Linolenic ácid—bioactive component of evening primrose oil widely used as a dietary supplement, astaxanthine—the food colouring material, methyl ketones—flavour molecules characteristic of blue cheese, .4-hydroxydecanoic acid used as a precursor of γ-decalactone—fruit flavour, dicarboxylic acids as pre polymers, interesterification of cheaper glycerides to more valuable forms - serve as cocoa butter replacements in chocolate manufacture, palm oil is inter esterification of cheap palm oil by the replacement of the palmitoyl substituents in the 1,3 position of the triglyceride with stearic acids, modification of vegetable oils at 2 position to have fat similar to human milk fat—for use in baby food formula feeds, lipid esters including isopropyl myristate—for use in cosmetics, monoglycerides—for emulsification crystal modification in food and pharmaceutical applications. The production of chiral chemicals by hydrolysis, esterification and transesterification using lipases is also becoming very important for the pharmaceutical industry and other parts of the chemical industry such as pesticides and insecticides and is likely to play a revolutionary role in the chemical industry. One interesting new technology uses lipases in the form of cross-linked enzyme crystals (CLEC). Lipases from *Candida rugosa* and *Burkholderia cepacia* are crystallized to give microcrystals approximately 50-100 μm in length which are subsequently cross-linked by the addition of gluteraldehyde and dried preferably in the presence of a detergent. These crystals had been used for chiral resolution of commercially important organic compounds by ester hydrolysis giving high enantiomeric excesses (ee): S-ibuprofen (95% ee), S-naproxen (97% ee) and (–)-menthol (95% ee). In addition, CLECs can catalyze esterification and transesterification reactions in low-water organic solvents to achieve chiral resolution of racemic alcohols and acids with ee values in the range of 95-99%. The enantioselective synthesis of chiral compounds is of rapidly increasing importance to the chemical industry. The total sales of therapeutics in 1995 has been estimated to be US $ 150 billion of which US $ 60 billion resulted from chiral compounds. Chiral compounds have been produced by chemical kinetic resolution using asymmetric catalysts or by biocatalysts identified through screening of large number of enzymes or organisms. X-ray crystal structures of lipases have been used successfully to predict enantiopreference and to engineer novel catalytic properties. Chiral drugs with a current sales volume exceeding US $1 billion include amoxycillin (an antibiotic), captopril (an angiotensin converting enzyme inhibitor) and erythropoetin (the haematopoietic growth factor). Efficient chemo- or biocatalysts have been identified. Transaminases with increased enantioselectivities towards a β-tetralone have also been identified upon screening of a mutant library; but a large number of problems remained unsolved. Nevertheless, a solution has emerged that may well continue a new direction in asymmetric catalysis: the creation of an enantioselective lipase by *in vitro* evolution. This technique had been used previously to develop enzymes with improved thermostability, activity in organic solvents or substrate specificity.

Thus the biotechnological potential of microbial enzymes is steadily increasing. Availability of new primary and 3D structures and novel molecular-biology techniques provide the tools to tailor the enzymes for a variety of different applications. However there are still only a few enzymes used at an industrial scale in organic synthesis, presumably because of difficulties in establishing cost-effective scaling up and downstream processing protocols. Growing concern for the environment has made detergent manufacturers take a close look at their formulation as certain ingredients have received a bad name. An example of the changes taking place is the field of automatic dishwashing detergents (ADDs) in the European market. Manufacturers have learnt that they can do without many of the questionable ingredients and still maintain the same performance by adding enzymes which have the advantage that they are 100% biodegradable. A new generation of ADDs in Europe has been introduced with no phosphates, no chlorine bleach and a safer, less caustic formulation. All of the new ADDs contain enzymes are likely to take up the whole world market in the coming years. Another industry where the enzyme preparations can help to reduce pollution is the pulp and paper industry. One of the biggest challenges the industry faces is to reduce the use of chlorine and chlorine compounds. These are used as bleaching chemicals, but, result in a group of troublesome waste products called chlorinated organic compounds. The pulp and paper industry has already invested heavily to reduce the chlorine usage and to improve the treatment of effluents. Neverthless, the regulations are becoming tougher, consumer pressure is growing and new markets have opened up for pulp which has been bleached using little or no chlorine. A lot of research input has been made in this direction with the use of enzymes, but, the commercial utilisation in industrial pulp bleaching is still in its infancy. The use of the new enzymatic process allows the same level of bleaching to be achieved but using upto one third less chlorine or chlorine compounds. That means a corresponding reduction in the chlorinated effluents. In the future, enzymes may be able to replace bleaching chemicals completely. It will be exciting to witness the future developments in basic research as well as biotechnological applications of various versatile microbial enzymes.

In conclusion, the discovery of new catalysts and molecules that rival natural enzymes in their substrate specificity and catalytic turnover is one of the major aims of biotechnology. The *de novo* design and synthesis of catalysts from organic molecules, the modification of existing catalysts such as enzymes or ribosomes and the creation of catalytic antibodies are now under examination. However, there are still a large number of organisms, particularly microorganisms, that have not been screened or examined for their ability to catalyse specific reactions, and the breadth of reactions catalysed by enzymes will certainly continue to expand. There have been attempts to screen microorganisms that, as yet, remain uncultivated. Here, microorganisms from natural habitats are concentrated by filtration, their DNA extracted without further cultivation and a variety of enzyme activities screened from their DNA-cosmid libraries. This approach enables the screening of enzyme activities in as-yet-uncultivated procaryotes and could represent a superior method to discover new, powerful enzymes. However, the enzyme is totally separated from its original function in the microorganism. To find a novel enzyme means to find a novel microbial function; in this respect, gene-library-screening misses original microbial function, although classical screening techniques including enrichment and isolation strategies, remain powerful tools, because they involve the natural environment of the microorganisms.

FURTHER READING

Gupta, J.K. and Soni, S.K. "Industrial uses of enzymes." *J. Pb. Acad. Sci.* **2**: (2000) 75.

Godfrey, A. "Production of industrial enzymes and some applications in fermented foods". In: Microbiology of Fermented Foods. (Ed. Wood, B.J.B.), Thomson Science, London. (1998) 622-657.

Godfrey, A. and West, S. Industrial Enzymology. Stockholm Press, New York (1996).

Srinivasan, M.C. In: Industrial Microbiology and Biotechnology. Proceedings of the Symposium on Biotech Industry—A challenge for 2005 (Eds. A.D. Chand and S.C. Jain) Replika Press, New Delhi, India. (1999) 179-187.

Ward, O.P. Microbial Enzymes and Biotechnology. Applied Science, London (1983).

Immobilized Enzymes and Microbial Technology in Agroprocessing and Agriculture

S.F. D'SOUZA

Nuclear Agriculture and Biotechnology Division
Bhabha Atomic Research Centre
Trombay, Mumbai-400085, India

BIOTECHNOLOGY IS CURRENTLY considered as an useful alternative to conventional chemical process technology. In a developing country like India, with its growing manpower and increasing pressure on natural resources; food production, processing and utilization of renewable resources are becoming important for sustainable economic growth and development. From a technological point of view, enzyme and microbial technology seems to be best suited and most relevant to meet this requirement. Enzyme and microbial technology has influenced the process industry significantly in the recent years by improvement of existing processes towards achieving a better quality product at reduced cost, as well as in the production of unique products. The current demand for better utilization of renewable resources and pressure on industry to operate within environmentally compatible limits has also been a stimulus to the development of new eco-friendly industrial bioprocesses. One of the techniques which has played a significant role, is the immobilization of enzymes and cells. Immobilized enzyme and microbial technology is creating newer avenues in bioprocessing, fermentation, bioremediation, agriculture and biosensor fields.

Biotechnology has influenced enzyme industry significantly in the recent years especially in the more efficient production of enzymes, their stabilization and economic reuse. The use of enzymes in industrial applications has been limited by several factors, mainly the high cost, their instability and availability in small amounts. Also the enzymes are soluble in aqueous media and it is difficult and expensive to recover them from reactor effluents at the end of catalytic process. This restricts the use of soluble enzymes to batch operations followed by their denaturation. Newer technological developments in the field of immobilized biocatalyst can offer the possibility of a wider and more economical exploitation of enzymes in agroprocessing. Immobilized enzymes, microbial, plant and animal cells and cellular organelles are being investigated for a variety of applications. The aspects of immobilized biocatalysts in agroprocessing with special reference to food industry and also the future potentials of this technology in agriculture are discussed in this chapter.

1. IMMOBILIZED ENZYMES AND CELLS

Immobilization means associating (fixing) the biocatalyst with an insoluble matrix, so that it can be retained in a proper reactor geometry for its economic reuse under stabilised conditions. Immobilized biocatalysts offer several other advantages, notable among them is the availability of the product in greater purity. Purity of the product is very crucial in food processing since contamination could cause serious toxicological, sensory or immunological problems. Basically now it is possible to use very crude enzyme preparations or cells or cell lysates in an immobilized form without contaminating the final product. The other major advantage includes greater control over fermentation or enzymatic reaction, as well as high volumetric productivity with lower residence time, which are of great significance in the food industry, specially in the treatment of perishable commodities. Inactivation of soluble enzyme by heat or chemicals can cause undesirable physicochemical alterations in protein including denaturation or reverting to insoluble form. This has been the problem often in food industry where thermolabile enzymes are not available. Such problems can be easily obviated using immobilized enzymes since the enzyme and the substrate are in two distinct phases and the enzyme can be separated by physical removal from the processing solution.

In addition to enzymes, whole cell immobilization has shown promise. Immobilized whole cells have been used either in a viable or nonviable form. Non viable cells which are often obtained by permeabilisation to allow for the diffusion of substrates/products across the cell membrane have gained importance as economical alternative for the otherwise expensive purified enzymes. A number of techniques have been developed for obtaining permeabilised cells containing catalase, invertase, alcohol dehydrogenase, amino acid oxidase fumarase and some enzymes from halophilic organisms.

One of the important challenges for the biotechnologists in the future is in improving the fermentation techniques. The classical fermentation suffers from various constraints such as low cell density, nutritional limitations and batch mode of operation with high down times. It has been well recognised that microbial cell density is of prime importance to achieve higher volumetric productivity. The major limitation in the development of continuous fermentation process has been the wash out of cells from the bioreactor. Use of flocculating strains, cell recycle and membrane reactors are being investigated to solve some of these problems. The immobilized viable cell technology can eliminate most of the constraints faced with the free cell systems. The remarkable advantage is the freedom to determine the cell density prior to fermentation. It also facilitates operation of fermentation on a continuous mode without cell wash out even at high dilution rates. The immobilized cell technology process also decouples microbial growth from cellular synthesis of favoured compounds. Immobilized viable cells have been extensively studied in the continuous production of fuel ethanol, amino acids, organic acids, enzymes etc. In addition to microbial cells the fermentation technology using bioreactors is also gaining importance for the production of high value compounds using plant and animal cell cultures. It has been shown that immobilization offers them stability against shear force when used in continuous stirred tank bioreactors.

Other future developments will be in integrating downstream processing with bioprocessing/fermentation and also in the development of appropriate bioreactors.

2. TECHNIQUES FOR THE IMMOBILIZATION OF BIOCATALYSTS

A large number of techniques and supports are now available for the immobilization of enzymes or cells (viable or non-viable). The choice of the support as well as the technique depends on the nature of the enzyme, nature of the substrate and its ultimate application. Care has to be specially taken to select supports and reagents used for immobilization that have GRAS status when their ultimate applications are in the food processing industry. Macromolecular, colloidal, viscous, sticky, dense or particulate food constituents also limit the choice of reactor and support geometries. The various techniques and supports used for immobilization have been extensively reviewed in several earlier publications.

Biomaterials can be immobilized either through adsorption, entrapment, covalent binding, cross-linking or a combination of all these techniques. Covalent binding, is a commonly used technique for the immobilization of enzymes and antibodies. It has not been useful however, for the immobilization of cells. One of the general problems with this technique is that the cells are exposed to potent reactive groups and other harsh reaction conditions thus affecting their viability. There may also be a loss in the structural integrity of the cell during continuous use, leading to loss of intracellular enzymes. Besides these, the other limitation is the very low cell loading that is achieved as compared to entrapment and other techniques. Novel techniques have been developed at Bhabha Atomic Research Centre (BARC) for the covalent binding of enzymes either through their amino or carbohydrate moiety (glycoproteins) on mucillagenous seeds.

Cross-linking using bifunctional reagents like glutaraldehyde has been successfully used for the immobilization of enzymes and cells in various supports. Of these, proteinic supports such as gelatine, collagen, albumin and hen egg white have been extensively used. The chemical cross-linking reagents used, often affect the cell viability. Thus cross-linking technique will be useful in obtaining immobilized non viable cell preparations containing active intracellular enzymes. Novel techniques have been developed for immobilizing enzyme and cells in hen egg white either in a powder or bead form. A highly porous sponge type cross-linked proteinic matrix has also been recently developed which allows the diffusion of even bacterial cells into the vicinity of the bound enzyme. Immobilization of enzymes through adsorption followed by cross-linking on cellulosic surfaces like cotton cloth, threads etc has also shown promise specially for use in the processing of eggs and sugar syrups. Cross-linking helps in the stabilisation of enzymes and has also been used for the stabilisation of cellular organelles to osmotic shock, prevention of lysis of extremely halophilic cells in low salt or salt free environments and the prevention of lysis of microbial cells by lytic enzymes present in the processing streams.

Entrapment is another useful technique for the immobilization of cells. It is not suitable, however, for the immobilization of cell free enzymes in view of their possible leakage from the entrapment matrix. Cells have been immobilized in a variety of synthetic and natural polymers like polyacrylamide, polyvinyl alcohol, polyuerthane foams, carrageenan, agarose, alginate and chitosan. Radiation polymerisation technique has been extended even to entrap cells in gelatine. Entrapment in alginate by ionotropic gelation using a variety of divalent and trivalent cations has found extensive use in immobilized viable cell technology. The major limitations of Ca-alginate gels is their destabilization and subsequent solubilisation by the Ca-chelators present in the processing solution or waste. A novel technique has been developed at BARC's laboratory for stabilising the alginate membranes and beads towards Ca-chelators by reinforcing them with gamma-ray polymerised polyacrylamide. A number of other techniques for the stabilisation of Ca-alginate have also been proposed.

One of the very useful and simple technique for the immobilization of cells is through adhesion. The major advantage is that the cells immobilized through adhesion are in direct contact with the liquid phase containing the substrate, even though the cell and the liquid phase are distinctly separate thus reducing or eliminating the mass transfer problems commonly associated with gel entrapment methods. Such biofilms have often been formed making use of the microbial cell's natural ability to adhere to surfaces. However, major limitation of such a natural adhesion process is the lengthy time required to build up a biofilm sufficient to carry out a bioconversion. Also, changes in the environmental conditions of growth or processing medium can often drastically alter the degree of cell adhesion. Author's laboratory was the first to demonstrate the possibility of immobilizing viable or non-viable cells through adhesion on a variety of polymeric surfaces including glass, cotton fabric and synthetic polymeric membranes using polyethylenimine.. The adhesion using this technique is found to be rapid and the cells adhere as a monolayer. Adhesion is also very strong and high ionic concentrations and extreme pH conditions, which normally disrupt the ionic interactions, fail to desorb the cells. Cells can be adhered by coating either the cells, the supports or both, with PEI. Viability of the cell was not affected by this treatment. This approach has recently been extended for the simultaneous filtration and immobilization of microbial cells from a flowing suspension thus integrating downstream processing and bioprocessing. The technique has been extended to obtain enzyme-cell co-immobilizates wherein a deficient enzyme is immobilized directly on the cell surface. The PEI technique is used in the fabrication of a microbial biosensors and a variety of processes including lactic acid fermentation.

Various techniques developed for the immobilization of enzyme and cells have been reviewed in 1998. In addition to these, recently novel techniques have been developed for the simultaneous isolation and immobilization of D-amino acid oxidase using hydrophobic supports and invertase using lectins. Others include magnetised immobilized biocatalysts.

3. APPLICATIONS OF IMMOBILIZED ENZYMES AND MICROBES IN AGROPROCESSING

The world-wide sale of industrial enzymes approximates to US $ 1.6 billion, which is expected to reach US $ 3.0 billion by the year 2008. Over 45% of the enzymes produced are being used in the food industry and the remaining is shared by detergent (34.4%), textile (11%), leather (2.8%), paper & pulp (1.2%) and other industries (5.6%) excluding enzymes for use in diagonostics and therapeutics. Food processing industry is the oldest and largest user of enzymes for imparting innumerable desired changes in the food and a horde of new potential applications are under investigation. Enzymes have formed an important part of the processing aids. They can improve extraction, aid filtration, change functionality, synthesize a variety of food additives, produce flavours, avoid undesirable chemical reactions during drying as well as storage and can also act as preservatives. Major applications have been especially in brewing, dairy, baking, fruit and vegetable processing, starch and sugar industries. Some of the applications of enzymes in the food industry are summarized in Table 1.

Table 1: Enzymes Used in Food Industry

Enzyme	*Applications*
Papain, bromelain, ficin	Meat tenderisation, haze removal, chillproofing
Rennin (chymosin)	Cheese
Amylases	Baking, brewing, sweetener production
Cellulases/xylanases	Biomass conversion, juice extraction
Glucanases	Brewing
Glucose isomerase	High fructose corn syrups
Glucose oxidase/ catalase	Desugaring of egg, oxygen removal
Invertase	Invert sugar, candy making
Catalase	Removal of excess H_2O_2 after cold pasteurization
Lactase	Low lactose dairy products, whey disposal
Pectinase	Beverage and juice clarification, juice extraction
Lipases	Texture modification, flavour generation
Mannuronic epimeras	Modification of alginate
Cyclodextrin glycosyl transferase	Production of cyclodextrins
Lysozyme	Preservative
Aminoacid acylase	Resolution of DL amino acid mixtures
Fumarase	Malic acid
Aspartase	Aspartic acid
Phytase, α-galactosidase, tannase	Detoxifying enzymes

India being the largest producer of milk in the world, and consequently the surplus availability of milk has triggered the food and dairy industry to convert the liquid milk into value-added products using microbial and enzymatic processes. It has now been well recognised that a major population of non-Caucasian origin, including Indians, suffer from lactose intolerance as they are lacking in sufficient quantities of β-galactosidase in the intestine thus leading to various complications of the digestive system. The dairy industry is interested in neutral (yeast) β-galactosidase which hydrolyses lactose to glucose and galactose. It is used to prevent crystallisation of lactose during concentration of milk products or during freezing and also to make milk sweeter without increasing the calories and to reduce lactose content in milk to be consumed by people who suffer from lactose intolerance. β-Galactosidases from a variety of sources have been successfully immobilized for their economic reuse. In order to further simplify the technology and make it even more competitive, a number of efforts have been made to utilize this intracellular enzyme in a crude form using permeabilized cells. A novel approach has also been developed for the complete removal of sugars from milk through heterogeneous fermentation using immobilized viable cells of *Saccharomyces fragilis*. A microbial cell-glucose oxidase co-immobilizate has been investigated for minimizing microbial contamination during continuous hydrolysis of lactose in milk through the initiation of lactoperoxidase system.

A waste product that is abundantly available is whey, a major by-product of cheese manufacturing and casein production. Whey represents an important environmental pollution problem because of high volumes produced and its high organic matter content exhibiting a BOD of 40000-60000 ppm. Cheese whey utilization has been the subject of much research. BOD reduction of higher than 75 %, with concomitant production of biogas, ethanol, single cell protein, organic acids, volatile flavouring compounds or other marketable products has been demonstrated. Actual market trends point to gradual increase in the production of cheese and allied dairy products globally as well as in our country thus in turn increasing the volume of whey. Better use of whey both in terms of decreasing its BOD as well as utilizing this nutritionally rich medium for development of newer value added products as well as processes will gain significance. Immobilized fungal β-galactosidase has been used in the preparation of lactose hydrolyzed whey thus converting a plentiful and inexpensive by product into a highly nutritious good quality food ingredient to be used in the confectionery, baking and condensed milk industries. Utilisation of whey for a variety of applications has been a major programme in the authour's laboratory.

Other major application is the use of renin in cheese making. Even though a number of attempts have been made to use immobilized renin/proteases for this process, industrial success has not been encouraging. Major limitation of immobilized enzyme technology in the treatment of macromolecular substrates like protein, starch, pectin etc. is the steric hindrance. Development of newer techniques in the future is essential to solve this problem. Other enzymes include the use of lipases, esterases and dipeptidases in cheese ripening and flavour development. Use of enzymes can drastically reduce time required for obtaining the same flavour and texture in a shorter time. Lipases in either aqueous or micro aqueous reaction environment will have an important impact in the food industry in the future with a great potential for producing uniquely flavoured products at low cost. In this direction considerable work has been carried out in the development of bioreactors containing immobilized lipases and other enzymes.

In the brewing industry, the major enzymes used are the amylases, glucanases, proteases and the xylanases for substitution of malt with domestic starch sources like the grains. Other enzymes which are gaining importance are the debranching enzymes like the pullulanase and isoamylases. These enzymes will gain importance in ingredient substitution (starch for molasses) in the distillery industry. Thermolabile amylases find applications in the baking industry. Xylanases present as side activity in crude amylase preparations have been shown to have a positive effect on baking quality. Pectinases are being extensively used in the fruit juice industry for the clarification of juices, to facilitate extraction of more juice and in liquefying fruit pulps like banana, guava etc. Catalase has a potential use in milk processing for the destruction of hydrogen peroxide employed in the cold sterilization of milk. The treatment of milk with hydrogen peroxide is interesting because it does not affect the organoleptic properties and does not destroy the temperature sensitive compounds of nutritional value. Major limitation, however, has been the high cost of catalase. Studies have shown that the catalase can be immobilized by entrapment of yeast cells. Currently we are assessing their stability characteristics for the removal of hydrogen peroxide used in the cold pasteurization of milk. The growing concern over nitrites and nitrates in foods related to the formation of carcinogenic nitrosamines, is fueling the use of lysozyme in place of nitrates as preservatives specially in the cheese industry. Its future potential will depend on its production in large quantities at reduced cost. A novel biospecific technique based on cross-linked *Micrococcus lysodeikticus* cells as the affinity ligand has been recently developed for the isolation of lysozyme. An interesting possibility of immobilizing lysozyme by directly cross-linking the raw egg white which retains its bacteriolytic activity has also been demonstrated.

Food processing means avoiding undesirable chemical changes especially like oxidation of unsaturated fatty acids or browning due to Maillard reaction. Problems related to these have become more acute since the use of sulphur dioxide has been discouraged. Enzymes can help in this direction eg. glucose present in the egg which is responsible for the browning during spray drying and storage can be removed using glucose oxidase. The application of this enzyme has been limited due to the high cost of the enzyme. For economical reasons many industries are removing the glucose from egg by fermentation with yeast. Even though this process is comparatively more economical, it results in an inferior product due to the presence of yeast in the final product. The studies in these directions have shown the possibility of removing glucose using either immobilized glucose oxidase or through heterogeneous fermentation using immobilized viable yeast cells or lactic acid cultures. Glucose oxidase is also being used for the removal of oxygen from canned food products. Another oxidase which may have potential is the D or L amino acid oxidase for the preparation of α-keto acid derivatives of essential amino acids for use in the diets of uremic patients. D amino acids have become important chiral-building blocks for many pharmaceutical and food products. Immobilized microbial hydantoinase has been investigated for this purpose.

Major problem of plant based food products is the presence of anti-nutritional stress factors like raffinose, stachyose, phytate, trypsin inhibitor, tannins, oxalate, gossypol, solanine, goitrin and limarin. Considerable work has been carried out in this direction by applying both the genetic engineering as well as plant tissue culture techniques to obtain plants with lower levels of these stress factors. However, practical application of these approaches have yet to be established. This is mainly because plants synthesize these metabolites mainly for their survival. The more practical approach would be to remove the stress factors through post harvest bioprocessing using enzymes like phytase, tannase, α-galactosidase, oxalate oxidase, debittering enzymes etc.

A number of processes have been successful in food industry especially for the production of food additives due to developments in the area of enzyme immobilization. One of the important applications of enzyme technology which has revolutionized the sweetener (sugar) industry, the world over, is the efficient use of the enzyme glucose isomerase in an immobilized form for the production of high fructose syrups (HFCS) from starch hydrolysate. In tropical countries like India such syrups can be more easily obtained by the hydrolysis of sucrose using invertase. A comprehensive study demonstrated the possibility of using the periplasmic invertase of yeast by immobilization of whole cells and optimization of various process parameters and has resulted in an industrial technology for the preparation of invert sugar syrups. Other products which have become possible as a result of immobilization technology are the production of aspartic acid using aspartase, malic acid using fumarase, resolution of amino acids using amino acid acylase and fermentative production of lactic acid, ethanol etc.

4. IMMOBILIZED MICROBIAL CELLS FOR AGRICULTURAL APPLICATIONS

Immobilized cells are being investigated as an alternative technology for a variety of environmental applications in agriculture, biocontrol, pesticide application and pollutant (eg pesticide/xenobiotics) degradation in contaminated soils. Microbial inoculants have been investigated for soil applications such as enhancement of symbiotic or associative nitrogen fixation, biological control of soil-borne plant pathogens, biological control of frost injury and biodegradation of xenobiotic compounds. The use of toxigenic fungal strains has also been suggested as a generic approach to biologically reduce aflatoxins in all crops including cotton bolls, maize etc. One of the major limitations for the application of microbial consortium into soil is their survival under biotic and abiotic stresses like drought (wet- dry cycles), salinity, phage attack, heavy metal toxicity etc. Both abiotic and biotic factors play critical roles in determining microbial survival. Soil moisture content, temperature, pH, texture, oxygen availability, rate of oxygen diffusion and nutrient availability have been suggested as abiotic factors controlling survival of introduced bacteria in soils. Biological factors include predation by protozoans, a lower level of starvation, resistance of the introduced bacteria and lack of suitable soil niches for extended cell survival.

Immobilization of the microbes through encapsulation or entrapment in certain defined polymers has been shown to afford protection to cells under such adverse conditions. The physical soil environment is heterogeneous and changing environmental conditions can result in various alterations of the soils. Encapsulation provides not only protection, but a more stable microenvironment for the entrapped microbial cells. The microbial cells entrapped in alginate are reported to remain stable after a few drying/wetting cycles in soil, whereas free cells under similar conditions were reduced by about log 2 CFU/g.

The entrapment techniques discussed earlier have been modified to include in addition to the microbial consortia, nutritional amendments like milk proteins, oil seed meal etc and fillers like clay. Such immobilization techniques help in devising newer strategies in the future in the introduction of microbes into the soils. Using this approach it is now possible to provide the microbial consortium with a more defined microenvironment, quite different from that encountered when directly in contact with soil environments. In this context encapsulation process adds a medium of control, potentially becoming a miniature reactor in the environment. Immobilized cells can also act as synthetic inoculation carriers for the slow release of plant growth related organisms like *Rhizobium* into soils. Microenvironment in the bead may initially protect cells from the soil microenvironment. Microorganisms are released after adaptation to prevailing environmental conditions. This may enable cells to overcome the numerous changing conditions in soil and increase microbial survival. Entrapment in gel-matrix offers a stable, defined, consistent, protective environment, without the immediate release of large number of cells, where cells can survive and metabolic activity can be maintained for extended periods of time. In general entrapment technique is being increasingly investigated for the preservation of cultures as well as a source of continuous inoculum for a variety of fermentation applications.

Other advantage observed in the case of immobilized viable cells is the increased plasmid stability in view of their controlled growth. Immobilization can provide additional benefits for commercial purposes especially in terms of ease of storage and transportation and also in terms of biosafety features that limit contamination and bioaerosol formation. Some work in this area is in progress at BARC especially in developing encapsulation materials with protective and magnetic properties. Entrapped cells have also been extensively investigated for polluatant degradation with an emphasis on *in situ* bioremediation of chemically contaminated soils. The approach has also been used for the restoration/rejuvenation of degraded ecosystems (metal and organic pollutant) especially in and around a variety of mining areas in India. These areas are being converted into thick vegetation cover with a variety of trees. More research is required in the future, to establish the potential effectiveness of immobilization technology for use in the environment in varied soil systems and in understanding the plant and rhizosphere microbe interactions.

5. ENZYME AND MICROBIAL BIOSENSORS

To meet the present and emerging requirements of productivity, quality and reliability new analytical devices are being developed for quick and specific analysis. One of the rapidly emerging area in the field of analysis is the biosensors which exploits the highly specific molecular recognition capability of certain biological materials. A biosensor consists of a biological sensing element such as an enzyme, antibody, cell, cellular organelle, chemoreceptor, tissue, chemical molecule in close contact with a transducer such as an ion specific electrode, thermistor, optical fibre, piezoelectric crystal, silicone wafer, conducting polymer etc. The biological signal arising due to its interaction with the analyte can result in an electronic or allied signal that can be easily measured, amplified and documented. Major advantages of biosensors will include simplicity of operation, low operating cost and could provide faster real time analytical data. This makes it relevant and cost-effective technology for small and medium enterprises that have to be modernised rapidly to be effectively integrated with the global economy.

Some of the commercial biosensor devices available for food industry based on enzymes include lactate, glucose, ethanol, lactose and a variety of other fermentation processes and products. From analysis of these routine food components enzymes have also played a major role in the fabrication of biosensors in the monitoring of highly complex parameters like freshness and aroma. Biosensors based on the estimation of hypoxanthine or amines using xanthine oxidase (fish), monoamine oxidase (meat) glucose oxidase (meat) have been developed. Biosensors based on non-enzymatic proteins like antibodies, receptors etc are also being investigated for the determination of microbial contamination. The olfactory parameters like taste and smell (coffee, tea, alcoholic beverages) are currently judged by sensory taste panels. In this direction biosensors could provide a tool for rapid, in-process analysis of organoleptic properties by using olfactory sensors like the electronic nose or the tongue. Volatiles as an indicator of fungal activity and differentiation between species, and the potential use of electronic nose technology for early detection of grain, milk and other food ingredient spoilage is currently gaining importance. Fabrication of biosensors using glucose oxidase-fabric membranes has also been reported.

Some of the segments where biosensors will play a major role in the future are in environmental monitoring and agriculture. Biosensors are being studied for the monitoring of pesticides, insecticides, herbicides, xenobiotics, heavy metals etc in agricultural products, processed foods and soils. In this direction genetically engineered microbes are being developed for monitoring the microbes released into the environment and also in monitoring the bioavailability of nutrients as well as contaminants. Genetically modified strains of *Psuedomonas fluorescence* that are good rhizosphere colonisers have shown promise.

6. CONCLUSION

Enzyme technology has influenced the agroprocessing industry significantly in the recent years by improvement of existing processes towards achieving a better quality product at reduced cost as well as in the production of unique products. Enzymes will play a major role in the future as processing aids and the search for novel enzymes should continue. A major aim of intensified research and development in the fields of immobilization and stabilization is an ever-large quantity of products from an ever-small amount of the enzyme. Future applications also call for a concerted basic research in understanding the structure function relationship of enzymes. Finally we need to keep in mind that unlike the success of biotechnology in pharmaceutical industry its success in agro industry will depend on cost-benefit constraints and also on consumer acceptance.

Acknowledgement

Author is thankful to all his colleagues who over the years have contributed to the work reviewed in this paper.

FURTHER READING

1. Bickerstaff, G.F. Immobilization of Enzymes and Cells, Human Press, Totowa, New Jersey, (1997).
2. Bilitewski, U. and Turner, A.P.F. Biosensors for Environmental Monitoring, Harwood, Amsterdam, (2000).
3. Broadent, J.R. and Kondo, J.K. Dairy Science and Technology Handbook 3, Application Science, Technology and Engineering.VCH Publishers, Inc., New York (1993).
4. D'Souza, S.F. Applications of Biotechnology in Molasses Based and Allied Industries: Trends and prospects, (Ed. V.K. Jain) Wisdom Publishing House, New Delhi, India, (1998).
5. D'Souza, S.F. Advances in Bioprocessing and r-DNA Technology, (Eds. V. Bihari and S.C. Agarwal), Modern Printers, Lucknow, India, (1999).
6. D'Souza, S.F. *Curr. Sci.* 77 (1999) 69.
7. D'Souza, S.F. Proc Symp on Emerging Trends in Biotechnology in the New Millennium. All India Biotech Association, Southern Chapter, Hyderabad, 26-27 Feb., (2000).
8. D'Souza, S.F. Yeast Molecular Biology and Biotechnology, (Ed. R. Prasad), Omega Science Publications, New Delhi, India, (1991).
9. Mulchandani, A. and Rogers, K.R. Enzyme and Microbial Biosensors: Techniques and Protocols, Human press, Totowa, New Jersey, (1998).
10. Neelakantan, S., Mohanty A.K. and Kaushil J.K. *Current Sci*, 77 (1999).
11. Rogers, K.R. and Mulchandani, A. Affinity biosensors: Techniques and protocols, Human Press, Totowa, New Jersey, (1998).
12. Wagner, G. and Schmid, R.D., *Food Biotechnol.* 4 (1990).

Biotechnological Approaches for Ethanol Production from Agro-Residues

S.K. SONI[1] *and* S.S. MARWAHA[2]

[1]*Department of Microbiology, Panjab University, Chandigarh 160 014*
[2]*Punjab State Council for Science & Technology, Chandigarh 160 019*

INDIA HAS ACHIEVED self sufficiency in most of the agricultural produce including cereals, pulses, fruits and vegetables. The sustainability in agriculture could be achieved with the introduction of high yielding varieties, improved agricultural practices and the increase in the area under cultivation.

In the recent years the country has become the second largest producer of fruits and vegetables in the world with total annual production of about 103 million tonnes.. However, upto 40% fruits and vegetables grown are wasted due to poor or no post harvest processing practices being followed. These products being highly susceptible to microbial attack due to the presence of various nutrients, required by the microorganisms get spoiled by microbial attack. Most of the cereal and pulses are also lost by spoilage or infestation after the harvesting and/or during the storage. Losses occur in all operations from harvesting through handling, storage, processing and marketing. The losses depend on factors such as the perishability of the commodity, ambient temperature and relative humidity. These factors determine the natural course of decay, fungal and bacterial decay, damage by pests, insects, rodents and birds; the length of time between harvesting and consumption; and practices of post harvest handling, storage and processing.

Biotechnology offers a very economical and efficient means of post harvest processing of the agricultural produce including cereals, pulses, vegetables and fruits by exploitation of the useful activities of microorganisms. Microorganisms have been used in the production of industrial chemicals from the agricultural produce since long and today a number of high value utilities and chemicals are produced through microbial technology. A number of different types of industries use different organisms in fermentation processes to produce high value chemicals (Table 1). These include a variety of organic acids, enzymes, chemicals, pharmaceuticals, biomass, beverages etc. In India, the industries based on microbial technology contribute a large sum to the national exchequer. Microbial technology is not only involved in producing high value industrial products, but, is also involved in processing industrial wastes to generate useful by products, and, to keep the environment clean. Among the various industries that use microbial technologies, the alcohol production is the major one and in this chapter the role of microbial technology in alcohol production is discussed.

Table 1: Chemicals from Microorganisms

Chemical	*Organisms involved*
Antibiotics	Actinomycetes, Fungi & Bacteria
Ethanol	Yeast
Organic acids	Fungi, Bactera
Enzymes	Fungi, Bacteria
Amino acids	Bacteria
Glycerol	Yeast
Acetone-Butanol	Bacteria
Gums	Bacteria
Sorbitol, Mannitol	Fungi
Gibberellic acid	Fungi
Steroid transformation	Fungi, Bacteria

1. ALCOHOL PRODUCTION

Alcohol production was the first fermentation known to mankind. Distilleries began to appear in Europe in the middle of the seventeenth century. At first alcohol was used only for human consumption. Lately the demand for alcohol as a universal solvent and chemical raw material developed.

The fermented alcoholic products can be classified in two main categories, fermented & non-distilled and fermented & distilled. The former includes two main sub-types, wines and beers which are prepared from fruit juices and cereals and have alcohol content of 12% and 4-6%, respectively. On the other hand the latter includes whisky, rum, brandy, gin and vodka which are prepared by distillation of the fermented liquids and have an alcohol content of 42.8%.

The wines are traditionally prepared from various varieties of grapes and are classified into different categories depending upon the colour, alcohol content, presence of residual sugar and carbonation etc. Some of the common wine categories are: white wine, red wine, table wine, dessert wine, sweet wine, dry wine, sparkling wine etc. The other fruit juices which are also used in wine preparation include apple, pear, mango, jambal, banana, guava, plum, orange, strawberry, raspberry, cherry, pine apple, date, apricot, litchi etc.

Beer is traditionally prepared from barley, but, other cereals including wheat, rice, corn can also be used in combination with barley. There are two main categories of beers depending upon the type of the microorganism used and the alcohol content: lager beer, fermented by bottom fermenting yeast and ale beer, fermented by top fermenting yeast .

The distilled alcoholic beverages can be prepared from a variety of raw materials—fruit juices, sugar cane juice, molasses, root tubers, cereal grains and lignocellulosic residues etc. The preparation of these involves an additional step, distillation which involves the separation of purified and concentrated alcohol i.e. ethyl alcohol, which is a high value product and can be used in various ways including the preparation of potable beverage blends, organic chemicals, drugs and pharmaceuticals, pesticides, plastics, textiles, agriculture and as a substitute of petrol.

Ethanol, ethyl alcohol, alcohol, spirit, denatured spirit, etc. there are myriad descriptions for this agriculture-based product. A globally traded commodity, Ethanol slakes the thirst of many in Europe and finds its way in pharmaceutical and chemical industries, across the world.

Large volumes of alcohol for use as chemical feed stock have been produced commercially since World War II by fermentation. After the war, the availability of low cost petroleum ushered in the growth era of the petrochemical industry. Ethylene and other petroleum-based compounds became building blocks for the chemical industry. It was not until the oil price stocks and energy shortages of 1973 and 1979 that the Carter Administration (USA) established government policies to provide incentives for the establishment of alternative energy sources, including ethanol-gasoline blends (gasohol). The availability of surplus corn from increased yields and the Russian grain embargo also provided impetus for establishing a fuel ethanol programme.

In 1978, only 10 million gallons of fuel ethanol were produced and sold from existing beverage and industrial grade ethanol plants. In 1979, private industry built the commercial scale fuel production ethanol facility. Fuel ethanol became a co-product of existing agro-processing plants as well as new facilities specifically designed to produce fuel alcohol. Unfortunately, the failure rate of the new facilities has been very high. Two out of the every three facilities built during the 1980s are presently shut-down.

In the early 1980s, ethanol was marketed as a gasoline extender. However, because of the Environmental Protection Agency's (EPA) programme to phase out lead in fuels, during the mid 1980s ethanol usage increased because of its ability to boost the octane rating of a gallon of gasohol by approximately 3 points. In the late 1980s, ethanol production came to be viewed as a way to increase demand for corn, stabilize commodity prices and reduce the dependence of the USA on foreign oil. In the 1990s the environmental benefits of oxygenated fuels such as ethanol blends were recognized as a way to lower carbon monoxide emissions from automobiles.

During the past 15 years, the industry has matured as the processing technology has improved and ethanol has gained a measure of public acceptance in the market place. Although the sense of urgency over the need to develop alternative transportation fuels has diminished because of the return of low-cost petroleum, future prospects for fuel ethanol remain bright. The technological advancements in the ethanol production process will make grain-based ethanol more competitive with gasoline from crude oil.

While the Americas account for over 66% of total alcohol production and Asia, a mere 19% (Fig. 1). A major part of the production is tilted towards the Indian sub-continent and North East Asia. In parts of Asia, the production and consumption of alcohol is affected by religious considerations, like in West Asia. Pakistan, which has a substantial molasses production, mainly trades in molasses instead of processing it. China is the largest producer of ethanol in the region, however, the industry remains fragmented, largely broken into small scale and state holdings across provinces. At the last estimate, there were about 431 units producing alcohol at 3,000 million liters per year. India is the second largest producer of Ethanol with an organized industry producing 1800 million liters per year. Japan, although the highest per capita consumer of alcohol, is the largest importer of alcohol/ raw spirit which it then redistills locally. Thailand and Indonesia are the other two countries in the region, which have seen a steady growth of alcohol producing industry in the past five years. Going by the production figures, Indonesia has achieved the highest rate of growth followed by Thailand. The reason for this growth lies in its export potential. Both Thailand and Indonesia export alcohol to Japan. Thailand is the largest exporter in the region with over 60 million liters exported in 1997. Indonesia and Thailand have a fairly organized sugar-alcohol link that makes it a very steady business with assured raw material supply.

The world's ethanol production in 2001 got a strong rebound after several years of decline. Overall output was 31.4 billion litres as compared to 29.9 billion litres

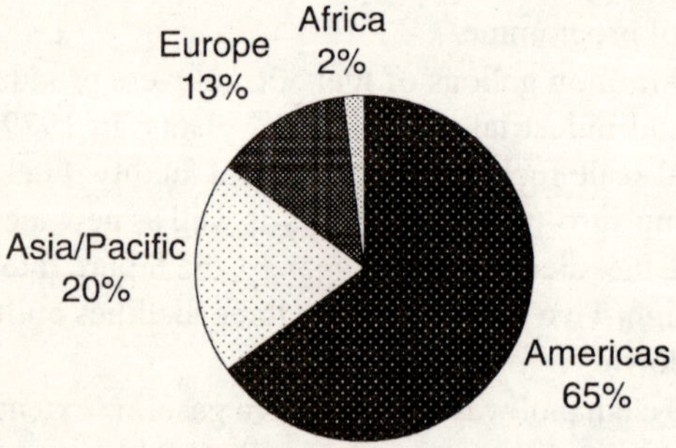

Fig. 1: World ethanol production by continent (2001)

in 2000 and 31.1 billion in 1999. Nevertheless, the world total is still below the all-time high reached in 1997, when a total of 33.0 billion litres were produced. Even though the bulk of production still comes from Brazil and the USA, two countries with an elaborate fuel ethanol program, there are interesting developments in other countries as well. There are some quarters that seek to establish fuel ethanol as an energy commodity, just like gasoline or diesel. There is some likelihood that this plan will be successfully implemented, once all these ambitious fuel ethanol programmes, which have been announced over the last two years, are put in place. In this case, the world market would have enough liquidity to facilitate a transparent price formation process and, as a result, attract the interest of the players involved. In fact, this instrument could have positive repercussions on use and production world-wide. If the burden of fuel ethanol production is shared by a sufficient number of producers, crop shortfalls in a specific country need not necessarily disrupt supplies, as these could be sourced from elsewhere.

India produces only about 1.4 billion litres ethanol from sugarcane molasses in about 300 distilleries largely by conventional batch fermentation process. This production in India is very low in terms of its size and per capita production (Table 2). India uses this alcohol in the ratio of 40:60 for potable and industrial purposes. As per reports at present actual production of alcohol is around 1.8 billion litres per annum against the installed capacity of 3.2 billlion litres in the country. India consumes about 1,200 million litres of alcohol. This leaves a surplus stock of 400 million litres which can be exported. Efforts are on at Govt of India's level to utilize this surplus as fuel-ethanol additive to gasoline. Govt is in process of framing a National Policy for alcohol production similar to the alcohol policy already excercised by Brazil and some advanced countries. In July-August 2000, there was sharp rise in crude oil prices. While there has been a drop in consumption of fuel world-over to counter the increase in prices, in India we saw a renewed interest in the promotion of ethanol production for use as fuel. Ministry of Petroleum announced its intention to introduce fuel ethanol programme. In October 2000 pilot projects were announced in the state of Maharashtra and Uttar Pradesh. Five percent addition of ethanol to gasoline would require approximately 350 million litres of ethanol annually at the current rate of consumption of motor fuel. An addition of 10% would mean 700 million litres which constitutes about 40% of the annual ethanol production. As a sharp departure from other, for fuel ethanol programmes, no subsidies are planned- neither for the producers nor for the oil companies. Full scale programme has been chalked out on the success of pilot projects firm schedule for the full scale programme has not been drawn yet, but, is in different stages of framing up for the implementation of the ambitious project on oil saving by substituting with fuel alcohol.

Table 2: Per Capita Production of Alcohol

Country	*Area Sq. Miles* $\times 10^6$	*Population (Billion)*	*Alcohol (Billion l/annum)*	*Major Substrate used*	*Per Capita Production (l)*
Brazil	3.28	0.10	12.0	Sugarcane juice	120.0
USA	3.56	0.25	7.6	Corn	30.4
India	1.26	1.0	1.8	Molasses	1.8

2. SUBSTRATES FOR ALCOHOL PRODUCTION

Ethanol production by fermentation requires sugar as raw material. This is available in various forms of agricultural origin. Depending on the source of sugar and other applications of that source crop, the availability and price of sugar for ethanol production, varies to a considerable extent. Being Agro-based product, production of fermentation alcohol or ethanol is guided by the availability of its primary raw material such as in the tropical region. However, sugar from cane can be used in more than one forms to produce ethanol viz. cane juice, cane syrup or cane molasses. Depending on cane price and processing costs, the advantages and economics of using these substrates varies from region to region or from time to time. In Australia and in certain other parts of the tropics, cane juice is used as a substrate.

The cane crop is the ruling factor in the entire chain as the same is primarily grown for sugar production. The world price of sugar has critical bearing on the feasibility of sugarcane to be used for ethanol production as a raw material, especially with a view on WTO. Companies presently producing ethanol in the world vary by size, type of feed stock, technology, byproducts and operating experience. Ethanol is generally produced by fermentation from biomass and its products like sugarcane, sugar beet, molasses, eight principal cereal grains: wheat, maize, rice, barley, sorghum, oats, millet, rye and root tubers. The substrates that have been classified into various categories are indicated in Table 3. Current economics dictate that the grains are the primary raw materials in various developed countries. Ethanol is produced by enzymatically converting the starch in grains to fermentable sugars, followed by metabolism of the sugars to ethanol by distiller's yeast.

Table 3: Various substrates used for ethanol production

Directly Fermentable	Molasses, Sucrose, Sugarcane juice, Fruit juices, Sugar beet juice, Sorghum, Honey etc.
Easily hydrolysable substrates	Starchy materials-grains, tubers
Difficult to hydrolyse substrates	Lignocllulosics

2.1. Ethanol production from sugarcane (*Saccharum officinarum*)

Sugar cane juice and two important co-products of sugar production, molasses and bagasse, are commonly used for alcohol production. Molasses is thick brown fluid containing sugary residue which is left behind after primary sugar recovery. Bagasse is the fibrous cane left over after the extraction of juice from the sugarcane. Of world's total sugar production of over 110 million tonnes, more than 65 million tonnes comes from cane. Cane sugar is produced in South and Central America, Africa, Asia and Oceania. Brazil has an area of 37,000 square kilometer under sugar cane plantation for ethanol production. Total annual sugar cane production in India is about 270 million tonnes. The juice from the crushed cane is weighed, adjusted for pH and clarified before evaporation and crystallisation, then centrifuged to separate the molasses from the sugar crystals. Cane juice contains 15-16% dissolved solids of which 85% is sucrose. The juice is partially evaporated before fermentation is carried out. The fermentation is highly productive, producing 75 litres of alcohol per tonne of cane or 4800 litres/ hectare of land. This is six times the amount of alcohol obtainable from molasses based on the same land area. Several ethanol tolerant yeast strains have been isolated which are able to produce and tolerate above 20% (v/v) alcohol during the fermentation of sugar cane syrup.

2.1.1. Molasses

The term 'molasses' is used for the viscous, sugary co-product of sugar refining, deriving from sugar cane, sugar beet and similar products. In refining cane juice for sugar production, 42 kg molasses (50% on sugar basis) are obtained from each tonne of cane or 2.2 tonnes/hectare. Typical molasses compositions are given in Table 4. Brazilian molasses has 60% fermentable sugars or 1.32 tonnes/ha which upon fermentation yields 11.5 litre of alcohol/tonne of cane or 730 l/hectare. Before fermentation, molasses may be clarified, refined and diluted. In addition to its use as a principal raw material for the production of alcohol, as used in various countries including India, molasses is also used in citric acid fermentation. Molasses has the advantage over the cane juice that it can be stored for longer periods, under optimal conditions for long enough to carry over one season to the next. After fermentation, the high ash content of molasses, which contains a substantial proportion of potassium, is of high value as fertilizer.

Molasses is the predominant raw material for alcohol production in the asian region with the exception of China and Saudi Arabia, which produce alcohol from grains and synthetic route, respectively. Japan which produces very small quantities of molasses is the largest importer of alcohol ~450 million liters per year. It is apparent that molasses supply becomes a very potent force with the sugar industry which is likely to play it to its advantage. In Philippines, this factor was dealt with head-on by a distillery which faced shortage of molasses supply. They have set up a cassava/cane juice/molasses triple mode distillery to handle the vagaries of molasses supply and price. In India also, this trend has begun. Currently dual route plant operating on molasses as well as grain as a feedstock,are being set up.

Table 4. Typical molasses composition

Component	*Range (%)*	*Average (%)*
Fermentable sugars	48-60	52
a) Sucrose	36	
b) Fructose	9	
c) Glucose	7	
Other organics	9–12	10.5
Inorganic ash	10–15	12.5
Solids	70–85	75
Water	15–30	25

Molasses production is showing gradual increase. The production levels in Asia for 1998/99 was 46.5 million tonnes, up from 45.3 million tonnes in 1997/98. India was more or less absent from the world's molasses market in first half of 1990's, when fortunes of the sugar industry turned with the steep increase in production which started in 1994/95. In 1996/97, India recorded as the world's largest supplier of molasses with exports of 1.5 million tonnes and production of 8.3 million tonnes with 42% share in the total Asian production (Fig. 2). There was a downturn in production in 1996/97 that continued in 1997/98 reducing India's role as a sizeable exporter. In 1997/98 India imported molasses. In 1998/99 estimated molasses production was 7 million tonnes, which is more than the requirement of the domestic market. Since the predominant raw material for ethanol production is cane molasses (only 8 units operate with other than molasses as raw material for beverage alcohol production in the country), it becomes pertinent to throw some light on the price movement and its effect on alcohol. Molasses prices were decontrolled around 1992. Consequently price shot up from US $ 6 per MT to US $ 60 per MT. This, of course, was an initial reaction. Subsequently, price settled down to around US $ 25-30 within a year. The price of molasses is again experiencing an upward trend which can be attributed to the rise in fuel prices. The price moved up from US $ 5-25 range last year to US$ 15-50 per MT this year.

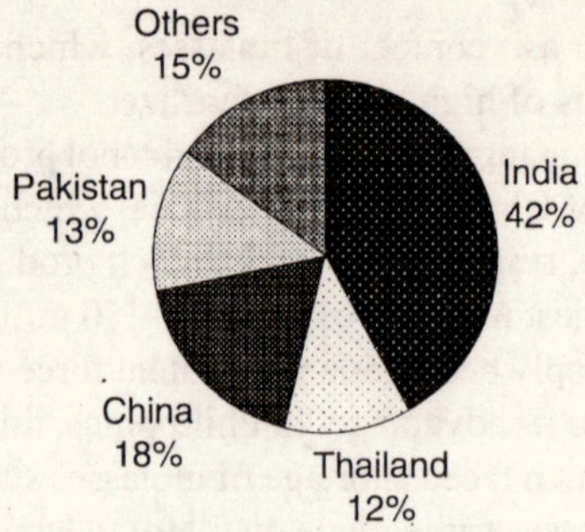

Fig. 2: Largest producers of molasses in Asia (1998-99)

2.1.2. Bagasse

Bagasse is the fibrous residue of extraction of juice from sugar cane. About 270 Kg raw, wet bagasses or 130 Kg dry matter is produced from each tonne of cane. Bagasse is a lignocellulosic residue with a substantial carbohydrate content. It typically consists of cellulose, 35%; pentosans, 25%; lignin, 20%; other organics and ash. In most sugar mills, bagasse is used largely for its fuel value. India produces 50 million MT of bagasse a year. Whilst bagasse based paper production was considered to be a significant revenue-earner over two decades ago, the importance of bagasse as fuel has overridden this importance. In fact, co-generation of energy is the new wave which is being promoted actively in certain states to counter the deficit of electrical power as well as to improve financial health of the industry.

2.2. Sugar beet

Sugar Beet is widely used in European countries for the production of sugar and alcohol. Tropical varieties, which are suitable to tropical climate, are being developed. These varieties of sugar beet crop are tested for high yield in India and other tropical countries. Sugar beet is grown extensively in UK and Europe. Annual world wide production of sugar beet is 300 million tonnes. Sugar content of sugar beet varies from 14-19%. The sugar containing juice extracted from the crushed beet is crystallised resulting in the formation of molasses. Beet molasses has similar use as that of cane molasses and generally serving as a feed stock for alcohol fermentation, the yield of which is 90-100 l/tonne. The short harvest season limits the use of sugar beet juice and its molasses for alcohol production. Two hundred and fifty kilograms of wet pulp is obtained from each tonne of sugar beet. Sugar beet pulp or bagasse is a cellulosic fibrous residue. It is not as yet, commercially processed for alcohol, but it is expected that the hydrolysis and fermentation of pulp can produce 40 l alcohol/tonne of beet. Several attempts have also been made to ferment the beet molasses by some alternative microorganims in place of the conventional distiller's yeast, *Saccharomyces cerevisiae*. The conversion of sugars present in sugar beet particles to ethanol by bacterium *Zymomonas mobilis* has also been reported in both conventional submerged and solid state fermentation. Ethanol yield of 0.48 g/g sugar, volumetric productivity of 12 g/l/h and final ethanol concentration of 130 g/l have been reported to be achieved in a solid state fermentation. The bacterium produced fewer by-products during solid state fermentation than in conventional submerged fermentation. An ethanol yield upto 95% of the theoretical value was obtained. Several studies have been carried out to recycle the spent wash which proved to be beneficial in beet molasses fermentation. The stillage from beet molasses is also recycled for alcohol production to lower the production costs by decreasing the energy requirements for waste water treatment. Some reports have indicated the detrimental effects of recyclings as there can be an in wort osmolarity leading to the inhibition of yeast growth. Moreover, the stillage recycling can lead to the accumulation of inhibitory substances which may be potentially toxic to microorganisms which are responsible for fermenting the sugars, although these inhibitory substances may act as a shield against the contaminants. The production of alcohol from sugar beet molasses has also been reported without heat or filter sterilization wherein n-butyl p-hydroxy benzoate has been used as an antimicrobial substance in the sugar beet molasses medium. The yeast cells, immobilized in calcium alginate, completely utilized molasses containing upto 25% of sugar with an ethanol yield of 10.58%, equal to 83% of the theoretical value. The addition of EDTA, potassium ferrocyanide and zeolite-X into sugar beet molasses medium improves ethanol production. The effect is more pronounced when the three substances are added to fermentation medium rather than to the growth medium.

2.3. Ethanol from Cereal grains

Annual world cereal production amounts to be more than two billion metric tonnes. Major cereal crops produced worldwide include wheat, rice, maize and barley. Other cereal crops produced include sorghum, oats, millet and rye (Fig. 3). Asia, America, and Europe produce more than 80% of the world's cereal grains. Wheat, rice, sorghum, and millet are produced in large quantities in Asia; corn and sorghum are principal crops in America, and barley, oats and rye are major crops in the former USSR and Europe. Apart from human and animal consumption, a major part of these cereal grains are used for the industrial production of starch and its derivatives, or for alcohol. The alcohol produced from these cereal grains can be used as a source of fuel and as potable alcohol. Saccharified cereal starches have been used as sweetners in food, the use of high fructose corn syrup (HFCS), which is made by enzymatic conversion of corn starch to fructose and glucose which is used as sweetner in non-alcoholic beverages. The major steps involved in processing cereal grains to ethanol include (i) feed stock preparation (milling) which, converts the starch bearing grains to a defined particle size and quality in a wet or dry mill. (ii) starch conversion (liquefaction and saccharification) to fermentable sugars using α-amylase and glucoamylase. (iii) establishing conditions for good yeast viability and fermentation of fermentable sugars by *Saccharomyces cerevisiae.* (iv) distillation of ethanol produced by fermentation and (v) removal and processing of the stillage and various by-products for animal feed. The world wide production of major cereal crops from 1961 to 1996 are presented in Fig. 3.

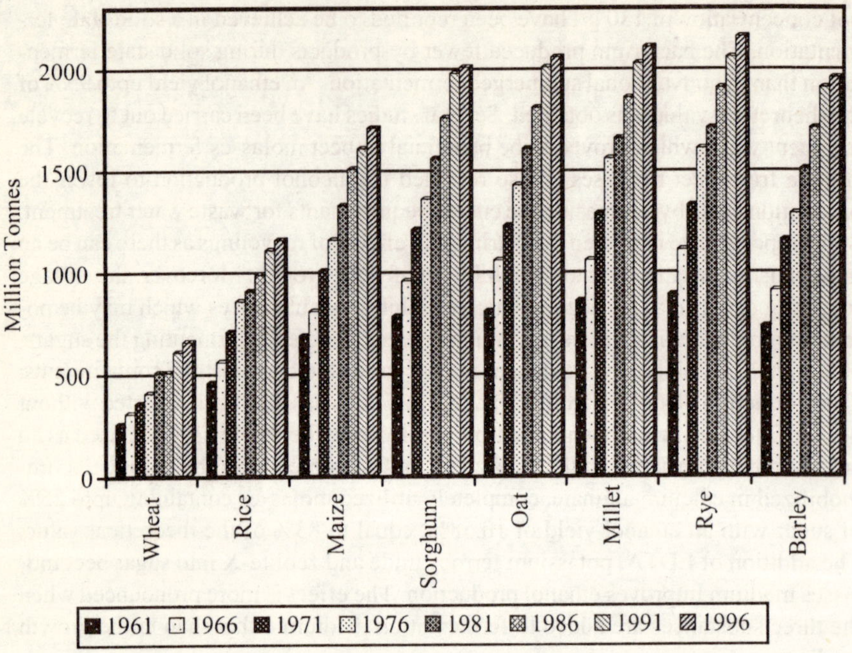

Fig. 3: Global production of major cereal crops from 1961-1996

Various common cereal grains employed for commercial production of ethanol in various parts of the world are discussed below:

2.3.1. Maize

Corn was originally cultivated in Central America and became the staple of the Incas of Peru, the Mayas and Aztecs of Mexico and early cliff dwellers of the American Southwest. Columbus brought corn back to Europe where it became a popular crop in the south. Different types of maize are classified on the basis of their protein content and the hardness of the kernel. These include pop, flint, flour, Indian and sweet corns. Much of the niacin in corn is in a bound form and this led to pellagra in areas where corn became the food staple. It was not until the late 1920s that pellagra was identified as a vitamin deficiency. The traditional practice by early American natives, of boiling corn in 5% lime or ashes, releases bound niacin making it available as a nutrient. Maize is a major cereal crop in Canada and USA and a minor crop in UK. Hybrid corn varieties produce about 210 million tonnes of corn in North America every year which is used for human consumption, manufacture of starch, corn syrup (glucose) and high fructose corn syrup (HFCS). The annual production of maize in India is approximately 10 million tonnes.

In a kernel of maize corn, the various constituents are starch, (72%); protein, (10%); oil, (5%); and fibre, (13%). For fuel alcohol production, the separation into various components is unnecessary, and is not widely practicised. For most single product plants such as dry ground corn alcohol plants, the kernel corns are first steeped to swell and then ground, saccharified by amylases and the resulting sugar is fermented.

For alcohol production, whole maize corn kernels are steeped in warm water (50°C) for 24 h before grinding. The swollen ground corn is then liquefied by α-amylase at 95°C for 1 h at pH 6.5. The corn is then heated to 105°C by direct steam injection for 10 min followed by heating to 140°C for 2 min for gelatinization, then cooling to 90°C and held at that temperature for 1 h. The treated corn is then cooled to 50-53°C, pH is lowered to 4.5-5.0 and 0.5ml/l glucoamylase is added for saccharification. The addition of glucoamylase sachharifies the starch in the kernels more or less completely within 1-2 h at 50-53°C. The slurry is then cooled and yeast is added. The alcohol yield from such a process is 450 litre per tonne of dry ground corn.

The researchers have investigated the fermentation of an enzymatic hydrolysate of ammonia fiber explosion (AFEX) pretreated corn fibre (containing a mixture of different sugars including glucose, arabinose, and galactose) by genetically engineered *E. coli* strain SL 40 and KO 11 and *Klebsiella oxytoca* P2 strain under controlled pH. Both *E. coli* strain utilized most of the sugars contained in the hydolysate and produced a maximum of 26.6 and 27.1 g/l ethanol respectively, equivalent to 90 and 92% of the theoretical yield. *K. oxytoca* P2 showed poor growth and ethanol production (maximum 20 g/l). The spent wash recycling in the fermentation of corn mashes increases the overall mashing and fermentation efficiencies and thus the alcohol yield.

2.3.2. Wheat

Wheat is one of the oldest of all cutivated plants. Today, there are more than 50000 cultivars of wheat in existence and as a result it can be grown in a relatively wide range of climatic conditions. Growing best in temperate climates, it is susceptible to disease in warm, humid regions and cannot be grown as far from the equator as can rye and oats. Wheat was brought to America early in the seventeenth century where it came to prominence in the Great Plains by 1855. Different types of wheat are classified based on planting season and endosperm composition. Wheat holds a special place amongst the cereals because upon mixing wheat flour with water, an elastic matrix called "gluten" required for the production of leavened breads is formed. "Hard wheats" tend to contain relatively high levels of starch and relatively low levels of protein, while the reverse is true for "soft wheats". High protein flours are best suited for pastas and breads, while flour from soft wheats is excellent for cakes and pastries, etc. Wheat is the second largest cereal crop of India mainly used as a major part of human diet with annual productivity level of about 60 million tonnes. Wheat flour usually contains a starch content between 74-81%. About 5-10% of this is recovered as 'B' starch, and the remainder constitutes the 'A' starch or prime starch. The value of the prime starch is very high. For the production of alcohol the feed stock of choice is 'B' starch, which commands a lower price. For cereals, where the gluten does not have high value, dry milling, saccharification and fermentation of the whole grain is the preferable route to alcohol.

For the bioalcohol production, wheat starch 'B' is slurried in water at 15% dry solids and pH adjusted to 7.0. Then α-amylase is added and liquefaction is carried out at 85°C for 30 min under constant stirring. After liquefaction the slurry is cooled to 55°C, the pH lowered to 5.0 and glucoamylase is added. After 40 h at this temperature, the sugar solution contains about 130 g/l of glucose. To this solution, nutrients and yeast are added and after 24 h fermentation, alcohol concentration rises to 7.3% (v/v), the yield being 500 l ethanol/tonne of starch 'B'. The wheat mashes contain growth limiting amounts of free amino nitrogen, so the fermentations by active dry yeast (*Saccharomyces cerevisiae*) is completed in 8 d at 20°C. Supplementing wheat mashes with yeast extract, casamino acids and glutamic acid stimulated the growth of the yeast and reduces the fermentation time. With 0.9% yeast extract as the supplement, the fermentation time is reduced from 8 d to 3 d and a final ethanol yield of 17.1% (v/v) is achieved. However, amino acids such as glycine reduces the cell yield and prolonged the fermentation time.

2.3.3. Barley

In the United States, barley is mostly used as feed, for brewing and alcohol production and only about 2% is used for human food. Barley flour is produced by abrasion dehulling, followed by milling of the "pearled" barley. In UK, 2 million tonnes of barley is used each year in brewing and distilling, while another 6 million tonnes is used as animal fodder. In a dual enzyme process, whole barley grains are steeped in water for 18 h and ground. The ground barley is stirred in water at 20% solids, pH adjusted to 7.0, 0.1% α-amylase is added and the slurry heated to 85°C for 45 min. The liquefied solution is then cooled to 50°C, pH adjusted to 5.0 and 0.1% glucoamylase added. The solution is held at 50°C for 48 h, pH adjusted to 5.5 and then cooled to 30°C and fermentated. Fermentation sugars after saccharification are around 147 g/l and the ethanol produced is 70 g/l or 9% (v/v), yield being 450 l/tonne of barley. When saccharification and fermentation are combined, the liquefied starch is treated with 0.1% glucoamylase and fermented using yeast at 37°C, the results on the alcohol yield are the same. When saccharification and fermentation are separate operations, 60 h are required, whereas, when the two steps were combined, alcohol production gets completed in 12 h.

2.3.4. Rice

The second most abundant cereal crop originated in the Indian subcontinent and Africa. Today, 90% of the world rice crop is grown in Asia. Its annual productivity level in India is over 80 million tonnes. Alexander the Great is credited with introducing rice to Europe around 300 BC. Growing rice requires more water than other cereal crops, although rice is a highly productive crop. There are several thousand rice cultivars which may differ in color, aroma and grain size. The main commercial distinction between rice types is the grain size, i.e. long, medium and short. Long grain rice, also called "Indian", tends to separate relatively easily on cooking and is dry and flakey. Short grain rice, also called "Japanese" is sticky, moist and firm when cooked. Unlike wheat, rice is most often consumed as grain rather than as a flour. Different grades of milling include brown rice (hull removed), unpolished rice (hull, bran and most of germ removed), and polished rice (aleurone layer removed from unpolished rice). Since polishing removes most of the lipid, the latter product is relatively stable during storage. The discovery that rice bran can alleviate beriberi led to the discovery of the vitamin thiamine. The traditional technique of parboiling rice in India and Pakistan (also called "converted rice") prior to milling improves the nutritional quality of the grain by allowing the B vitamins in the bran and germ to diffuse into the endosperm.

2.3.5. Rye and mixed grains

Rye is an impotant bread grain in Eastern Europe and Scandinavia. World production of rye is over 25 million tonnes. The main uses of rye is for bread and other foods, animal feed, and/or rye whisky. The use of rye in not reported for industrial or fuel alcohol production. It can, however, be used as a component of mixed grains. While its starch content is lower than that of corn or barley, it is less expensive.

2.3.6. Millet and Sorghum

Millet and sorghum are often grouped together because their growing conditions, processing and uses are similar. Millets are native to Africa or Asia and have been cultivated for more than 6 000 years. Millets grow well in regions with poor soils and are valued for their relatively high protein content among the cereals.

Sorghum originated in East Africa and today is an important food crop in Africa, Asia, India and China where it is made into porridge, unleavened bread ("roti" in India) and beer. Whole grain sorghum flour has a relatively short shelf-life. The production of low-fat sorghum flour requires removal of about 20% of the grain weight by abrasion. In North America, millet and sorghum are used primarily as livestock feed. *Sorghum vulgare* or *S. bicolor* is a member of grass family. It bears starchy seeds upon its stalks and has ability to withstand adverse conditions of draught and water logging. The annual production of sorghum is about 75 million tonnes, grown on 48 million hectare. The leading producers are USA (31%), India (16%), China (10%), Argentina (10%), Mexico (9%) and Nigeria (5%). Over 95% of the total food use of sorghum occurs in Africa and Asia, where it serves as a staple food and constitutes a major source of energy and protein. About 97% of the total use of sorghum in developed countries is for animal feed. Nevertheless, sorghum is also a substrate for producing industrial alcohol, distilled spirits, starch, dextrose, syrups and edible oil. Traditional alcohol and non-alcoholic beverages are brewed using sorghum in parts of Africa and Asia. The use of sorghum, as an adjunct or as a substitute for barley in beer brewing has also been reported by several workers. The stalks of sorghum contain 14% sugars. It is very fast growing and two crops per year are cultivated in the most favourable areas. Yields of 35 tonnes/hectare, in addition to about 3 tonnes of seeds is recorded. The sorghum crop of 1 hectare allows production of 2800 l ethanol. Sweet sorghum, similar to the normal sorghum varieties, is also used in various parts of the world for alcohol production. However, in addition to the grains, the sweet sorghum variety contains sweet juice in the stalk materials. This juice is very similar to cane juice in characteristics and has between 12 to 16 % (w/w) sugar (Sucrose). Sweet sorghum also contains a large amount of fibrous material in the stalks (Similar to sugarcane), which generates equivalent bagasse for energy generation. Direct conversion of sorghum carbohydrates to ethanol by a mixed microbial culture has been reported by several workers who produced ethanol from sweet sorghum in submerged as well as solid state fermentation. One of the study involved simultaneous saccharification and fermentation of sweet sorghum carbohydrates to ethanol by *Fusarium oxysporum* alone or in mixed culture with *Saccharomyces cerevisiae* or *Zymomonas mobilis*. The process achieved its maximum value by the mixed culture of the fungus and the yeast. Under optimum conditions, ethanol yields and concentrations as high as 29.7 g of ethanol per 100 g of dry sorghum stalk and 7.5% (w/v), respectively, were obtained. Another study involved the production of bioethanol from sweet sorghum carbohydrates which were simultaneously saccharified and fermented to ethanol by a mixed culture of *Fusarium oxysporum* and *Saccharomyces cerevisiae*. The optimum yield of bioconversion and ethanol concentration was 5.2-8.4 g ethanol/100 g of fresh sorghum and 3.5-4.9 % (w/v) respectively. In all experiments, the ethanol yield exceeded the theoretical level, based on soluble sugars, by 20.0-32.1% due to the bioconversion of polysaccharides to ethanol.

2.3.7. Cassava

Cassava, perennial woody shrub is an annual crop. This is a low cost oarbohydrate-producing crop, which finds applications in food as well as alcohol production. Brazil is the largest producer of cassava followed by, Thailand, Nigeria, Zaire and Indonesia. However, Thailand is a major exporter of cassava. The enlarged cassava roots contain maximum concentration of starch on dry weight basis among all food crops. The cassava is processed in different ways in different countries for food and for ethanol production.

2.3.8. Straw and Chaff

The annual production of straw and chaff is enormous. In addiiton to 1.5-4.0 tonnes of straw/tonne of grain, the chaff adds 100-150 kg/tonne. Wheat straw contains 73% polysaccharides, with about equal quantities of cellulose and hemicellulose. Wheat chaff contains 68% polysaccharides of which 45% is cellulose. If half of all straw and chaff were hydrolysed and fermented, alcohol production would amount to 480,000 million l/year equivalent to 3000 million barrels of oil, about 15% of all crude oil production.

The bioconversion of straw to fuel alcohol has been the subject of great deal of research in the past several years. There is a large pilot plant at Soustons in South Western France owned by "Institue Francais de Petrole, IFP" which is capable of processing 2-4 tonnes/h of straw to ethanol. After a steam pretreatment the fibres are enzymatically saccharified and the reducing sugars fermented. The fact that straws and chaffs being bulky materials, difficult to collect, transport and process. The use of straw and chaff as animal feed and farm yard manure has effectively hindered the development of industrial conversion of these to ethanol, though several reports have appeared indicating the production of alcohol from these, at laboratory scale. One study employed the fungus *Trichoderma viride* and *Pachysolen tannophilus* for the single batch bioconversion of wheat straw to ethanol. Another study reported production of ethanol from barley straw in a solid state fermentation using *Kluyveromyces marxianus*. After fermentation, the ethanol concentration increased to a maximum of 20 g ethanol/100 g of straw.

3. NOVEL MICROORGANISMS FOR ALCOHOL PRODUCTION

Alcoholic fermentation is the basis for the production of several useful products including industrial alcohol, fuel alcohol as well as potable alcoholic beverages - wine, beer, whisky, rum, gin, vodka etc. Efficient ethanol production depends upon the choice of microbes (which can adapt to the production procedures employed), the composition of the feedstock. The micro-organism selected should have the following:

- High yield of product per unit substrate assimilated.
- High fermentation rate.
- Substantial tolerance for alcohol.
- Stability under adverse environment conditions, including low and high pH, temperature etc.
- Inability to produce undesirable products
- Large scale easy cultivation on cheap and simple substrates

Although ethanolic fermentation is common to a wide variety of microorganisms (Fig. 4), with a few exceptions. Only yeasts have traditionally been used for industrial scale production of alcohol or alcoholic beverages. The near-exclusive application of yeasts is due, in part, to their homofermentative mode (EMP pathway) which yields mostly a single product, ethanol, often in relatively high concentrations. *Saccharomyces cerevisiae* and *S. uvarum* (*S. carlbergensis*), for example, produce 12% ethanol, whereas in a slow fermentation *S. sake* produce up to 20% ethanol. Under anaerobic conditions, these yeasts normally produce small amounts of byproducts such as glycerol, methanol and fusel oil, which require that specific measures be taken for their separation during distillation of alcohol.

One of the advantages of using yeasts has been their ability to ferment a wide spectrum of substrates. For example, when whey is used as substrate for ethanol production, the yeasts *Torula cremoris, Kluyveromyces fragilis, K. lactis* and *Candida pseudotropicalis* may be employed. Substrates containing low molecular weight oligomers of various sugars, dextrins, can also be fermented directly after partial hydrolysis. *K. marxianus* and *K. fragilis* can ferment inulin, a polyfructan directly, whereas *S. diastaticus* can produce ethanol from a partial digest of liquefied starch (maltodextrins) and *Torulopsis wickerhamii* has been shown to give good yields of ethanol from depolymerized cellulose (cellodextrins) while *Pichia stipitis* has been shown to be useful for the production of ethanol from sugars derived from the hydrolysis of lignocellulosic biomass.

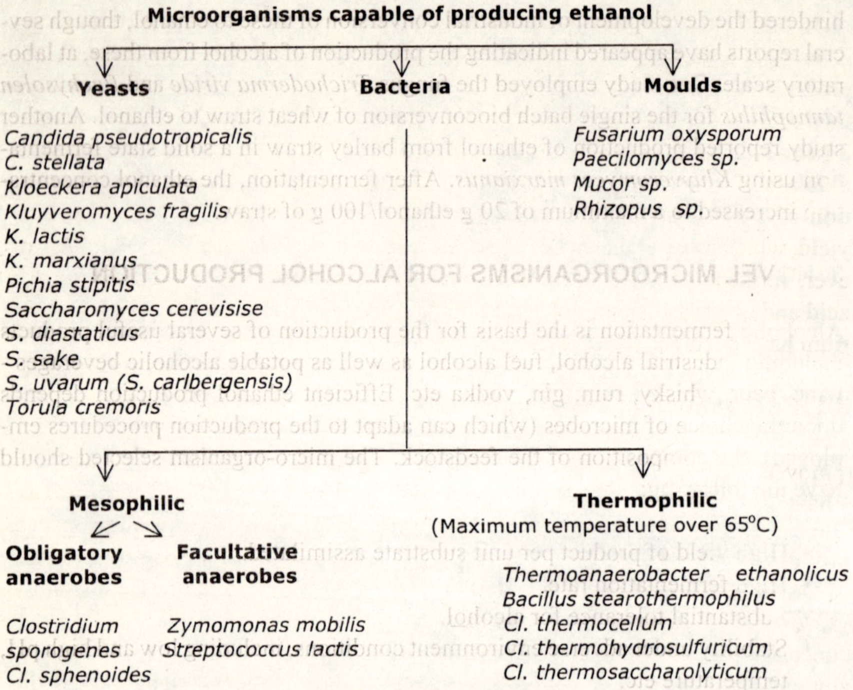

Fig. 4: Ethanol producing microorganisms

One disadvantage of ethanol-producing yeasts has been their inability to tolerate high temperatures as temperature above 40°C usually inhibit growth and fermentaion. This means that during large-scale fermentation a cooling system is an absolute requirement. It has been reported that certain thermotolerant yeasts may overcome this handicap. The strains of *Pichia stipitis* have been reported to carry out fermentation well at 41°C, which results in reducing the cooling costs.

Some bacteria are capable of producing ethanol under both aerobic and anaerobic conditions as efficiently as yeasts, and therefore have attracted attention for their possible application as an alternative organism for alcohol production. However, in contrast to yeast, bacteria characteristically generate multiple end products in addition to ethanol. These include other alcohols (butanol, isopropanol), glycols (2-3-butanediol), organic acids (acetic, propionic, butyric, formic, lactic), polyols (arabitol, glycerol, xylitol), ketones (acetone), and gases (methane, CO_2, H_2). Of a number of bacteria capable of producing good yields of ethanol, only a few may be considered for a alcohol production at industrial scale. Within the mesophilic group only the Gram-negative facultative anaerobic bacterium *Zymomonas mobilis* can be regarded as an efficient ethanol producer. It offers advantageous that makes it a potential organism for industrial scale production of alcohol. This bacterium is used in tropical regions of America, Africa and Asia in the fermentation of plant juices and production of native alcoholic beverages such as pulque, palm wine and cactus wine and there are reports of its adaptation for fuel alcohol manufacturers in the United State because of its high fermentation efficiency.

In contrast to the yeasts, which use the classical glycolysis (EMP) pathway, *Z. mobilis* uses the less common Enter-Doudoroff (ED) pathway for anaerobic production of ethanol from glucose. In this pathway glucose is phosphorylated and then oxidized or dehydrogenated to 6-phosphogluconate which loses H_2O to form 2-keto-3-deoxy-6-phosphogluconate (KDPG). This is then cleaved by KDPG aldolase. The net yield is 2 mol of pyruvate from 1 mol of glucose and the generation of 1 mol of ATP. *Zymomonas mobilis* achieves 92.5-97.5% of theoretical yield, which is better than most yeast fermentation yields, typically 88-90%. However, fermentation with *Z. mobilis* also generates acetylmethyl carbinol, acetic acid and glycerol. The biology, taxonomy, physiology and genetics of this bacterium have been very well reviewed.

3.1. Comparison of ethanol production by bacteria and yeast

Bacteria *Zymomonas mobilis* and *Thermoanaerobacter ethanolicus*, have been shown to offer some advantages over traditional yeast for ethanol fermentation and are finding attention as alternative organisms to yeast for commercial scale production of ethanol.

Zymomonas mobilis grows rapidly under anaerobic conditions and, unlike yeasts, does not require the presence of oxygen to maintain viability at high cell concentration. In contrast, yeast fermentation requires addition of a controlled amount of oxygen for cell wall & lipid synthesis and general maintenance of cellular health . However, oxygen also leads to an increase in biomass concentration (aerobic growth) and a decrease in ethanol production due to the Pasteur effect. With *Z. mobilis*, under strictly anaerobic conditions, higher ethanol and lower biomass production are obtained. Lower cell production is also a consequence of the reduced energy available for growth (1 mol of ATP/mol of glucose consumed via the ED pathway versus 2 mol of ATP/mole of glucose via the EMP pathway with yeasts). Also *Z. mobilis* takes advantage of uncoupled growth metabolism in which glucose fermentation is continuous in the absence of cell growth. The fermentation rate of *Z. mobilis* is higher than that of yeasts. The key fermentation kinetic parameters indicating the superiority of this bacterium include higher specific growth rate, μ (2.4 times high than yeast), high specific ethanol producitivity Qp, (g/g/1) 2.9 times greater and high specific glucose uptake rate Qs, (g/g/h) 2.6 times greater than for *S. uvarum*.

Zymomonas mobilis is also reasonably ethanol (over 7%, v/v ethanol), glucose and thermotolerant. Another advantage of *Z. mobilis* is that it being a prokaryote, genetic manipulation can be carried out with greater ease than in yeasts. *Z. mobilis* was discovered in 1936 and despite its several favourable attributes and high production effectiveness, it has not yet worked its way into an industrial alcohol process for the following reasons:

- *Zymomonas mobilis* being a temperamental, it requires some skill to work with; unless fermentation is closely controlled acetaldehyde tends to accumulate in the beer.
- *Z. mobilis* converts ethanol to acetate, in the presence of even traces of oxy-

gen, through the Krebs respiration cycle, thus resulting in reduction in ethanol yield. Moreover, beer containing acetate can cause technical separation problems during distillation.
- *Z. mobilis* grows optimally at pH values above 5.0, which is more than the pH of most yeast fermentations. Higher pH increases the chances of contamination by other organisms, especially at the beginning of the fermentation when the ethanol concentration is low.
- *Z. mobilis* ferments a very limited range of substrates: glucose and fructose and only a few strains decompose sucrose. It cannot be utilized to ferment some classical substrates of the alcohol industry such as lactose, maltose, mannose, galactose, starch, or lignocellulosics.

The thermophilic bacteria, *Themonoanerobacter ethanolicus* and similar thermophiles have the following advantages over yeast for ethanol production:

- Grow at high temperatures, exhibit rapid metabolic activity and a high fermentation efficiency with high product output. The ethanol yield by *T. ethanolicus* approaches that of yeasts. Fermentation at a high temperature would favour continuous operation and at the same time assist in ethanol recovery directly from the fermenter.
- Use of thermophiles minimize the chances of contamination and need for cooling.
- Viscosity of growth medium decreases at high temperature, thereby the energy requirement to maintain agitation is reduced.
- Reduced solubility of oxygen and other gases at high temperature assists maintenance of anaerobic conditions.

However, the thermophilic bacteria have the following disadvantages:

- Majority of thermophilic bacteria follow the EMP pathway, as do yeasts, but they also synthesize large amounts of acetic and lactic acid, thereby lowering yield of ethanol. The energy relation is 2.5-3.0 mol ATP/mol of glucose. High yields of organic acids and ATP diminish the amount of substrate available for the formation of ethanol. e.g. *T. ethanolicus* produces 1.8 mol of ethanol, 0.1 mol of acetic acid and 0.1 mol of lactic acid per mol of glucose. *Clostridium thermohydrosulfuricum* produces only 0.5-0.6 mol of ethanol and 0.5 mol of each acid.
- For *Cl. thermocellum* and *Thermoanaerobium brockii*, the rates of ethanol production are below 1 mol alcohol/mol glucose. Technical problems are experienced in the separation of the acids from the alcohol.
- Thermophilic bacteria require specific substrates and growth conditions, for instance *Cl. thermohydrosulfuricum* produces large amounts of ethanol at a starting pH of 9.5 which decreases during the course of fermentation to 6.9 or even less. Also, the medium needs a significant amount of yeast extract as a growth factor for ethanol production.
- Thermophilic bacteria have low ethanol tolerance. Yeasts are known to tol-

erate 6-12% ethanol without a significant decrease in the growth rate, but, in contrast the thermophilic bacteria are able to tolerate only up to a maximum of 3-7% ethanol. Strains of *Cl. thermocellum* and *Cl. thermnohydrosulfuricum* are reported to be inhibited by only 1% ethanol in the fermentation medium.

- Bacterial strains are susceptiblie to viral infection (bacteriophages), while the yeasts are not amenable to viral contamination.

It has become evident that the bacteria do not at present meet the criteria for industrial alcoholic fermentation as an alterative to yeasts. In order to use them as alternative to yeast for industrial alcohol production an improvement through genetic transformation is necessary. It is most unlikely that yeast will be replaced by other microbes for industrial scale production of ethanol in the near future, however, *Z. mobilis* may be adopted in some specific situations, especially where glucose is available as main sugar.

4. FACTORS INFLUENCING ETHANOL PRODUCTION

A brief review of various process variables and conditions that influence the yeast growth and their efficiency of industrial ethanol production is presented.

4.1. pH

Hydrogen ion concentration has a significant influece on industrial fermentation due to its importance in controlling bacterial contamination and has effect on yeast growth, fermentation rates, and by product formation. In an uncontrolled batch fermentation of a highly buffered medium, the best ethanol yields are generally obtained at pH 4.5-4.7. At higher pH more glycerol and organic acids are formed at the expense of ethanol.

In lightly buffered media, the optimum starting value is pH 5.5. At the completion of fermentation pH falls to about 3.5. If low pH is used initially in a lightly buffered medium, the final pH tends to fall enough to slow the fermentation rate. This difficulty can be overcome to a certain extent by the use of larger inocula. Yeasts survive in a pH range of pH 2.0-8.6.

4.2. Choice of organism

The ability to ferment rapidly and efficiently the sugars present in the fermentation medium containing high concentrations and sometimes unusual sugars, ethanol tolerance and ability to remain stable and viable under prevailing fermentation conditions determines the yeast to be used for industrial scale ethanol production. Presently, large-scale ethanol producing industry use *Saccharomyces* strains. The yeasts, however, have limitatios in fermenting lignocellulosic biomass, due to the presence of cellobiose, xylose, arabinose and other wood sugars. Moreover, the yeasts are not able to ferment inulin from Jerusalem artichoke, an attractive renewable resource. However, a xylose-fermenting yeast like *P. stipitis* or an inulin-fermenting yeast such as *K. marxianus* can be used to ferment such feedstocks.

4.3. Sugars fermented by yeasts

Yeasts can ferment a wide variety of oligosaccharides and sugars in addition to glucose. Different yeasts have different capabilities in this regard due to the presence or absence of enzymes which are capable of converting these different sugars to substaces which appear in the metabolic pathways of alcoholic fermentation.

With very few exceptions, L-sugars, including the common L-arabinose, are not fermented by yeasts, Some yeasts are known to ferment other pentoses, but, not methyl pentoses although the latter may be utilized in the respiration of aerobic yeast. The substrates for alcoholic fermentation are, thus, primarily hexoses, xylose and oligosaccharides.

D-hexoses and oligosaccharides fermented most often by yeasts are glucose, mannose, fructose, galactose, maltose, sucrose, lactose, melibiose, trehalose, and raffinose. The last, a trisaccharide, is partially fermented by some yeasts, but is assimilated completely by many more species during aerobic growth. This behaviour toward raffinose differentiates the bottom and top fermenting brewers' yeasts *S. cerevisiae* and *S. uvarum*.

Certain generalizations can be made regarding the fermentation characteristics of yeast on the basis of following rules or guidelines:

- Yeast unable to ferment D-glucose will not ferment other sugars.
- If D-glucose is fermented, so is D- mannose.
- If maltose is fermented, lactose is not, and vice versa (with a few exceptions, such as *Brettanomyces clausenii*, an ale yeast that ferments both sugars).
- If sucrose is fermented, so is raffionse.
- All sugars fermented, are also utilized aerobically.
- No L-sugars are fermented.

The advantage of using yeast strains *P. stipitis* is that in addition to sugars utilized by *S. cerevisiae*, this yeast can also ferment xylose and cellobiose. These sugars are normally present in lignocellulosics. Also the ability to ferment cellobiose and perhaps cellodextrins offers advantages in using partially hydrolysed lignocellulosic substrates.

4.4. Sugar concentration

The concentraion of sugar which can be fermented most efficiently depends to some extent on other components of the medium. Most industrial processes utilize a fermentation solution containing 12-20% sugars by weight. The advantages of a high concentration feed solution include reduced water requirement, suppression of osmosensitive contaminants and reduced distillation cost if ethanol inhibition can be limited.

4.5. Temperature

The rate of alcohoic fermentation increases with temperature and is optimum between 30° and 40°C. Both optimum and maximum temperature tolerance for growth and fermentation are strongly strain-dependent. However, for most ordinary yeasts, if the ethanol concentration reaches 8-9% by volume prior to completion of fermentation, the fermentation may stop if temperatures are held above 33°C. This effect has been attributed to intracellular ethanol being produced more rapidly than it can be transported through the cell membrane, causing inhibition of intracellular fermentation enzymes. Thus, in the course of fermentation, the temperature should not exceed 32°C, especially once the ethanol begins to accumulate. The temperature optimum for the growth of yeast is about 25°C. This temperature can be increased slightly by use of a rich nutrient medium. Since yeast growth and fermentation both produce heat, cooling may be necessary to maintain the desired temperature.

4.6. Ethanol

A limitation of ethanolic fermentation is the capacity of the microbe to tolerate this solvent, because ethanol inhibits alcoholic fermentation, which limits the concentration of ethanol which can be produced by a given strain of yeast. The maximum concentration of ethanol which can be produced by yeast varies with species and a maximum upto 20% ethanol by volume can be produced. The results of mating experiments with yeasts indicate that ethanol inhibition is a polygenic phenomenon, that it is a result of multiple enzyme and cell physiological functions. The degree of inhibition is also related to other environmental factors, in particular high sugar concetration and high temperature which cause the inhibition to be more severe. Brewers' yeast (*S. cerevisiae* var. *carlsbergensis*) ceases to ferment at about 6% ethanol by volume, whereas bakers' yeast (*S. cerevisiae* var. *diastaticus*) stops fermentation at about 12% ethanol by volume, and wine yeast (*S. cerevisiae* var. *ellipsoideus*) at about 15% ethanol by volume. Ethanol which is produced during fermentation (autogenous ethanol) is more inhibitory to cell growth than that from an exogenous source. Addition of ethanol to actively fermenting yeast cultures results in rapid reduction of fermentation and growth rates due to its influence on protein synthesis and irreversible denaturation of enzymes. Ethanol has also been shown to exhibit non-competitive inhibition of yeast growth, that is, independent of other inhibitors which may be present. The toxic effect of ethanol has also been attributed to damaging the cell membrane or changing its properties. The extent of ethanol tolerance of some yeasts is highly strain dependent and appears to be related to the unsaturated fatty acids and the fatty acyl composition of the plasma membrane. The intolerance exhibited by some yeasts appears to be related to low cellular lipid content and the inability of the plasma membrane to maintain during high rates of ethanol production, a high ethanol flux through the plasma membrane and out of the cell to avoid intracellular accumulation.

4.7. Nutrients

Most yeasts grow well on a variety of amino acids, purines, and pyrimidines as the sole source of nitrogen. Mixtures of amino acids are better than any single amino acid. About 15 mg of assimilable nitrogen per 100 ml of fermentation solution is sufficient to assure good fermentation rates. The nitrogen can also be supplied in the form of ammonium salts, urea, corn steep liquor or distillers' malt. The distillers' malt have the advantage of containing vitamins and minerals necessary for maximum yeast growth. Yeasts require trace amounts of biotin, thiamine, pyridoxine, calcium pantothenate, and inositol for maximum growth and fermentation rates. These vitamins also regulate yeast metabolism, vitamins being generally either coenzymes or precursors for fully active enzymes. Vitamin requirements are strain dependent. *Candida utilis* and *Hansenula anomala* are vitamin independent and synthesize all their own needs, whereas biotin and pantothenate are essential for all strains of *Saccharomyces*.

Phosphorus in the form of phosphate is an important ionic factor determining the rate of fermentation. It is required at a concentration of about 0.16 mM/g cells for optimum fermentation performance. Yeast growth also depends on compounds containing potassuim (K), sulphur (S), and traces of zinc (Zn), iron (Fe) and copper (Cu), required in concentrations of 0.1-1 mM. For anaerobic growth of yeasts, ergosterol and unsaturated fatty acids are beneficial additive while some thermophilic yeast strains have requirements for choline or leucine.

4.8. Yeast inoculum

Provided that temperature, pH, sugar concentration, assimilable nitrogen, vitamins and minerals are at their optimal values, a controlling factor is the amount of yeast added in the inoculum, or the "pitch" as it is often called in industrial operations. Using a larger inoculum of yeast will decrease the time required for growth, and alcoholic fermentation will set in sooner. For commerical production of alcohol, yeast is usually added to provide a starting population of 7-10 million cells/ml, about 0.2g of dry yeast/l of broth. At this inoculation level, fermentation is usually completed within a day or two, depending upon sugar concentration. Much higher inocula, up to 10 or more g cells/l, result in much more rapid fermentation, and is also beneficial in resisting inhibition and suppressing contamination.

4.9. Dissolved oxygen

Ethanol fermentation is not a purely anaerobic process. Dissolved oxygen concentration is an important variable for industrial alcohol production. In many instances, active starter cultures are grown under aerobic or semiaerobic conditions to improve yeast yields and growth rates. Yeasts are unable to multiply for more than four to five generations in solutions containing less than 1 ppm oxygen, even if sufficient nutrients, such as sugar, nitrogen, and vitamins, are present. It is the dissolved oxygen in the solution which limits the population of yeast in the fermentation medium. Once the dissolved oxygen is depleted during fermentation, the yeast growth slows and eventually ceases. Also during the time the metabolism of yeast switches from aerobic respiration to anaerobic alcoholic fermentation, and the synthesis of alcohol occurs until the sugar is depleted or until a limiting ethanol concentration is obtained.

Oxygen is utilized by yeasts for at least two separate functions:

- as the ultimate electron acceptor in oxidative phosphorylation, that is for growth and respiration; and
- as an essential nutrilite in lipid, ergosterol and nicotinic acid biosyntheis.

During respiration yeasts have a high demand for dissolved oxygen which promotes complete oxidation of glucose to carbon dioxide and water via the tricarboxylic acid cycle at the expense of ethanol accumulation via fermentation (Pasteur effect). However, the extent of this respiration is tightly controlled by the concentration of fermentable carbohydrates in the medium. In the presence of high sugar concentrations, ethanol is formed during aerobic growth and the phenomenon is known as 'aerobic fermentation'. This happens due to the operation of the so-called reverse Pasteur, or Crabtree effect which results in the repression of both the synthesis and the activity of the respiratory enzymes, notably some Krebs cycle enzymes under fully erobic conditions.

The Pasteur and Crabtree effect are illustrated in Fig. 5. Conditions 3 and 4 are carried out exclusively for alcohol production, whereas, condition 2 is the mode for fodder yeast production, and condition 1 is practised during production of bakers' yeast. The Crabtree effect is also applied in brewery operation, where due to high initial sugar concentration (condition 1) a substantial degree of aeration of the wort prior to pitching, up to about 20% oxygen saturation, is practised to get maximum yeast yield, high viability, growth rate, and overall fermentation rate.

Oxygen tension	Glucose concentration: High (1)	Glucose concentration: Low (2)
High	Aerobic alcohol fermentation + Aerobic cell growth CRABTREE EFFECT	No alcohol Aerobic cell growth PASTEUR EFFECT
	3	4
Low	Alcohol synthesis Anaerobic environment No growth	Alcohol synthesis Anaerobic environment No growth

Fig. 5: The influence of oxygen on glucose metabolism

The effect of dissolved oxygen is critical in continuous yeast growth or single cell protein production, where it is usually the limiting factor. It is also important in continuous fermentation, in order to maintain yeast health as well as a satisfactory fermentation rate. The exact level of the minimal oxygen required is strain-dependent and varies with environment. At oxygen concentrations below 1 ppm in the broth, the addition of sterols (ergosterol) and unsaturated fatty acids (oleic acid) may be required during anaerobic growth, since oxygen is required for the synthesis of unsaturated fatty acids, essential components of yeast membranes. It is for these reasons that low concentrations of oxygen are generally maintained in industrial alcohol fermentation.

4.10. Inhibitors

Several trace elements including silver (Ag), arsenic (As) and mercury (Hg), with concentrations greater than 10-100 μM, are toxic to yeasts. In addition, anionic detergents, sorbic acid, diethyl pyrocarbonate (DEP) and the antibiotics cyclohexamide, antimycin A and nystatin completely inhibit the growth of most common yeasts. In fermentation of lignocellulosic hydrolysates, acetic acid, furfural and lignin-derived phenolics are inhibitory.

4.11. Carbon dioxide

At pressures greater than atmospheric, carbon dioxide inhibits both aerobic growth of yeast and anaerobic fermentation and this effect being enhanced markedly by low pH and high concentrations of ethanol. The inhibitory effect of CO_2 is due to its influence on the permeability and the composition of the cell membrane resulting in altered enzyme activity and changes in solute transport. A major inhibition due to dissolved CO_2 is to be expected in deep fermentations, for example in towers, where pressure may reach upto two atmospheres or more and with fermenters having dissolved CO_2 levels as high as 0.02 M.

4.12. Microbial contaminants

The two most common types of contaminants encountered in alcoholic fermentation are acetic acid bacteria (*Acetobacter* spp.) and lactic acid bacteria, principally species of *Lactobacillus* and *Streptococcus*. Lactic acid bacteria are a more serious problem, since they metabolize and multiply under low pH and anaerobic conditions. Acetic acid bacteria metabolize ethanol in the presence of oxygen and thus appear as a gross infection in the early stages of yeast growth, when the fermentation liquid has been exposed to air for long periods. Strict precautions of cleanliness, sterilization of media and the addition of a high inoculum usually ensures that the yeast population will predominate in a batch fermentation. However, regular microscopic examination of the fermentation broth is necessary.

As fuel ethanol is not for human consumption, it is possible to introduce antibiotics to fermentation media to eliminate, or at least slow down, microbial contamination. Penicillin works well in this regard. Antibiotics are usually unnecessary in batch fermentation, however, in continuous fermentation, where contamination may be more difficult to get rid of, there use may be beneficial.

5. CURRENT STATUS AND FUTURE PROSPECTS OF ALCOHOL PRODUCTION IN THE WORLD

America continues to be the leading ethanol producing continent and it seems likely to continue in the future also. Total production in 2001 was 20.6 billion litres, up from 19.5 billion in 2000. This is equivalent to around 66% of world output (Fig. 1; Table 5). The world leader continues to be Brazil, where output for 2001 was 11.9 billion litres compared with a revised estimate of 11.4 billion for the year 2000. The downturn in production in 2000 was mainly due to a sharp decline in cane crop production. Alcohol prices, on the other hand, were quite attractive. The value of cane, if used for fuel alcohol production, was at par with sugar produced for the domestic and the export markets until February 2001. After that, alcohol values on the domestic market weakened and, as a result, the value of cane for alcohol production also dropped. It is not yet clear what consequences arise from this deterioration in the economic position.

Canada is also in the process of expanding its ethanol production capacity. The total production for 2001 was 238 million litres which is unchanged from last year, but, more than twice the amount produced in 1997. For 2002, the Canadian Renewable Fuels Association (CRFA) forecasts a rise to 358 million litres, as several new projects are expected to start operating by then. Longer term output could rise to as much as 750 million litres, boosted by the government's Action Plan 2000 on Climate Change. The government supports the ethanol industry with a C$0.10 per litre-exemption on federal excise taxes for the ethanol portion of blended fuels.

Spain is expected to become the largest bioethanol producer in the EU by 2004, when its third and largest plant will come on stream in Salamanca. So far, one bioethanol plant is up and running in Cartagena with a total capacity of one million hl ethanol production a year. The second one started operations in the first quarter of 2002 in Galicia. Production capacity there is 1.26 million hl. Finally, the Salamanca plant will have a production capacity of two million hl. This plant will use biomass as feedstock and, if successful, would be the largest ethanol producing plant, employing such a production process world-wide.

In France, the prospects for bioethanol production have dimmed, at least for the short term. While the Farm Ministry in September 2000 raised the production ceilings for bioethanol by an extra 115,000 tonnes, The decision by a European court has challenged the current practice of support. Up to now, biofuel plants in France were granted tax exemptions on the grounds that these were pilot projects that developed more environmentally friendly fuels. However, the court said the plants benefiting from the lower rates were operating established processes, and that the aid was thus for economic and industrial rather than technological purposes.

Production in Germany rose slightly in 2001, due to higher output in the fermentation sector. In the wake of the reforms, some industrial distilleries (molasses based) could opt out of the current market regulation, taking advantage of rather favourable transition rules. Figures so far suggest that fermentation alcohol production could rise by some 100,000 hl in the coming years.

In Italy, production will fall due to the exit of a major molasses market in the alcohol sector, as well as lower wine alcohol production.

Production in central and eastern Europe is expected to grow only marginally. The Russian alcohol market is the largest in Europe, even though its size is vague because of the big black market. According to official figures, total alcohol output in 2000 reached around 800 million litres, the largest part of which was beverage alcohol. In Ukraine, the government is trying to put the industry on a sounder footing. A fuel alcohol program was implemented in summer 2000, but, so far production has remained below expectations with total beverage alcohol production around 190 million litres.

Ethanol production in Africa is concentrated on the southern tip of the continent, with the Republic of South Africa accounting for approximately 70% of the total production of ethanol. The production of high purity ethanol has been growing in recent years, with the total in 2001 reached 126,000 tonnes, against 97,000 tonnes in 2000. Besides synthetic alcohol, South Africa also produces increasing amounts of fermentation ethanol, with molasses being the major feedstock. The production levels reached 40 million litres in 2001 as compared with 38 million in 2000.

One of the most interesting import markets in Africa in recent years was Nigeria. The country stopped producing ethanol in 2001, after cheap world market imports and a difficult domestic feedstock situation had undermined the viability of the domestic sector. The total market volume in Nigeria is estimated at around 90 million litres, the largest part of which is now supplied by South Africa, Brazil and Spain.

Among the Asian countries, China is one of the latest countries to announce a comprehensive fuel ethanol program. In late 2000 it was announced that the country plans to build a fuel ethanol plant in the province of Jilin. The facility is designed for an annual production of 800,000 tonnes of fuel alcohol from corn, and operations are to start by the end of 2002. Earlier, a beverage ethanol plant in Heilongjiang was refitted so that it could produce fuel ethanol with a capacity of 100,000 tonnes a year. In addition, the Henan Tianguan Group started to produce fuel ethanol at a 200,000 tonne facility in March in 2002. Finally, a 500,000 tonne plant has been announced for the cane growing province of Guangdong. All plants combined would raise fuel ethanol production capacity in China to 1.6 million tonnes. China sees fuel ethanol as an opportunity to utilise its grain surpluses more efficiently and to reduce the amount of oil imports. In Thailand the fuel ethanol program, announced with much fanfare in late 2000, is getting under way. Local and international investors are looking at the possibility of building plants with an overall capacity of 2 million litres per day, with sugar cane and tapioca being used as feedstocks. As in the case of China, Thailand aims to reduce its oil import bill and to create new outlets for farm produce. So far, two oil companies have started distributing fuel ethanol blends, but have run into difficulties sourcing sufficient quantities of feedstocks. Fuel ethanol blends are sold at prices which are somewhat lower than those for premium gasoline and demand had been brisk. India is the third country in the region which has started to produce anhydrous alcohol for fuel purposes. At present there are three pilot plants operating in the country producing ethanol for sale in fuel blends containing around five per cent alcohol. India is heavily dependent on oil imports with the total around 75 million tonnes in 2001

Table 5: Country-wise total ethanol production

	World Ethanol Production (1000hl)				
Country	*2001*	*2000*	*1999*	*1998*	*1997*
France	8 000	8 120	7 540	7 788	7 671
Germany	2 950	2 850	3 400	3 640	3 750
Italy	1 900	2 056	2 009	2 251	2 297
Spain	2 250	1 450	1 250	1 400	1 360
UK	4 300	4 350	4 100	4 220	4 100
Other EU	2 221	1 871	1 853	1 876	2 020
EU	**21 621**	**20 697**	**20 152**	**21 175**	**21 198**
Czech Republic	900	900	900	1 050	1 090
Hungary	510	553	481	494	531
Poland	1 580	1 600	1 700	2 080	2 400
Russia	11 700	11 500	12 800	12 000	11 800
Ukraine	2 200	1 960	1 740	1 560	2 470
Other Europe	3 023	3 032	2 952	3 045	3 587
Europe	**41 534**	**40 242**	**40 725**	**41 404**	**43 076**
Argentina	1 530	1 710	1 735	1 766	1 610
Brazil	119 000	114 000	129 821	141 221	154 934
Canada	2 380	2 380	2 000	1 500	1 500
Cuba	850	840	800	795	1 100
Ecuador	627	375	321	313	263
Guatemala	600	600	450	450	500
Mexico	701	671	562	531	532
USA	75 800	70 500	66 050	64 500	58 860
Other Americas	4 180	4 042	3 867	3 828	3 770
Americas	**205 668**	**195 118**	**205 606**	**214 904**	**223 069**
China	30 900	29 700	28 600	28 000	26 900
India	17800	17 200	16 900	16 881	16 470
Indonesia	1 650	1 600	1 500	1 650	1 740
Japan	1 360	1 100	1 040	1 020	1 040
Saudi-Arabia	3 900	4 100	3 900	3 700	3 900
Thailand	1 500	1 000	3 200	2 650	3 750
Other Asia	2 485	2 613	2 653	2 820	2 763
Asia	**59 595**	**57 313**	**57 793**	**56 721**	**56 563**
Australia	1 540	1 200	1 000	900	850
New Zealand	174	195	176	190	178
Other Oceania	80	80	80	80	80
Oceania	**1 794**	**1 475**	**1 256**	**1 170**	**1 108**
Malawi	120	120	122	119	154
South African CU	3 852	3 800	3 900	4 100	4 300
Zimbabwe	293	257	263	221	252
Other Africa	1 059	1 036	1 048	991	1 089
Africa	**5 324**	**5 213**	**5 333**	**5 431**	**5 795**
World	**313 915**	**299 361**	**310 713**	**319 630**	**329 611**

valued at $17.5 billion. At the same time, the ethanol industry in India is suffering from undersupply, with total capacity now put at 3.2 billion litres, while actual production is around 1.8 billion. However, the argument is not as straight forward as it may seem. The country's large alco-chemical industry fears that production of fuel ethanol could raise prices of molasses, thus increasing production costs for them as well. The industry predicts that many facilities could face bankruptcy as a result of higher feedstock costs. However, so far the government seems to be determined to push ahead with the scheme.

6. STATUS OF ALCOHOL PRODUCTION IN INDIA

The ethanol industry in India has been a traditional industry. However, winds of change are blowing in India and so is it for the ethanol industry. This is very evident from the fact that molasses, which had been the most controlled commodity, has been decontrolled across all states with only a few states still exercising partial control. To predict the future trends in the ethanol industry we need to study some of the key trends in the market over the past decade. India is 28 states in the context of one. Each state has different laws for controlling taxes and movement of molasses and ethanol. In some states, like in the northern state of Uttar Pradesh, use of molasses for industrial and potable use—local liquor and for superior grade liquor category—is still controlled, whilst in other states total decontrol is being followed. There are about 300 distilleries in India which operate with average capacity of 40,000 to 50,000 litres per day, producing 1.8 billion litres per annum. The ethanol industry is closely related to the sugar economy and its cycles. With over 18 million MT of sugar produced, about 8 million MT of molasses and 1.5 billion litres of ethanol was produced in the year 1999-2000. Since the predominant raw material for ethanol production is cane molasses (only 8 units operate with other than molasses as raw material for production of beverage alcohol), its availability and price affects the alcohol production. Molasses prices were decontrolled around 1992. Consequently price shot up from US $ 6 per MT to US $ 60 per MT. This, of course, was an initial reaction. Subsequently, the price settled down to around US $ 25-30 within a year. May be due to the increasing interest in the use of alcohol as biofuel the price of molasses is again experiencing an upward trend.

In terms of quality of molasses, there has been a significant change in the composition of molasses. As the sugar mills are improving their extraction efficiency, composition of molasses is adversely affecting the distillery industry. Content of fermentable sugar has dropped from 47-50% to as low as 35% by weight. This, of course, has notable impact on the yield and efficiency. Content of volatile organic acids, caramel and sludge has been on the increase as the extraction efficiency has been increasing. Volatile organic acids and caramel pose severe problems to activity of yeast, thereby affecting not only efficiency, but, also quality of alcohol. The high content of sludge results into deposition in distillation column adversely affecting uptime. This has put a new demand on technology—'How to tackle molasses with adverse composition?'

India has around 300 alcohol producing distilleries with an installed capacity nearly 3.2 billion litres of ethanol per annum. However, capacity utilization generally does not exceed more than 60% and as a consequence, alcohol production has not exceeded 1.8 billion litres per annum. This amount is small compared to the figure of nearly 12 billion litres of alcohol produced by Brazil during 2001 while considering the fact that India is World's largest producer of sugar and sugar cane. Most distilleries in India use cane molasses as the substrate and the process involves diluting the cane molasses to about 14-15% of fermentable sugars, which is then inoculated with a yeast and allowed to ferment an ambient temperature for 24-48 h after which the wash is steam distilled to recover the alcohol. Ethanol yields of 6-8% are average and alcohol recoveries from a tonne of cane molasses are generally around 230 l per tonne. There is no control over the quality of molasses and as a consequence the recoveries are largely variable. The major substrate for ethanol production in India is cane molasses. Generally about 50-60% of the cane produced in India is processed in sugar mills and the yield of cane molasses i.e. around 30% of the amount of the sugar produced annually. With the price of sugar on the steady rise, there is tendency to recover the last crystal of sugar from the cane and as a consequence, harsh methods of processing are used. As a consequence, the quality of molasses produced is extremely poor. Table 6 shows the quality of Indian cane molasses vis a vis the cane molasses from other countries. Indian cane molasses contain the lowest amount of fermentable sugars and the highest amount of ash material. Further, the quality of molasses in India also varies from factory to factory.

Table 6: Average composition of Indian cane molasses vis a vis other cane producing countries

Component	*India*	*Brazil*	*Philipines*	*Thailand*
pH	4.5-5.0	—	5.3	5.2
Total solids g%	75	—	73	72
Total ash g%	12	7	6	7
Total sugars %	42-45	56	57	53
Invert sugars %	15	—	20	19
Unfermentable sugars %	12	5	6	7
CaO %	2	0.5	0.7	0.5
Total Nitrogen %	0.1	0.3	0.2	0.6

Table 7 shows the composition of molasses obtained from different distilleries in India. The composition not only varies with regard to the fermentable sugars but also with the nutrients available. As a consequence, each batch of molasses needs to be examined for nutrient supplementation particularly nitrogen and phosphate, to ensure satisfactory recoveries. For efficient production of ethanol, a number of factors are involved. These include the substrate, the yeast strain, the fermentation conditions, distillation conditions and finally the effluent treatment process. All these factors individually and collectively determine the final efficiency of the fermentation process. In recent years, a number of limitations have been faced by the Indian alcohol distilleries at each of these stages.

Table 7: Composition of cane molasses from different Indian distilleries

Distillery No.	*% Total sugars*	*% Fermentable sugars*	*pH*	*% Ash*
1	51.8	47.2	5.5	13.0
2	54.2	49.6	4.8	9.0
3	50.6	44.2	5.0	13.6
4	48.5	43.9	5.3	11.6
5	50.6	45.6	5.0	18.2
6	49.3	44.2	4.8	13.8
7	48.5	4.5	5.1	13.0
8	42.4	38.4	4.8	12.0

Ethylene, being cheaper, was used earlier (World War II) for ethanol production by chemical synthesis. As now, the ethylene cost is high, production of ethanol from agricultural renewable sources is preferred. Due to short supply of sugarcane molasses, its rising cost, deteriorating quality and technological advancement in Indian sugar industry, the industries have shown keen enthusiasm to utilize other alternate substrates for ethanol production. However, the cost of production of ethanol from such substrates is always more than that of the cost of ethanol production from molasses (Table 8).

Table 8: Approximate cost of production of ethanol from different substrates by fermentation

Substrate	*Cost/ton Rs.*	*EtOH yield Lit/t*	*Appoximate Cost/l EtOH (Rs.)*
Sucrose	14,000	520	27
Jaggery	10,000	400	25
Molasses	3,000	225	13
Wheat and other starchy materials	5,000	350	15

Other than cane molasses there are a number of additional substrates which can be employed for ethanol production (Table 3) in India, but, due to some limitations, many have not been exploited yet on commercial scale. In the directly fermentable category of substrates, the substrates other than cane molasses are not available in adequate quantities in India for use in alcohol production. However, considering the recent increase in cost of cane molasses as a result of price decontrol, sugarcane juice appears to be an attractive substrate for ethanol production. Use of this substrate does not need any change in the fermentation conditions and further will create no problems of effluent disposal. Average yields of 50-60 L of ethanol per tonne of cane can be expected. At the current production of nearly 250 million tonnes of cane annually, a part of this can be diverted for the production of ethanol by direct fermentation. Production of ethanol by direct fermentation of cane juice has also been examined in India and yields of 7-8% ethanol in the fermented wash have been reported (Table 9). Economical production of ethanol from direct fermentation of sugarcane is now possible mainly because of the changed price structure of cane molasses.

Table 9: Ethanol production from cane juice

Fermentation time (h)	*%(v/v) Ethanol in Farm cane juice*	*% (v/v) Ethanol in sugar mill juice*
6	1.7	1.6
24	5.9	4.9
30	7.8	5.0
36	8.3	5.1

The other directly fermentable substrates available are sweet sorghum juice, jaggery, sugar beet etc. The use of these substrates have some limitations. Sweet sorghum appears to be an excellent substrate since the stems in addition to yielding the juice, can also yield valuable grain and fodder. However, difficulties in its successful cultivation in India, has placed restrictions on its use. Should easy cultivation be possible, sweet sorghum appears to be an interesting crop for ethanol pro-' duction. The fermentation of the juice does not need any changes in the fermentation conditions and as discussed later, the grain can also be used for ethanol production after hydrolysis.

Substrate such as Jaggery is being used for ethanol production especially when molasses is in short supply. Fermentation of this does not need any changes in fermentation procedure. Infact, use of jaggery or even concentrated cane juice appears to be best alternative for producing ethanol from cane molasses. These substrates are directly fermentable and moreover, pose no serious effluent disposal problems.

Another substrate that has been examined in India for the production of ethanol is sugar beet. This crop is next only to sugarcane in using the solar energy efficiently. However, it is not grown at large scale in India except in Ganganagar region of Rajasthan and in the Sunderbans of West Bengal. With slight modification of fermentation conditions, both the sugar beet as well as beet juice can be efficiently used for the production of ethanol.

The second group of substrates available for ethanol fermentation in India include the starchy substrates. These substrates need to be first hydrolysed before they can be fermented by yeast to produce ethanol. The four major starchy substrates available in adequate quantities in country are rice, wheat, corn and sorghum. Since the first three substrates are a part of staple diet of majority of people of India and are not available in surplus the only potential substrate that can be exploited for commercial production of alcohol in India, in future, appears to be sorghum. This is already in use by a one of the largest alcohol producing companies of India, but, the yields are not sufficient. The procedures used by this company are kept secret which needs to be looked into to improve the ethanol prodution.

The technology of producing alcohol from starchy substrates is well documented. Briefly, the process involves gelatinisation of starchy material using bacterial α-amylases and fungal glucoamylases. The hydrolysate is supplemented with nutrients and fermented by strains of *Saccharomyces cerevisiae*. Recently, high gravity wheat hydrolyssate fermentations have been reported and alcohol yields as high as 18% have been recorded. The main problem in using starchy substrates as alternate substrates for alcohol production is that the thermostable microbial enzymes for the hydrolysis of starch are not produced in India in sufficient quantities, though, Biocon (India) Limited, Bangalore is the main producer of such enzymes, which are generally used in the hydrolysis of barley malt, carried out in the temperarture range of 60-65°C. However, the high-temperature amylases, required for the liquefaction of wheat, rice, corn and sorghum, are generally procured from Novo Nordisk, Denmark, the leader in the world enzyme market. According to a recent survey, a largest alcohol industry of Northern India, (the third largest alcohol group of India) imports the thermostable amylases worth rupees 1.0 crore annually for the hydrolysis of various cereal grains for ethanol production. The main advantage of this category of substrates is that they does not lead to the production of an effluent that is difficult to handle. Infact, the beer after the alcohol is removed, which can be processed to a high value low alcohol beverage rich in yeast nutrients. While in the western world, the fermented grains are the main high value products of the grain based distilleries. In India, the effluent can be the main revenue earner with ethanol becoming a by product. Microbiological techniques to quantitatively convert the starch present in the cereal grains to alcohol need to be only standardized and the development of some efficient, in-house amylase preparations is the requirement of the present scenario, if the cost of ethanol production from starchy biomass is to be made comparabele to that of molasses. At the moment, the cost of sorghum and other cereal grains including wheat in India is comparable or even less than that of molasses in some parts the country, but the cost of ethanol production from starchy biomass is very high as compared to that from molasses (Table 9).

At present the hydrolysis of starch is required prior to its fermentation, since as yet yeasts that can ferment starch directly to yield high levels of ethanol are not available. Strains capable of fermenting starch substrates directly to ethanol have been derived by genetic manipulation, but, these do not produce significantly high levels of ethanol for use in distilleries. Also currently available technologies for the production of ethanol from starchy substrates need high temperature gelatinisation of the substrate before enzymatic hydrolysis. Fungi producing enzymes that hydrolyze uncooked starch have been reported and their use in raw starch hydrolysis would substantially reduce the cost of alcohol production from starchy materials.

The third group of substrates include the difficult to hydrolyse lignocellulosics. Although available in plenty in India, and is of renewable type, there are still difficulties in quantitatively converting this substrate to fermentable sugars. Also, hydrolysis of this material leads to the production of a hydrolysate containing both hexoses and pentoses and as yet there is no organisms that utilize both substrates efficiently to produce alcohol. Technology for acid hydrolysis of the cellulosic wastes and to produce alcohol is well documented. However, acid hydrolysates contain products that are inhibitory to yeast growth. Enzymatic hydrolysis of lignocellulosics has been investigated for a long time, all over the world, but as yet no efficient method is reported and commercialized. Attempts are also being made to derive yeasts that can ferment the cellulose present in lignocellulosics directly but as yet, not much success is reported. Recently much research has been focussed on the utilization of lower value substrate such as lignocellulose and agrowaste for fuel ethanol production, because cost of the substrate constitutes 55-70% of the total process costs. Thus the low value lignocellulose offers a great potential for reducing the production cost and increasing the use of ethanol as a fuel additive. The biological process of ethanol fuel production utilizing lignocellulose requires:

- delignification to liberate cellulose and hemicellulose from their complex with lignin.
- depolymerization of the carbohydrate polymers (cellulose and hemicellulose) to produce free sugars and
- fermentation of mixed hexose and pentose sugars to produce ethanol.

The development of the feasible biological delignification process is possible if lignin degrading microorganisms, their ecophysiological requirements and optimal biorector design are effectively coordinated. Some thermophilic anaerobic and recombinant bacteria have advantageous features for direct microbial conversion of cellulose to ethanol i.e. the simultaneous depolymerisation of cellulosic carbohydrate polymers with ethanol production. Few workers have used a verified mathematical model to examine the most critical biochemical engineering aspects of ethanol production from cellulose. Extensive simulations of the simultaneous saccharification and fermentation of cellulose were conducted to identify the effect of operating conditions, pretreatment effectiveness, microorganism parameters and enzyme characteristics on ethanol production. The simulation studies have shown that the biomass-enzyme interaction play a dominant role in determining the performance of solid state fermentation in batch and continuous operating modes. The digestibility of the substrate and the cellulase enzyme dosage, specific activity and composition had a profound effect on ethanol yield. Some researchers have developed a low cost fermentation medium for ethanol production from biomass. The low cost medium containing 0.3% corn steep liquor and 2.5 mM $MgSO_4.7H_2O$ alongwith pretreated poplar was similar in performance to a nutrient rich medium. Besides its low cost, the medium consisted of components that are avilable on a commercial scale. Bioconversion of pentose sugars (Xylose etc.) and rice straw hydrolysate has also been studied using free and immobilized cells of *Candida shaetae* NCL 3501 in batch, fed batch and continuous culture conditions. *C. shaetae* NCL 3501 effectively utilized the sugars present in rice straw hydrolysate. Ethanol yields were higher with immobilized cells in all reactor types, i.e. batch, fed batch and continuous processes. Few workers have reported an increase in ethanol production by metabolic modulation of cellulose fermentation in *C. thermocellum*. Significant quantitative difference in ethanol yield alongwith repression in acetic acid production were observed in *Clostridium thermocellum* SS21 and SS22 in the presence of H_2, acetone and sodium azide. The use of a thermotolerant strain of *Kluyveromyces maxianus* IMB-3 has been studied in solid state fermentation of cellulose to ethanol. At maximum achievable cellulose concentration, ethanol production increased to 6.6 g/l at 45°C, representing 21% of the theoretical yield.

7. THE INDIAN ETHANOL INDUSTRY—IN TRANSITION

Although China is the largest producer of alcohol in the region, it is India which has the largest number of significantly sized plants in Asia that produce ethanol of 95-96% (v/v) concentration. The fate of the ethanol industry in India is linked to the sugar economy. Almost 50% of the distilleries are attached to sugar factories, with as many as 70% of them in the co-operative sector. In India, the distillery industry has shown remarkable resilience and progressiveness by adopting modern technology and trying to meet the environmental norms. The fact that almost 90% of the continuous fermentation plants in Asia are operating in India, is in itself a testimony to the aspiration of the industry to move to an international arena.

Cane Molasses is the single largest raw material. India produces almost 25% of the world's molasses. State levies and control factors often affect the availability and the price of molasses thereby creating a lopsided demand-supply situation. The cost of production is driven higher by these factors. Marketers around the world have understood that India is in fact 28 countries in the context of one. A number of controls are exercised by the individual states (provinces) on cross-border movement of molasses and ethanol. The tax on ethanol is determined by the state and not the Federal Govt., leading to vastly different tax levels in even neighboring states. This creates some regional imbalances within the country. For example, the states of Maharashtra and Uttar Pradesh have surplus molasses which is trading within the states at US$ 10 to 20 per MT. On the other hand, in the states of Madhya Pradesh and Kerala which are deficit in molasses, the price of molasses is as high as US$ 70 to 90 per MT. This made it prudent for the distillers in these states to put up ethanol plants based on grain and cassava.

Production methods across the region are traditional. Fermentation is done in batch mode. India took a lead in successful introduction of continuous ethanol feremention a decade ago. About 100 continuous plants have been set up all over the country during the last decade. The major reasons for this shift have been:

- Easier operation
- Higher efficiency of conversion,
- Consistent quality of product and
- Lower consumption of steam due to higher concentration of ethanol in the mash.

Lower steam consumption and consistently high qualities are the issues in distillation. The Asian region makes its liquor largely from molasses-based ethanol. With competition in the liquor industry, the requirement of a better quality of ethanol as base spirit is going up day by day. This has necessitated the installation of better distillation systems. Earlier, the level of instrumentation and controls in a distillation plant was fairly low. However, this has undergone a change and plants with better controls to achieve better and consistent quality are being put up. Several refining or re-distillation plants have been put in the last decade.

Out of the 900 million litres being consumed as beverage alcohol, 60% goes as cheap liquor and the balance for fine liquor production. Tax structure in each state shapes the supply of alcohol within the state and the price thereof. Beverage alcohol commands a premium in price of 20%-40% over industrial alcohol. Thus, the value-addition is higher for sugar mills that go in for beverage-grade alcohol. Considering the downstream blending and bottling unit, the return on investment is higher than the alcohol-based chemical manufacturing industry. However, State control over production of liquor restricts the growth of this alternative for sugar mills.

The size of the alcohol-based chemical industry is estimated to be in the region of US$ 1 billion. Through licensing and price controls on feedstock, Central and State governments encouraged the alcohol-based chemicals industry prior to 1993. In June 1993, the Govt. of India decontrolled movement and pricing of molasses to encourage this industry.

At present, there are over 200 alcohol-based products manufactured in India, the single largest alcohol-based chemical being acetic acid. Acetic acid itself is the building block for the manufacture of several other alcohol-chemicals like ethyl acetate (EA), butyl acetate (BA), acetic anhydride, etc. Ethyl alcohol is also used to produce ethylene and its downstream products. The Table 10 lists out the capacities of major alcohol based chemicals in the country:

Table 10: Alcohol based chemicals production in India

Alcohol chemicals	*Capacities (MT per annum)*
Acetic acid	290,000
Acetic anhydride	78,000
Vinyl acetate monomer	52,000
Pyridines, picolines and other derivatives	15,000
Ethyl acetate	65,000
Mono ethylene glycol	25,000
Amines, ethoxylates and others	50,000
Pentacrythritol	15,000
Ethyl/butyl acrylates	12,000
Glyoxal	8,000

Fuel ethanol, power alcohol are buzz words today. There are many reasons to justify the use of alcohol as a blend in petrol for automobiles. Environmental causes will prevail in utilizing one of the best sources of oxygenating agents in gasoline. Countries that are short of hard currency and/or have to import crude petroleum, the oxygenating agents see more reasons in doing so. In India, there are missions working on this aspect. The Ministry of Non-conventional Energy Sources, the regional Pollution Control Boards in India are following the leads in this regard. Ethanol can have a brighter future if the fuel-ethanol develops in the region. More stringent norms for environment could increase the cost of production and could restrict volumes of production. The issue needs to be addressed today. Molasses as a raw material, although available in plenty, could be leveraged for bringing down the cost of production and, hence, there could be a stronger linkage between sugar producing mills and distilleries for consistent molasses supply. Other raw materials are being looked at. As the trend for alcoholic beverages across the region is shifting to 'white liquors', the need to switch to 'grain-spirit' is greater. This will also balance out the use of molasses for industrial purposes. A review of local levies and taxes on this commodity could alter the scenario significantly. Technological upgradation could take a driver's seat in bringing down the cost of production and in ensuring improved and consistent quality of the product.

Production is expected to reach 1.9 billion litres this year (Table 11). The growth has been consistent at 10%. As the movement of molasses is restricted by various constraints and barriers, almost 60% of production of ethanol is limited to 4 states, that of Uttar Pradesh, Maharashtra, Tamil Nadu and Karnataka. Another state which is fast catching up is Andhra Pradesh.

Table 11: Ethanol Production in India

Year	*Production, million liters*
1999	1569
2000	1652
2001	1739
2002	1832
2003	1929

Almost 50% of the ethanol produced is consumed by the industrial sector for production of ethanol-based chemicals and the remaining 40-50% is utilized by the potable sector. It is expected that the pattern of utilization of ethanol may change only slightly in the coming years. However, with the Government removing control on price and allocation of molasses, the ethanol-based chemical industry is facing problems as the procurement price of alcohol has increased substantially in the last two years. One probable reason for the drop in industrial consumption is realisation of lower price for chemicals as a result of heightened competition. Also, due to depressed oil prices in the last decade, the petroleum route had been quite attractive. With crude prices rallying around US$ 30 per barrel, it is now competitive to use ethanol as feedstock. A spurt is already seen in this area. Consumption of ethanol is expected to rise significantly (Table 12), provided the growth in sugar production keeps pace and fuel ethanol programme takes shape.

Table 12: Ethanol Consumption in India

Year	*Consumption, million litres*
2000	1240
2001	1290
2002	1342
2003	1396
2004	1451

Some concerns which need to be kept in perspective are: Vagaries of nature could seriously influence the availability of ethanol impacting the prices adversely. This, in turn, would influence the downstream applications. The use of fuel ethanol would require production competitiveness amongst producers of ethanol.

FUTURE OUTLOOK

The "fuel ethanolization" of the world alcohol industry is set to continue. If all recently announced ethanol projects are implemented, total fuel ethanol production world-wide could grow to 31 billion litres by 2006 against somewhat below 20 billion in 2001 (Fig. 6). As a result, the share of fuel ethanol in 2005 could jump to almost 70% compared with around 63% in 2001.

Such a change is likely to have far reaching implications for the world alcohol sector. At a national level, such a growth can only be facilitated if plant sizes increase tremendously. This, in turn, would have repercussions on the feedstock market as well, where infrastructural adaptations would have to take place. At the same time, distribution systems would have to change. In countries where fuel ethanol receives financial assistance from the state (in the form of tax incentives or direct subsidies, etc.), fuel ethanol plants are likely to crowd out those producers that traditionally supply the industrial or beverage sector. This is so because fuel ethanol producers can fully utilise economies and are also able to cross-subsidise production for other markets. The USA is a good example of what to expect if fuel ethanol production takes off on a large scale. Today, fuel ethanol producers also dominate the beverage and industrial markets. However, the structure of the distilling industries in other countries is different and therefore the development path needs not necessarily be the same. However, the forces at work will be the same and

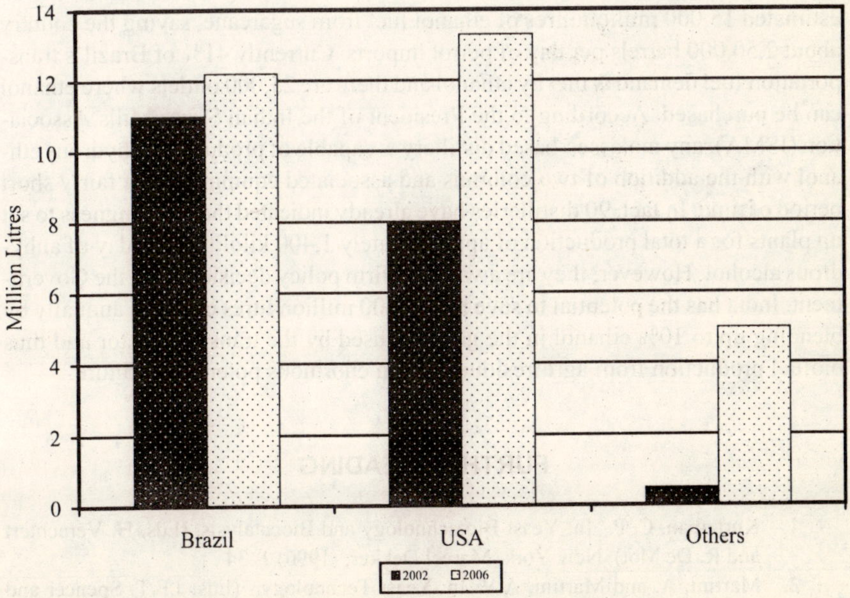

Fig. 6: World fuel ethanol production (2001 vs 2006)

something similar can therefore be expected. At an international level, the increased production and utilisation of fuel ethanol will make necessary an international exchange mechanism, which can help to stabilise the market at times of regional production shortfalls. This will be a very difficult operation. After all, fuel ethanol production in most countries is being subsidised in order to support domestic farmers. If large scale imports occurred, foreign farmers and ethanol industries would benefit from these support mechanisms. Nevertheless, an international exchange will be the only way to firmly establish large-scale fuel ethanol programmes in any one country. Biofuels are produced from renewable resources, supplies of which are dependent on the weather. This is a fundamental difference to non-renewables, such as oil and gas. As a result, the level of production cannot be fully controlled by mankind. On the other hand, being produced from agricultural crops, fuel ethanol production need not be concentrated in a handful of countries, as is the case in the oil sector. Therefore, it is unlikely that a typical importer/exporter relationship will develop. Instead, international trade will occur in order to compensate for temporary production shortfalls and, as such, will remain at a rather limited level relative to world production. There could be some countries that will start to regularly import more environmentally friendly fuel ethanol to be sold in the domestic market, However, these countries will remain the exception. Fuel ethanol, besides its environmental value, is and will remain first and foremost an instrument to support farmers. It is they, who in the first place, will profit from fuel ethanol programs and it is they who primarily lobby for the production and the utilisation of fuel ethanol. Roughly two-thirds of the world's ethanol production is used as fuel. As against this, despite being the world's fourth largest ethanol producer, India has not been able to use even one third of this production as transport fuel. This is unlike Brazil, which during 1998-99 produced an estimated 15,000 million litres of ethanol fuel from sugarcane, saving the country about 2,50,000 barrels per day of petrol imports. Currently 41% of Brazil's transportation fuel demand is met by ethanol and there are 25,000 outlets where ethanol can be purchased. According to the President of the Indian Sugar Mills Association (ISMA), any molasses-based distillery is capable of producing anhydrous ethanol with the addition of two columns and associated equipment in a fairly short period of time. In fact, 90 distilleries have already indicated their willingness to set up plants for a total production of approximately 1,400 kilolitres per day of anhydrous alcohol. However, they are awaiting a firm policy decision from the Government. India has the potential to save nearly 800 million litres of petrol annually by blending up to 10% ethanol in the gasoline used by the transport sector and thus biofuel production from agro-residues has an enormous potential in future.

FURTHER READING

1. Kurtzman, C. P. In: Yeast Biotechnology and Biocatalysis. (Eds. H. Verachtert and R. De Mot) New York, Marcel Dekker. (1990) 1–34 .
2. Martini, A. and Martini, A.V. In: Yeast Technology, (Eds. J.F.T. Spencer and D.M. Spencer) Berlin, Springer-Verlag. (1990) 105–123.

3. Soni, S.K. and Sandhu, D.K. Microbiology of Fermentation's. In: Biotechnology: Food Fermentation Microbiology, Biochemistry and Technology. (Eds. V.K. Joshi and A. Pandey) Volume I: Basic. Educational Publishers and Distributors, New Delhi, India (1999). 25-85.
4. Wayman, M. and Parekh, S.R. Biotechnology of biomass conversion, Fuels and chemicals from renewable resources, New Jersey, Prentice Hall. (1990).
5. Wood, W.A. Trends in the biology of fermentation for fuels and chemicals, New York, Plenum Press. (1981).

Alcohol Production from Cereals: A Case Study

V. SHEORAIN, R. BANKA *and* M. CHAVAN

Seagram R & D Centre, 128/3 Telco Road, Mohan Nagar, Chinchwad, Pune 411019

INDIA IS THE THIRD largest producer of cereals and it ranks second in wheat, rice and sorghum production in the world. In 1998, India produced 192.4 million tons of grain, half of which was rice. Coarse cereals contribute on average about 10% of the total quantity of cereals produced. This is largely due to less demand of these cereals both for human consumption and other applications. The situation is further compounded, because, affluent segment of the society consumes mainly wheat and rice. Therefore, coarse cereals such as sorghum and millets are neglected both by the farmers as well as agricultural scientists. In addition to this, a large quantity of some of these grains is damaged due to unseasonal rains, post harvest handling and poor storage facilities (Table 1).

Table 1: Wastage of grains from farm to consumer

Cereal	*MT / Year*	*% Damaged grain*
Rice	834,000	1.0
Sorghum	631,000	7.0
Millets (Bajra & Ragi)	567,000	5.4

There is hardly any demand for the damaged grains which leads to lower price realisation. The damaged grain can be processed for value added products, to partly cover the losses. The R & D Centre of M/s Seagram has developed a process of making good quality potable alcohol from damaged sorghum and work is in progress for the processing of other grains for ethanol production. In order to ensure constant supply of good quality raw material, the group has undertaken collaborative projects to increase the cultivation and productivity of coarse cereals. Seagram's R & D Centre is working with key National Agricultural Research Institutes (NARIs) and International Agricultural Institutes on breeding programmes to develop high yielding hybrids. While India may be the largest producers of grains, it ranks very low in terms of yield (Table 2).

As can be observed from Table 2, India is lowest among both the advanced countries as well as developing countries like China and Egypt. This means that there is a large scope to increase grain production in the country. Higher grain production will not only feed our population but will also be available for industrial applications.

Table 2: Grain yields of different countries

Country	*Average Yield (Kg/hac)*
France	6460
Germany	6110
USA	4620
Egypt	5610
China	4130
India	1750

1. USE OF GRAINS FOR ALCOHOL PRODUCTION

Alcohol can be produced by fermentation of any material that contains sugars either in the free form as in case of molasses or in the form of starch as in case of grains. However, grain starch will have to be hydrolysed to free sugars before it can be fermented to alcohol. This step is achieved by a combination of cooking the grain slurry and use of enzymes. Grain alcohol is much superior in quality compared to alcohol produced by fermenting molasses. However, the molasses alcohol contains high levels of sulphates and aldehydes. The former gives sulphurous odour and bad taste and latter may have deleterious effects on health if consumed in large quantities. Therefore, in developed countries, grain alcohol is used for potable purposes. However, in India, molasses alcohol is being used for manufacture of all types of alcoholic beverages despite its poor quality. World over, only rum is produced from molasses alcohol. The balance quantity is used by chemical industries as raw material for the production of more than 200 different organic chemicals and polymers including their downstream products. These products cater to a wide range of industrial and consumer products such as plastics, rubber, synthetic fibres, drugs and pharmaceuticals, pesticides, footwear, perfumes, flavours, dyestuffs, adhesives etc. Use of molasses alcohol for potable purposes makes it unavailable for industrial purposes and adversely affects industries using alcohol as resource material.

Molasses contains sucrose, which is easily fermented by yeast to produce alcohol, whereas, for alcohol production from grains, the starch present in the grain is hydrolysed using costly enzymes and heat, to get fermentable sugars. This additional step of starch hydrolysis and low cost of molasses as compared to grains, makes alcohol produced from molasses cheaper by 25-30%. Although molasses alcohol is cheaper to produce, it is very difficult to treat its by-product thus it pollutes the environment. Typical yield of alcohol from molasses is 225 l/tonne. Considering molasses contains 45% sugars, only 70-75% of the sugars get fermented to alcohol. So remaining sugars go into spent wash, thereby, effluent with high organic matter. There are more than 250 distilleries in India with installed capacity of 2000 million litre alcohol. They generate about 10,000 million litres of spent wash which has BOD in the range of 40,000 to 60,000 mg/l. Most of this acidic spent wash is being disposed of without proper treatment. Even in the distilleries with modern waste treatment plants, although BOD and COD is reduced, the effluent remains black in colour and has foul odour.

On the other hand, in a typical grain fermentation the alcohol yield is more than 360 l/tonne of grain, containing 60% starch which corresponds to 85% of conversion of sugars to alcohol. Moreover, the remaining sugars/starch do not go into the effluent because the stillage is dried. The dried residue is called Distiller's Dried Grain and Solubles (DDGS) which is an excellent ingredient for animal feed. Seagram's state-of-the-art grain distillery at Nashik is zero polluting.

2. PROCESS OF ALCOHOL PRODUCTION USING GRAINS

Ethanol production from grain involves milling and hydrolysis of starch grains to release fermentable sugars followed by inoculation with yeast. Chemically starch is a polymer of glucose. Yeast can not use starch directly for ethanol production. Therefore, grain starch has to be completely broken down to glucose by a combination of two enzymes viz. amylase and amyloglucosidase before it is fermented by yeast to produce ethanol. The biochemical reactions and process involved in starch hydrolysis and fermentation are shown in Figs. 1 & 2. Alcohol so produced is distilled from fermented broth. The remaining stillage is processed to produce DDGS (Fig. 3).

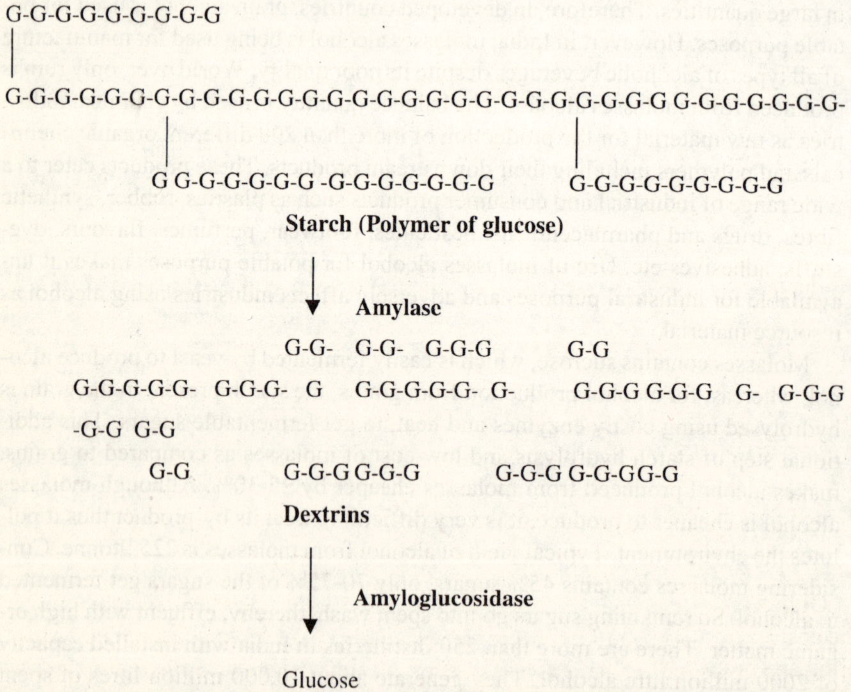

Fig. 1: Enzymatic hydrolysis of starch to glucose

$$(C_6H_{10}O_5)_n + n\,H_2O \longrightarrow n\,C_6H_{12}O_6$$

Starch Water Glucose

$$C_6H_{12}O_6 \longrightarrow 2\ C_2H_5OH + 2\,CO_2$$

Glucose Ethanol

Fig. 2: Chemical equation for conversion of starch to glucose and glucose to ethanol.

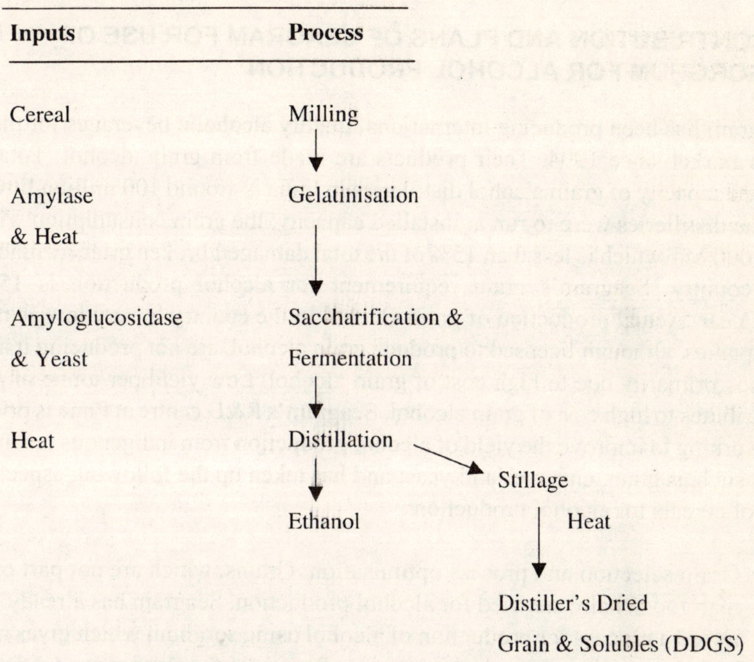

Fig. 3: Flow chart of ethanol production from cereal

Any cereal can be used for alcohol production. However, the yield of alcohol production varies depending on the starch content of the grain (Table 3).

Table 3: Yield of alcohol from various cereals

Cereal	*% Starch*	*Alcohol Yield (l/MT)*
Sorghum	63	447
Rice	72	511
Broken Rice	60	426
Bajra	58	411

Grain cost contributes about one third to the cost of grain alcohol. Therefore, the cost and availability of grain close to the distillery is very important for considering any proposal for grain alcohol distillery. Table 4 gives suitability of grains in different regions of the country.

Table 4: Availability of cereals for alcohol production in different regions of the country

Region	*Available grains*
North	Broken rice & Bajra
West	Sorghum
South	Broken rice & Ragi
East	Broken rice

3. CONTRIBUTION AND PLANS OF SEAGRAM FOR USE OF SORGHUM FOR ALCOHOL PRODUCTION

Seagram has been producing international quality alcoholic beverages for the Indian market since 1994. Their products are made from grain alcohol. Total installed capacity of grain alcohol distilleries in India is around 100 million litres. If all the distilleries were to run at installed capacity, the grain consumption will be 285,000 MT which is less than 15% of the total damaged/broken grian available in the country. Seagram's grain requirement for alcohol production is 15,000 MT/Year. Actual production of grain alcohol in the country is very low as many companies, although licensed to produce grain alcohol, are not producing it at all. This is primarily due to high cost of grain alcohol. Low yield per tonne of grain contributes to high cost of grain alcohol. Seagram's R&D centre at Pune is primarily working to improve the yield of alcohol production from indigenous raw materials such as grain, enzymes and yeast and has taken up the following aspects for use of cereals for alcohol production:

- Grain selection and process optimisation: Grains, which are not part of human food chain, are tried for alcohol production. Seagram has already optimised a process for production of alcohol using sorghum which gives much higher yield than the industry average. Processes for alcohol production using broken rice is being optimised by the group.
- Backward linkages with farmers: Farmers are encouraged to grow grains suitable for fermentation. Seagram along with seed companies and agriculture research organisations plans to educate the farmers to use good seeds and agronomic practices to increase the yield. The company plans to buy grains directly from farmers or their co-operatives to pass on maximum benefit to them.
- Breeding of new varieties/hybrids: In an attempt to further increase the output of Sorghum cereal, the company has initiated a collaborative project with National Research centre for Sorghum (NRCS) and International Crop Research Institute for Semi Arid Tropics (ICRISAT) to undertake breeding programmes.
- The company is also working for the upgradation of key by-products. This is important to reduce the gap between cost of grain alcohol and molasses alcohol.
- The company has already filed three patent applications for the processes developed for alcohol production from cereal grains.

Cereals have good potential for being used as a raw material for ethanol production. The process for alcohol production using cereals although little expensive is still preferable as it is environment friendly and quality of produce is better as compared to the produce from molasses. Use of cereals for ethanol fermentation will create market for damaged grain which will help farmers realise a good price for their produce. Ethanol producing companies, research institutions and the Government can co-ordinate with farmers to strategically develop value added utilisation of damaged and coarse cereals.

FURTHER READING

1. Galliard, T. and Bowler, P. Morphology and Composition of starch. In: Starch: Properties and potential. (Eds. T. Galliard), John Wiley & Sons. UK. (1987).
2. Gandi, G., Chandnani, K., Dhale, A., and Mukhopadhyay, S. Pollution control in sugar industries: An overview. *Chemical Weekly*, **44**, (1999) 167.
3. Paterson, A. Production of fermentable extracts from cereal & fruits. In: Fermented beverage production. (Eds. A.G.H. Lea and J.R. Piggot), Blackie Academic & Professional, London. (1995).
4. Patser report . Technology Evaluation in Alcohol Industry. DSIR, Govt. of India. (1994).
5. Powers, W.J., Van Horn, H.H., Harris, B. Jr. and Wilcox, C.J. Effects of variable sources of distillers dried grains plus solubles on milk yield and composition. *J. Dairy Sci.*, **78**, (1995) 388.

Ethanol Fermentation Technology of *Zymomonas Mobilis*

P.S. PANESAR[1], S.S. MARWAHA[2] and J.K. ARORA[3]

[1] *Department of Food Technology*
Sant Longowal Institute of Engg. & Technology
Longowal-148 106, Punjab (India)
[2 & 3] *Punjab State Council for Science & Technology*
Chandigarh-160 019, India

THE INCREASE IN THE oil price and possibilities of supply shortfalls have prompted an extensive evaluation of alternative technologies for the production of liquid fuel. Microbial processes have been found to be useful processes for the production of alternate energy products from renewable resources. Ethanol is being considered to be most promising to use it as an alternate liquid fuel, since it can be readily produced by fermentation of agriculture based renewable materials like sugarcane juice, molasses, potatoes, corn, barley etc. As early as 1890, the value of the use of ethanol as substitute for gasoline was indicated.

Today, Brazil, a developing country is meeting its substantial energy requirements by using bioethanol and getting technological independence from oil. During 1989, in Brazil, 40% of vehicles were operating on ethanol (95% ethanol, 5% water) and 60% of vehicles on a blend of 78% gasoline and 22% ethanol as their fuel. Currently, yeast *Saccharomyces cerevisiae* is used all over the world as the major ethanol producing organism. Despite its extensive use, it has number of disadvantages such as high aeration cost, high biomass production, low temperature and ethanol tolerance. Therefore, a number of microorganisms are being tested for the production of ethanol during the past few years. Out of these, *Zymomonas mobilis* has emerged as a potential alternate to presently used yeast for ethanol production. The comparative attributes for ethanol production by *Zymomonas* and yeast are given in Table 1. The studies have demonstrated that it has following advantages over yeast, a conventional organism used for alcohol production:

- higher specific rate of sugar uptake and ethanol yield,
- lower biomass production as compared to yeast
- higher volumetric sugar uptake and ethanol productivity
- non- requirement of controlled addition of oxygen to maintain the cell viability.
- amenability to genetic improvement

Zymomonas mobilis is a facultative anaerobic gram negative bacterium, first isolated in tropical countries from alcoholic beverages like African palm wine, the Mexican "pulque" and also as a contaminant of cider (cider sickness) or beer in European countries. It is one of the few facultative bacteria degrading glucose by the Enter-Doudoroff (ED) pathway usually found in strictly aerobic microorganisms such as *Pseudomonas*. Although it is classified as an anaerobic microorganism, *Zymomonas* grows well under aerobic conditions also.

The first report on the great potentialities of *Z. mobilis* for ethanol production was published by an Australian group of scientists. At that time, the petroleum crisis had led the developed countries to search for alternative fuel from renewable resources. The Australian group has demonstrated the advantages of the bacterium over the yeasts traditionally used for alcoholic fermentation.

Table 1: Comparative attributes for ethanol production by *Zymomonas* and yeast.

	Z. mobilis	*Yeast*
Conversion of sugar to ethanol (%)	96	96
Maximum ethanol concentration (%)	12	12
Ethanol productivity rate ($gg^{-1}h^{-1}$)[a]	5.67	0.67
Volumetric ethanol productivity ($gg^{-1}h^{-1}$)[b]	200	29
ATP yield (per mole glucose) (Enter-Doudoroff vs. Embden-Meyerhof)	1	2
Sugar tolerance (%)	>40	>40
pH range for ethanol production	3.5-7.5	2-6.5
Optimum temperature (°C)	25-30	30-38

[a]Batch fermentation utilizing 10% glucose
[b]Continuous culture with cell cycle

1. SUBSTRATES FOR ETHANOL PRODUCTION BY *ZYMOMONAS MOBILIS*

1.1 Glucose and Fructose

D-glucose and D-fructose have been reported to be transported by facilitated diffusion in *Z. mobilis*, but, to attain rapid growth, there is need to metabolize considerable amount of sugar. An NMR study has shown that the rate limiting steps in the sugar metabolism is the conversion of glucose 6-phosphate to 6-phosphogluconate and of 3-phosphoglycerate to 2-phosphoglycerate. Pyruvate decarboxylase (pdc) found mainly in yeast and molds, has also been reported to occur in *Z. mobilis*. It catalyzes the non-oxidative decarboxylation of pyruvate to produce acetaldehyde and carbon dioxide. Two alcohol dehydrogenase (adh) isozymes are reported to be present in *Z. mobilis* which catalyze the reduction of acetaldehyde to ethanol during fermentation, accompanied by the oxidation of $NADH^+$ to NAD^+. In *Z. mobilis*, adh I is a zinc dependent enzyme, whereas adh II an unusual enzyme was found to contain iron and not zinc.

Fructose is phosphorylated by a constitutive kinase which is highly specific for fructose and ATP. The phosphorylation of fructose has been reported to be regulated by the concentration of glucose . The enzyme gets strongly inhibited by glucose and to a lesser extent by glucose-6-phosphate. The V_{max} value of fructokinase is reported to be similar to that of glucokinase. The ethanol yield obtained from fructose in batch fermentations was generally lower (90%) than that from glucose (95%). The formation of the byproducts such as dihydroxyacetone, mannitol, glycerol, sorbitol, acetoin, acetaldehyde, lactic acid and acetic acid may be the cause for the reduction in the ethanol yield. The formation of mannitol and sorbitol followed the uptake of fructose, while dihydroxyacetone and glycerol were reported to accumulate during the later stage of the fermentation. Moreover, the decreased cell yield is also characteristic of its growth in fructose.

Most of the fermentation work using *Z. mobilis* has been done on glucose. The potential of *Z. mobilis* over traditional yeast fermentation and in a continuous cell recycle system, a volumetric ethanol productivity of 120 g/l/h has been reported. Further, the effect of temperature and ethanol concentration on the maintenance and yield co-efficient of *Z. mobilis* has also been studied. The most important breakthrough was achieved by the use of flocculent strains in continuous reactors. Very high ethanol productivities were obtained with these systems, usually at an ethanol concentration around 50 g/l. High cell concentration in the reactor was obtained by recycling the flocculent cells with a simple internal or external setter. The optimum temperature of 37°C and broad pH range of 4.4-6.0 are reported optimal for maximum ethanol productivity and ethanol yield in continuous system using immobilized *Z. mobilis* on glucose .

The effect of the initial concentration of glucose, fructose and sucrose over a range of 100-250 g/l on the fermentation kinetics in batch cultures of *Z. mobilis* NRRL B-4286 at an adapted initial inoculum has been reported. It is observed that at increasing concentrations of substrate, parameters related to growth are more rapidly and strongly affected than those related to ethanol production.

The optimization of fermentation conditions using free and immobilized cells of *Z. mobilis* ATCC 10988 have also been reported. The temperatures of 30° and 35°C have been observed optimum for free and immobilized cells, respectively, whereas, a pH 7.0 for both the systems. Moreover, the immobilized cells could tolerate higher concentrations of sugars than the free cells. The role of different nitrogen sources on ethanolic fermentation of glucose by *Z. mobilis* has been studied. Among several organic and inorganic nitrogen sources tried, urea has been reported to be better nitrogen source than the ammonium sulphate.

1.2 Sucrose

The ability to hydrolyze sucrose is not a common feature of *Z. mobilis*, but it appeared to be strain dependent. The use of sucrose for ethanol production by *Z. mobilis* present two different problems. The first at the stage of cleavage of the sucrose molecule by the enzyme levansucrase and the second in the direct oxidation of glucose linked to the reduction of fructose to form sorbitol, or the polymerization of fructose to polyfructo-oligosaccharides. These problems result in lowering the ethanol yield from sucrose compared with that obtained with the conversion of starch or glucose.

During fermentation of sucrose by *Z. mobilis*, three types of transfructosylations occur which result in the formation of free fructose, oligosaccharides and higher polymers of fructose and levan, where, water, sucrose and levan or polyfructose, respectively, could act as an acceptor. In fermentation of sucrose to ethanol, *Z. mobilis* has been reported to produce as much as 11% sorbitol, 70% small molecular weight oligomer and 2% levan from original carbon source. Levan is a β (2-6) linked fructose polymer with a molecular weight of 10^7. The temperature and pH were observed to have a profound effect on ethanol production by *Z. mobilis*. It has also been reported that in the fermentation of sucrose to ethanol by *Z. mobilis*, the ratio of fructose to sorbitol varied significantly with culture conditions. A pH around 5.0 favoured sorbitol formation whereas a pH around 6.0 supported fructose release.

The fermentation pattern of *Z. mobilis* ATCC 39676 at high sucrose concentration (200-400g/l) revealed that the efficiency of sucrose hydrolysis dropped from 94 to 78.6%, whereas, the efficiency with which the hydrolysed products were converted to ethanol decreased from 94 to 43%.

The formation of the non metabolizable by-products levan, sorbitol, and fructo-oligosaccharides from sucrose based substrates have hampered efficient ethanol production by *Z. mobilis*. The neutralization of levan formation by increasing the fermentation temperature to 35°C have been reported. However, the problem of formation of sorbitol and oligosaccharides existed as such.

Further, the coupled microbe - immobilized invertase system for the sucrose hydrolysis has been reported better than the two stage (separate hydrolysis and fermentation) system for ethanol production.

1.3 Sugarcane and Beet molasses

Molasses, the noncrystallizable residue remaining after sucrose purification has additional advantages as it is readily available and cheap feedstock for ethanol production. However, a number of components in the molasses can cause inhibition of growth and fermentation. It has been postulated that herbicides and some minor components of molasses such as phenol and indole, thermal degradation products and/or some acids, nitrites and chlorphenolate which accumulate in molasses through the extended use of chemical fertilizers are responsible for inhibition of ethanol production by *Z. mobilis*. The relatively high levels of some inorganic salts, especially the chlorides of potassium, calcium and magnesium also exhibit significant inhibitory effects.

The use of membrane technology to desalt the molasses and the use of desalted molasses for ethanol production by *Z. mobilis* has been reported. Fermentation studies using desalted molasses showed faster rates of ethanol production with overall higher yields. It has been reported that the fermentation of sugarcane syrup and A-molasses by *Z. mobilis* require the addition of small amounts of magnesium sulfate and diammonium phosphate. Further, it has been observed that an increase of B-molasses or C-molasses from 250 g/l to 300 g/l in the fermentation medium lead to decreased fermentation efficiency. This reduction in ethanol yield to 45.7% was resulted due to the increase of the Cl^{-1} concentration from 5.91 g/l to 6.75 g/l.

Potassium, sodium, magnesium and calcium exhibit lesser effects. The higher molasses concentrations (300 g/l) also lead to fructose accumulation.

The potential of four hydrolyzed sugar beet substrates from the sugar industry (syrup, crystallizer effluent 1, crystallizer effluent 2 and molasses) for ethanol production using *Z. mobilis* SBE 15 have also been reported . All substrates at 20% (w/v) concentration has been reported to show nearly complete sugar conversion (94.9%), good growth and ethanol production (40 g/l). Moreover, nutrient supplementation have shown that reproducible growth and ethanol production could be achieved on the four substrates supplemented with only magnesium sulfate or additionally with ammonium sulphate.

The ethanol fermentation kinetics of *Z. mobilis* ZM4 on sucrose, cane juice and molasses show that cane sugar is fermented more efficiently, yielding 0.50 g/g ethanol at 185 g/l sugar concentration, whereas, lower ethanol yields (0.39 g/g) from molasses medium at similar sugar concentration are achieved.

The effect of nutrient addition, molasses clarification, sugar concentration (55-145 g/l) and temperature (3 0-40°C) on growth rate and ethanol production using *Z. mobilis* CP4 in batch fermentation process has also been studied. The best yields are obtained using clarified and supplemented molasses. Growth rate and ethanol production are reported to be maximum at 34°C and sugar concentration of 110 g/l. Further, the studies on the fermentation of unclarified molasses by *Z. mobilis* have depicted 30°C as optimum temperature for efficient fermentation. The higher temperatures (35-40°C) have been reported to decrease the fermentation efficiency as well as the ethanol tolerance of the bacterium.

1.4 Other substrates

Attempts have been made to widen the substrate range of *Z. mobilis* for ethanol production. It has been reported that the metabolism of mannitol by *Z. mobilis* is oxygen dependent and results in limited production of ethanol, but, increases production of lactic acid. The ethanol production from wood-derived cellulose hydrolysates has also been reported. It has been observed that the presence of lignin and suspended solids did not inhibit the ethanol fermentation by *Z. mobilis*. With this bacterium, the enzymatically saccharified aspen having 50g/l glucose was fermented in a batch culture with 96% ethanol yield.

Various attempts have been made to prepare *Z. mobilis* for replacing yeast in the ethanol industry using renewable resources such as corn/maize, potatoes, maltrin, wheat, milo, cassava and sago. From these attempts, two processes of ethanol production by *Zymomonas* have been developed, the "Glucotech Process" by the Australian research group at the University of Queensland and the "Bio-Hol Process" by the Canadian research group at the University of Toronto.

The ethanol production from corn using inoculum cascading of *Z. mobilis* ATCC 39767 has been reported. The industrial trials have been carried out using wet and dry milled corn as well as corn, milo, and wheat mixtures as feedstocks. Corn is found easier to ferment to ethanol than milo. At production rates in excess of 0.25% /h. ethanol, fermentation was completed within 36 to 40 h in comparison to a typical yeast fermentation of 52 to 70h. A mixed culture of *Endomycopsis fibuligera* NRRL 76 and *Z. mobilis* ZM4 for direct fermentation of cassava starch to ethanol has also been tested. It was found that mixed culture could directly and more efficiently ferment cassava starch (22.5%,w/v) to ethanol (10.5%,v/v) than the monocultures.

2. FACTORS AFFECTING ETHANOL PRODUCTION BY *ZYMOMONAS MOBILIS*

2.1 Sugar concentration

It has been reported that the extracellular osmotic pressure of the glucose solution may rapidly be balanced by corresponding intracellular sugar concentrations. The reduced water activity was found to have only minor effects on the E D pathway enzymes. About 40 strains of *Z. mobilis* were found to grow in 20% glucose and most strains grow in media containing 30 or 40% glucose. However, the salt tolerance of *Z. mobilis* is rather low; 71% of the strains tested grew in the presence of 1% NaCl, whereas, none could grow in the presence of 2% NaCl. High sugar concentrations decrease the total water concentration and exerted osmotic pressure.

2.2 Ethanol concentration

Ethanol has been reported to affect the integrity of biological membranes by intercalation into the lipid bilayer and the leakage of co-factors and co-enzymes through the plasma membrane as a consequence of ethanol damage. The ethanol induced leakage is more rapid at elevated temperatures. The leakage of nucleotides has also been reported to occur in the presence of increased concentration of ethanol. The damage to the barrier properties of the cell membrane appeared to be of primary importance in the inhibition of growth and inhibition of ethanol fermentation. Further, it has been observed that *Z. mobilis* could continue to grow in the presence of 7-8% ethanol at 30°C, while at higher temperatures (37-40°C), growth is inhibited by 2-3% ethanol. Thus the basis for decreased ethanol production at elevated temperatures appeared to be the increased potency of ethanol at these elevated temperatures.

2.3 Temperature

The membrane composition of *Z. mobilis* has been reported to be changed dramatically in response to growth temperature. The increase in temperature, resulted in the decline of the proportion of vaccenic acid with an increase in myristic acid. It has been further reported that the leakage of magnesium, nucleotides (A_{260}), and proteins increased with increasing assay temperature in cells grown at 30°C. The leakage of proteins into buffer at 45°C roughly correlate with the thermal inactivation of cells grown at different temperatures and provided a possible mechanism for cell inactivation. The protein loss from cells is interpreted as a disruption of membrane integrity. Moreover, protein leakage at 45°C is most rapid in cells grown at 41°C, consistent with their acute sensitivity to thermal inactivation. The minimal and intermediate protein leakage is observed in cells grown at 30°C and 20°C, respectively.

The optimal temperature for ethanol fermentation by *Z. mobilis* is around 30°C, whereas the higher temperatures are found to reduce the fermentation efficiency and the maximal level of ethanol production. The maximum level of ethanol produced at 40°C is found half vis-à-vis at 30°C. It has been reported that an increase in temperature above 30°C, increases the sensitivity of cells to ethanol. The reverse is true because ethanol accumulates the thermal tolerance of *Z. mobilis* rapidly declines. Thus, it has been reported that temperature and ethanol concentration have synergistic effect on ethanol production by this bacteria.

2.4 Aeration

Aeration of *Z. mobilis* cultures lead to diversion of reducing power and a consequent decrease in ethanol production (Table 2). The growth of *Z. mobilis* is strongly inhibited when aeration rates are increased . *Z. mobilis* was found to posses a NADH and NADPH-oxidase located in the cell membrane which catalyze the oxidation of NAD(P)H. Therefore, it was assumed that the oxygen gets reduced to water by this reaction. In addition to the catalase, super-oxide dimutase and peroxidase are also found in *Z. mobilis*. The transfer of electrons via the respiratory chain has not been coupled with oxidative phosphorylation. Under aerobic conditions, a switch over from ethanol plus carbon dioxide to the formation of acetaldehyde, acetoin, and other byproducts has been observed.

Table 2: Influence of oxygen on growth and product formation of *Z. mobilis*.

*Culture conditions**	*Ethanol (g/l)*	*Acetaldehyde (g/l)*	*Growth yield (g cell dry wt/mol glucose)*
Anaerobiosis	63.4	0.24	2.3
Aerobiosis	20.1	1.31	1.4

* Continuous cultures with 14% (w/v) glucose.

2.5 Product inhibition

Various theories exist with respect to product inhibition, as *Z. mobilis* produces two major products in ethanol and carbon dioxide. The most common has been reported to be ethanol inhibition theory. It has been reported that ethanol inhibition is responsible for the final ethanol obtainable, as the same results in decrease in the cell growth of *Z. mobilis* with the increase in ethanol concentration. It results in decrease in biomass formation and increase in residual substrate concentrations. An extractive fermentation system has been reported to greatly improve the performance of *Z. mobilis*. The, ethanol concentration as reported earlier causes damage to the barrier properties of the cell membrane which adversely effect the ethanol production capability of the microbe.

The second product and its behavior with respect to high ethanol productivity has been neglected owing to the dominance of the ethanol inhibition theory. However, It has also been suggested that the bicarbonate ions (due to CO_2 aqueous) are responsible for the destabilization of the membrane in complexing with the strongly acidic proteins and possibly charged phopholipids or lipopolysaccharides head groups, whereas CO_2 (aqueous) influenced the fluidity. The combined action caused the inhibition in substrate utilization efficiency by mainly blocking glucose uptake.

3. STRATEGIES FOR STRAIN IMPROVEMENT OF *ZYMOMONAS MOBILIS*

Industrial strain improvement plays a central role in the commercialization of microbial fermentation processes. The strain improvement depends upon the nature, quality and quantity of the product desired. *Z. mobilis,* being a prokaroyte, is more amenable to genetic manipulations as compared to yeast. The following strategies of strain improvement have been applied to *Z. mobilis* to overcome its various limitations so that the bacterium can be used for ethanol production at large scale.

3.1 Mutagenesis

Several mutagenic agents have been tested for their general effectiveness on *Z. mobilis* to obtain high ethanol yielding mutants. The various studies reveal that UV light and N-Methyl-N'-Nitro N-Nitrosoguanidine (NTG) are effective mutagens for *Z. mobilis*. The isolation of auxotropic mutants of *Z. mobilis* by using NTG and the minimal medium for their isolation have been reported. Through mutagenesis with nitrosoguanidine, an ethanol tolerant strain ZM 481 has been developed. and used in continuous cell recycle to increase the steady-state ethanol level from 65 to 85 g/l. With the same chemical mutagen, another strain of *Z. mobilis* ZM 482 has been isolated which could tolerate higher concentration of Na^+, Mg^+ and Cl^- ions. The resultant strain helped in preventing the desalting of such ions from fermentation media and made the fermentation process more economical.

A *Z. mobilis* mutant deficient in the utilization of fructose for the growth and ethanol formation is isolated with the use of NTG. The fructose negative mutants of *Z. mobilis* are of potential use for the enrichment of fructose in high fructose corn syrup (HFCS) or invert sugar. Another successful use of this mutagen lies in the isolation of an osmotolerant mutant SBE 15 of *Z. mobilis* ZM4, which showed successful growth and ethanol production when cultivated on hydrolyzed sugar beet molasses.

A stable thermotolerant mutant of *Z. mobilis* ZM 4 is isolated and found to possess almost normal cell morphology and better ethanol tolerance than parent strain. The mutagenesis with NTG and isolation of mutants which showed tolerance to high temperature, salt and alcohol concentration have been developed.

The isolation of mutants defective in levansucrase and found to produce more ethanol than parent strain have been reported. Glucose-negative and fructose-negative mutants are also reported to be isolated by D-cycloserine-mediated enrichment method. A mutant of *Z. mobilis* is reported to use mannitol as the sole carbon source due to the presence of a newly evolved mannitol dehydrogenase. This is shown to have evolved from mutation in the *adh* A gene, thus expanding the substrate range of the wild type to include mannitol.

3.2 Recombinant-DNA technology

Large scale ethanol production by *Z. mobilis* has been restricted, because of its substrate range limited to glucose, fructose and sucrose. By genetic approaches, efforts have been made to increase its substrate range to include industrially attractive feed stocks, such as whey, starch and cellulose.

The occurrence of natural plasmids in *Zymomonas* strains has been reported by several groups. Although not many functions have been assigned to any plasmid, yet a high degree of homology has been demonstrated between plasmids from strains isolated from widely differing locations. Native plasmids of *Z. mobilis* strains ZM3 and ZM4 has been found identical and approximately 70 and 30 kb in size. These plasmids are also found in strains ZM1 and Ag11. The strains of ZM1 have also been found to contain three low molecular weight plasmids approximately 1.7, 1.9 and 2.5 kb. The presence of three 1.7 kb species is also reported in *Z. mobilis* B806.

Suitable cloning vectors for *Zymomonas* have been developed which are categorized into three broad groups : broad host range plasmids, fused *E. coli- Z. mobilis* plasmids and modified broad host range vectors which further incorporate *Z. mobilis* features, such as promoter functions. The low molecular weight plasmids of *Z. mobilis* have been used to develop cloning vehicles for *Zymomonas*. They have been fused with small *E. coli* plasmids such as pACYC184 and pBR 325 to produce novel shuttle vectors. A set of expression vectors are constructed for cloning and expression of heterologous genes in *Z. mobilis* under the control of the pdc promoter.

Several *Zymomonas* strains can act as recipients for a number of broad host range plasmids. However, cloning vehicles commonly used in *E. coli* such as pBR 322 and phage vectors can not be transferred and/or maintained in *Z. mobilis*. The plasmids most commonly used for cloning in *Zymomonas* are a series of broad host range plasmid cloning vehicles derived from plasmids RSF 1010 and RP 4, but, their stability in general vary. The broad host range, high copy number of RSF 1010-derived vectors such as pkT 210 and pkT 230 have been found to be more stable in *Z. mobilis* than RP-4 derived plasmids. A method of transformation of plasmid, pNSW601 by modifying the $CaCl_2$ method has been developed and a frequency of 5 x 10^3 transformants/μg DNA has been recorded. However, this method failed to work for other plasmids. The other transformation systems using calcium chloride and electroporation technique have been developed and a high transformation frequency of 1.0 x 10^7/μg using electroporation technique has been reported.

Attempts to transfer appropriate hydrolase genes, in order to grow on novel substrates such as starch, cellulose, lactose have also been made. The lac operon encoded by transposon Tn 951 present on the plasmid pGC 91.14 is expressed in *Z. mobilis*. A new strain capable of producing ethanol from lactose is reported although the very low rate of ethanol production has been recorded. The endogluconase gene from *Bacillus subtilis* is introduced into *Zymomonas*, but no endoglucanase activity detected in the culture. However, the cellulase gene from *E. chrysanthemi* coding for endoglucanase cloned into *Z. mobilis* with the help of plasmid RP 4, expressed 40% cellulase activity in culture media. The cloning of glucoamylase gene from *Aspergillus niger* is also attempted, but, stable conjugants are not obtained in *Z. mobilis*. Also α-amylase with pkT 210 and the recombinant plasmid pGNB 6 had been transferred into *Z. mobilis*.

D-xylose catabolic genes from *Xanthomonas* are subcloned and the plasmid, pND90 is introduced into *Z. mobilis*. The levels of expression for D-xylose permease and D-xylose isomerase is 12-fold and 2-fold lower, respectively, whereas, xylulokinase is approximately the same when compared to the expression levels in *Xanthomonas*. The successful cloning of the β-carotene biosynthetic gene into *Z. mobilis* has also been reported. The substrate fermentation range of *Z. mobilis* ATCC 39676 is expanded to include L-arabinose by inducting genes encoding L-arabinose-isomerase, L-ribulokinase, L-ribulose-phosphate-4-epimerase, transaldolase and transketolase of *E. coli.* The engineered strain with plasmid pZB 206 grew on arabinose as C-source and produced ethanol at 98% of theoretical yield. A plasmid pZB5 carrying genes encoding enzymes for xylose metabolism into *Z. mobilis* has also been transferred. Recently, studies have been reported on the physical mapping of the genome of *Z. mobilis* ZM4 and localization of genes on the map has been carried out. Despite these reports on the introduction of catabolic genes into *Z. mobilis* , the expression of these foreign genes, however, is not high enough to allow growth on these substrates at a reasonable rate. Attempts to improve gene expression in strains of *Z. mobilis* is in progress.

Thus it may be summed up that the energetics of alcohol improvement strategies are advancing rapidly as the main inhibitory factors causing hindrance in alcohol production by *Z. mobilis* have been identified and studied. Ethanol, elevated temperatures and high substrate concentrations have been regarded as synergistic in their inhibition of ethanol fermentation. The *Z. mobilis* being a prokaryote and amenable to advanced engineering techniques, efforts are on to improve its substrate range to make it an alternative to yeast for commercial scale production of ethanol.

FURTHER READING

Benschoter, A.S. and Ingram, L.O., Thermal tolerance of *Zymomonas mobilis* temperature induced changes in membrane composition. *Appl. Environ.* Microbiol, **51** (1986) 1278.

Browne, G.M., Skotinicki, M.L., Goodman, A.E. and Rogers, P.L., Transformation of *Zymomonas mobilis* by a hybrid plasmid. *Plasmid*, **12** (1984) 211.

Buchholz, S.E., Dooley, M.E. and Eveleigh, D.E., *Zymomonas*—an alocoholic enigma. *Trends Biotechnol*, **5** (1987) 199.

Doelle, H.W., Kirk, L., Crittcnden, R., Toh, H. and Doelle, M.B., *Zymomonas mobilis*—science and industrial application. *CRC Crit. Rev. Biotechnol*, **13** (1993) 57.

Esser, K. and Karsch, T., Bacterial ethanol production: advantages and disadvantages. *Process Biochem*, **17** (1984) 116.

Gunasekaran, P., Karunakaran, T., Kamini, N.R. and Mukundan, A.G., Current status and prospects of an ethanol producer, *Zymomonas mobilis*. *Indian J Microbiol*, **30** (1990) 107.

Ingram, L.O., Carey, V.C., Dombek, K.M., Holt, A.S., Osman, Y.A. and Walia, S.K., Biochemical and genetic improvement of *Zymomonas mobilis*. *Biomass*, **6** (1984) 131.

Lam, C.K., Mullan, P.O. and Eveleigh, D.E., Transformation of *Zymomonas mobilis* by electroporation. *Appl Microbiol Biotechnol*, **39** (1993) 305.

Rogers, P.L., Lee, K.J., Scotnicki, M.L. and Lee, J.H., Recent developments in the *Zymomonas* process for ethanol production. *CRC Crit Rev Biotechnol*, **1** (1984) 273.

Singh, A. and Jain, V.K., Fermentation kinetics of *Zymomonas mobilis* on sucrose and other substrates. *Indian J. Microbiol.*, **34** (1994) 205.

Viikari, L., Carbohydrate metabolism in Zymomonas. CRC *Crit Rev Biotechnol*, **7** (1988) 237.

Biotechnology in Agro Industries: Role of Safety and Quality

S. SAREEN

Bureau of Indian Standards (BIS)
Plot No. 58, 1st Floor, Neelam Batra Road,
Faridabad, Haryana.

WITH INCREASING WORLD POPULATION, there is a growing demand for food products. Need is there to grow and process more food supply to meet the consistently growing demand. A number of technologies have been developed through the years to cater to increasing demand for foods. One of the currently developing technologies is 'biotechnology'. Exploitation of microorganism for production of food stuffs had been done since time immemorial when yeasts had been employed to make wine and beer (6000 BC) and leavened breads (4000 BC). However, the use of microbes was a "hit and miss affair" due to lack of knowledge and skills on understanding the technology and manipulating it to an advantage. Now with advancements in biotechnology, the term is commonly used to refer to foods derived through genetic engineering and this is gaining importance in devicing new ways to increase food production, improve nutrient content and provide better processing and storage characteristics. As food or food component produced through this technology results in a new food, its safety and quality aspects become important.

Safety and quality have played an important role since the past few years for both protection of health of consumer and facilitation of international trade in foods. However, since the last 4 to 5 years, with the establishment of World Trade Organisation (WTO) and signing of the Sanitary and Phytosanitary Measures (SPS) and Technical Barriers to Trade (TBT) Agreements, the international scenario is rapidly changing, with safety and quality gaining further significance. In the area of biotechnology, safety becomes imperative with safety evaluations and assessments playing a significant role. Quality of food derived through biotechnology becomes equally important due to significant technical changes in the property of food as well as nutritional implications resulting from changes in composition of the food. Labelling is also an important issue as it enables the consumer to be informed and thereby make a conscious choice.

With this latest understanding of biotechnology, a lot of interest and concern on the subject has been generated world wide with significant discussions and debate at both national and international levels. In the succeeding text, the scenario which is emerging in the light of the new environment created by the SPS and TBT agreements has been highlighted. The concepts of risk analysis to include risk assessment, management and communication have been covered as also the importance and role of standardization, both product and systems. The work done at both international and national level in the area of standardization as related to biotechnology has also been covered.

1. WORLD TRADE AGREEMENTS

The basic rules that govern international trade in food were agreed upon during the Uruguay Round of Trade Negotiations and apply to member governments of the WTO. With respect to food safety matters, these rules are set out in the agreement on application of sanitary and phyto-sanitary measures (SPS Agreement), whereas other food standard and quality issues are covered under the agreement on technical barriers to trade (TBT Agreement).

1.1 Agreement on the Application of Sanitary and Phyto-Sanitary Measures

The SPS agreement confirms the right of the WTO member countries to apply measures necessary to protect human, animal and plant life and health provided that such measures have a scientific justification, are not applied in a manner which could constitute a means of arbitrary or a disguised restriction on international trade and are established and maintained in an open and transparent manner. It requires that with regard to food safety measures, WTO members base their national measures on international standards, guidelines and other recommendations adopted by the Codex Alimentarius Commission where they exist. With regard to animal and plant health, the standards, guidelines and recommendations adopted by the FAO International Plant Protection Convention Secretariat (IPPC) and the Office International des Epizooties (OIE), have been referred to. The SPS Agreement further states that any measures taken which conform to international standards, guidelines or other recommendations should be appropriate, necessary and non-discriminatory. The SPS agreement also calls for a programme on harmonization of national requirements based on international standards.

In the area of human health, therefore, the work of Codex Alimentarius Commission is of great significance and is being used as a "benchmark" or "yardstick". Governments of most of the countries are either adopting codex standards for both domestic and international trade or are basing their legislation on codex documents as far as feasible. WTO members who are applying stricter national requirements than Codex for food imports will be required to provide scientific justification for the same. Codex standards thereby assume great significance for both domestic and international trade in foods.

1.2 Agreement on Technical Barriers to Trade (TBT)

The objective of this agreement is to prevent the use of national or regional technical requirements, or standards in general, as unjustified technical barriers to trade. It covers all types of standards including aspects of food standards, except requirements related to sanitary and phyto-sanitary measures and includes a very large number of measures designed to protect the consumer against deception and economic fraud. The aspects of food standards, it covers, relate specifically to quality provisions, nutritional requirements, labelling and methods of analysis. The agreement basically provides that all technical regulations must have a legitimate purpose and that the impact or cost of implementing the standard must be proportional to the purpose of the standard. The agreement further stresses on transparency in laying down standards and requires that the code of good practice for preparation, adoption and application of standards should be followed for both mandatory and voluntary standards. It also places emphasis on international standards and refers to the work of the International Organization for Standardization (ISO).

2. RISK ANALYSIS AS RELATED TO BIOTECHNOLOGY

The use of biotechnology processes, particularly genetic modification, may result in the development of new foods or food components. However, with the development of new foods, the potential human health and safety risk needs to be considered. Foods derived through biotechnological techniques, therefore, need to be assessed for their safety through risk analysis techniques.

Risk analysis is a key element of any comprehensive food safety programme. The SPS agreement requires that health and safety measures be based on sound scientific risk assessment and take into account risk assessment techniques developed by relevant international organizations. The risk analysis approach has basically three components, namely, assessment of risk followed by approaches to manage the risk and further communication of the risk to various interested parties. A primary purpose of risk assessment in food safety is evaluation and documentation of scientific evidence, including consideration of severity and uncertainty to produce a risk estimate for use by risk managers. Risk assessment is a major aspect in assuring that sound scientific principles are used for establishing standards related to food safety to ensure consumer protection as well as facilitation of international trade.

Risk management is a process of weighing policy alternatives in the light of the results of risk assessment and, if required, selecting and implementing the appropriate control options, including regulatory measures. Risk communication is an interactive exchange of information and opinion concerning risk among risk assessors, risk managers, consumers and other interested parties.

The concept of substantial equivalence has been an approach advocated by Organisation for Economic Cooperation and Development (OECD) to address the concept of risk assessment for foods/food components derived through biotechnology. Through this dynamic, analytical exercise, an assessment of the safety of new foods relative to an existing food is carried out.

3. STANDARDIZATION IN THE AREA OF BIOTECHNOLOGY

Standardization, like in any other industry, plays an important role in biotechnology also. However, because of the unique nature of this industry as it relates to living organisms and genetic modification, standardization is somewhat more complicated. Standardization in this area basically encompasses two aspects;

(i) Standards of the food products—whether derived through traditional means or through biotechnology.
(ii) Labelling aspects to provide consistent information to the consumer about foods obtained through biotechnology to enable him/her to make an informed choice.

Work has been done in this area at both international and national level.

3.1. International Standardization

At the international level, it is the Codex Alimentarius Commission known in short as Codex which is responsible for standardization work. Codex was established as a joint FAO/WHO Intergovernmental Commission in 1962 with the aim of developing international standards in the area of food to protect the health of the consumers and to facilitate international trade. The Codex has more than 160 governments as its members. Codex operates through a series of committees. It has presently 17 commodity committees and 8 subject committees which are responsible for developing documents. In addition, it has expert groups to guide consultations in various areas.

Codex has considered two aspects of biotechnology; i) safety of the foods and ii) labelling aspects of these foods. With regard to safety, "joint FAO/WHO expert consultations on biotechnology and food safety" have been held in 1990 and 1996 to address the issue of safety aspects related to the use of biotechnology in both food production and processing. The general approach decided at these consultations is that safety assessment based on the concept of substantial equivalence be applied in establishing the safety of food and food components derived from genetic modifications. Substantial equivalence embodies the concept that if a new food or food component is found to be substantially equivalent to the existing food or food component, it can be treated in the same manner with respect to safety, i.e. the food or food component can be concluded to be as safe as a conventional food or food component.

With regard to labelling, the issue has been under discussion since a number of years and a document "Proposed Draft Recommendations for Labelling of Foods obtained through Biotechnology" has been developed which is presently at Step 3 of the "Adoption Procedure" of Codex. A distinction has been made in this document between genetically engineered foods and those derived through other types of biotechnology. It has generally been agreed that for any new product, produced through genetic modification, it is necessary to provide complete information about it to the consumer. For the purpose of labelling recommendations, the document defines products obtained through biotechnology as

> "Foods composed of or containing genetically modified organisms, defined as organisms whose genetic material has been altered in a way which does not occur naturally through multiplication and/or natural recombination."

Genetic modification techniques include:

- recombinant DNA (r-DNA)techniques which use vector systems.
- techniques involving the direct introduction into the organism of hereditary materials prepared outside the organism including micro-injection, macro-injection and micro-encapsulation.
- cell fusion or hybridization techniques in which living cells with new combinations of heritable genetic material are formed through the fusion of two or more cells by means of methods which do not occur naturally.

The document is also proposing mandatory labelling requirements for foods obtained through biotechnology which are no longer substantially equivalent to the corresponding existing food or food ingredient as regards composition, nutritional value or intended use.

The following mandatory labelling requirements are being included:

- if the nutrient content is significantly modified (relevant/comprehensive), nutrient declaration should be provided in conformity with the guidelines for Nutrition Labelling.
- if the mode of preparation is significantly different from that for the equivalent food, clear instructions for use should be provided.

When a food produced by biotechnology is not substantially equivalent to any existing food in the food supply and no conventional comparator exists, the labelling shall indicate clearly the nature of the product, nutritional composition, intended use, (the method by which it was obtained) and any other essential characteristic necessary to provide a clear description of the product.

3.2. National Standardization

At the national level, BIS, the national standards body of India, is responsible for standardization and quality control activities in the country. BIS was established in 1947 (then known as ISI) with its main functions being formulation of Indian Standards for products and processes and promoting their implementation; operation of a certification scheme for both product and system including gold certification and certification of foreign manufacturers and imported products; establishing, maintaining and recognising laboratory for standardization and quality control; undertaking research for formulation of Indian standards; offering consultancy and training services and harmonizing standardization and quality control activities in the country.

Standardization in the area of agriculture and food was started by BIS in 1956. More than 1000 standards have been published in this area. In the area of biotechnology, although not much work has been done, particularly in the area of genetically engineered foods. However, standards have been brought out on bio-based products which can broadly be covered under biotechnology. These are being used as processing aids in various industries, for improving nutritional content of foods, as sugar substitutes etc. Some of the standards brought out in these areas include brewer's yeast (IS 10759:1983), baker's yeast (IS 1320:1988), food yeast (IS 3839:1966), papain (IS 3543:1966), alga spirulina marketed as a health food (IS 12895:1990), canned mushrooms (IS 14266:1995), monosodium L-glutamate (IS 9505:1992), α-amylase (*Aspergillus oryzae*), Food grade (IS 9605:1992), raw seaweeds both agarophytes and alginophtes (IS 12342:1988) and aspartame (IS 13657:1993).

4. ROLE OF SYSTEMS IN THE AREA OF BIOTECHNOLOGY

There has been a recent shift in trend from end point inspection to safety and quality assurance and management systems approach. These systems focus on building up quality or safety right from the production or procurement of raw materials till the product reaches the consumer. The focus is on preventive measures for ensuring health, safety and quality aspects. There are basically three systems which are internationally recognised and important in the food sector including biotechnology.

(a) Safety management systems
(b) Quality assurance and management systems
(c) Environmental management systems

4.1. Safety Management Systems

These have been recognized and published as the Hazard Analysis and Critical Control Point Systems by the Codex. As in the case of any other industry, even in the area of biotechnology, these systems play an important role in ensuring the safety of the final product produced.

The hazard analysis and critical control point system known in short, as "HACCP' is a science-based risk analysis and management system which assures safety of food products at all levels of food handling right from the production of raw material till the food reaches the final consumer. It is a tool to assess the hazards (biological, physical and chemical) which may occur in the food chain and establish control systems that focus on prevention of these hazards rather than relying on eliminating these hazards through end-product testing.

The system works on identification of significant hazards based on the risks involved; determination of critical points at which controls need to be applied to control the hazards; establishment of acceptable, measurable limits for each critical point of control; monitoring to see that the limits are not violated and in case of violation, taking pre-planned corrective actions. HACCP basically consists of 7 principles and 12 steps in application of the principles for implementation of the system.

Any HACCP system is capable of accommodating changes such as advances in equipment design, processing procedures or technological developments including biotechnological aspects. Successful application requires multi-disciplinary approach and should include expertise from relevant areas such as agronomy, veterinary health, production, microbiology, food technology, environmental health, etc including experts in genetic engineering, if required.

Although, the concept of HACCP developed in the late 60's, and at that time, was used in the US space programme as a preventative tool to control hazards at all stages of food production. It is only as recent as 1993 that it has been adopted at the international level by the Codex Alimentarius Commission. The concept is, therefore, not new, but, its significance has increased since its international adoption, especially after establishment of the WTO and signing of the SPS and TBT agreements. The Codex guidelines on design, operation, assessment and accreditation of food import and export, inspection and certification systems which have been recently adopted have recognised HACCP and other quality assurance systems for achieving quality and safety and recommend that the due recognition should be given to these systems in inspection and control programmes by governments of various countries. Accordingly, the HACCP system is being gradually included to form a base for official food control in a number of countries.

Countries including USA, Canada, Australia and many members of the EU have already started insisting on implementation of such systems both for their domestic and overseas trade, especially for the more risk prone products such as marine, meats, poultry, dairy and processed fruit and vegetable products. It is only a matter of time before these systems would be insisted upon for other products.

In India, institutional mechanisms for the implementation and certification of these systems have been developed. A number of organizations, including BIS have been involved in helping industry build up HACCP systems or to certify the units implementing these systems.

4.2. Quality Assurance and Management Systems

The quality assurance and management systems standards have been published by the International Organization for Standards (ISO) as the ISO 9000 series of standards. ISO is a worldwide Federation of National Standardization Bodies and has more than 100 members (the national standards bodies of different countries). The standards have been developed through the technical committee ISO/TC 176. The aim of the standards is to provide the company with an organizational model on the basis of which it can, by involving all its functions, organize itself in a way so as to manage quality and ensure that its product conforms to the level of quality it has targetted itself. A company is free to set a quality standard or level for its products on the basis of marketing considerations and customer's requirement (including legislation of the country). The ISO 9000 standards help the company to plan and constantly achieve the requisite product quality standard.

The ISO 9000 standards define a system for managing quality in generic terms and apply to all products and services including those related to the area of biotechnology. The standards provide a basic frame work around which the company builds up its quality management system.

The ISO 9000 series consists of two broad categories of standards-the "Core Standards" and the "Supplementary Standards". There are five core standards which are as follows:

- ISO 9000–1:1994, Quality management and quality & quality assurance standards—Part 1: Guidelines for selection and use. This standard clarifies principal quality-related concepts and provides guidance for the selection and use of the ISO 9000 family of standards for quality assurance and management.

- ISO 9001:1994, Quality systems—Model for quality assurance in design, development, production, installation and servicing.
- ISO 9002:1994, Quality systems—Model for quality assurance in production, installation and servicing.
- ISO 9003:1994, Quality systems—Model for quality assurance in final inspection and test.
- ISO 9004-1:1994, Quality management and quality system elements—Part 1: Guidelines. This standard is meant for internal use by organizations and provides guidance in designing and implementing a quality system so that the organization can meet their market needs and achieve overall success.

The quality assurance models set out in ISO 9001, ISO 9002 and ISO 9003 represent three distinctive quality management requirements for two-party contractual performance and also form the basis for third party certification. These standards are presently under review.

In addition, there are a number of supplementary guidance standards related to guidance for generic application, process materials, quality improvement, quality plans, development of quality manuals, auditing etc.

For establishing quality management and assurance system, 20 elements need to be addressed by the company through a documented system consisting of quality manuals, procedures and work instructions. Implementation of this documented system results in producing consistent quality of the product, the quality being that which the system has been designed for certification is being done against ISO 9001, 9002 & 9003 by various certification bodies in the country including BIS.

4.3 Environmental Management Systems

A major issue of concern related to the use of biotechnology is the environmental aspect due to the release of genetically engineered product into the environment. The ISO 14000 is a standard on environmental management systems which addresses the issue of significant environmental aspects of the activities of any manufacturing or other facility and lays down requirements of a system that would help an industry in evaluating its own environmental performance for continual improvement. Certification is being issued by various bodies including BIS against this standard.

Biotechnology is now gaining increasing importance in devising new ways to increase food production, improve nutrient content and provide better processing and storage characteristics. As per Dennis Avery, the Hudson Institute's Director of global food issues, U.K, "Biotechnology has been considered as the biggest piece of knowledge on humanity's shelves which has so far not been fully exploited.' However, as the food or food component produced through this technology results in a new food, it becomes necessary to carry out a safety or risk assessment, prior to its marketing. This has become a truly global issue after the establishment of the WTO, and it has become imperative to follow certain rules for both safety and risk assessment, standardization as well as labelling to allow consumers to make an informed choice.

FURTHER READING

1. Bhat, R.V. and Vasanthi, S. Meeting international requirements of agricultural produce by a quality evaluation approach. *Ind. Food Industry*, **19**, (2000) 135.
2. FAO. Codex Alimentarius. Secretariet of the Joint FAO/WHO Food Standards Programme. Food and Agriculture Organisation, Rome. (1997).
3. Mehta, A., Ranote, P.S. and Bawa, A.S. Indian Fruit Processing Industry: Quality control aspects. *Ind. Food Industry*. **21**: (2002) 37.
4. Soni, R., Gupta, J.K. and Marwaha, S.S. Food Safety and Standards. *In* Food Processing: Biotechnological Applications (Eds. S.S. Marwaha and J.K. Arora). Asiatech Publishers Inc. New Delhi. (2000) 317–347.
5. Srilakshami, B. Food Science. New Age International Pvt. Ltd., New Delhi. (1997).

Agrobiotechnology: Misconceived Concerns About Consumer and Environmental Safety

P.N. VISWANATHAN

2-A, Way Lane
Lucknow

THE LAST CENTURY WITNESSED phenomenal rise in chemical technology and natural resource utilization, for overall better quality of life, but also the considerable side effects on man, materials and environment. Since biotechnology is the technology of this century, it is natural to be apprehensive about its long term impact, sustainability and safety. Fortunately man has learned from his earlier mistakes and safety can be built into the technology unlike in the chemical revolution so that only benefits are felt. With the high precision of the powerful tools of molecular biology, selection of specific genes with the desired quality, their transfer in a safe, and stable way and monitoring field performance through specific markers could be the right approach. Agro-biotechnology has especially gained from the experience of quality and safety assurance guidelines, test protocols and surveillance methods evolved to achieve bio safety. PDA guidelines for substantial equivalence that any natural toxicant present in the parent should not increase by more than 10% and that the principal nutrients should not decrease by more than 20% as compared to parent and selection of gene, donor and acceptor based on familiarity, helps to build in safety with the technology.

It would be self contradictory and scientifically unjustifiable, if we accept readily, foods produced by the traditional methods of genetic modification that entail thousands of recombination events, most of which are uncertain, unpredictable and unknown, on the one hand, and on the other be apprehensive of accepting agro-biotechnology products which are much more precisely designed and controlled. As nature is the biggest set up in genetic engineering, it is also the biggest and most objective, time tested, rigid and effective safety assurance facility. Apart from ease of availability and wholesomeness, safety was also a consideration, collectively acquired by local society during evolution of food preferences. When selective breeding, optimization of production, storage and household use and industrial processing, underwent evolution for better product, the acceptability of the modified material was not questioned in terms of safety so long as final product was desirable.

1. APPROACHES

The approach for safety assessment of whole food items or major ingredients, in food such as proteins is quite different from that for toxicity evaluation of occupational and environmental chemicals and substances added in small amounts as food additives, cosmetic ingredients and pharmaceuticals. The influence of the food component administered in usual amounts on growth and gross physiological processes as compared to a suitable control diet and not a placebo, is followed. In the case of proteins, protein related specific parameters such as catalytic activity, special effects like haemagglutination and enzyme inhibition, allergenicity, unusual and toxic amino acids, membrane damaging and antinutrient effects and influence on essential trace elements also have to be checked in addition to biological value and digestibility. With novel bulk proteins, such as petroleum based single cell proteins or algal proteins, all these testings are easy. In the case of transgenic food proteins, unless expressed in bacteria, testing is difficult and unrealistic in relation to diet composition. However, sequence homology comparisons for predicting effects are possible. All these approaches are needed in the safety assessment of specific genetically engineered proteins in food crops and the safety of the whole crop is to be evaluated, rather than the protein alone.

Wholesomeness and substantial equivalence have been conclusively established in case of all the transgenic crops approved such as glyphosate tolerant cotton and soybeans, Bt expressed cotton, tomato and soybeans, canola, delayed ripening tomatoes and others. A widely used bacterial pesticide Bt protein, when engineered into a plant, the plant actually becomes a new delivery system for an approved pesticide and not a new plant product. As in the case of chemical pesticides, if the added ingredient is within safety level for residues, the whole product can be generally recognized as safe.

The safety of the engineered Bt protein through multispecies testing and sequence homology analysis was proved beyond doubt. Detailed molecular characterization, digestibility, acute minimum level toxicity test, and other mandatory data were also made available. With microbial pesticides like Bt, extensive studies on toxicity were done before field release and the information will be applicable for Bt engineered crops also. Further to establishing of substantial equivalence and wholesomeness, it is essential to quantify all potential risk factors including allergens, toxicants and anti-nutrients. Whereas, highly allergenic ovalbumin and B lactoglobulin were stable even at 60 min, Bt insecticidal protein and most other engineered proteins were digested within 30 sec. The high target specificity of Bt protein is due to the fact that only the target insects cannot digest the protein and all mammals completely break it. If the donor of the gene has no allergenic history and the gene product is not allergenic, or has no homology with known allergens, the chances of food allergy are remote for transgenic product. It is considered that if a protein is expressed at levels more than 1% of total protein, allergy testing is needed, while engineered proteins like Bt are below 0.01%.

2. SAFETY EVALUATION STUDIES

With transgenic tomatoes rendered insect resistant by engineering Bt. Cry A-b insecticidal protein and its marker NPT II, detailed studies were conducted on rodent model. The test material was expressed in *E. coli* to facilitate production of large amounts. In Brown Norway rats, no binding with gastrointestinal tract or any histological lesions, were seen in target larvae. Immunochemical methods did not show any receptor for the protein. Digestibility was also established in the Brown Norway rat fistulated ileum model and on simulated gastrointestinal *in vitro* models. The dose of protein administered when computed to whole tomato contents was around 2000 kg daily. Even at this dose there was no toxicity. No peptide capable of eliciting reaction could be detected to alter digestion. Oral feeding to NMRI mice through water daily for 28 days and New Zealand white rabbits fed for 31 days also did not show any treatment-related toxicity. Detailed chemical analysis, including toxic (α-tomatin) showed no significant variation with non-transgenic crops. A 90 days rat study with lyophilized tomato (transgenic and control) also did not show any adverse effects using all standard toxicity indicating parameters. Issues regarding any influence of post translational modification (tomato vs *E. coli*) may not be serious since glyphosate resistance protein expressed in soybean and *E. coli* had same sequence and 3D structure.

Bacillus thuringiensis based transgenic crops have been thoroughly studied on various models even at very high doses for any oral, dermal, pulmonary toxicity as well as all mandatory ecotoxicological tests. There were no evidence of human or other non-target risks. The proteins disappear in environment rapidly. These studies along with comparative studies on feeding Bt expressed and control seeds in different animal models prove the absence of any risk. Detailed studies in a 90 day feeding trial on goats did not show any treatment related effect on food intake, weight gain, behaviour, growth and biochemical, hematological and histological parameters. Thus a science backed approach, can remove apprehensions about safety of transgenic crops.

3. ENVIRONMENTAL CONCERNS

A genuine concern for any adverse impact on environment and biodiversity, created by lack of awareness, got exaggerated by emotionally and ethically sensitized persons and sensationally publicized by mass media. The removal of gnosophobia by mass awareness generation, as done in controlled field trial of transgenic crops with free accessibility to society, is the responsibility of all concerned with technology development, propagation, mass production, sale regulation and surveillance. The initial hostility is gradually subsiding and societal needs are considered along with hypothetical ethical issues. All viable agrotechnologies so far are based on already familiar crops and gene donors and no new life is created only better survival and performance while assured, which man was always doing progressively since early agriculture. The chance of the gene from decayed plants getting taken up by soil microfiora deleting the plant promoter and adding its promoter and expressing is difficult. Stable spontaneous specific genetic exchanges and recombinations are not as frequent as imagined by some, in nature. If so, evolution would have been a much faster process and the biodiversity and its time tested interrelations would have been in disarray. Genetic outcrossing through pollination, resistance development and weediness are also not as serious as conceived and the experience with random cross breeding success stories will tell us that in organized agriculture, than can be anticipated and avoided. The same applies to non-target effects as evident from the current data on monarch butterflies. In over 25000 field trials, so far no adverse environmental impact has been encountered because risks had been anticipated and safety precautions built in. Further, the release of millions of tons of more hazardous, less predictable manufactured biocide chemicals can be avoided. Biotechnology also has potential for rehabilitation of damaged sites through cleanup, protection of threatened germplasm and assisting in better health. On the whole, biotechnology has consolidated all the benefits of natural productivity such as quality; quantity, stability; acceptability and environmental friendliness built into it through efficient, well-defined, expressed, amplified and maintained trait improvement. Like artificial intelligence, biotechnology is also speeding up natural process and not replacing it for desired needs from nature to provide better health, food and overall quality of life for all in the new millennium.

FURTHER READINGS

1. Cook, R.J. Agricultural Biotechnology and the Poor. In: Proc. International Conference, (Eds. G.I. Persley and M.N. Lantin). Washington DC, October 1999, (2000) 123–130.
2. James, C. Global Status of Transgenic Crops in 1999. ISAAA: Ithaca, New York (1999).
3. Rai, M. and B.M. Prasanna. Transgenics in Agriculture. Indian Council of Agricultural Research: New Delhi (2000) 144.
4. University of Illinois (1999). The Economics and Politics of Genetically Modified Organisms in Agriculture: Implications for WTO 2000. USA: University of Illinois Bulletin 809, November 1999.
5. US National Research Council. Genetically Modified Pest Protected Plants: Science and Regulation, National Academy Press: Washington, DC, USA, (2000).

Strategies for Improvement in Production of Secondary Metabolitis through Plant Cell Technology: Case Studies

A. KUMAR

Global Biotech
130-A, Golf Course Scheme
Jodhpur 342 011 (Rajasthan)

DRUGS OF PLANT ORIGIN FORM an important segment of modern pharmacoepia and our dependence on plants for natural pharmaceuticals is likely to continue as some of these compounds are structurally too complex to be synthesized chemically, serve as model compound for the chemical synthesis of more potent analogues and the increasing consumer preference for natural products over their synthetic counterpart. Plants have traditionally been the source of pharmaceuticals and fine chemicals. Many plants are still being screened for new compounds . In course of biochemical and physiological studies, plant cells/tissues have also been discovered to accumulate many of the compounds characteristic of a particular plant. Many of these naturally occurring cytochemo compounds have become the source of large numbers of industrial products including agro-chemicals, pharmaceuticals and food additives (Table 1).

Most of these valuable plant products are defined as secondary metabolites, often complex in structure, and are the end products of long biochemical pathways. These secondary products are not essential for growth and have a minor role in the basic life processes of the plant, but, do often confer some advantage to the plant as attractants of pollinators and chemical defense against microorganisms, insects, higher predators etc.

Plant tissue culture technique offers an avenue to evaluate and exploit the metabolic potential of higher plants for the bio-production of useful secondary metabolites of therapeutic/pharmaceutical importance. The cultures can be an additional source of production for a number of useful and valuable secondary products if the product:

- has high demand
- gets affected by diseases
- supply gets affected by climate
- accumulates in extremely low concentration in the plant of origin
- is of rare plant origin
- can not be synthesized chemically

Table 1: Natural Plant Products of Industrial Importance

Category	*Examples*
Pharmaceuticals	
(a) Alkaloids	Ajmalicine, atropine, berberine, codeine, reserpine vincristine, vinblastine
(b) Steroids	Diosgenin
(c) Cardenolides	Digitoxin, digoxin
Food and flavours	
(a) Sweetners	Stevoside, thaumatin
(b) Bittering agent	Quinine
(c) Pigment	Crocin
Pigments and perfumes	
(a) Pigments	Shikonin, anthocyanins, betalins
(b) Fragrances	Rose oil, jasmine oil, lavender oil
Agrochemicals and fine chemicals	
(a) Agrochemicals	Pyrethrin, salanin. azadirachtin
(b) Fine chemicals	Proteases. vitamins, lipids, latex oil

The regular supply of some medicinal plants has been difficult recently due to several factors like their ruthless exploitation, extinction and decimation of the plants in ecosystem (Table 2), lack of conservation of the environment, increasing labour cost, erratic and changing seasonal patterns, lack of organized cultivation and other economical and technical problems. The tissue culture based production system can extend solutions to some of these problems and can offer an alternative route of production of secondary metabolites.

The utility of plant tissue culture for the production of natural compounds was visualized in the early 1940's and the first patent on plant cell culture was taken by JB Routin and LG Nickel in 1946 (US Patent 2,747,334, 1956). However, it took more than a decade to establish that plant cells in culture can produce metabolites in reasonable concentrations. Since then, there has been an increasing interest among scientists to standardize the production of high value natural products by cell cultures which can overcome many of the problems associated with industrial production by extracting field grown plants (cultivated or naturally occurring). The cultures offer factory-type controlled, round the year production, unaffected by seasonal behaviors, natural calamities/hazards and puts no threat to our biodiversity or ecosystem. Cell culture system has not only been used for de-novo synthesis of target compounds, but also as a reactor for bioconversion of cheaper molecules into expensive compounds. Sometimes, cell cultures have been reported to produce compounds that are novel and not reported in intact plants (Table 3). At least 85 novel compounds including 23 alkaloids, 19 terpenoids, 30 quinones and 11 phenyl compounds have been isolated from some 30 different plant

Table 2: Endangered /Vulnerable medicinal plants

Commiphora wightii (Guggul)
Swertia chiraita (Chiraita)
Coptis teeta (Mamri)
Ferula faeskeona (Jangli hing)
Saussarea sacra (Jogi badshah)
Atropa belladonna (Indian belladonna)
Gentiana kurroo (Karu)
Picrorhiza kurroa (Kutki)
Nardostochys grandiflora (Jatamansi)
Aconitum chasmanthum (Aconites)
Rheum australe (Tukshu)
Podophyllum hexadrum (Bankakri)
Rauvolfia serpentina (Sarpgandha)
Panax pseudo-ginseng (Indian ginseng)
Dioscorea deltoidea (Singli-mingli Kins)
Aristolochia brae tea ta (Kiramar)
Chlorophytum arundinaceum (Safed musli)
Ephedra gerardiana (Somlata)
Dactylorhiza hatagirea (Orchis latifolia) (Salampanja)
Enlophia compestris (Salam misriO
Inula racemosa (Puskarmool)
Nepenthes khasiana
Onosama hracteatum (Shankhahuli)
Bunium persicum (Kalzeera)
Colchicum luteum (suranja-talka)
Piper cubeba (Kabab chini)
Thymus serphyllum (Banajwain)
Malaxis muscifera (Javan Rishbhak)
Podygonatum cirrhifolium (Maha meda)
P. verticillatum (Meda)
Roscorea-procera ralpinu (Kakoli)
Lilium polyphyllum (Shera kankoli)
Lavateria khashmiriana (Shera kankoli)
Anchusa strigosa (Gaozaban)
Coscinium fanestratum (Jahri-haldi)
Holostemma annularis (Kasturi manzil)

culture systems. During the last four decades, after the conception of tissue culture production of natural compounds, many technological advances have been made and in few cases, cell cultures have shown higher production of the product of interest than the intact plants of their origin (Table 4).

Table 3: Some compounds produced exclusively in culture but absent in the corresponding intact plants

Compounds	*Species*
Echrosine	*Ochrosia elliptica*
Dehydrodiconiferyl alcohol–y-B-D glucoside	*Plagiorhegmci dubium*
Paniculid A	*Andrographis paniculala*
Pericine	*Picrulima nitida*
Rutacultin	*Ruta graveolens*
Tarennosid	*Gardenia jasminoides*
Voafrine A and Voafrine B	*Voacanga africana*

Table 4: Cell Cultures producing higher or equal amount of natural product than whole plants

		Yield (% dry weight)	
Product	*Plant species*	*Whole Plant*	*Cell Culture*
Ajmalicine	*Catharanthus roseus*	0.3	1
Anthraquinones	*Morinda citrifolia*	2.2	18
Berberine	*Coptis japonica*	2.4	13.4
Caffeine	*Coffea arabica*	1.6	1.6
Catharantheine	*Catharanthus roseus*	0.0017	0.005
Diosgenin	*Dioscorea deltoidea*	2.4	7.8
Ginsenoside	*Panax ginseng*	4.5	27
Rosmarinic acid	*Coleus blumei*	3	23
Serotonin	*Pagamim harmala*	2	7
Serpentine	*Catharanthus roseus*	0.26	2
Shikimic acid	*Galium mollugo*	2-3	10
Shikonin	*Lithospermum erythrorhizon*	1-2	15-20
Trigonelline	*Trigonella foenum-graecum*	0.4	5
Tripodiolide	*Tripterygium wilfordii*	0.01	0.2
Vomilenine	*Rauwolfia serpentina*	0.004	0.214

Out of several natural products of plant origin, pharmaceuticals have received maximum attention as far as tissue culture production is concerned and the maximum contribution to this field has been by two countries, Japan and Germany.

This chapter deals with the strategies for better secondary metabolite production with increased yield and a few case studies for better learning and analysis.

1. PRINCIPLES

The production of secondary metabolites is the result of morphogenesis including metabolic differentiation and cytodifferentiation. Secondary metabolites are also formed as a result of infection or stress due to certain gene activation and are referred to as phytoalexins. Other secondary metabolites, produced during stress and disease by the stimulation of enzymes catalyzing regular metabolism and are referred to as phytoanticipins.

1.1 Secondary Plant Metabolites

1.1.1 Products of differentiation

The induction of secondary metabolites in tissues grown *in vitro* was recognized after the heterogeneity of cultured cells was described. Hence, pigment, flavour, and fragrance or tannin production was interpreted as the result of cell differentiation. Lack of alkaloids in poppy or terpenoids in carrot cultures could be readily explained with the lack of laticifers or resin canals respectively. Failure to demonstrate vindolin and vinblastine in cell cultures of periwinkle and the recurrence of vindolin and vinblastin in cell cultures with shoot formation highlights the point. And finally, menthol in mint *(Mentha piperatci)* cell cultures with shoot formation was a strong confirmation of morphogenetic regulation of secondary metabolite production.

The concept of secondary metabolite as differentiation product is supported by the following arguments:

- secondary metabolites occur *in vitro* during the stationary phase of plant cell cultures
- during maturation of cells and cyto-differentiation
- during organogenesis

1.1.1.1 Stationary phase of plant cell cultures

Kinetics of growth and secondary metabolite synthesis have long and repeatedly been shown to occur in the stationary phase of cell cultures. Profitable management of cell culture for industrial production initially by boosting growth and the metabolite accumulation adds convincing evidence. Still, not all experimental results are in accordance with this concept.

1.1.1.2 Hormonal regulation

Hormonal regulation of metabolite formation and accumulation has been well established for many systems. Thus, deacceleration of growth because of deletion or reduction of auxin in culture media prompts the appearance of products. Cytokinins, on the other hand, have been demonstrated to stimulate production.

1.1.1.3 Cell maturation/Cytodifferentiation

Cell culture deprivation of hormones leads to organogenesis (differentiation) and subsequently to an increased metabolite concentration. Concurrent microscopic examination clearly reveals correlation between the appearance of thevaine. thevetin and laticifer like cells. The relation between cell morphology and metabolite synthesis and accumulation came to the fore when cells of protoplast derived periwinkle clones were investigated. The degree of differentiation of a given tissue certainly makes for differences in the spectrum of products.

Maturation and specialization of cells and product formation may be achieved by technology that allows cells to assemble in a mat, a pseudo-tissue (i.e. immobilization of surfaces of a matrix). Cell to cell contact and nutrient gradients in cell layers may indeed convey physiological attributes of maturation.

1.1.1.4 Organogenesis

The competency of *in vitro* cultures for metabolite biosynthesis and accumulation, after roots or shoots are formed, has been established (Table 5). The ultimate in

Table 5: Examples of secondary products accumulated by root and shoot cultures

Culture	*Product*
Root cultures	
Atropa belladonna	Tropane alkaloids
Catharanthus roseus	Ajmalicine
Papaver bracteatum	Thebane
Papaver somniferm	Codeine
Senecio spp	Pyrrolizidine alkaloids
Shoot cultures	
Atropa belladonna	Tropane alkaloids
Rauwolfia serpentina	Ajmalicine
Digitalis purpurea	Digitoxin
Pelargonium fragrans	Pinene
Cinchona ledgeriana	Quinine
Centranthus macrosiphon	Valpotriates
Linum flavum	Coniferin
Coleus forskohii	Forskolin

morphological differentiation for accumulation of secondary metabolite is the formation of trichomes in shoot cultures. Their structures can be rather complex because they provide effective compartmentalization of products that are generally toxic. Development of transgenic shoots *in vitro* is followed by differentiation of hair that accumulate menthol at levels comparable with those in field grown control plants. The most definite answer to metabolite production by organized tissues and organs is presented by hairy roots and shoots. This material is known to grow profusely while retaining histological integrity. Recently, co-cultivation of

Papaver somniferum hypocotyls cells with *Agrobacterium rhizogenes* gave rise to transformed cell cultures and subsequently to adventitious shoots which accumulated ten-fold higher levels of morphenine alkaloids, especially, thevaine and codeine than controls.

1.1.2 Products of Stress

Secondary metabolism has been viewed as a result of differentiation until the concept of chemical defense was established, and elicitors were employed for metabolite production *in vitro.* The following are stress factors having remarkable effect on secondary metabolism in plant cell tissue and organ culture.

1.1.2.1 Osmotic stress

The relation between osmotic stress caused by high levels of sucrose in nutrient media and production of secondary metabolites has been known for some time and collaborative observations have been made. The products include not only anthocyanins and alkaloids, but, also terpenoids. Increasing the carbohydrate concentration to 6% (3% sucrose + 3% mannitol) and addition of 25uM abscissic acid significantly increased the accumulation of monoterpenes in shoot cultures of *Lavendula latifolia.*

1.1.2.2 Ultra-violet radiation

The response of plants of high altitudes to UV irradiation by an increase in anthocyanin biosynthesis is a common observation. Pigmentation in cell cultures of parsley on treatment with UV light has not only improved conditions for metabolic production, but, also has significantly affected the molecular biology of induction, transduction and gene expression of the plant metabolite biosynthesis. UV light has not been applied for production of secondary metabolites *in vitro* as much as had been expected.

1.1.2.3 Biotic and abiotic stress

Cell cultures have been exposed to very different kind of stress, both biotic and abiotic. Most radical was the use of fungal homogenates and extracts in nutrient media as elicitors for alkaloid synthesis and accumulation in plant cell cultures. The list of materials successfully treated with fungal elicitors for production of secondary metabolites has grown rapidly and continues. Investigations of elicitor induced production of benzophenanthridine alkaloids in *Sangunaria canadensis* cell cultures were extended to include the effect of pretreatment with a specific calcium chelatant and a calcium channel inhibitor. Product formation through fungal elicitors was stimulated by heavy metal ions such as vanadyl ions. Cell cultures of various plant species could be elicted to produce secondary metabolites by supplementing nutrient media with methyl jasmonate. Secondary metabolites resulting from elicitation is not a function of cell maturation or differentiation. Rather highest yield will be achieved when elicitors are being introduced to *in vitro* systems at a time when carbohydrates are exhausted and when the cells are in the deacceleration mode of growth.

1.1.3 Products of Genetic Engineering

In general, the productivity of organisms directly isolated from nature is not satisfactory for industrial production. Mutagenesis and/or breeding is used to enhance the output of the targeted product. Recombinant DNA (r-DNA) technology now provides new ways of modifying the genomes of organisms in a directed manner. Genetic incompatibility of a species, a restriction in classical genetics has been overcome by this new technology. As a result, naturally occurring organisms can be endowed with additional genetic information from other sexually incompatible organisms. Thus, construction of novel organisms which do not exist in nature with new metabolic capability is now routine. The manipulation of metabolism in plants through DNA technologies for enhanced production of secondary metabolites is in part curiosity driven. Recombinant DNA technologies are also pooled by the resource and processing industries owing to prospects of marketable and profitable botanicals. Generally, the strategy is aimed at enabling transgenic plants to synthesize new and novel products by

- Completion of partial pathways
- Amplification of regular pathways
- Blockage of competing pathways
- Interdiction of regular pathways
- Revising metabolic regulation
- Minimizing response cascades.

1.2 Preservation and Multiplication of Plant Resources

High yielding cell lines selected from wild or transgenic populations comprise unique germplasm that require preservation, best accomplished by cryo-technology using liquid nitrogen. Cryopreservation is well developed. Equally important is the cryopreservation of germplasm of wild medicinal or industrial plants classified as rare and threatened. Modern concepts to remedy the situation comprise germplasm preservation by cryotechnology as mentioned and micropropagation of remaining stock for rehabilitation for domestication and establishment of plantations.

Clonal or micropropagation of elite plants is a viable technology in plant tissue culture. Several industries have standardized protocols for the multiplication of various types of plants but, no serious attempt has been made to clone medicinal plants on industrial scale. By standardizing protocols for clonal propagation of high yielding clones, it will be possible to achieve an 8-10 fold increase in the natural products per unit area of cultivation.

2. METHODOLOGY FOR YIELD IMPROVEMENT

Biotechnology for enhanced production of secondary metabolites in plant cell cultures need to be complimented by methods that are applicable in the laboratory. Various publications have covered the production of secondary metabolites in cell cultures in transgenic plants emphasizing methodology Strategies for enhancement of secondary metabolite production include the following:

- Selection of source material
- Selection of superior cell lines
- Optimization of culture conditions
- Elicitation and stress
- Scale-up
- Recombinant DNA technologies
- Genetic transformation
- Cryopreservation
- Micropropagation

2.1 Selection of source material

The initiation and maintenance of the callus and cell culture is the subject of several laboratory manuals. Species limitation, therefore, is no longer a problem when searching for source material to produce secondary metabolites *in vitro.* Plants known to be rich in the product under investigation would naturally be considered the best starting material. Given a specific natural product, explants should be taken from as many source species as possible. The screening can be done from different parts of one plant, different plants, different cultivars and different plant species. However, it must be appreciated that the source of the explant has only a limited effect on the yield of secondary products in some cases.

2.2 Selection of superior cell lines

High yielding cell lines are generally a result of screening and selection process. For obvious reasons, this process is more easily accomplished with visible products i.e. pigments and flavours such as vanilla. Extraction and chemical analysis are the traditional procedures. Radioimmunoassays (RIA) and enyme-linked immunosorbent assay (ELISA) tests are being applied whereever possible. Non-destructive analysis by nuclear magnetic resonance (NMR) may soon become the method of choice. Generally, the methods applied for screening have been chemical analysis of plant material by chromatography and more recently by NMR over a substantial time. Such procedures have included hairy root cultures, as demonstrated for *Datura stramonium* resulting in stable production of scopolamine that was 2-fold higher than from source plant.

2.3 Optimization of Culture Conditions

2.3.1 Light

Among physical parameters of culturing plant cells, light has been a significant factor in some instances. Sometimes consideration may have to be given to specific light requirements. In shoot cultures of *Catharanthus roseus* for vinblastine production, synthesis was stimulated by visible light closer to UV light.

Light is required for growth of shoot cultures and for production of metabolites in leaves, such as essential oils in glands and hair or idioblasts. More complex is the interrelation of metabolite synthesis and accumulation with chloroplasts. Light will lead to greening of callus and cell cultures and may hinder or interfere with the detection of desirable products. It has been experienced with paclitaxel formation in *Taxus* spp.

2.3.2 Culture Medium

Growth of cells and tissues cultured *in vitro* and production of secondary metabolites demand the specifications of nutritional requirements for each species, and also for different genotypes within each species. The general procedure is to permutate the components of the nutrient media (i.e. mineral salts, vitamins, organic supplements, carbohydrates and hormones). A general approach begins with testing growth and product formation of tissues in the presence of different media. An expedient approach would be to optimize conditions by directing cells towards the production of biomass by exposure to a growth medium, followed by production of secondary metabolites by substituting the growth medium with a production medium as introduced for alkaloids (serpentine and ajmalicine) and for naphthaquinones (shikonin). For production media, a decrease in the level of auxin and nitrate and an increase in the level of ammonium salts and carbohydrates are generally recommended. Numerous efforts to optimize individual components of production media have been rewarded by high yields and have been incorporated into scale-up procedures.

Determining the quality and quantity of hormone supplements to nutrient media has played a major role towards increasing product synthesis and accumulation in cell cultures. As a rule, the concentration of auxin has been reduced to minimum or eliminated altogether and cytokinins have been adjusted to levels of 0.1-1.0 mg/1. The effect of gibberellic acid and abscissic acid has generally been judged ambivalent.

The feeding of precursors is common practice for biochemist when elucidating biosynthetic pathways. Supplementation of limited media with substantial amount of precursors is aimed at increasing yields of metabolites. Such a procedure must satisfy two concerns i.e. the concentration of the precursor must be tolerated by the cells under investigation and the precursor must be catalysed to the desired end product and not be diverted to the synthesis of metabolites on branched pathways.

The feeding of secondary metabolites to cell cultured *in vitro* raises the question of tolerance or the mechanism of protection from compounds that are potentially toxic. Recently, ajmaline was fed to cell cultures of *Rauvolfia serpentina.* The cells tolerated lg/1 of added alkaloid. The process, however, led to produce several previously unknown compounds. Instant extra and intra cellular catalysis may be the response by cells. Optimization should also be achieved by decreasing catabolism of desirable products. Concept of how to achieve this remains a question. The addition to nutrient media with undefined supplements may underscore the degree of empiricism that governs some tissue culture experimentation.

2.3.3 Elicitation

The term elicitation is used in its widest definition as induction and stimulation of synthesis of secondary metabolites by stress agents regardless of whether metabolites are phytoalexins or phytoanticipins. The interest is the identification and characterization of more powerful elicitors for further improvement of product formation including manipulation of signals and transduction mechânism. Elicitation parameters to be optimized to improve the production of secondary metabolites in plent cell cultures are the selection of cell line and the type of elicitor, the concentration of elicitor, the timing of elicitation and the general culture conditions during elicitation.

2.3.4 Immobilization

Immobilization is meant to aggragate cells in a tissue like manner, but, exposed to the medium for continuous culture and metabolite production *in vitro*. Most methods recommend the entrapment of cells in a polymeric matrix be it alginate, agarose, carrageenan. polyacrylamide gels, hollow fibres, reticulate polyurethane foam in membranes or stainless steel screens. Further development will depend on the methods of forced release of products and recovery by adsorption to resins. Usage appears to have decreased over the last few years.

2.4 Genetic Transformation

More recently, transformed roots (hairy roots) have also been shown to accumulate secondary products (Table 6). Plants are induced to form roots at the point of infection by *Agrobacterium rhizogenes,* the causative agent of hairy root disease. The infection process is complex but leads to the insertion of DNA from the Ri plasmid in the bacterium into the plant genome. The expression of the genes coding for auxin synthesis results in root formation at the site of infection. Transformed roots are characterized by a high degree of genetic stability and retain their biosynthetic stability for long periods. The rate of growth is high, compares well with plant cell suspension and is generally faster than normal roots in a medium

Table 6: Examples of phytochemicals produced by Hairy Roots

Compound	*Culture*
Shikonin	*Lithospermum erythrorhizon*
Quininc	*Cinchona ledgeriana*
Nicotinc	*Nicotiana tabacum*
Ajmalicinc	*Catharnthns roseus*
Hyoscyaminc	*Datura stramonium*
Saponin	*Panax ginseng*

lacking growth regulators. Secondary product accumulation reflects the plant and exhibits the kinetics for suspension cultures.

Shoots transformed by co-cultivation of stem segments with *Agrobacterium rhizogenes have* become the prime source material for production of essential oils.

2.5 Product release and Adsorption

Cell cultures arriving from a phase of growth at the stationary phase will mature and sometimes will show signs of deterioration and lysis. As a consequence, secondary metabolites will appear in the nutrient medium which can be exploited for downstream processing. Release of berberine from cultured cells of *Thalictrum minus* in to the medium was temperature dependent and was suppressed by such inhibitors of membrane bound ATPase as vanadate and diethylstilbestrol. Active release of products by cells may be the major challenge for future manipulation *of in vitro* systems. The release of metabolites triggers the question of whether these compounds will not be toxic to cells generating them or whether they may be resorbed. Uptake has been investigated and led to the observation that products indigenous for a given species will indeed be taken up from the medium and stored in vacuoles, but, this is not so always.

Forced release of secondary metabolites from cells and the harvesting of these products from the medium by adsorption would be a breakthrough and at the same time will facilitate continuous culture processes that may reduce cost of production, purification and allow one to economize immobilized systems, as well as hairy roots and shoots cultures. The release of desirable products from the vacuoles without inferring damage to the cells remains the actual challenge to be met.

Passive diffusion, unmitigated or stimulated by elicitation or enhanced by permabilization with dimethylsulphoxide (DMSO) or other agents, and subsequent extracellular accumulation by adsorption of metabolites has been demonstrated. Since the inception of the idea, several solid and liquid adsorbents of secondary metabolites have been introduced in the nutrient media, thus, adding an extraction phase to a growth phase. Such a procedure would increase the product yield by adding an additional source, but, more importantly, by suppressing the end product inhibition, by suppressing enzymatic or non-enzymatic metabolism of extracellular products and by capturing volatile products. It requires adsorbents that are non-toxic, do not interact with nutrient media, bind-in as specific manner as possible to desirable metabolites and release metabolites during subsequent extraction. A combination of agents that permeabilize cells, such as DMSO and of adsorbents may enhance both release and the productivity of cell suspension by scavenging even trace amounts of chemicals from the medium, thus, avoiding any kind of feedback inhibition that may hinder product formation.

2.6 Cryopreservation

Technology for successful cryopreservation of cells, tissues or organs of medicinal plants has been well documented which basically involves the viable freezing of cells and tissues, their subsequent storage at ultra low temperatures (preferably liquid nitrogen, -196°C) and their eventual retrieval in a genetically unaltered condition.

2.7 Micropropagation

Domestication and establishment of plantations with cuttings or with plants regenerated from *in vitro* grown cells, tissues, and organs have become the goal when rare and high yielding genotypes are to be maintained. The simpler method will be given preference and the plant material which qualifies for micropropagation of "elite' clones, wild mutants or transgenics and rare plants threatened by extinction. The procedures to follow have been amply and repeatedly described in manuals, books and reviews.

3. CASE STUDIES

3.1 Monoterpenes—Menthol

Menthol is the main constituent of the essential oil of *Mentha xpiperata*. It is derived from pulegone via menthol. Functionally, menthol and other terpenoids of the essential oil of mint are phytoanticipins. a chemical defense mechanism of the plant against insect pests. Worldwide demand for mint oil as over-the-counter medication and food additive is met by plantations in many countries. The challenge is to develop technology that will stimulate essential oil accumulation in cells and organs cultured *in vitro*.

Undifferentiated, finally dispersed cell suspension cultures from a wild variety of aromatic plants typically accumulate only insignificant amount of monoterpenoids. Thus, the production of essential oils by Mentha spp. cultures has been very low (0.1% FW), whereas, intact plants yield oil at 0.1-5.0% FW. The lack of accumulation of terpenoids in undifferentiated cells of mint may not necessarily be due to lack of expression of genes encoding enzymes in terpenoid biosynthesis, rather these enzymes may be expressed only in differentiated cells and tissues i.e. trichomes. Also, essential oils may be subjected to catabolism potentially at a greater rate than synthesis. Finally, the toxicity of oil may become limiting if it exceeds 0.1% FW. With the production of transgenic shoots using disarmed strains of *Agrobacterium tumefaciens,* mint oil resembles that of parental material i.e. menthol, but menthone levels were lower and menthfuran levels were higher than in parent plants.

3.2 Diterpenes—Paclitaxel (Taxol)

Taxol and taxanes are complex oxygenated diterpenes initially isolated from the stem bark of Pacific yew (*Taxus brevifolia*). Little is known about the biosynthesis of paclitaxel and the possibility if its biosynthesis remains frustratingly remote.

Paclitaxel has been effective in the treatment of refractive ovarian cancer, metastatic breast cancer, malignant melanoma and other solid tumours. The world-wide demand for paclitaxel will quickly exceed 100 kg/year. The bark required to yield 1 kg of paclitaxel is about 7000 kg. This situation has mandated the development of alternative source of this substance. The demand for paclitaxel has prompted several biotech laboratories to develop alternative methods for its production. Total synthesis of paclitaxel has been accomplished rather recently. However, it is unlikely that total synthesis will ever contribute significantly to its supply as the process is rather uneconomical. Production by semi-synthetic route from precursor taxanes will, however, replace the bark of *Taxus brevifolia.* Taxanes are readily obtainable from needles harvested from *Taxus* spp. grown in plantations. For the bioengineering of paclitaxel production, prospects of success are slim at the moment as its biosynthetic pathway is undefined. As it stands, 1 kg paclitaxel will require the harvest of 500-1000 bioreactors of 500 L, each filled with a heavy suspension of *Taxus* cells. It is a great achievement that callus and cell suspension cultures of *Taxus* have been established and the problem of browning and deterioration of tissue/callus in culture has been overcome. The production of paclitaxel by some cell lines cultured *in vitro* has equaled, if not surpassed that of *Taxus brevifolia* bark. Remarkable is the enhancement of production by media supplements of phenylalanine. Also, a very encouraging effort for the enhanced production of taxol has been carried out using fungal elicitation.

3.3 Indole Alkaloids—Vinblastin

Vinblastin, a dimeric compound composed of catharanthine and vindoline, two monoterpene indole alkaloids, is found in *Catharanthus roseus.* The biosynthetic pathway has been elucidated. In plants, vinblastine occurs at very low concentrations (5g/2 tonnes FW). Due to its cyto-toxic effect, it may well function as phytoalexin, a chemical defence agent. Ever since 1958, when vinblastin and its analogue vincristine were recognized as anti-cancer agents and established as important drugs in the chemotherapy of Hodgkin's disease and acute leukaemia, its supply has been of great concern. Plantations of *C. roseus* could match the demand, but, for the extreme cost associated with extraction and purification. Alternative sources of higher value are the goals targetted by scientists. The development of cell cultures of *C. roseus* has been established as method in plant physiology to understand the basic processes of its biosynthesis. More recently, these efforts have been complimented with DNA technologies. Despite many years of efforts and remarkable success in increasing the production of catharanthine in cell suspensions, vinblastine has eluded all laboratories. The problem is lack of vindoline synthesis and accumulation in cell suspension cultures. Still, vindoline recurs in shoots regenerated from callus and accumulates in multiple shoot cultures. Hairy root cultures generated by employing *A. rhizogenes* have shown remarkable levels of catharanthine and traces of vinblastine. Enhanced production of catharanthine and vindoline in plants and *in vitro* is expected to be achieved by genetic manipulation of their biosynthesis or by genetically directed changes of the cyto environment. Cryopreservation of C. *roseus* cells has been successfully demonstrated. Finally, given sufficient yields of catharantheine from cell cultures and vindolines from shoot cultures or plants, vinblastine can be obtained through dimerization catalyzed by cell-free extracts of cells cultured *in vitro,* by enzymes of such cultures, and by horseradish peroxidase with 3' 4'-anhydrovinblastine as an intermediate.

3.4 Phenylpropanoids—Vanillin

Vanillin is widely used for flavouring and in the perfumed industry. It is produced through the fermentation of unripe fruits of *Vanilla planifolia* (Orchidacea). Global production of natural vanilla rests on a very narrow genetic base. Tissue culture technology is being used to conserve vanilla germplasm collected from various places. Vanilla is a vine orchid, difficult to grow and fruit set depends on hand-pollinating skills. Hence, there is a need for alternative production of vanillin. Biotransformation of eugenol to vanillin has recently been improved by using microbial catalyst. The biology of vanillin production has not been studied. The surge in demand for natural vanillin may change this situation and drive R & D technology to increase its production. There has been little effort towards breeding of *Vanilla* and as the vanillin occurs not only in beans, but, in several other parts of the *Vanilla* plant, tissue culture is being utilized to produce vanillin *in vitro.* The pathway has not been fully studied yet ferulic acid has been found to be the best precursor for the biosynthesis of vanillin using axenic aerial roots of *Vanilla* cultured in hormone-free medium as the biocatalyst. The aerial root material could be reused several times.

3.5 Flavonoids—Anthocyanins

Anthocyanins are wide-spread metabolites among flowering plants and most conspicuous as pigments in flowers and fruits. These are potential products as marker in experimental biology and as colorant in the market. Because anthocyanins have been discovered to readily accumulate in vacuoles of cells cultured *in vitro*, it was expected that increasing its yield would only be a matter of superior cells selection and optimizing culture conditions. Culture conditions tested for improving anthocyanin synthesis and accumulation are the composition of the medium including hormones, pH and light. Remarkably, cell biomass and anthocyanin content in carrot cultures were highest on media containing a combination of galactose and glucose. An equimolar mixture of both carbohydrates, however, led to a lower performance. Stress clearly affects anthocyanin formation in plants. *In vitro* cultured material exposed to UV irradiation with supplementation of media with carbohydrate showed upward anthocyanin synthesis. With phenylalanine ammonia-lyase (PAL) and chalcone synthase (CHS) identified as key enzymes in anthocyanin biosynthesis and with genes encoding both these enzymes available as cassette, genetic manipulation of cells for improved quality and yield of production has been applied to petunia plants. PAL and CHS transgenes have not yet been expressed in cells cultured *in vitro* for production of anthocyanins.

4. CONCLUSION

Living organisms are source of many valuable products of diverse chemical structures and biological activity. This is a special period of rapid progress in biotechnology. Molecular biology is finally contributing towards the development of new commercial products and new biological methodologies have spawned a new industry known as biotech industry. Research and development of biotechnologies for the production of secondary metabolites has received a major impetus by the very positive response of cells and tissues, roots and shoots, cultured *in vitro.* Levels of production in selected cases have been very substantial, many times the level of the source material (Table 7). Synthesizing a new compound on a laboratory scale is only the first step in a long trek towards producing and marketing a new product. To realize full potential, it must be optimized and scaled up, clinical trials performed to seek approval of appropriate governmental regulatory agencies, and finally the product packaged and marketed. All of these steps require several years and enormous financial costs to complete. Attention and efforts are now being shifted from enhanced production of secondary metabolites in *in vitro* systems to genetic manipulations of productions in plants, accompanied by expectations of directing secondary metabolism towards desirable compounds. Much of the support for R & D has been considered an investment for the human well-being and this century will indeed be dominated by safe bio-products for the benefit of life, the environment, and the survival of human society.

Table 7: Potential Pharmaceuticals from Plant Cell Cultures

Product	*Plant spp.*	*Application*	*Yield (%)*	*Price $/Kg*
Shikonin	*Lithospermum erythrorhizon*	antibacterial (Japan)	12.4	4500
Berberine	*Thalictrum minus*	drug (Japan)	10.6	3250
Diosgenin	*Dioscorea deltoidea*	steroid	3.8	1000
Morphine	*Papaver somniferum*	drug	0.025	340,000
Sanguinarine	*P. somniferum*	antibiotic	2.5	4800
Taxol	*Taxus brevifolia*	anticancer	0.06	0.06M
Vincristine	*Catharanthus roseus*	antileukemic	R&D	20M

FURTHER READING

1. Balandrin, M.F. and Klocke, J.A. In Biotechnology in agriculture and forestry, 4, Medicinal and Aromatic Plants I (Ed. Y.P.S. Bajaj). Springer-Verlag, Berlin, (1988) 3–36.
2. Bhojwani, S.S. and Razdan, M.K. Plant Tissue Culture: theory and practice, a Revised Edition, Elsevier-Amsterdam, (1996) 537–562.
3. Kokate, C.K. Adv Hort vol 11- Medicinal and Aromatic Plants (Eds. K.L. Chadha and R. Gupta) Malhotra Publishing House, New Delhi. (1995) 223–245.
4. Konowicz, K. and Janick, 1. *Hortsci* **19**: (1984) 58.
5. Smith, N.J.H., Williams, J.T., Plucknett, D.L., and Talbott, J.P. Tropical Forests and their Crops, Cornell University Press, Ithaca. New York. (1992).

Cereal Products: Biotechnological Approaches for their Production

S.K. SONI[1] *and* S.S. MARWAHA[2]

[1]*Department of Microbiology, Punjab University, Chandigarh-160 014, India.*
[2]*Punjab State Council for Science & Technology, Chandigarh-160 019, India.*

CEREAL GRAINS ARE the fruit of plants belonging to the grass family (Gramineae). The sustenance provided by cereals is frequently mentioned in the Bible, and they are by many other criterea the most important group of food crops produced in the world. Cereal crops are energy dense, containing 10 000-15 000 kJ/Kg, about 10-20 times more energy than most succulent fruits and vegetables. Nutritionally, they are important sources of dietary protein, carbohydrates, the B complex vitamins, vitamin E, iron, trace minerals, and fiber. It has been estimated that global cereal consumption directly provides about 50 percent of protein and energy necessary for the human diet, with cereals providing an additional 25 percent of protein and energy via livestock intermediaries. Some cereals, notably wheat, contain proteins that form gluten, which is essential for making leavened bread. Although dried cereal grains constitute living cells that respire, when kept in an appropriate environment, whole grains can be stored for many years. In 1996, world cereal production amounted to more than two billion metric tons. Major cereal crops produced worldwide include wheat, rice, maize and barley. Other cereal crops produced include sorghum, oats, millet and rye. Asia, America, and Europe produce more than 80 percent of the world's cereal grains. Wheat, rice, sorghum, and millet are produced in large quantities in Asia; corn and sorghum are principal crops in America, and barley, oats and rye are major crops in the former USSR and Europe.

Cereals have a variety of uses as food. Only two cereals, wheat and rye, suit to the preparation of leavened bread. The most general usage of cereals is in cooking, either directly, in the form of grain, flour, starch, or as semolina, etc. Another common usage of cereals is in the preparation of alcoholic drinks such as whiskey and beer (barley; sorghum), vodka (wheat), American bourbon (rye), Japanese sake (rice), etc. A variety of unique, indigenous fermented foods, other than leavened breads and alcoholic beverages, are also produced in regions of the world that rely mainly on plant sources of protein and calories. In developed countries, that obtain most of their protein from animal products, cereals are increasingly used as animal feed. More than 70 percent of the cereal crop produced in developed countries is fed to livestock, whereas, in developing countries, 68-98 percent of the cereal crop is used for human consumption. There is evidence that the people of Uadi el-Natuf Tell of Southeast Asia were the first grain cultivators at about 7 800 B.C (Table 1). By 5 000 BC, wild animals became rare, forming only 5 percent of the diet, while cereals and farmed animals provided a sizeable part of human food. Early wild wheat (*Triticum*) and barley (*Hordeum*) species were diploid, carried few seeds, and abscissed from the plant at maturation, thus making harvest difficult. Polyploid plants can originate in nature, but, have little chance for self-propagation without cultivation. The cultivation and irrigation of cereals allowed the expansion of polyploid grains. The polyploid grains exhibit less genetic variation, since each gene is represented in several copies, causing more genetic uniformity and considerable increase in stability and yield. The first stable lines of polyploid cereals were identified as early as 6 000 BC.

Table 1. Estimates of origin and early cultivation of cereals

Cereal	*Time*	*Location*
Wheat	7000 BC	Near East
Barley	7000 BC	Near East
Rice	4500 BC	Asia
Maize	4500 BC	Central America
Millet	4000 BC	Africa
Sorghum	4000 BC	Africa
Rye	400 BC	Europe
Oats	100 AD	Europe
Triticale	1930 AD	USSR, Europe

Cereals are consumed as such or after being acted upon by microorganisms and thus are known as fermented foods. In general, animal or plant tissues subjected to the action of microorganisms and/or enzymes to give desirable biochemical changes and significant modification of food quality are referred to as fermented foods. The reports about the origin of fermented foods are lacking, but, it is believed that the discovery of fermentations was purely by chance. The transition from hunting and gathering food to organized food cultivation and production is believed to have taken place some 10000 to 15000 years ago in the Middle East. It is believed that it was purely by chance that the fermentation became a method of preservation. Fermentation is the oldest known form of food biotechnology. Records of barley conversion to beer date back more than 5000 years. Microbial activity in foods makes them more attractive in terms of appearance and flavours. Such foods are more appetizing, easily digestible and form an important sector of food industry. The traditional fermentation of foods serves several functions including:

- the enrichment of the diet through development of a diversity of flavors, aromas, and textures in food substrates,
- preservation of substantial amounts of food through lactic acid, alcoholic, acetic acid, and alkaline fermentations
- enrichment of food substrates biologically with protein, essential amino ac-

ids, essential fatty acids, and vitamins
- detoxification during food fermentation processing
- decrease in cooking times and fuel requirements
- improvement of the digestibility of the ingredients for human consumption and protection against the food poisoning and transmission of pathogenic microorganisms
- Fermented foods add to the variety of foods available

The biological agents involved in fermentations include filamentous fungi, bacteria and yeasts occurring in succession or in combination. In general, filamentous fungi bring about the saccharification and proteolysis of the starting materials, bacteria cause acidification and leavening while yeasts ferment the sugars to produce alcohol and esters which impart desirable flavours to the products. A review of published reports shows that filamentous fungi belonging to *Aspergillus, Mucor, Neurospora* and *Rhizopus*, bacteria belonging to *Leuconostoc mesenteroides*, lactobacilli, *Streptococcus faecalis*, *S. pediococci* and yeasts belonging to *Saccharomyces cerevisiae*, *Candida* spp, *Debaryomyces hansenii* and *Hansenula anomala* are most commonly associated with the traditional fermentations. These organisms are generally contributed by the ingredients, workers and the surroundings. Since the source of inoculum is not apparent, the study of biological agents actually associated with fermentations has attracted considerable attention over the years. Ordinarily the microbes developing during the initial soaking of ingredients and later are sufficient to bring about the fermentations. Presoaking of the ingredients release the free sugars and non-protein nitrogen which support the growth of lactic acid bacteria. Spices, as added in some ingredients, also enhance the food fermentations by stimulating growth and proliferation of certain lactic acid bacteria and accelerating lactic acid production. Seasonal variations affect qualitative and quantitative prevalence of bacteria and yeasts in different traditional fermented foods with summers favouring the former and winters favouring the later.

The content and quality of cereal proteins may be improved by fermentation. Fermentation of cereals increases their relative nutritive value and available lysine. Bacterial fermentations involving proteolytic activity increase the biological availability of essential amino acids more so than yeast fermentations which mainly degrade carbohydrates. Starch and fiber tend to decrease during fermentation of cereals. Although fermentation does alter the mineral content of the product, the hydrolysis of chelating agents such as phytic acid during fermentation, improves the bioavailability of minerals. Changes in the vitamin content of cereals with fermentation vary according to the process and the raw material used in the fermentation. B group vitamins generally show an increase on fermentation.

Generally the traditional methods of preparing indigenous fermented foods are simple and inexpensive. However the ancient methods of making such foods are changing rapidly through modern microbial technology. Some of the fermented products, widely used throughout the world, including alcoholic products, bread & related products, dairy products, coffee and cocoa have been studied extensively for many decades and their production methods have been modernized keeping pace with the advancement in modern science. During the last few years much interest has been generated in the fermented foods of Asian and African countries where such foods are still being manufactured according to the traditional, technologically less advanced methods using simple equipment and are produced at cottage industry scale by means of natural microflora from the staples and the surroundings. The technology of producing many indigenous fermented foods from cereals remains a household art in these countries.

1. COMMERCIALIZED FERMENTED CEREAL PRODUCTS OF THE WORLD

Wheat, rice, corn, sorghum and rye are the principal cereals being widely used throughout the world for the preparation of some fermented cereal products which have already been studied in depth and have been commercialized. There are two categories of such products including alcoholic beverages and non-alcoholic staple breads.

1.1. Alcoholic beverages

Alcohol fermentation was the first fermentation known to mankind. Distilleries began to appear in Europe in the middle of the seventeenth century. At first alcohol was used only for human consumption. Lately the demand for alcohol as a universal solvent and chemical raw material developed. Ethanol is the largest volume (liquid) industrial fermentation world-wide and is gaining significance due to its potential use as an alternate liquid fuel in place of gasoline. The fermented alcoholic products can be classified in two main categories, fermented & non-distilled and, fermented & distilled. The former includes beers, having an alcohol content of 4-6% while the latter includes whiskies, gin and vodka which are prepared by distillation of the fermented liquids and have a much higher alcohol content of 42.8%. Beer is traditionally prepared from barley but other cereals including wheat, rice corn can also be used in combination with barley. There are two main categories of beers depending upon the type of the yeast used and the alcohol content: lager beer, fermented by bottom fermenting yeast and ale beer, fermented by top fermenting yeast. The distilled alcoholic beverages can be prepared from a variety of cereals grains and the preparation of these involves an additional step, called as distillation which involves the separation of purified and concentrated ethyl alcohol which is a high value product and can be used in various ways including in the preparation of potable beverage blends, organic chemicals, drugs and pharmaceuticals, pesticides, plastics, textiles, agriculture and as a substitute of petrol. The major steps involved in processing the cereal grains to ethanol include (i) feed stock preparation (milling—converts the starch bearing grains to a defined particle size and quality in a wet mill or a dry mill), (ii) starch conversion (liquefaction and saccharification) to fermentable sugars using α-amylase and glucoamylase, (iii) establishing conditions for good yeast viability and fermentation of fermentable sugars by *Saccharomyces cerevisiae*, (iv) distillation of ethanol produced by fermentation and (v) removal and processing of the stillage and various byproducts for animal feed.

1.2. Bakery products

Baked foods are produced and consumed in most of the countries of the world as the staple food. Considerable variation exists in the type of baked foods, that are made from country to country and among regions within a given country.

1.2.1. Bread

Bread is one of the most commonly used baked food and has traditionally been an important factor in human nutrition. In many countries bread consumption accounts for a large part of the diet. Bread was first baked in Egypt over 6000 years ago and from that time and place, travelled west, varying in ingredients, flavour, texture, and form. European and Middle East immigrants brought recipes for their native breads and these have been handed down to second and third generations. There are various types of bread available in the world and these can be classified in different ways, some of which are summarized as follows:

Leavening: Breads may be unleavened such as chappatti, leavened with yeast, such as pan breads, chemically leavened, such as Irish soda bread; or leavened with bacteria, such as salt rising bread.

Formulation: Many breads around the world are made with lean formulas comprising wheat flour, water, yeast and salt, while some breads such as certain premium breads also contain considerable amounts of other ingredients such as fat, eggs and milk.

Shape and size: A virtually endless variety of shapes and sizes have evolved; some breads are baked in pans while many are baked in open hearth.

Specific volume: This factor can vary from the low-density, high volume white pan breads in North America to the dense, dark breads of Central Europe.

Crust characteristics: Some breads, such as Vienna loaves, have thin crispy, while others, such as German pumpernickel breads have thick crusts. Colours range from quite light to dark colours.

Crumb Characteristics: Continuous mix breads in the United States have a very fine, uniform crumb structure, while other breads such as French baguettes have coarse, irregular structures. Some Middle East breads have little crumb structure at all, being comprised largely of crust because of the high baking temperatures that are used.

The bread dough is usually made from white wheat flour and leavened by bread yeasts which ferment the sugars in the dough and produce mainly carbon dioxide and alcohol. However, other actively gas forming microorganisms such as wild yeasts, coliform bacteria, saccharolytic *Clostridium* spp., heterofermentative lactic acid bacteria and various naturally occurring mixtures of these organisms have been used instead of bread yeasts for leavening. The actual baking process is really the last and most important step in the production of bakery products. Through heat, an unpalatable dough mass is transformed into a light, porous, easily digestible and most appetizing product. The biological activities that have been taking place in the dough are stopped by baking, with the destruction of microorganisms and enzymes present.

1.3. Bread like products

A large number of other bread-like, leavened and baked foods are also produced in different parts of the world. Some of them are discussed below:

1.3.1. Sourdough bread

This is usually made from a mixture of rye and wheat flour containing upto 80% of wheat flour. The acid taste may be obtained by adding mixtures of lactic and acetic acids. Such breads are usually leavened with yeast and the action of lactic acid bacteria. In Europe and for some American rye sour breads the use of sourdough starters are preferred because it gives the full flavour which depends upon the microbial souring action. The use of such rye sour cultures predates recorded history. It is relatively easy to produce such starter by mixing rye flour and water in equal proportions and permitting it to ferment for several days at 30°C. This forms the basis of the culture which can then be maintained by further additions of rye flour and water. Sourdough starter cultures are also available commercially and these can be used to inoculate the doughs. Such cultures usually contain mixtures of lactic acid bacteria. Ordinary commercially available baker's yeast (*S. cerevisiae*) usually contains 10,000 to several million lactic acid bacteria per g and its use will result in souring of the dough if the fermentation is carried out for prolonged periods of time and at relatively elevated temperatures. A white sourdough bread produced only in the western United States is the San Francesco sourdough bread. It is of special interest because it is still produced by propagation with sponges from the previous day's production. The characteristic yeast of this process is *Torulopsis holmii* and the bacterium producing both lactic and acetic acids has been identified as *Lactobacillus sanfrancisco* . The sponges contain only water, flour and microorganisms and require about 8 h at 25°C for full development. The final pH may be as low as 3.9. About 20% of the starter sponge is used in the dough.

1.3.2. Rye bread

This is usually made from rye flour with or without a starter or sour, The old method of preparing sour bread depend upon the bacteria naturally present in a mixture of rye flour and water. The mixture is allowed to ferment for 5 to 10 h; then more flour and water are added and the fermentation is continued for additional 5 or 6 h. Half of the sour is incorporated in the sponge or dough for the bread and the rest is carried over to start a new sour. This sour was modified by some bakers by the addition of yeast and lactic acid bacteria from Cultured buttermilk or Bulgarian buttermilk to a sour that was made anew daily. Modern methods involve addition of considerable amounts of cultures of acid-forming bacteria to the dough mass to be used as a sour and controlling the time of fermentation (18-24 h) and the incubation temperature (about 25°C). An excessively high incubation temperature, for example, 32-35°C, favours the growth of undesirable gas-formers, such as coliform and butyric bacteria. Some bakers prefer to use low-temperature lactic acid bacteria, whereas, others use high-temperature lactics, and adjust the incubation temperature accordingly. The growth of heterofermentative lactic acid bacteria is considered desirable. The starter imparts a desired sour flavour to the rye bread.

2. INDIGENOUS FERMENTED FOODS FROM CEREALS

Indigenous fermented products form an important part of human diet in Indian subcontinent, the Middle East, Africa and Latin America. People of Asia are the pioneers in the development of means of fermenting vegetable proteins to produce meat like flavours. Indonesians developed the fermentation methods for introducing meat like texture into vegetable products. Koreans are credited for the production of acid fermented vegetables. People of Egypt developed wheat breads leavened with yeasts while Indians discovered methods for souring and leavening cereal-legume batters. There are several varieties of fermented cereals commonly used in various geographical regions of the world differing from each other on the basis of the substrates employed, the method of the preparation, the nature and uses of various products.

2.1. Indigenous Fermented cereals of the Asian-pacific countries

The Asia-Pacific region is characterized by its tropical and subtropical humid climate, which is suitable for rice paddy cultivation and mold growth. The consumption of rice as a staple food, and the high population density which limits animal husbandry practices in this region, has resulted in a typical food processing technology—cereal fermentation with molds. Molds and other microorganisms convert unpalatable carbohydrates of low digestibility and proteins into palatable sugars and amino acids, respectively, with a high conversion efficiency. Cereal consumption in the Asia-pacific region varies in accordance with geographic and climatic conditions. Inhabitants of the tropical South-Eastern regions consume primarily rice, while those in subtropical and temperate zones of the North-Eastern region including Northern China, Korea and Japan consume wheat, barley, corn, millet and soybeans in addition to rice. These cereals are the basis of a large number of indigenous fermented foods in Asia which are classified in accordance with the raw materials used, type of fermentation, usage of the product, and the area of production. In general, fermented products are classified according to usage of the products and the major fermentation process taking place; e.g. alcoholic foods and beverages, vinegars, breads, fermented porridges and snacks.

2.1.1. Alcoholic Food and Beverages

Alcoholic and the acidic beverages are the important fermentation products of cereals in the Asian region. Alcohol fermentation is more important than acid fermentation in this region in terms of the relative amounts of cereals used for fermentation and the varieties of the products produced. There are two main categories of the indigenous alcoholic beverages. Clear products which are generally referred to as *shaosingjiu* in China, *chongju* in Korea and *sake* in Japan, contain at least 15% alcohol and are designated as rice-wine, while turbid beverages, such as *takju* in Korea and *tapuy* in the Philippines which contain less than 8% alcohol along with suspended insoluble solids and live yeasts, are referred to as rice-beer. Examples of alcoholic beverages prepared from cereals in Asia-Pacific region are listed in Table 2.

Table 2: Examples of cereal alcoholic beverages prepared in the Asia-Pacific Region

Product name	*Country*	*Major Ingredients*	*Microorganisms*	*Appearance & Usage*
		Ice Wine		
Shaosinghjiu	China	rice	*S. cerevisiae*	clear liquid
Chongju	Korea	rice	*S. cerevisiae*	clear liquid
Sale	Japan	rice	*S. sake*	clear liquid
		Rice Beer		
Takju	Korea	rice, wheat	Lactic acid bacteria *S. cerevisiae*	turbid drink
Tapuy	Philippines	rice, glutinous rice	*Saccharomyces, Mucor Rhizopus Aspergillus Leuconostoc L.Plantarum*	sour, sweet liquid, paste
Brembali	Indonesia	glutinous rice	*Mucor indicus Candida*	dark brown liquid sour, alcoholic
Jaanr	India Himalaya	finger millet	*Hansenula anomala Mucor rouxianus*	sweet-sour alcoholic paste; mix with water
		Alcoholic Rice Paste		
Khaomak	Thailand	glutinous rice	*Rhizopus, Mucor Saccharomyces Hansenula*	semisolid, sweet alcoholic
Tapai pulut	Malaysia	glutinous rice	*Chlamydomucor Endomycopsis Hansenula*	semisolid, sweet alcoholic
Tape-ketan	Indonesia	glutinous rice rice	*A. rouxii E. burtonii E. fibulinger*	sweet-sour alcoholic paste
Lao-chao	China	rice	*Rhizopus A. rouxii*	paste
Bhattejaanr	India Sikkim	glutinous rice	*Hansenula anomala Mucor rouxianus*	sweet-sour alcoholic paste
		Alcoholic Rice Seasoning		
Mirin	Japan	rice, alcohol	*A. oryzae A. usamii*	clear liquid seasoning

The process of cereal alcohol fermentation using mould starters was well established in the year of 1000 BC, and 43 different types of cereal wines and beers were described with detailed processing procedures in Chi-Min -Yao-Su (AD 530-550). Millet appeared to be the main ingredient for alcohol fermentation. Among the 43 product types described, 16 were prepared from millet, 11 from rice and 12 from Indian millet. Figure 1 compares the processing methodology for the preparation of *quing-chu-jiu* as appeared in Chi-Min-Yao-Su with that for traditional *chong-ju* prepared in Korea. Cereal wines are generally prepared at low temperatures around 10°C which improve the taste and keeping quality of the wine, commonly known as *Shaosinghjiu* in China and *Chongju* in Korea. The dried and powdered starter is mixed with water and steamed. The steamed grains are then fermented for 2-3 weeks or up to 5-7 months depending on the brewing method. The traditional method of rice-wine brewing was industrialized by Japanese brewers in the late 19th century, who adopted pure starter culture manufacturing technology from Europe and transferred it to Korea and China.

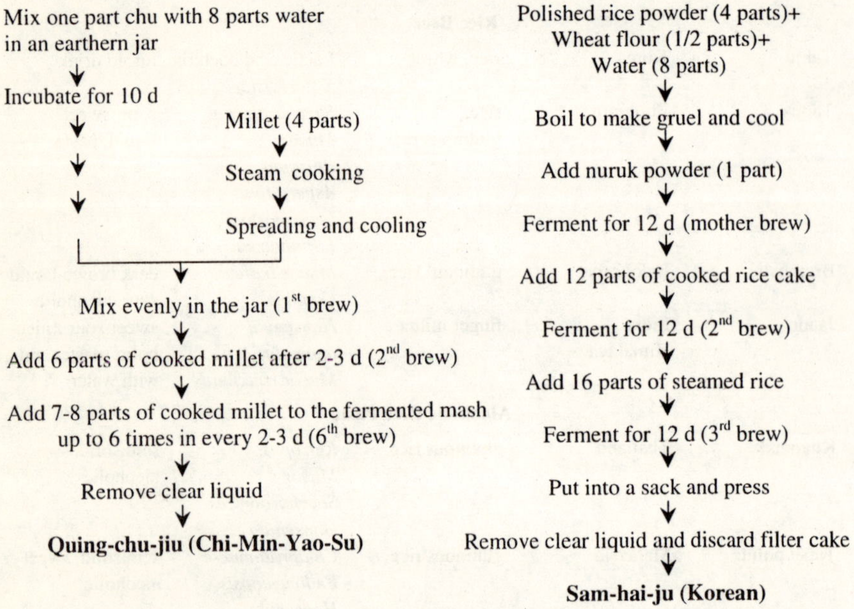

Fig. 1: Flow sheet for the production of rice-wines in China and Korea

Figure 2 shows the Japanese process for preparing rice-wine. Cereal beers are produced at a higher fermentation temperature around 20°C. The fermentation starter powder is mixed with cooked cereals, incubated at approximately 20°C for 2-3 d, following which it is filtered through a fine mesh sieve or cloth. Figure 3 compares the preparation of Korean *takju* with that of Philippine *tapuy* Similar products such as *brem bali* in Indonesia and *jaanr* and *bhatte jaanr* in Sikkim, India are also prepared in other Southeast Asian countries.

The alcoholic fermentation is also used for the preparation of paste-type food products including *tape ketan*, an alcoholic rice paste, *Mirin* is a traditional Japanese alcoholic seasoning prepared from rice and *koji.* The fermentation starter also referred to as *chu* in China, *nuruk* in Korea, *ragi* in Southeast Asia and *bakhar ranu* or *marchaar* (*murcha*) in India. The digestion of rice by native enzymes in *koji* is

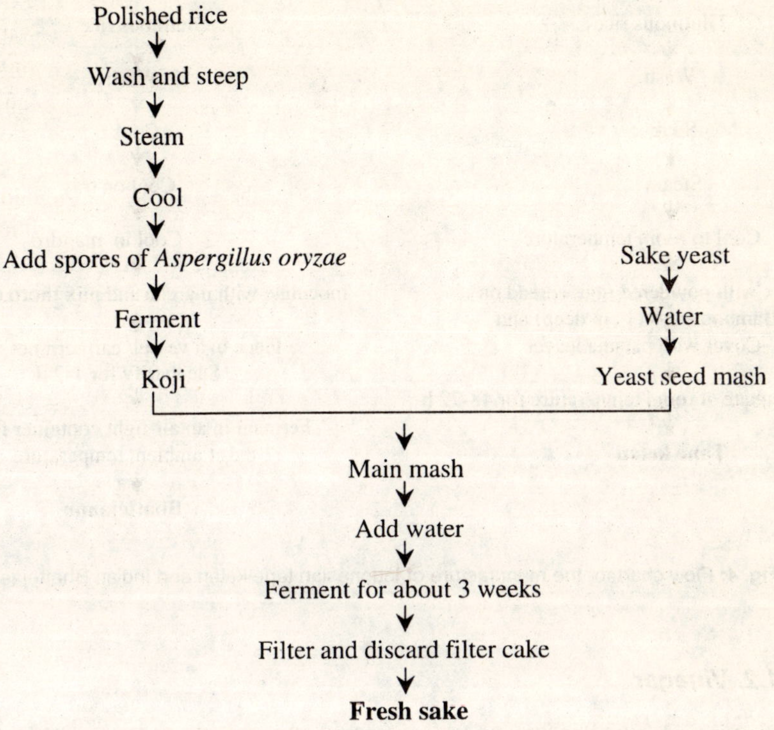

Fig. 2: Flow sheet for the Japanese sake brewing process

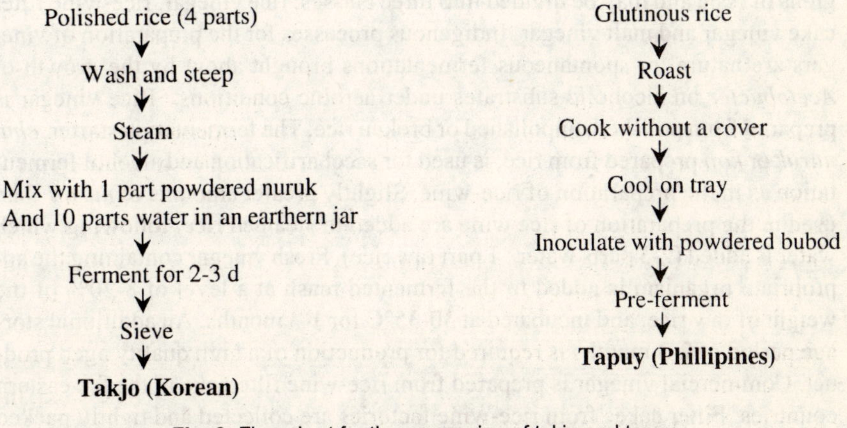

Fig. 3: Flow chart for the processing of takju and tapuy

carried out in ethanol in order to prevent microbial contamination. *Mirin*, the final product, is a clear liquid containing approximately 40% sugar, which is produced as a result of starch hydrolysis by *koji* enzymes. Figure 4 outlines the processing procedure for the manufacture of *tape ketan* in Indonesia and *bhattejaanr* in East Sikkim, India.

Glutinous rice
↓
Wash
↓
Soak
↓
Steam
↓
Cool to room temperature
↓
Mix with powdered ragi; spread on Bamboo trays (1 cm deep) and Cover with banana leaves
↓
Incubate at room temperature for 48-72 h
↓
Tape ketan

Glutinous rice
↓
Wash
↓
Cook
↓
Cool on tray
↓
Cool in mandro
↓
Inoculate with marcha and mix thoroughly
↓
Place in a vessel/ earthern pot and Saccharify for 1-2 d
↓
Ferment in an air-tight container for 2-8 d at ambient temperature
↓
Bhattejaanr

Fig. 4: Flow chart for the manufacture of Indonesian tape ketan and Indian Bhattejaanr

2.1.2. Vinegar

Vinegar production is as ancient as is alcoholic fermentation, since acetic acid is produced in any natural alcoholic fermentation upon exposure to air. Vinegar prepared from cereal alcoholic fermentations is widely used in North-Eastern regions of Asia and may be divided into three classes: rice vinegar, rice-wine filter cake vinegar and malt vinegar. Indigenous processes for the preparation of vinegars are natural or spontaneous fermentations brought about by the growth of *Acetobacter* on alcoholic substrates under aerobic conditions. Rice vinegar is prepared from polished, unpolished or broken rice. The fermentation starter, *chu, nuruk* or *koji* prepared from rice, is used for saccharification and alcohol fermentation as in the preparation of rice-wine. Slightly greater amounts of *nuruk* than used in the preparation of rice wine are added to steamed rice, following which water is added (2-3 parts water : 1 part raw rice). Fresh vinegar containing the appropriate organism is added to the fermented mash at a level of 8-20% of the weight of raw rice, and incubated at 30-35°C for 1-3 months. An additional storage period of 2-3 months is required for production of a high quality aged product. Commercial vinegar is prepared from rice-wine filtercake in the far-eastern countries. Filter cakes from rice-wine factories are collected and tightly packed into a storage tank for 1-2 years. The filtercake contains large amount of unused carbohydrates and proteins, which are further hydrolyzed by inherent microorganisms and enzymes during storage, converting them into alcohol and other nutrients and flavour substances. The cake is slurried in two to three volumes of water prior to filtration. The filtrate is fermented with *Acetobacter* to produce filtercake vinegar.

2.1.3. Fermented Breads, Noodles and Porridges

Large quantities of acid leavened bread and pancakes are consumed daily in India, Sri Lanka, Pakistan, Nepal, Sikkim, Tibet and neighboring countries. *Idli*, *dosa* and *dhokla* are produced primarily in South India and Sri Lanka, and *jalebies* and *Kulcha* are consumed throughout India, Nepal and Pakistan. Various acid-leavened bread and noodles prepared in the Asia-Pacific region are described in Table 3.

Table 3: Examples of acid-leavened bread and noodles used in Asia-Pacific region

Product Name	*Country of Use*	*Major Ingredients*	*Microorganisms*	*Appearance & Usage*
Idli	South India Sri Lanka	Rice grits; black gram powder;	*L. mesenteroides* *S. fecalis* *T. candida* *T. pullulans*	Steamed cake
Dosa	India	Rice flour; black gram powder;	*L. mesenteroides* *S. faecalis* *T. candida* *T. pullulans*	Griddled pancake
Dhokla	India	Rice bengal gram	*L. mesenteroides* *S. faecalis* *T. candida* *T. Pullulans*	Steamed cake
Jalebies	India Nepal Pakistan	Wheat flour	*S. bayanus*	Pretzel-like confection
Mantou	China	Wheat flour	*Saccharomyces*	Steamed cake
Kichudok	Korea	Rice, takju	*Saccharomyces*	Steamed cake
Puto	Philippines	Rice, sugar	*L. mesenteroides* *S. faecalis* *yeast*	Steamed cake
Brem	Indonesia	Glutinous rice		Cake
Mungbean starch	China Thailand Korea Japan		*L. mesenteroides* *L. casei* *L. cellobiosus* *L. fermenti*	Noodle
Khanomjeen	Thailand	Rice	*Lactobacillus sp.* *Streptococcus sp*	Noodle
Me	Vietnam	Rice	Lactic acid bacteria	Sour food ingredient

2.1.3.1. Idli and dosa: Idlis are small, white acid-leavened and steamed cakes prepared by bacterial and yeast fermentation of a thick batter prepared from carefully washed coarsely ground rice and finely ground dehulled black gram dhal. Acidified and leavened foods like Indian idli, constitute a very interesting group of cereal-pulse based foods which have considerable potential in the developing and under-developed world. Idli is closely related to sour dough bread of western world, but, does not depend upon wheat or rye as a source of protein to retain CO_2 during leavening which is produced by bacterial rather than yeast activity.

Dosa batter is very similar to idli batter, except that the rice and black gram are finely ground. Following fermentation, the dosa is quickly fried as a thin, fairly crisp pancake and eaten directly. Dhokla is similar to idli except that dehulled Bengal gram dhal is used instead of black gram dhal in its preparation. The fermented batter is poured into a greased pie pan, and steamed in the open rather than in a covered idli steamer. Figure 5 shows the flow chart for idli production.

Both bacteria and yeasts have been reported to participate in Idil/dosa fermentation. However, acid and gas production has been found to be mostly dependent upon the bacterial growth. Some people add yeasts and buttermilk as the inocula, but, Idli/dosa prepared with added yeast and/or sour butter milk inoculum is generally similar to that prepared without inocula. Bacteria identified as part of the microflora responsible for the production of good Idli include *Leuconostoc mesenteroides, Lactobacillus coryneformis, L. delbrueckii, L. fermentum, L. lactis, Streptococcus faecalis* and *Pediococcus cerevisiae,* while the yeast flora generally involved in fermentation include *Torulopsis candida, Trichosporon pullulans, Candida cacaoi, C. fragicola, C. kefyr, C. glabarata, C. troipicalis, C. sake, C. krusei, Debaryomyces tamarii, Rhodotorula graminis, Wingea robertsii, Issatchenkia terricola, Hansenula anomala,* and *Torulopsis holmii. Leuconostoc mesenteroides* and *Streptococcus faecalis* develop concomitantly at the soaking stage, and then continue to multiply following grinding. *L. mesenteroides* is considered to be the microorganism essential for leavening of the batter and also responsible, along with *S. faecalis*, for acid production in idli, dosa and related products.

Black gram Dhal	White polished rice
↓	↓
Wash and soak for 5-10 h	Wash and soak for 5-10 h
↓	↓
Grind finely in a mortar	Grind coarsely in a mortar

↓

Combine the slurries into a thick batter and mix well

↓

Add salt for seasoning (approximately 1%, w/v)

↓

Incubate overnight in a warm place (30-32°C)

↓

Pour batter into small cups in idli cooker

↓

Steam for 10 min

↓

Ready for consumption

Fig. 5: Flow chart for the production of traditional Indian Idli

Two more significant changes during *Idli* fermentation are leavening and acidification of the batter and these have been used as criteria for judging the progress in fermentation. During fermentation, the volume of *Idli* batter increases and the pH decreases. Soluble solids and non-protein nitrogen generally show a slight increase during fermentation, whereas soluble nitrogen , amino nitrogen and reducing sugars have been found to decline. *Idli* is generally considered a nourishing and easily digestible food. Fermentation improves the nutritional value and digestibility of proteins of rice and black gram mixture . Feeding of fermented *Idli* reduces the liver fat contant of persons fed with a high fat and low protein diet. Fermentation also brings about an increase in soluble solid contents, the reducing sugars, soluble nitrogen and free amino acids. Availability of essential amino acids, choline, methionine, cystine, water soluble group-B vitamins including thiamine, riboflavin, cyanocobalamin , folic acid and ascorbic acid (vitamin C) also increase significantly.

2.1.3.2. Bhatura: This is a leavened, deep fried bread made from white wheat flour, consumed as a breakfast food and snacks with masala chana in Northern India. Dough is prepared by mixing white wheat flour with water and kneading for 20 min. at room temperature. The dough is supplemented with 1% common salt and inoculated with a portion of ripe batter from previous batch and kept in open to undergo fermentation for 8-14 h. The dough is then made into balls, which are then flattened by hand using vegetable oil to 12-18 cm in diameter before deep frying in oil.

2.1.3.3. Kulcha: This is a popular North Indian bread eaten with vegetables and meat and is generally used as a staple diet. Dough preparation is usually made by the straight dough method. White wheat flour, sugar, portion of the dough from previously fermented batch and water are mixed (100:1:10:50) at room temperature, kneaded by hand and allowed to ferment overnight in earthern pots. Some times gram flour (*Cicer aerietinum* L.) is also added at 20% level. After fermentation the dough is worked into balls, placed on a smooth surface, lightly sprinkled with flour and flattened by hand to 15-20 cm in diameter. Smoothly flattened dough is transferred onto a metallic baking tray or scooped in a special oven, made of clay. The baking is carried out for 30-40 min to get the final product. In some areas of Pakistan the dough is raised with low level of inoculum but fermentation is prolonged which gives characteristic flavour and taste to the product whereas, in some parts of India commercial baker's yeast and some chemical leaveners are also added to raise the dough.

2.1.3.4. Nan: This is leavened flat bread with central pouch. Nan is generally consumed as a staple food by the people of India, Pakistan, Afganistan and Iran. Its dough is raised by using white wheat flour, salt, inoculum from previous batch and water (100:2:10:50) which are then kneaded with hand at room temperature and allowed to ferment in open for 6-8 h. Sometimes the dough is prepared by sponge dough method when yoghurt is added as inoculum alongwith baker's yeast and some chemical leaveners. The flattened dough is then transferred onto a circular pad of cotton cloth and plastered onto the inner wall of a clay oven called tandoor. Before baking, some spices like caraway seeds (*Carum carvi*), coriander (*Coriandrum sativum*) and kalaunji (*Bochanania lauzan*) are sprinkled on it. The finished product is obtained by baking at 120-150°C for 10-15 min.

2.1.3.5. Jalebies: These are pretzel like, syrup-filled confections prepared from deep-fried-, fermented wheat-flour dough. Mixed fermentation is by lactic acid bacteria, *Lactobacillus fermentum, L. lactis and L. buchneri with S. lactis* and *Streptococcus faecalis.* The yeast *Saccharomyces cerevisiae* is also active, producing CO_2 and causing about a 10% increase in volume during fermentation.

2.1.3.6. Miscellaneous breads: Other types of breads prepared primarily by acid fermentation of rice flour dough include Korean kichudok and Philippines puto. These products are leavened, steamed rice cakes which are similar to Indian idli, except for the fact that they do not contain any legumes. Kichudok is prepared at the household level in Korea and is consumed on special occassions, while puto is normally consumed as a breakfast and snack in Philippines. Puto is a common food of the lower-income group, but, varieties containing added cheese, egg etc., are consumed as delicacies by higher-income groups. Figure 6 outlines methodologies used for the preparation of kichudok and puto.

Thai rice-noodle khanom-jeen, is also prepared from acid-fermented rice. Soaked rice is drained and fermented for at least 3 d prior to grinding. *Lactobacillus* species and *Streptococcus* species are involved in the acid fermentation.

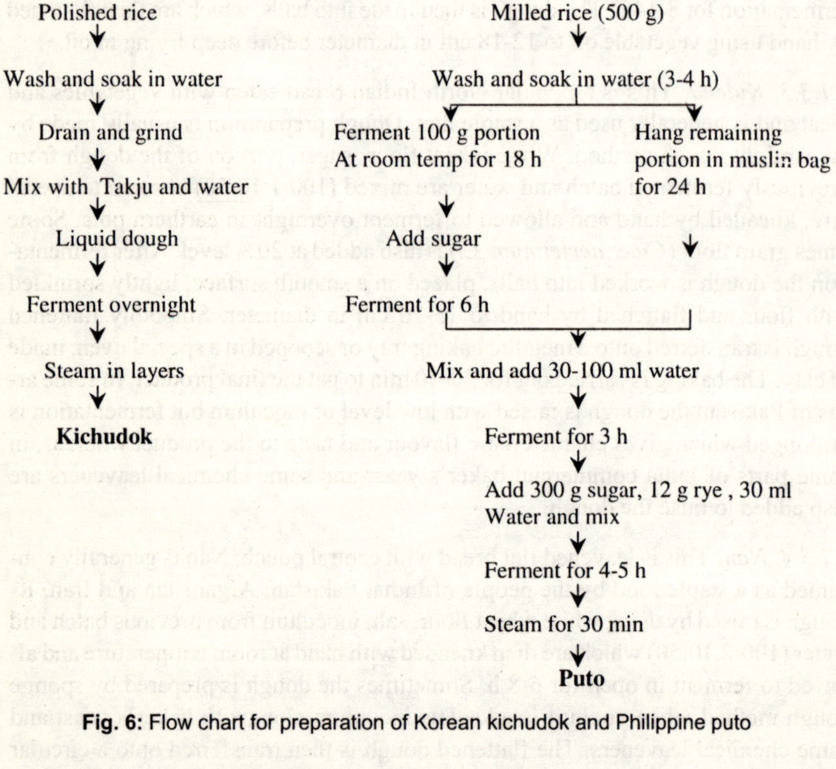

Fig. 6: Flow chart for preparation of Korean kichudok and Philippine puto

2.2. Indigenous Fermented cereals of African countries

Africa is one of the lowest producers of cereals globally, but, these are more widely utilized as food in African countries, than in the developed world. Major cereals grown in Africa include maize, rice, sorghum and millet A majority of traditional cereal-based foods consumed in Africa are processed by natural fermentation. Fermented cereals are particularly important as weaning foods for infants and as dietary staples for adults. Some are used as beverages and breakfasts or snack foods while a few are consumed as staples and weaning foods. There are three main categories of fermented foods in Africa, the fermented gruels, breads and pancakes (Table 4) and beverages (Table 5).

Table 4: Fermented gruels, breads and pancakes in Africa

Product name	*Area of production*	*Substrate*	*Microorganisms involved*	*Textural characteristics of product*
Ogi	Nigeria, Benin	Maize, sorghum or millet	*Lactobacillus* sp. and yeasts	Soft or stiff gel
BogobeKoko and kenkey	Botswana, Ghana	Sorghum, maize,	*Lactobacillus* sp. and yeasts	Porridge Dough
Mawe	Dahomey	Sorghum or millet	*L. fermentum*, *L. cellobiosis*, *L. brevis*, *Candida Krusei* and *S. cerevisae*	Dough
Mahewu (magou)	South Africa	Maize	*Streptocoocus lactis* *L. delbrueckii*, and *L. bulgaricus*	Liquid
Uji	East Africa	Maize sorghum or millet	*Lactobacillus* sp.	Liquid
Kisra	Sudan	Maize, sorghum or millet Sorghum	Unknown	Dough
Injara	Ethiopia	Sorghum	*Candida guilliermondii*	Dough

Table 5: Alcoholic beverages produced from cereals in Africa

Product name	*Area of Production*	*Substrate*	*Starter*	*Textural characteristics of product*
Burukutu	Ethiopia Nigeria (north)	Guinea corn and cassava	Yeasts and lactic acid bacteria	Liquid
PitoKaffir beer	Northern Ghana	Guinea corn and maize	Moulds, yeast and *Lactobacillus* sp.	Liquid
Busaa (maize beer)	Nigeria (Bendel) Ghana	Kaffir corn (or maize)	*Lactobacillus* spp. and yeasts	Liquid
Malawa beer	South Africa	Maize	Yeasts and *Lactobacillus* spp.	Liquid
Zambian opaquemaize beer	East Africa, Uganda, Zambia	Maize	*Candida krusei*	Liquid
Merissa	Sudan	Sorghum	Yeasts, Lactic acid bacteria, acetic acid bacteria	Liquid
Seketeh	Nigeria (south)	Maize	Unknown	Liquid
Bouza	Egypt	Wheat or maize	Unknown	Liquid
Talla	Ethiopia	Sorghum	Unknown	Liquid
Kishk	Egypt	Wheat and milk	*Lactobacillus* spp., yeasts and *Bacillus* spp.	Liquid

2.2.1. Fermented gruels

2.2.1.1. Ogi: This is a porridge prepared from fermented maize, sorghum or millet in West Africa. It is a staple of that region, and serves as a weaning food for infants. The traditional preparation of ogi involves soaking of corn kernels in water for 1 to 3 d followed by wet milling and sieving to remove bran, hulls and germ (Fig. 7) . The pomace is retained on the sieve and later discarded as animal feed while the filtrate is fermented for 2-3 d to yield ogi, which is a sour, white starchy sediment. Ogi is often marketed as a wet cake wrapped in leaves or transparent polythene bags. It is diluted to a solids content of 8 to 10% and boiled into a pan, or cooked and turned into a stiff gel called "agidi" of "eko" prior to consumption. Lactic acid bacterium *Lactobacillus plantarum*, the aerobic bacteria *Corynebacterium* and *Aerobacter*, the yeasts *Candida mycoderma*, *Saccharomyces cerevisiae* and *Rhodotorula* and molds *Cephalosporium*, *Fusarium*, *Aspergillus* and *Penicillium* are the major organisms responsible for the fermentation and nutritional improvement of ogi. *Lactobacillus. plantarum* is the predominant organism in the fermentation responsible for lactic acid production. *Corynebacterium* hydrolyses corn starch to organic acids while *S. cereviseae* and *Candida mycoderma* contribute to flavour development.

Corn
↓
Clean for 2 to 3 d
↓
Wet mill
↓
Sieve and discard pomace
↓
Ferment filtrate and allow to sediment
↓
for 1-3 d
↓
Ogi

Fig. 7: Flow diagram for the preparation of ogi

2.2.1.2. Banku: This is a popular staple consumed in Ghana. It is prepared from maize and/or from a mixture of maize and cassava. Preparation procedures for banku are summarized in Fig. 8. This involves steeping the raw material(s) in water for 24 h followed by wet milling and fermentation for 3 d. The dough is then mixed with water at a ratio of 4 parts dough to 2 parts water; or 4 parts dough to 1 part cassava and 2 parts water. Continuous stirring and kneading of the fermented dough is required to attain an appropriate consistency during subsequent cooking. The predominant microorganisms involved in fermentation include lactic acid bacteria and moulds.

2.2.1.3. Kenkey: This is a fermented maize dough which is also popularly consumed in Ghana. During the production of kenkey, the dough is divided into two parts: one part, the 'aflata' is cooked into a thick porridge, while the other un-

Corn
↓
Clean
↓
Steep for 24 h
↓
Wet mill
↓
Ferment for 3 d
↓
Dough formation
↓
Stir and knead
↓
Cook
↓
Banku

Fig. 8: Flow diagram for the preparation of Banku

cooked part is later mixed with the 'aflata'. The resulting mixture is moulded into balls and wrapped in dried maize husk or plantain leaves, after which it is steamed (Fig. 9). It is interesting to note that kenkey varieties vary widely throughout Ghana. In northern Ghana, sorghum is sometimes used instead of maize for preparation of the dough. Both yeasts and the moulds play a significant role in the pro-

Maize
↓
Clean
↓
Steep (24-48 h)
↓
Mill
↓
Dough formation
↓
Ferment for 72 h
↓
Cooked dough   Raw dough

Mix (Aflatasation)
↓
Mould into balls
↓
Wrap with maize husks
↓
Boil for about 3 h
↓
Kenkey

Fig. 9: Flow chart for the traditional preparation of kenkey

duction of the fermented maize dough. A homogenous group of obligately heterofermentative lactobacilli related to *L. fermentum and L. reuteri* play a dominating role during kenkey production in addition to *Candida*, *Saccharomyces*, *Penicillium*, *Aspergillus* and *Furasium spp.*

2.2.1.4. Mahewu: This is a fermented maize meal commonly consumed as a staple among black South Africans. It is traditionally prepared by adding one part of maize meal to 9 parts of boiling water. The suspension is cooked for 10 min, allowed to cool and then transferred to a fermentation container. At this stage, wheat flour (about 5% of the maize meal used) is added to serve as a source of inoculum. Fermentation occurs in a warm sunny place within 24 h. Mahewu is currently produced on an industrial scale (Fig. 10) as a dry food product which is marketed as a pre-cooked ready-mix powder. *Streptococcus lactis* is the main fermenting organism in traditionally prepared mahewu which offer some advantages over Ogi in that the initial fermentation by wild fungi, etc. is eliminated by boiling both the maize meal and water for steeping. Furthermore, it is pre-cooked and requires only mixing prior to consumption. Mahewu consists of coarse maize particles while Ogi contains very fine pasty maize particles.

Maize meal

↓

Mix in warm water to give 8% solids content

↓

Cook at 121°C for 15 min

↓

Cool

↓

Inoculate
(5% wheat flour or an adapted pure culture of *Lactobacillus delbrueckii*)

↓

Incubate at 30-50°C for wheat innoculum, or at 45°C for *L. delbrueckii* innoculum

↓

Ferment for 36 h with mixing only at the beginning of fermentation

↓

Heat for 10-15 min under pressure (7 psi)

↓

Spray or drum dry

↓

Mahewu

Fig. 10: Industrial preparation of mahewu

2.2.1.5. Mawe: This is a sour dough prepared from partially dehulled maize meal which has undergone natural fermentation for one to three-days period. Quantitatively mawe is less important than Ogi, but, is suitable as a basis for the preparation of many dishes in Dahomey, including those prepared from Ogi. Mawe is produced using both a traditional (home) process (Fig. 11) and a commercial process. The main difference between the traditional and the commercial process of mawe production is that hulls and germs are removed during the commercial processing

Maize grains

↓

Clean and wash

↓

Crush

↓

Screen and dehull; discard hull

↓

Soak in water for 2-4 h and drain

↓

Grind

↓

Knead to form a dough

↓

Ferment for 1-3 d

↓

Mawe

Fig. 11: Flow diagram of the home process of mawe production

of mawe. Commercial mawe is whiter in appearance than home-produced mawe and has better swelling and thickening characteristics, but, is of lower nutritional value. Dominant microorganisms in mawe preparation include lactic acid bacteria (*Lactobacillus fermentum* and its biotype *L. cellobiosis*, *L. brevis*) and yeasts (*Candida krusei* and *Saccharomyces cerevisiae*).

2.2.2. Bread and Pancakes

2.2.2.1. Injera: This is the most popular baked product in Ethiopia. It is a fermented sorghum bread with a very sour taste and is the undisputed national bread of Ethiopia. The baked product is referred to by different names depending on the locality of production in Ethiopia. It is referred to as 'bidena' in Oromigua, 'taeta' in Giragigua, and 'solo' in Walaytigna. The sorghum grains are dehulled manually or mechanically and milled to flour which is subsequently used in the preparation of injera (Fig. 12). On the basis of production procedures three types of injera are distiguishable: (i) thin injera which results from mixing a portion of fermented sorghum paste with three parts of water and boiling to yield a product known as 'absit' which is, in turn, mixed with a portion of the original fermented flour (ii) thick injera, which is reddish in colour with a sweet taste, is a 'tef' paste that has undergone only minimal fermentation for 12-24 h; (iii) komtata-type injera, which is produced from over-fermented paste, and has a sour taste. The paste is baked or grilled to give a bread-like product. Yeasts are the major microorganisms involved in the fermentation of injera.

2.2.2.2. Kisra: This is a thin pancake like leavened bread prepared from whole sorghum flour. It is a dietary staple in Sudan. This fermented sorghum bread has a very sour taste. It is prepared by mixing sorghum flour with water to give a thick paste which is allowed to ferment for 12-24 h, following which the paste is thinned to a desirable consistency with water just prior to baking (Fig. 13). There is a slight

Sorghum flour
↓
Mix with water, 4:1(*w/v*)
↓
Knead to form a dough
↓
Mix with starter
(Fermented yellowish liquid saved from previously fermented dough)
↓
Knead
↓
Add water
↓
Ferment for 48 h
↓
Add water and allow to stand for 1 h
↓
Bake on hot
greased clay griddle metal till holes begin to form on top
↓
Injera

Fig. 12: Flow diagram for the preparation of injera

Sorghum flour
↓
Mix with water (60:40, *w/v*)
↓
Ferment for 12 – 24h
↓
Add water to form a thin slurry
↓
Bake on hot oiled pan
↓
Kisra

Fig. 13: Flow diagram for the preparation of Kisra

increase in protein and fiber contents and an appreciable decrease in carbohydrate (starch and sugars) content during the fermentation of kisra.

2.2.2.3. Kishk: This is a fermented product prepared from parboiled wheat and milk (Fig. 14). It is consumed in Egypt and in most Arabian countries.

During the preparation of kishk, wheat grains are boiled until soft, dried, milled and sieved in order to remove the bran. Milk is separately soured in earthen containers, concentrated and mixed with the moistened wheat flour thus prepared, resulting in the preparation of a paste called hamma. The hamma is allowed to ferment for about 24 h, following which it is kneaded and two volumes of soured salted milk are added prior to dilution with water. Alternatively, milk is added to the hamma and fermentation is allowed to proceed for further 24 h. The mass is thoroughly mixed, formed into balls and dried. Kishk is a highly nutritious food,

Wheat grains covered with water
↓
Heat slowly to boiling and simmer until soft
↓
Wash with cold water
↓
Dry on mats
↓
Grind
↓
Remove seed coats by sieving
↓
Place in pots and moisten with lightly salted boiling water →

Sour milk by churning in skin bags
↓
Concentrate
↓
↓
↓
Mix to form a paste
↓
Ferment for 24 h
↓
Mix and add whey
↓
Dilute with milk or water to give a syrupy consistency
↓
Ferment for 24 h
↓
Mix and form small balls
↓
Place on mats and sun dry
↓
Kishk

Fig. 14. Flow diagram for the preparation of kishk

having a protein content of about 23.5%. It is of a high digestibility, and high biological value. Microorganisms responsible for fermentation include *Lactobacillus plantarum*, *L. brevis*, *L. casei*, *Bacillus subtilis* and yeasts.

2.2.2.4. Bogobe: This is a sorghum porridge prepared in Botswana from fermented and non-fermented sorghum (Fig. 15). Fermented bogobe is a soft porridge, known

Sorghum grains
↓
Wash with water
↓
Dehull (mechanical or manual)
↓
Discard bran and grind to a coarse meal
↓
Add lukewarm water (1:1, w/v) to form a slurry
↓
Allow to ferment in a closed environment for 24 h
↓
Cook in boiling water for 12 – 15 min.
↓
Bogobe

Fig. 15. Flow diagram for the preparation of bogobe from sorghum

as *ting* while the non-fermented bogobe is a thick porridge called *monokwane*. Information relevant to microorganisms involved in the feremention of bogobe, and the nutritional changes which occur during fermentation is still scanty.

2.2.3. Fermented Beverages

2.2.3.1. Kunu-Zaki: This is a millet-based non-alcoholic fermented beverage widely consumed in the Northern parts of Nigeria. This beverage is however becoming more widely consumed in Southern Nigeria, owing to its refreshing qualities. The traditional process for the manufacture of kunu-zaki involves the steeping of millet grains, wet milling with spices (ginger, cloves, pepper), wet sieving and partial gelatinization of the slurry, followed by the addition of sugar, and bottling (Fig. 16). The fermentation which occurs briefly during steeping of the grains in

Dehulled millet grain
↓
Clean
↓
Steep
↓
Wet mill with the addition of spices
↓
Wet-sieve
↓
Allow to settle
↓
Decant supernatant and retain slurry
↓
Slurry in cold water + Slurry in boiling water
↓
Add sweetener and mix
↓
Bottle
↓
Kuni-zaki

Fig. 16: Flow chart for the Traditional Process of Kunu-zaki

water over a 8-48 h period is known to involve mainly lactic acid bacteria and yeasts. The product has a shelf-life of about 24 h at ambient temperature, which can be extended to 8 d by pasteurization at 60°C for 1 h and storage under refrigeration conditions.

2.2.3.2. Burukutu: This is a popular alcoholic beverage with vinegar-like flavour, consumed in the Northern Guinea savanna region of Nigeria, in the Republic of Benin and in Ghana. The preparation of burukutu involves steeping sorghum grains in water overnight, following which excess water is drained. The grains are then spread out on to a mat or tray, covered with banana leaves and allowed to germinate. During the germination process, the grains are watered on alternate days and turned over at intervals. Germination continues for 4-5 d until the plumule attains a certain length. The malted grains are spread out in the sun to dry for 1-2 d, following which the dried malt is ground into a powder. Gari, (a farinaceous fermented cassava product) is added to a mixture of the ground malt and water in a ratio of one part gari to two parts malt and six parts water. The resulting mixture is allowed to ferment for 2 d, following which it is boiled for approximately 4 h and allowed to mature for further 2 d. The resulting product is a cloudy alcoholic beverage. Sorghum malt contains primarily yeasts and moulds as the indigenous microflora. Microorganisms associated with the fermentation include yeasts mainly *Saccharomyces cerevisiae* and *S. chavelieri* and the bacteria, *Leuconostoc meseteroides*. Boiling prior to maturation eliminates lactics and other yeasts. At the termination of the 2-d maturing period *Acetobacter sp.* and *Candida sp.* are the dominant microorganisms.

2.2.3.3. Pito: This is the traditional beverage drink of the Binis in the mid-western part of Nigeria. It is now very popularly consumed throughout Nigeria and Ghana owing to its refreshing nature and low price. The preparation of pito involves soaking cereal grains (maize, sorghum or a combination of both) in water for 2 d, followed by malting, and allowing to settle for 5 d in baskets lined with moistened banana leaves. The malted grains are ground, mixed with water and boiled. The resulting mash is allowed to cool and later filtered through a fine mesh basket. The filtrate thus obtained is allowed to stand overnight, or until it assumes a slightly sour flavour, following which it is boiled to a concentrate. A starter from the previous brew is added to the cooled concentrate which is again allowed to ferment overnight. Pito, the product thus obtained, is a dark brown liquid which varies in taste from sweet to bitter. It contains lactic acid, sugars, amino acids and has an alcohol content of 3%. Organisms responsible for souring include *Geotrichum candidum* and *Lactobacillus* sp. while *Candida* sp. are responsible for the alcoholic fermentation.

2.2.3.4. Merissa: This is an alcoholic drink which is widely consumed in Sudan. It is prepared from sorghum and millet by a relatively complex process. Brewing takes place in three distinct phases (i) 'ajeen' fermentation, a lactic souring of sorghum, (ii) 'debosa' fermentation, a starter activating phase and (iii) merissa fermentation, an alcoholic fermentation. Ajeen fermentation is accomplished by lactic acid and acetic acid bacteria and yeasts at a pH of about 4.0. An alcoholic content of 1% and lactic acid content of 2.5% are formed at end of this fermentation. At the final stage of merissa fermentation, however, the alcoholic content increases to about 6%.

2.2.3.5. Bouza: This is a fermented alcoholic beverage produced from wheat in Egypt and has been known by the Egyptians since the days of the Pharaohs. It is a thick, pasty yellow beverage with an agreeable taste and produces a sensation of heat when consumed. It is prepared by coarsely grinding wheat grains, placing a portion of them (3/4) in a wooden basin and kneading them with water into a dough. The dough is cut into thick loaves which are very lightly baked. Meanwhile, the remainder of the grains are moistened with water, germinated for 3-5 d, sun-dried, ground and mixed with the loaves of bread which are soaked in water in a wooden barrel. Bouza from a previous brew is added to serve as an inoculum.

The mixture is allowed to ferment at room temperature for 24 h, following which the product is sieved to remove large particles and diluted with water to a desired consistency. Like other opaque beers, bouza has a very short shelf-life and is expected to be consumed within a day. Its pH decreases to between 3.9-4.0 and its alcoholic content to between 3.8-4.2% within 24 h period.

2.3. Indigenous Fermented Cereals of Latin American countries

Cereal crops, in particular maize which has its origin in Mexico, are very important in Latin America and have been consumed in the fermented form for hundred of years. A number of traditional products based on maize are developed by the populations of Mexico and Peru in the Pre-colombian era. These fermented products are utilised as stimulants, traditional medicine, as well as in religious ceremonies. Fermented beverages and porridges consumed in the Latin American region are presented in Table 6. Of the products listed, chicha (maize beer) is the most important traditional fermented beverage in the region and also the most widely studied.

Table 6: Fermented Cereal Products Consumed in Latin America

Name	*Description*	*Country*
Abati	Alcoholic beverage based on maize	Paraguay, Argentina
Acupe	Beverage based on germinated maize, fermented and sweetened	Venezuela
Agua-agria	Non alcoholic beverage based on ground maize and water	Mexico
Arroz requemado	Fermented rice grains	Ecuador
Atole	Non alcoholic porridge based on maize dough	Mexico
Atole agrio	Non alcoholic porridge based on black maize dough fermented 4-5 d	Mexico
Cachiri	Fermented beverage based on maize, manihot or fruits. It is produced in clay pots	Brazil
Champuz	Fermented beverage based on maize or rice	Colombia, Peru
Chica	Alcoholic beverage based on pineapple, barley steep liquor and black maize dough. It is fermented for 4 d, following which brown sugar, cinnamon and clove are added	Mexico
Charagua	Alcoholic beverage based on "pulque" syrup, chili and toasted maize leaves, heated slowly and fermented	Mexico
Fubá	Germinated maize grains fermented in water	Brazil
Jamin-bang	Bread based on maize fermented for 3-6 d and cooked as a cake	Brazil
Napú	Beverage based on germinated,ground and fermented maize	Peru
Ostoche	Alcoholic beverage based on maize juice and "pulque"or brown sugar	Mexico
Pozol	Non-alcoholic acidic beverage based on maize liquor. Balls prepared from fermented dough are enveloped in banana leaves	Mexico

Name	Description	Country
Quebranta huesos	Alcoholic beverage based on maize juice toasted maize and pirú fruits (Schinus molle)	Mexico
Sendechó	Alcoholic beverage (beer-like) based on germinated maize and red chili. Dough is resuspended in water, boiled, bestowed, cooled and inoculated with Sendechó	Mexico
Sora	Alcoholic beverage based on germinated, ground, cooked and fermented maize	Peru
Tepache	Alcoholic beverage based on maize grains, brown sugar and water	Mexico
Tesgüino	Alcoholic beverage (beer-like) based on germinated maize, gound and cooked with fragments of plants that serve as enzyme sources	Mexico
Tocos	Dessert based on maize fermented for 2-3 months and cooked	Peru
Zarzaparrilla bark wine	Alcoholic beverage based on maize beer and zarzaparrilla bark	Mexico
Zambumbia	Alcoholic beverage based on toasted barley and water; fermented for 3-4 d, following which brown sugar is added	Mexico

2.3.1. Chicha

This is the most important traditional fermented beverage consumed in Latin America. Chicha is a clear, yellowish, effervescent, alcoholic beverage prepared from maize and is also known an maize beer. It has a flavour similar to that of cider. Chicha has been consumed by the Andean Indians for centuries. When prepared from pigmented maize varieties, its colour varies from red to purple. The alcoholic content of chicha varies between 2 and 12% (*v/v*). The traditional production of chicha is a somewhat unique fermentation process in which saliva serves as the source of amylase for converting starch to fermentable sugars. Malting (germination) of maize kernels to produce the amylase required for starch conversion is an alternative procedure which is widely used in modern day processing. Frequently, salivation is combined with malting to yield chicha.

It is produced in the Andean regions and sometimes in the lower altitude regions of countries such as Argentina, Bolivia, Brazil, Colombia, Ecuador and Peru. Chicha manufacture is a significant household or communal activity and the beverage is consumed mainly by the Indians during religious and agricultural festivities and during important family and social events. A special annual festival designated the Kayova (Cayua) is dedicated to chicha production in Peru. This festival takes place in January at the beginning of the maize harvest.

Chicha is traditionally produced in an earthen pots about 75 cm high and 85 cm in diameter which is filled to one-third of its capacity with maize flour, obtained from salivated and sun-dried maize or germinated and sun-dried maize. The pot is then filled with water and heated to approximately 75°C. The flour and hot water are thoroughly mixed for 1 h and then allowed to settle and cool, following which three layers are formed: a top liquid layer called upi; a middle jelly-like layer, and a bottom layer which contains coarse particles (hanchi). The upi is scooped out with a gourd and is transferred to another earthen pot. The middle layer is placed in a shallow pan, heated, and concentrated to a sugar-like product. The hanchi is pressed and filtered and the filtrate thus obtained is added to the upi. The upi is simmered for several hours until it becomes caramelized. This caramelized product, referred to as *misqui kheta*, is allowed to cool. Additional upi may be boiled for as long as 3 h and then combined with the misqui kheta. After an incubation period of four days, the mixture of upi and misqui kheta begins to ferment and bubble violently. In the highlands, fermentation is complete and bubbling ceases in approximately 6 d, while in the hot lowlands, fermentation may be complete within 2 d. When bubbling ceases, the chicha is transferred to narrow-mouth pots and is ready for consumption.

After the chicha has been consumed, a layer of sediment remaining in the pot is filtered to obtain *sutu*, which resembles chicha in colour but lacks carbonation and the typical tangy flavour associated with cider. Sutu is reportedly of a higher alcohol content than chicha. Forty litres of chicha yield about 1 litre of sutu. Yeasts, particularly *Saccharomyces cerevisiae*, and bacteria of the genus *Lactobacillus*, are the primary fermenting organisms in chicha produced in northern Argentina. Various yeasts (*S. cerevisiae*, *S. pastorianus* and *Mycoderna vivi*, *Oidium lactis* and *Monilia candida*), bacteria of the genera *Leuconostoc*, *Lactobacillus*, *Acetobacter*, and various moulds including *Aspergillus* and *Penicillium* are present and presumably active in Colombian chicha. The natural mixture of organisms found in chicha has a greater ability to utilize starches, dextrins, and sugars than do pure cultures of *S. cerevisiae.* Chicha organisms produce a total alcohol content of 13.4 % (*v/v*) when fermenting a 30% solution consisting of 15 % starch and dextrin and 15 % fermentable sugars. Ninety percent of the total solids are fermented within 10 d, and the total acid produced is 0.8%. Chicha is also rich in the B vitamins. The riboflavin concentration is doubled, while thiamine and niacin remain fairly constant during the fermentation of chicha.

2.3.2. Pozol

This is a fermented maize dough formed into balls of various shapes and sizes ranging from 10 to 12 cm in length, 5 to 8 cm in width and 70 to 170 g in weight. Pozol is consumed by Indian and mestizo populations, mainly in the Southeastern states of Mexico, such as Chiapas, Tabasco, Campeche and Yucatan. Balls of freshly prepared pozol, or pozol at various stages of the fermentation process are diluted with water to produce a whitish porridge which is consumed in the uncooked state as a basic food in the daily diet of large communities. The ratio of pozol dough to water varies between 1:2 and 1:3. Salt, toasted ground chilli pods, sugar, or honey may be added. The beverage prepared from pozol is consumed particularly by low-income individuals during working hours, at meals, or as a refreshment at any hour of the day. Two basic types of pozol are distinguishable: a traditional-type prepared by the indigenous Indians and a mestizo-type, characterized by additional cooking of the dehulled grains.

Pozol is prepared either domestically for consumption or on a small commercial scale according to traditional procedures handed down from generation to generation. In the production of pozol, 1 to 1½ kg of kernels obtained by shelling cobs of maize (preferably white *Zea mays* L.), are boiled for 1 h in a pot containing 1 to 2 ℓ of an approximately 10% (w/v) calcium hydroxide solution. During boiling, swelling of the kernels takes place, thus allowing the pericarp to be relatively easily peeled off the kernels. The kernels are cooled, rinsed with water, and drained resulting in what may be described as *nixtamal*. The nixtamal is ground in a manual metal mill to obtain a coarse dough which is manually shaped into balls. The balls are then wrapped in banana leaves to prevent desiccation, and fermented for 1 to 14 d or more depending on consumer preference and prevailing circumstances (Fig. 17). During the initial 24 h of pozol fermentation, bacteria outnumber yeasts and moulds and are probably responsible for the majority of acid produced. Lactic acid bacteria, the predominant microflora of pozol include strains of *Leuconostoc mesenteroides*, *L plantarum*, *Lactobacillus confusus*, *Lactococcus lactis* and raffinolactis. Yeasts associated with pozol include *Candida krusei*, *C.*

Corn

↓

Boil in water containing calcium hydroxide
(this process is known as *nixtimalization*)

↓

Grind to obtain a dough

↓

Shape dough into balls and envelope with banana leaves

↓

Allow to ferment for 1-14 d

↓

Pozol

Fig. 17: Flow chart for the preparation of Pozol

guillermondii, *C. parapsilosis*, *C. tropicalis Trichosporon cutaneum*, *Hansenula fabiani*, *Kluveromyces fragilis*, and *S. cerevisiae*. Molds such as *Cladosporium clados-porioides* or *C. herbarum*, *Monilia sitophola*, and *Mucor rouxianus* or *M. recemosis* are also common in pozol balls as their surface progressively dries and their pH in lowered. Two essential changes that occur in maize dough during pozol fermentation are the development of an acidic flavour and a characteristic aroma which impart the refreshing properties of the product on ingestion. Pozol is richer in protein, niacin, riboflavin, lysine, tryptophane, and some other nutrients.

2.3.3. Tesgüino

This is a slurry-like, alcoholic beverage prepared by fermentation of germinated maize or maize stalk juice. Tesgüino prepared from maize stalk juice, is referred to as *paciki* and tesgüino prepared with the incorporation of the bark of certain species of *Rubiaceae*, as *batari*. The beverage is also referred to as sugiki. The beverage plays an important role in the everyday life of indigenous groups. It is the preferred beverage at any celebration, religious ritual, funeral, sporting game, or in the so-called *tesgüinadas*, which are some of the most important events in Tarahumara´s life. Tesgüino when diluted with water, is commonly fed to babies and infants, and is also drunk at meals during family reunions.

Methodologies used in the preparation of *tesgüino* vary among ethnic groups. The most common method for preparing tesgüino is summarised in Figure 18. For the preparation of tesgüino, about 10 kg of dry maize kernels are soaked in water

Corn
↓
Germinate in the dark
↓
Grind to obtain a dough
↓
Boil in water
↓
Transfer liquid portion to a clay pot and add catalysts
↓
Allow to ferment (2-3 d)
↓
Tesgüino

Fig. 18: Flow chart for the preparation of Tesgüïno

for several days, drained, and placed either in baskets in the dark or in a hole in the ground in order to allow them to germinate. The germinating kernels are protected from light, in order to prevent the formation of green and bitter sprouts. The germinated kernels are ground in a manual metal or stone mill and then boiled in water until the mixture turns yellow (about 8 h). The liquid portion is then transferred to a clay pot, and catalysts are added. The most common catalysts include bark (*batari*) or *kakwara* (*Randia echinocarpa*, *R. watsoni,* and *R. laevigata*) and *kaya* (*Coutarea pterosperma*), which are chopped, ground, and boiled for many hours prior to being added to the tesgüino. At higher altitudes where pine trees grow, the catalysts used are leaves of *roninowa* (*Stevia serrata*), *rojisuwi* (*Chimaphila maculata*), and *ubitakuwari* (*Datura meteloides*); stems of *basiawi* (*Bromus arizonicus*), roots of *gotoko*, *otoko*, or *goto* (*Phaseolus metcalfei* and *Plumbago scandens*); *rawici kitakame* or "mouse´s ear" (*Hieracium fendleri*). The mixture is allowed to ferment for several days prior to consumption. Several bacteria, yeasts and molds have been isolated from tesgüino. These include *Lactobacillus*, *Streptococcus*, *Leuconostoc*, *Pediococcus*, *Saccharomyces*, *Candida*, *Cryptococcus*, *Hansenula*, *Brettanomyces*, *Pichia*, *Geotrichum* and *Penicillium*.

2.3.4. Atole

This is a sour porridge-type product prepared from maize by members of the Tzotzil ethnic group in Southern Mexico. It is produced by steeping maize grains in water for 4 d, milling and allowing to stand for 1 d. During atole production, lactic acid fermentation commences during steeping and continues in the milled product. Lactic acid bacteria present in the atole form diacetyl, which contributes to the characteristic sensory properties of the product.

2.4. Future outlook and conclusions

Some advantages of traditional fermentations are that they are labor-intensive, integrated into village life, familiar, utilize locally produced raw materials, inexpensive, add interest and tradition to local consumers. From this perspective, research leading to new fermentation technologies should be sensitive to social and economic factors in developing countries. Rapid displacement of traditional foodstuffs with technologies developed in more affluent countries may result in centralized production, distribution problems, less local involvement in food processing, less employment in some areas, less nutritionally adequate substitutions in raw materials, displacement of traditional arts, loss of unique local know-how, dependence on importation of equipment and materials, initially require the use of outside consultants, and may otherwise not meet local needs as fully as traditional fermented products. On the other hand, indigenous fermentations may have a number of problems, i.e., they are uncontrolled and often unhygienic, labour intensive.

The microbiology of many of these fermentations is undoubtedly quite complex and the specific information on microflora is lacking for several indigenous fermented cereal products.. Many indigenous cereal fermentations involve the combined action of bacteria, yeast and fungi. Some microflora may participate in parallel while others may participate in a sequential manner with a changing dominant flora during the course of the fermentation. The specific microflora involved may vary somewhat from village to village and from family to family within the same village. The identification of specific microflora involved is needed to amplify and control such positive factors as well as to minimize or prevent negative factors such as growth and metabolism of pathogenic and toxinogenic bacteria. Identifying and providing a practical means of using appropriate starter cultures is advantageous due to the competitive role of microorganisms and their metabolites in preventing growth and metabolism of unwanted microorganisms. A strong starter may reduce fermentation times, minimise dry matter losses, avoid contamination with pathogenic and toxigenic bacteria and moulds, and minimize the risk of incidental microflora causing off-flavour, etc. Optimization of starter cultures may be achieved by either conventional selection and mutation, or by recombinant-DNA techniques to result in increased levels of safety. Relatively litle is known of the contribution of microflora to the formation of desired flavour during such fermentations. Genes for flavour and other beneficial enzymes that come from incidental microflora may be incorporated into starter bacteria to facilitate more subtle and ancillary aspects of the fermentation along with primary events such as lactic acid production, thus preserving the distinctive nature of products made in different regions.

The contribution of specific enzymes to indigenous cereal fermentations is perhaps even less understood than that of microorganisms. It is likely that there is considerable synergy between complimentary enzymes from the cereal itself and from the microorganisms. Partial germination of cereal in the field and contamination with otherwise innocuous fungal contaminants contribute enzymes, notably amylases and proteases, that aid these fermentations. Today, essentially all beer production and continuous breadmaking in the West is achieved with the aid of added enzymes. A similar situation may occur in developing countries, i.e. as improvements in cereal handling are introduced to minimise postharvest losses and mycotoxin formation, the otherwise improved crop may be less suitable in some ways for traditional fermentations. Hence, basic information is needed on the contribution of cereal enzymes and other constituents to indigenous fermentations. With this information in hand, consideration can then be given to use of enzyme supplements and other additives to improve the rate and quality of fermentations.

In conclusion, traditional fermentations are likely to remain an important part of global food supply; many may evolve into fermentations involving the use of starter cultures, enzyme additives and controlled environmental conditions, and others may benefit from genetic modification of the cereal or starter bacteria. Further research should be directed towards identifying the benefits and risks associated with specific indigenous fermented cereals; elucidating the contributions of microorganisms, enzymes and other cereal constituents in the fermentation process; and developing starter cultures, unique microbial strains for nutritive improvement and detoxification, and testing of new cereal varieties for their suitability as fermentation substrates.

FURTHER READING

1. Haard, N.F., Odunfa, S,A., Lee, C.H., Ramirez, R.Q., Quinones, R.Q., Radarte, C.W. Fermented Cereals. A global Perspective. FAO Agricultural Services Bulletin no. 138. Agriculture Organization of the United Nations Rome. (1999).
2. Sankaran, R. Fermented foods of the Indian subcontinent. In: Microbiology of fermented foods. Vol 2. (Ed. Brian J.B. Wood). Blackie Academic and Professional. UK. (1998). 753–789.
3. Soni, S.K. and Arora, J.K. Indian Fermented Foods: Biotechnological approaches. In: Food Processing: Biotechnological Applications. (Eds. S.S. Marwaha and J.K. Arora). Asiatech Publishers Inc., New Delhi, India. (2000). 143–190.
4. Soni, S.K. and Sandhu, D.K. Fermented cereal products. In: Biotechnology: Food fermentation Microbiology, Biochemistry and Technology. Vol. 2. (Eds. V.K. Joshi and A. Pandey). Educational Publishers and Distributors, New Delhi. (1999). 895–950.
5. Stainkraus, K.H. Handbook of Indigenous fermented foods. New York, Marcel Dekker, Inc. (1995).

Emerging Technologies for Commercial Applications in Horticultural Sector

A.S. DHATT *and* K.S. SANDHU

Punjab Horticultural Postharvest Technology Centre
PAU Campus, Ludhiana 141 004 (Punjab)

FRUITS AND VEGETABLES ARE HIGHLY perishable commodities. Their spoilage takes place due to continuous respiration, loss of moisture and microbial growth particularly moulds and rots. It is estimated that about 30-50% of the total produce is lost during transit in India. There is great need to develop the techniques to reduce the quantum of losses to increase their availability in fresh form. Refrigeration or cold storage is commonly used to prolong their shelf life. Other techniques are coming up commercially to supplement the effect of refrigeration to extend the shelf life of these commodities further. Among these techniques, edible coatings and modified atmosphere packagings and controlled-atmosphere storage are gaining popularity.

1. COATINGS

Natural waxes present on the surface of fruits and vegetables are softened by the adverse effect of atmosphere. Fruits in the field may contain dust, dirt, mould, bacterial spores and pre-harvest spray residues. For these reasons, the first step in a commercial packing line is the thorough washing of fruit over revolving brushes. The washing process results in removal of natural waxes of the cuticle leaving microscopic injuries.

The removal of natural wax coating during the washing operation has following two undesirable side effects that limit their market life:

- The loss of water by transpiration from the fruit surface is increased and susceptibility to infection is increased.
- Rind permeability to gases is increased resulting in increased rate of respiration. This higher rate of respiration results in a decrease in sugar & ascorbic acid content, loss of flavour, poorer texture and quality.

The coatings prolong the market life of the produce. Food grade coatings are used in fruits and vegetables for a number of reasons:

- To reduce shrinkage due to water loss.
- To provide a barrier to free gas exchange.

- To improve cosmetic appearance, mechanical properties and marketability by application of a shiny, sweat-resistant film.
- To encapsulate aroma compounds, antioxidants, pigments, ions that stop browning reactions and nutritional substances such as vitamins.
- To provide a carrier for decay-control fungicides and growth regulators.

Edible films used as coating or for packaging should have the following features:

- Good sensory quality
- High barrier and mechanical efficiencies
- Enough biochemical, physico-chemical and microbial stability.
- Free of toxic and safe for health
- Simple to apply
- Non-polluting/Ecofriendly
- Low cost

Various edible films/packagings such as Citrashine, Brilloshine, Stayfresh, Stafresh-451, Semperfresh, Shellac, Zein, Chitosan, Sugarcane and Carnauba waxes etc. which by and large fulfill the above listed criteria are used commercially to save the post harvest losses of fruits and vegetables.

There are several methods of applying coatings:

- Spray fluidization
- Falling and pan coating
- Spraying
- Dipping
- Brushing/Sponging

These processes are usually followed by drying steps for aqueous products or by cooling for lipid-based coatings or films. Their use and application rates are governed by legislation in the country where the coating is being applied or in the country where the fruits and vegetables are to be exported. The manufacturer's recommendations should be strictly adhered to for use. Coatings have been used for various fruits and vegetables including orange, lemon, grapefruit, apple, pear, cherry, banana, guava, mango, litchi, peach, grape, carrot, cucumber, root crops, sweet corn, egg plant, pepper, tomato, asparagus, radish, potato, turnip etc (Table 1). In Punjab, at present four waxing and grading plants situated at Kang Mai (Hoshiarpur), Badal (Faridkot), Baluana and Mojgarh (Ferozepur) are being used for coating the kinnow fruits. However, other round fruits and vegetables can also be waxed with same arrangements. Figure 1 show the citrus packaging using both water and solvent wax alternatives.

2. MODIFIED AND CONTROLLED-ATMOSPHERE STORAGE

Air contains about 20.9% oxygen, 78.1% nitrogen, 0.9% argon and 0.03% carbon

Table 1: Specific coating applications

Coating material	*Fruit/Vegetable*	*Effect of coating*
Tal Prolong	Banana	Delays ripening, builds-up of CO_2 & depletes O_2 causing effect of controlled atmosphere
Semperfresh	Banana	Delays development of crown rot caused by *Colletotrichum musae*
Semperfresh with organic acid	Banana	-Do- with increased effect
Ban-seel	Banana & plantains	Protects & extends storage life
Tal Prolong, Semperfresh and Apple Wax	Apple	Reduction in disease level, Delay colour, weight & firmness changes
Nutri-Save	Golden Delicious apple	Reduces C_2H_4 production
Semperfresh	Granny Smith apple	Ripening delayed
Brilloshine	Apple, avocado, melons & citrus fruits	Protects, increases shine & extends shelf life
Nu-coatflo, Brilloshine and Citrashine	Citrus fruits	Protect, increase shine & extend shelf life
Semperfresh	Guava	2 months storage at 10°C better in quality than untreated
Palm oil	Guava	-Do- Better & cheaper
Vapor Gard	Mango	Antitranspirant, maintains firmness, retains ascorbic acid, inhibits malic enzyme & increases polygalacturase activity
Chitosan	Pepper, cucumber, strawberry & raspberry	Antifongic, reduces water loss & maintains quality
N, O-carboxymethyl chitosan	Fruits	Selectively permeable to gases such as CO_2, O_2 & C_2H_4
Zein	Tomato	Delays colour, weight & firmness changes

dioxide and trace amounts of other gases including neon, helium, methane and water vapours. Fruits and vegetables get oxygen from atmosphere for respiration and produce carbon dioxide and moisture is given off in the surroundings along with generation of heat. These physiological processes are accelerated with increase in temperature. The chemical, enzymatic and microbial deterioration result in reduced shelf life at elevated temperature. The shelf life of fruits and vegetables can be prolonged by slowing down the rates of respiration, ethylene production, moisture loss and other deteriorative changes by modifying or controlling of carbon dioxide & nitrogen, increased relative humidity and lowered temperature in their surroundings.

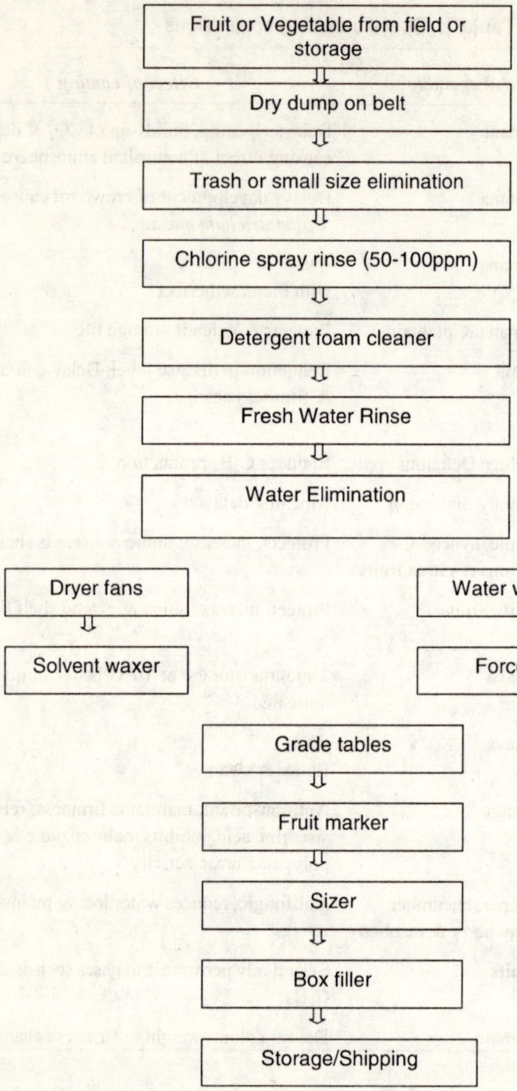

Fig. 1. Flow Diagram showing citrus packaging using both water and solvent wax alternatives.

2.1. Modified atmosphere packaging

In the modified atmosphere packaging (MAP) and storage, the atmosphere is generally modified by the commodity itself by consuming the oxygen and releasing the carbon dioxide in surrounding atmosphere. MAP requires the use of flexible packaging films specific to the requirement of each commodity alongwith temperature management to increase the shelf life. It is relatively expensive and may be cost effective only for high value produce such as berries and fresh-cut products. MAP is only a supplement to refrigeration and not a replacement for temperature management. Packaging films have been developed for specific produce to maintain the desired composition of atmosphere. MAP can be designed from the variables such as product mass (weight), respiration rate, film area, film thickness, desired O_2 and CO_2 concentrations, oxygen transmission or carbon dioxide transmission rates of films. Oxygen transmission rate or carbon dioxide transmission rate can be computed from the following formula.

$$\left.\begin{matrix}\text{OTR}\\(\text{cc/mil/cm}^2\text{/h}\end{matrix}\right\} = \frac{\left\{\begin{matrix}Wt(\text{kg})\times\text{ respiration rate (ml O}_2\text{kg / h)}\\ \times\text{ film thickness (mil)}\end{matrix}\right\}}{\left\{\begin{matrix}\text{Film area (cm}^2)\times\text{desired external O}_2\,(\%)\\ -\text{ desired internal O}_2\,(\%)\end{matrix}\right\}}$$

If the composition of desired atmosphere is known, then the type of film to meet that particular requirement can be selected. Oxygen transmission rate and carbon dioxide transmission rate values are given by the manufacturers of films. MAP films are available under various trade names like "Exact', 'Affinity' (Mettalocene films), 'P-Plus' (Microperforated films) and 'Fresh Hold' (a combination of microporous film and flexible oxygen impermeable film).

2.1.1. *New Trends in MAP*

Now the films have been developed which can change the transmission rate of gases with change in respiration rate affected by temperature. These are called active packagings like 'Intellimers' or smart films. Another breakthrough in the area of packing material is the incorporation of antimicrobial compounds to develop edible films/packaging materials such as 'Zeolite' film which can reduce the microbial population on contact with coated fruits and vegetables.

Currently, the efforts are in progress to safely incorporate ethylene and carbon dioxide absorbers directly into the packaging films. 'PeakFresh' is a polyethylene film impregnated with mineral particles to absorb ethylene.

2.2. Controlled-atmosphere

In controlled-atmosphere (CA) storage, composition of gases and other conditions are controlled precisely to avoid the undesirable physiological effects. Any atmosphere qualifies to be defined as CA for the storage of fruits and vegetables to improve their shelf life which fulfills the following criteria.

Provision for :

- air circulation
- maintenance of high humidity
- constantly monitor the levels of oxygen and carbon dioxide with in ± 0.1% limits.
- removal of excess of carbon dioxide by using CO_2 scrubbers like hydrated lime or activated charcoal.
- controlling O_2 levels by ventilation with cold air.
- injecting nitrogen to speed up the process.

- ethylene scrubbers such as molecular sieves, activated carbon, hydrated aluminium silicate or potassium permanganate can be used to remove ethylene from the store.
- atmosphere generators to quickly create the desired atmosphere in CA stores

The approximate cost of 1000 cu. ft. CA store is about 12 lacs with storage capacity of 5-6 tonnes of fruits and vegetables.

The different fruits/vegetables require different CA rooms for their storage. For long time storage, it is desirable that fruits harvested at the right stage of maturity and quality be immediately stored in CA rooms after harvesting. Following points need attention during the operation of CA rooms for the effective storage of fruits and vegetables:

- temperature should be maintained between 0.5°C to 0°C during loading of store.
- bring the store quickly to desired storage temperature before sealing the room.
- O_2 and CO_2 concentrations of atmosphere should be checked twice daily and. record should be maintained.
- air purifiers should be installed in the store rooms to pick up objectionable odours from containers and fruits.
- when the room is opened for sale, temperature should be lowered to 0°c.

The storage conditions can vary considerably not only for different crops but also for the same species. These may depend on the condition under which the crop is grown. The conditions recommended for storage of selected crops are listed in Table 2.

Table 2: CA Requirements for Different Commodities

Name of the commodity	*Temperature*	O_2	CO_2
Apple:	0.5°C	1.6%	2.3%
—Golden Delicious (shelf life 7-11 months)	0°C	1.6%	1.8%
— Red Delicious (shelf life 6-11 months)			
Apricot Royal	0°C	2 – 3%	2.5 – 3%
Asparagus	0–5°C	21%	5 – 10%
Egg Plant	7–10°C	3 – 5%	0%
Avocado	5–3°C	2 – 3%	2 – 3%
Banana	12–16°C	2 – 5%	2 – 5%
Blackberries	0–5°C	5 – 10%	15 – 20%
Broccoli	0–5°C	1 – 2%	5-10%
Brussels sprouts	0–5°C	1 – 2%	5-10%
Cabbage	0–5°C	3 – 5%	5 – 7%
Capsicum	8–12°C	3 – 5%	0%

Name of the commodity	*Temperature*	O_2	CO_2
Carrot	2°C	1-2%	0%
Cauliflower	1°C	1%	2.5%
Celery	0–5°C	2–4%	0%
Cherry Sweet	0–5°C	3–10%	10–15%
Chillies	8–12°C	3–5%	0%
Chinese Cabbage	0–5°C	1–2%	0–5%
Cucumber	8–12°C	3–5%	O%
Fig	0–5°C	5–10%	15–20%
Garlic	0°C, RH 75%	1–2%	0%
Grapes	0–5°C	2–5%	1–3%
Grapefruit	10–15°C	3–10%	5–10%
Lemon	10–15°C	5–10%	0–10%
Lettuce	0–5°C	1–3%	0%
Litchi	5–12°C	3–5%	3–5%
Mango	10–15°C	3–7%	5–8%
Melon:	3–7°C	3–5%	10–15%
Cantaloupe type	10–12°C	3–5%	0%
Honey dew type			
Okra	8–10°C	3–5%	0%
Onion	0–5°C, RH 75%	1–2%	0%
Orange	5–10°C	5–10%	0–5%
Papaya	10–15°C	2–5%	5–8%
Peach	0–5°C	1–2%	3–5%
Clingstone/Freestone			
Pear:	0–5°C	1–2%	0–0.5%
Bartlett (Shelflife 5-9 months)	0–5°C	0.5–3%	0–1%
20th Century			
Peas	0°C	5–10%	5–7%
Pineapple	8–13°C	2–5%	5–10%
Plum	0–5°C	1–2%	0–5%
Strawberry	0–5°C	5–10%	15–20%
Tomato:	12–20°C	3–5%	0%
Mature Green	8–12°C	3–5%	0%
Partially ripe	10–15°C	3–5%	3–5%
Ripe			
Lime	10–15°C	5%	0–10%
Mushroom	0–5°C	3–21%	5–15%
Pomegranate	5-10°C	3–5%	5-10%

2.3. Benefits of CA over Conventional Cold Stores

By controlling the atmosphere along with increased RH and lowered temperature, the shelf life of highly perishable commodities can be extended more than that at ordinary refrigeration storage. In case of pears and apples, shelf life is about 40-60% longer and that of peas is doubled as compared to ordinary refrigeration storage. As it is possible to maintain higher RH (95%) in CA storage than normal cold storage, the fruits retain their original crispiness and lose less weight and are thus superior in quality.

In CA stored fruits ripening is slowed down after taking the fruits out of store and they have longer shelf life after the fruit is returned to room air. Moreover, CO_2 level used in CA has anti-microbial properties and also insects, mice and rats which sometimes infest fruit stores cannot survive in CA rooms. Also, the low temperature disorders of apples like core flush and internal browning do not occur in CA storage if concentration of gases is precisely controlled.

These technologies are the future needs for value addition, extension of availability and to meet the challenges in competitive global trade of fruits and vegetables in the domestic and export markets. Edible coatings are very promising systems of future for the improvement of fruit quality and preservation. They can be used where plastic films cannot be used. At present, their cost is about 10 to 50 folds higher than the polyethylene or polypropylene films, but, they are used in very low quantities for achieving specific goals. Though the FDA has approved a number of coating materials for specific applications, yet in India the use of edible coatings is not permitted under PFA. It is appropriate time that the government should take initiative in approving the use of edible coatings. Similarly, the controlled atmosphere storage of apple and pear has been widely studied and commercially used in the developed countries. The work is at experimental stage in India. The efforts should be intensified to work out the feasibility of commercial adoption of this technology. Also there is need to develop biotechnological processes to make these technologies economically more viable to augment the supply of fresh horticultural produce to the consumer.

FURTHER READING

1. Adsule, P.G. Existing Packaging Systems followed for fresh horticultural produce and then improvement using newer technologies. *Ind: Food Industry*, **20**, (2001) 58.
2. Cuq, B., Goritard, N. and S. Guilbert . Edible films and coatings as active layers. In: Active Food Packaging (Ed. M.L. Rooney) Blackie, Glasgow, (1995) 111–192.
3. Han, J.H. Antimicrobial Food *Food Technology*, **54,** (2000) 58.
4. Katz, F. New research in packaging. *Food Technology,* **52** (1998) 56.
5. Matche, R.S. New Trends in fresh produce packaging. *Ind. Food Industry*, **20**, (2001) 58.
6. Thompson, A.K. Post-harvest Technology. Blackwell Science Ltd., London. (1996).

Mushroom Production: Potential and Opportunities

P.K. KHANNA

Department of Microbiology
Punjab Agricultural University
Ludhiana-141 001 (Punjab)

MUSHROOMS ARE ECOLOGICAL wonders and have long been appreciated for their flavour and texture. In the past they were mentioned only for their condiment values, but, with more information available on the chemical composition of some edible mushrooms as well as on their medicinal and tonic attributes, they are now recognized for their nutritive value. This recognition has generated much interest not only towards the cultivation of more number of edible varieties, but, also taking the production of some of them to the commercial level. Broadly speaking mushrooms are the fruiting bodies of higher fungi which belong to the class Basidiomycetes or Ascomycetes and can be divided into four major categories:

(a)	Edible Mushrooms	*Agaricus*, *Pleurotus*, *Volvariella*, etc.
(b)	Medicinal Mushrooms	*Ganoderma*, *Coriolus*, *Lentinus edodes, etc.*
(c)	Poisonous Mushrooms	*Amanita phalloides*
(d)	Miscellaneous category	include a large number of edible fungi.

Some mushrooms fall in more than one category, e.g., they may be edible and also possess medicinal properties. Of the approximately 69,000 described species of estimated 1,500.000 species of fungi, there are about 10,000 known species of fleshy macro fungi, but, only a handful of them are lethal. Today mushrooms are recognized as a nutritionally rich food which contain reasonable amounts of high and good quality proteins, unsaturated fatty acids, minerals and vitamins and rank between low grade vegetables and high grade meats. They are low in nucleic acids, cholesterol and in calories. Therefore, mushrooms can serve as an additional dietary supplement to make up for the protein energy malnutrition (PEM) of the Indian population in addition to their dietary factors.. Some of the nutritional attributes of two most commonly cultivated edible mushrooms are reflected in Table 1.

Besides their food value, mushrooms also have medicinal potential and have traditionally been used in China and Japan for their medicinal /tonic properties (Table 2). In addition to the value of world mushroom crop estimated to be 7.5 billion dollars in 1990, a lot of money is also being generated from mushroom products, e.g in 1987, the value of pharmaceuticals developed from three mushrooms,

Table 1: Percent composition of *Agaricus bisporus* and *Pleurotus florida*

Component	*A.bisporus*	*P. florida*
Moisture	90.4	91.5
Protein(Nx4.38	26.8	19.1
Lipids	3.9	5.8
Crude fibres	8.2	9.5
Minerals(ash)	10.9	9.2
Carbohydrates	46.8	53.3
Kcal/100g	265.6	284.1

Table 2: Medicinal attributes of mushrooms

- Possess anti-fungal, anti-bacterial and anti-viral properties
- Possess anti-tumor qualities
- Possess hypocholesterolemic potential
- Show remedial allergenic reactions
- Cosmetic and health beverages can be prepared
- Pharmaceuticals and proprietary products including health drinks & foods can be obtained.

Coriolus versicolor, I.entinus edodes and *Schhizaphyllum commune* was reported to he 769 million dollars in Japan. This is in addition to several mushroom products from other mushrooms e.g. *Ganoderma,* manufactured and used in China, Taiwan and Korea that are also worth many million of dollars. Some of these pharmaceutical e.g., PS-K, PS-P, OK-432 etc. rank amongst the top selling anti- tumor drugs in Japan.

The awareness of mushrooms as a component of diet having varied medicinal applications has been growing during the past few decades and their place in the Indian dietary system becomes all the more important because of their ability to convert nutritionally waste agri-residues into edible protein food.

1. GLOBAL PRODUCTION SCENARIO

Presently, mushrooms are being cultivated in about 100 countries of the world. However, the production of mushroom is mainly concentrated in three geographical regions, 55% in W. Europe, 27% in N. America, 14% in Asia and 4% in the rest of the world (Fig. l). Netherlands, France, Italy, U.K., Germany and Poland are the major producers in European Union whereas China, Japan, Indonesia and Taiwan are the major producers in Asia. Of the 2000 edible species, 4-5 are produced on commercial scale throughout the world. *Agarcus bisporus, Lentinus edodes, Pleurotus spp, Auricularia spp* and *Volvariella volvacea* have been the most common cultivated types constituting about 85.3% of the total world production of mushrooms (Table 3) which has already crossed 5 million tonnes and this production figure is expected to reach 7 million tonnes in the next decade. In the developed countries of Europe and in North America, mushroom farming has attained

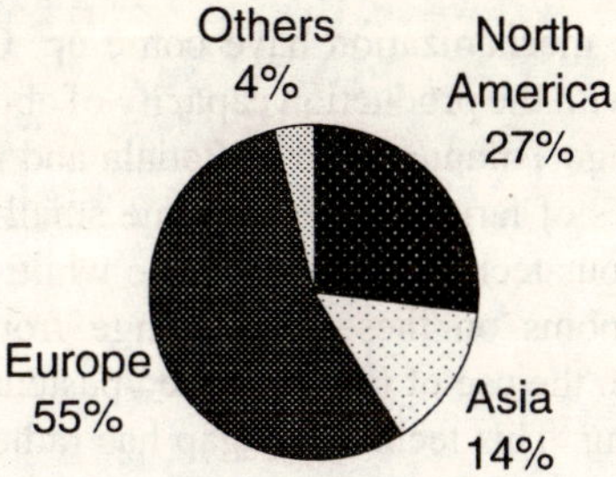

Fig 1: Global Distribution of Mushroom Production (1997)

Table 3: World production of major cultivated types of mushrooms

Variety	*Production (× 1000 MT)*	*Percent share*
Agaricus bisporus	1846.0	
Lentinus edodes	826.2	16.8
Pleurotus spp.	797.4	16.3
Auricularia spp.	420.1	8.5
Volvariella volvacea	298.8	6.1
Others	720.0	14.7

the status of the high tech industry with ample mechanization and automation especially for growing white button type although its share is declining in the production chart with a corresponding diversification to other cultivated species and increase in the production of specialty mushroom from China and Japan. Today China is the leading producer of Oyster mushroom and Japan of Shiitake mushroom with production figure of more than 80% each, respectively. India is yet to achieve a land mark in any of these mushrooms despite possessing all the attributes required to achieve this status.

2. INDIAN PRODUCTION SCENARIO

Mushroom farming in India started about 4 decades ago. but after a sluggish start it has shown an upward trend of growth in the last decade. However the production figures for such a vast country are dismal. Only 40,000 tonnes of mushroom were produced in India in 1997 that comprised of more than 90% of only button type and the major states producing this mushroom are Punjab, Haryana, H.P., J & K and some South Indian states. Oyster mushroom is the other variety becoming popular in North Indian and some South Indian states like, Tamil Nadu and Karnataka. The paddy straw mushroom which is being grown in some southern parts of the country has not picked up because of poor unpredictable yields No serious attempt has been made in popularizing other varieties of edible mushrooms whose cultivation is much simpler and does not involve much investment. Much of the 90% of white button mushroom production is in the hands of seasonal growers although during the last decade, a few Export Oriented Units using imported/improved cultivation technology with ample mechanization have come up. One such unit, the Agro Dutch Industries Ltd.. with the production capacity of about 15,000 tonnes per annum is situated in Village Tofapur, District Patiala and is working profitably. In between these two types of farms, there are some small climate controlled units which practice indigenous technology to produce while button mushrooms. The productivity of mushrooms on these firms range from 8-10Kg/qtl. to 25-30 Kg./qtl. depending upon the use of unpasteurized/pasteurized substrates and seasonal/controlled growing. This technology gap has rather hampered the over all development of mushroom production process in the country.

3. CONSUMPTION OF MUSHROOMS

While European countries lead in the production of mushrooms, the U.S.A. is the largest consumer of this protein rich delicacy followed by Netherlands and Germany. Group 6 countries comprising of U.S. A., Germany , U.K.. France. Italy and Canada consume 85% of the total mushroom produced in the world (Table 4), the remaining 15% is consumed by the rest of the world. Average per capita consumption in G-6 countries is between 2-3 Kg compared to 20-25g in India. Awareness about the nutritional and medicinal values of mushroom have dispelled many socio-religious beliefs and production along with productivity of mushroom has improved ever since, but, not to the level that makes our presence felt in the international market. There are various reasons for the low consumption of mushroom in our country that include low availability, high cost of commodity, poor purchasing power, social and religious taboos and little awareness about the potential use of mushrooms.

Table 4: Per capita consumption of mushrooms

Group-6 countries	*Kg/capita*	*Percentage*
U.S.A.	NA	30.0
Germany	3.83	17.0
U.K.	3.35	11.0
France	3.03	11.0
Italy	2.6	10.0
Canada	NA	6.0

NA: Figures not available

The world production of mushrooms has been increasing steadily during the past decade at an annual growth rate of 7-10%, however, the demand for both the fresh and the processed mushrooms has been more than the actual supply. There was an over production to the extent of 1,85,315 tonnes of mushrooms in 1993 which got reduced to 35,175 tonnes in 1996. Mushroom production is expected to register a sustained growth in demand in the next decade and is expected to touch a demand to supply gap of 6,53,454 tonnes by the year 2006 creating thereby, an enormous potential and oppurtunities for production both for domestic use as well as for its export (Table 5).

Table 5: Global demand, supply and demand-supply gap for button mushroom and projected estimates

Year	*Total demand (MT)*	*Total supply (MT)*	*Demand-Supply gap*
1993	1,816,073	2,001,388	1,85,315 (+)
1996	2,043,671	2,079.046	35,175 (+)
2000	2,394,486	2,193,127	2,01, 359 (–)
2003	2,696,494	2,285,192	4,1 1,302 (–)
2006	3,036,702	2,383,248	6. 53,454 (–)

4. POTENTIAL AND OPPORTUNITIES

The growth of mushroom industry in India has been very sluggish in the 70's and 80's. With the coining up of some export oriented units in the 90's, the production capacity has increased to around 25,000 TPA for export to Europe and North America . However, in the unorganized sector which comprised of more than 95% seasonal growing and caters to the need of domestic market, the growth has been rather very slow. This is despite the fact that India has all what is required to make it into a leading mushroom producer like abundant supply of agri-residues and other inputs, low-cost labour, favourable climate etc. In addition to this, there have been many reasons for the slow growth of this unorganized sector which have been listed in Table 6.

Table 6: Reasons for slow growth of mushroom production in unorganized sector

- Seasonal cultivation and low productivity
- Non availability of quality spawn
- Dependence on single raw material
- Obsolete production technology
- High cost of production and processing
- Poor domestic base and consumption
- Fragile marketing system
- Inadequate extension education

Inspite of various reasons listed above for the slow growth of mushroom production in the country, there seems to be enormous potential of developing mushroom industry into remunerative activity because of the following inherent advantages:

- Availability of basic raw material in abundance
- Availability of suitable technology and environment
- Availability of low- cost labour
- Huge potential for export
- Quality protein for a largely vegetarian population suffering from PEM

- Intensive use of land
- Recycling of organic residues causing environmental pollution

5. BASIC RAW MATERIAL

For mushroom cultivation, basic raw materials invariably are cereal straw, other lignocellulosics and organic residues which are available abundantly and continuously. These residues are generated annually through agriculture, forest and food industry operations. More than 325 million MT of agri-residues were available in India in 1998-99 and sufficient quantity of these are available to the farmer at his door step at no cost. Most of these residues particularly paddy straw is burnt in those niches where intensive crop rotation is followed and more preferred nutrient feeds are available . The state of Punjab alone burns more than 10 million tonnes of paddy straw annually loosing nitrogen worth many million of rupees besides causing enormous environmental pollution. Presently this vast resource of raw material is not being utilized at the desired level. Mushroom culture offers a low cost by conversion of straw into edible food which in an important need of the country for meeting the requirement of the exponentially growing population besides solving the problem of PEM.

6. TECHNOLOGY AND ENVIRONMENT

The development of technology suited to varied agro- climatic conditions of the country and its adoption has enabled states like Punjab, Haryana, J & K, H.P. to establish mushroom Industry. At present mushroom cultivation of one variety or the other is being carried out at more than 15 states of the country following low-cost cultivation technology. In addition, technology for controlled mushroom production is also available and is being practised by some small climate controlled farms as well as export oriented units. Many of these farms are producing their own compost, spawn and mushrooms and processing them as canned product. The availability of quality and productive strains have provided them another edge to compete in the global market. All these developments related to technology upgradation have resulted in increased mushroom production.

7. LOW-COST LABOUR

Cheep and skilled labour along with the availability of raw material is the main strength for mushroom growth in the country. According to one analysis of cost break-up in mushroom cultivation, the labour cost averages 7% of total production cost in India compared to 47% in the Western world (Table 7). Since mushroom cultivation is a labour intensive process, it simultaneously creates new avenues for employment. Mushroom cultivation in most part of the country is restricted to the rural areas because of cheap and easily available raw material and low-cost indoor space, therefore, it can be taken up as a self-employment and income generating activity to improve the economic standard of the masses besides enriching their diet with quality protein. Winter season is the lean period for the farmers when mushroom cultivation can easily be taken up as a profitable venture.

Table 7: Average production cost break-up (%)

Component	*In Europe*	*In India*
Raw material	26.0	20.0
Labour	47.0	7.0
Energy	3.0	17.5
Packing/marketing	6.0	23.0
Maintenance/depreciation	9.0	12.5
Interest of loans	3.0	18.0
Miscellaneous	6.0	2.0

8. EXPORT POTENTIAL

Liberalization of policies and globalization of trade has opened new vistas of export of this commodity. Although a highly competitive field ,mushroom export is a potential means to earn foreign exchange. There has been a tremendous increase in mushroom export from 1990-91 to 1997-98 and the value has exceeded 50 crores of rupees. Earlier the major export from India to Switzerland, France and Germany was of dried mushroom collections of morels (gucchi) and oyster mushroom from forests. With the inclusion of preserved mushrooms in the exportable quota, the export of canned button mushrooms has increased substantially during the last 5-6 years (Table 8). Bulk of export in the past years has gone to USA where the major

Table 8: Export* of mushrooms from India (1991 1998)

Year	*Fresh/chilled*		*Dried*		*Preserved*		*Total Value*
	Qty	*Value*	*Qty*	*Value*	*Qty*	*Value*	
1991-92	37.6	20.05	49.17	1222.40	1174.76	381.19	1623.64
1992-93	110.53	80.78	37.70	903.32	3294.20	1081.80	2065.90
1993-94	41.53	19.03	39.40	859.76	4811.48	1514.73	2393.52
1994-95	289.69	173.73	77.11	1995.72	6568.38	2368.83	4538.28
1995-96	1559.63	639.91	92.42	1895.09	6883.50	2593.27	5128.27
1996-97	2343.67	861.01	106.03	2099.44	5119.88	1931.67	4892.12
1997-98	5711.18	2013.13	148.37	3512.63	520.56	171.81	5697.57

*80% export to U.S.A. Qty: in MT Value: Rupees in Lakhs

competition has been from China, the largest exporter of mushroom from Asia. With the levying of anti-dumping duty on China, its presence is likely to suffer which could be an emerging opportunity for India to increase its export. India at the moment does not have any significant status either as a mushroom producer or a consumer or an exporter. With further reduction in the cost of production and government intervention in acquiring EEC quota, freight subsidy etc. there is every possibility of further boosting export of mushroom from India in order to earn valuable foreign exchange.

9. SOURCE OF FOOD PROTEIN

We are largely a vegetarian population and live on grains, tubers and small quantities of pulses and vegetables all of which are a poor source of protein quantity as well as quality. Vegetables provide hardly any dietary proteins and proteins from cereals, the staple food, are generally deficient in essential amino acids. Pulses on the other hand are deficient in other essential amino acids like, tryptophan. Consequently, protein energy malnutrition is rampant in Indian population. Mushrooms protein which is highly digestible (Table 9) and has a nutritional score which is intermediate in quality between low grade vegetables and high grade milk/meat protein can be of considerable significance in filling the protein gap especially of the rural masses because of the close location of mushroom production centers. Being rich in minerals, their usefulness becomes still more important .Edible mushrooms are apparently innocuous and this has been confirmed both at cellular and biochemical levels. With the increasing awareness in the production of mushrooms , their availability can be increased which would in turn improve per capita consumption.

Table 9: Some nutritional attributes of edible mushrooms

Parameter	*A.bisporus*	*P. florida*
Essential amino acid index (EAA)	51.10	46.30
Biological value (BV)	45.09	38.77
Nutritional index (NI)	13.69	12.59
Digestibility (%)	76.43	79.07
Relative nutritive value (RNV)	91.87	97.62

$$EAA = \sqrt{\frac{Cys_p \times Lys_p \times Meth_p \times Trp_p}{Cys_s \times Lys_s \times Meth_s \times Trp_s}} \quad NI = \frac{(\text{EAA INDEX} \times \%\ \text{PROTEIN})}{100}$$

$$BV - 1.09\ (\text{EAA Index}) - 11.7$$

10. MINIMUM LAND USE

Land is of relatively little importance for mushroom cultivation and is a bonus point where intensive agriculture is practised due to reduction in the land holdings. Comparatively speaking, growing mushrooms for food protein resources requires less land than producing protein from animal and other plant sources. It has been estimated that many white button mushroom growers throughout the world currently can produce 6730-7826 Kg of dry protein/hec/y compared with 78 Kg of dry animal protein /hec/y and 673 Kg of dry protein/hec/y from fish farming. The production of over 620 Kg of *Agaricus* mushrooms from a sq. m. bed area based on 5 crops/y with five shelves in a growing room with a yield of 24.4 Kg/sq.m. has been recorded. Similarly sq.m. of land can produce 586 Kg of fresh *Volvariella* mushroom/y based on 2 crops/month with five shelves in a growing house with a yield of 4.88 Kg/sq.m. Thus, less land required for growing mushrooms is very relevant in the context of intensive farming being practised due to reduction in land holdings.

11. RECYCLING WASTE MATERIALS

The availability of organic residues has increased tremendously during the last decade which are partly fed to the cattle, partly used as fuel in rural India but mostly burnt. Mushroom growing represents the only commercially viable process where these wastes can be converted through SSF process into quality food protein. Some varieties of mushroom (oyster type) also add to the feed value and manurial potential of the spent residues, which can either be fed to the cattle or returned to soil to enrich the land as these contain more concentrated, easily usable and soluble forms of organic and inorganic nutrients.

The foregoing account provides an evidence that mushroom production has a great prospect and potential to be developed into a viable remunerative industry. The country's strength lies in its cheap raw material availability, cheap and skilled man power and strong technical base .In addition, globalization of trade, liberalization of economy, anti- dumping duty against China, the major competitor, has made the climate extremely favorable for India to increase its production and export. If we can work to our strengths, overcome our weaknesses and exploit our advantages where we have an edge over other countries, then there is every likelihood that Indian Mushroom Industry can turn the comer and emerge as a major player in mushroom production and export.

FURTHER READING

Buswell, J.A. and Chang, S.T. Edible mushrooms: Attributes and applications. In: Genetics and Breeding of Edible Mushrooms (Eds. S.T. Chang, J.A. Buswell and P.G. Miles), Philadelphia: Gordon and Breach Science Publishers (1993) 297–324.

Chang, S.T. Cultivation of *Volvariella* mushrooms in south east asia. In: Tropical Mushrooms:Biological Nature and Cultivation Methods (Eds. S.T. Chang and T.H. Quimio), Hong Kong Chinese University Press (1982).

Chang, S.T. Mushrooms and Mushroom Biology. In: Genetics and Breeding of Edible Mushrooms (Eds. S.T. Chang, J.A. Buswell and P.G. Miles) Philadelphia: Gordon and Breach Science Publishers (1993) 1–13.

Chang, S.T. and Miles, P.G. Mushroom Biology—A new discipline *The Mycologist,* **6** (1992) 64.

Garcha, H.S., Khanna, P.K. and Soni, G.L. Nutritional Importance of Mushrooms. In: Mushroom Biology and Mushroom Products (Eds. S.T. Chang, J.A. Buswell and S.W. Chui), Hong Kong Chinese University Press (1993) 227–236.

Hawksworths, U.L. The fungal dimension of biodiversity—magnitude, significance and conservation *Mycological Research* **95** (1991) 641.

Hayes, W.A. Mushrooms, microbes and malnutrition. *New Scientist,* **44** (1969) 450.

Kaul, P.L. Scope of/export of mushrooms from India—potential markets. In: Conference souvenir on mushroom industry in India—A decade of achievements and future perspective. New Delhi. Dec. **3**, (1999) 17–33.

Kendrick, B. The Fifth Kingdom. Waterloo; Mycologue Publications (1985) 364.

Schroeder, N.E., Schisler, L.C., Snetsinger, R., Craw ley, V.E. and Barr, W.L. Transport of MTDF trays at casing to conventional mushroom doubles. *Mushroom* **22** (1974) 15.

Verma, R.N. National and international scenario of mushroom production and trade. In: Conference souvenir on mushroom industry in India— A decade of achievements and future perspective, New Delhi. Dec. 3. (1999) 5–8.

Fruit and Vegetable Processing: Canning

M. KALIA

Himachal Pradesh Agricultural University
Palampur 176 062 (Himachal Pradesh)

THE TERM 'PROCESSING' USED IN canning industry refers to heating or cooling of canned foods to inactivate bacteria. Absolute sterilization is difficult to attain, as many bacteria can form very minute and highly heat resistant bodies called spores, which can be killed/suppressed only by very high or very low temperature treatment. Such a drastic treatment, however, determines the quality of the canned product. Processing consists in determining just the temperature as well as the extent of cooking that would suffice to eliminate all possibilities of bacterial growth. Overcooking should be avoided, as it spoils the flavour as well as the appearance of the product. Since fruits and vegetables vary considerably in their composition and texture, it is difficult to lay down the temperature and time required for processing. Generally, almost all fruits can be processed satisfactorily at a temperature of 100°C, i.e. in boiling water. Further, they do not thrive in heavy sugar syrups, which are normally used in canning the fruits. Vegetables, except the more acidic ones like tomato, however, require to be processed at higher temperatures of about 115°C to 121°C. It is required that the contents of the can in center must reach these high levels. The processing time in such cases need to be maintained for a sufficiently long period to destroy the spores of the more resistant bacteria. The temperature and time of processing, therefore, vary with the size of the can: the larger the can the greater the processing time and *vice versa*.

1. HEAT PENETRATION IN CANS

Before a process for any particular product can be adopted, it is necessary to study the rate of heat penetration in the can. This is done by means of a thermocouple fixed to the center of the can. Some of the scientists have also done useful work in determining the time for which a given size of a can of a particular food should be processed at a temperature. For any canned product, temperature can be worked out, making use of heat penetration data, either by the simpler integrated graphic method or by the more sophisticated mathematical procedure based on thermal death time (TDT) of a typical spoilage organism practically similar to *Clostridium botulinum*. Details of these methods can be found in advanced books dealing with the subject. However, to give a fair idea of the technique and procedure followed, the subject has been dealt with in a succinct manner.

Some scientists have investigated the role of acidity in vegetable canning. They have shown that brines, acidified with a small amount of acetic or citric acid, greatly reduced the resistance of the heat-resisting spores of bacteria, and that in acidified brine, the time and temperature required to sterilize vegetables could, therefore, be reduced considerably. This process, however, is not to be recommended for home-canning of vegetables, as there is always an element of uncertainty in the strict application of the conditions prescribed.

2. PROCESSING METHODS

There are generally three types of cookers used for canning of fruits. These are:

2.1. Open Cookers

These cookers are very simple in construction and consist essentially of wooden tubs or galvanized iron tanks of any desired capacity. The sealed cans are placed in crates of galvanized iron and immersed in the tank containing boiling water, which is kept boiling by letting in jets of steam through perforated pipes placed underneath the false bottom of the tank. An improved type of this cooker is a galvanized iron tank with a cover; false bottom of perforated galvanized plate and perforated steam pipe for jets of steam.

2.2. Continuous Non-agitating Cookers

In these cookers, the cans travel in boiling water in crates carried by overhead conveyors in a single file, on a continuously moving belt. These cookers are fitted with temperature controlling and recording devices also.

2.3. Continuous Agitating Cookers

These cookers are of various designs and are generally used in big production units. The sealed cans move in the non-agitating cookers, but are at the same time rotated by special mechanical devices to agitate the contents of the cans. By this technique, the processing time is reduced considerably. These cookers may either be filled three-fourths with water, which is kept hot by steam or processing may be carried out in steam itself. These cookers have an advantage over other types in that they accelerate the output to a considerable extent.

Due to lack of acid, texture and proximity to soil, which may infect them with spore-bearing organisms, vegetables require much severe sterilization than processing of fruits. The vegetables are usually processed at temperatures much above 100°C, under steam and pressure in closed retorts. The retorts, generally of batch type, vary in shape and size, with capacities ranging from 50 cans to 3,000 cans. The more modern ones, which are known as automatic pressure cookers, operate continuously . The retorts are either vertical or horizontal. To the large commercial retorts, automatic steam and temperature and time controlling devices equipped with recording equipment are fitted.

In India, small vertical stationary retorts are generally used. They are made of cast iron cylinder of varying sizes and are fitted with a steam tight lid. The retorts are also equipped with necessary steam and water-feed inlets, air vent, drain cock, safety valve, pressure gauge and thermometer. The sequence of operating these retorts is as follows:

The sealed cans are arranged in crates, which are then placed inside the retorts. The lid is placed in position and tightened. The vent and drain cocks are opened and steam is let in until it blows freely through the vent-hole, thereby removing all the entrapped air. The vent and the drain are then partially closed till the temperature reached 100°C. The exits are then closed, leaving only a small opening for a fine jet of steam to escape, and the steam pressure and temperature are allowed to rise to the desired extent. During the cooking periods, the thermometer and pressure gauge records should show close agreement. For instance, if the cooking temperature is 115°C, the steam pressure should be a little above 0.7 kg/sq. cm. If at this temperature, the pressure is higher than 0.7 kg/sq. cm., the presence of entrapped air is indicated. On the completion of the cooking period, steam is cut off and the released valve at the top is opened gradually to let of steam; any sudden releases being avoided as it is likely to burst the cans inside. When the pressure becomes normal, the retort is opened and the cans are removed for subsequent cooking to prevent their overheating.

3. FACTORS AFFECTING PROCESSING

3.1. Effect of altitude on Processing Time

The boiling point of water decreases with an increase in altitude. The processing time for different kinds of fruits and vegetables, (as standardized at places about 150 metres above sea level) has to be relatively increased with increase in the altitude. As a rule, for every increase of 150 metres in altitude, the boiling point of water decreases by about 1°C and the normal processing time has to be increased by about two minutes. Thus, an A 2 ½ size can of pears which requires processing for 30 min at sea level, will require a total processing time of 30+25=55 min at an altitude of about 1850 metres above sea level.

3.2. Effect of Altitude on Processing Pressure and Temperature

As the altitude above sea level increases, the pressure required to maintain the specified processing temperature also increases. Since process-time schedules are based on the exposure of the canned food to the specified temperature in steam, it may be necessary to maintain higher pressure if the processing is done at an altitude substantially greater than 300 m above mean sea level.

3.3. Effect of Acidity on Sterilization

Acids in solution owe their acidic properties to the hydrogen ion concentration, which is a measure of the intensity or potential of acidity rather than of its quantum. It is expressed as the pH value. The pH value of a solution is equal to the logarithm of the number of litres of solution, which contains 1 gram of hydrogen ion. The pH value of water, which is neutral, is 7. The pH values rising above 7 represent progressive increase in alkalinity and those falling below 7 indicate increased acidity. Hence, the more acidic the substance, the lower is its pH value. The acidic property or sourness of fruits is due to the presence of hydrogen ions.

The pH value has a great influence upon the destruction of microorganism. The lower the pH the greater is the ease with which a product can be processed or sterilized. Based on the pH value, the foods are classified as:

- **Low Acid–pH 5.0 and Higher:** This class includes peas, beans, corns, asparagus, etc. These are subject to spoilage by non-sporing aciduric types, butyric anaerobes of the *Clostridium pasteurianum* and Wingogradskyi types and thermophilic anaeobes.
- **Highly Acid–pH 3.7 and Below:** This group includes sauerkraut, pickles, berries, citrus juices, grapefruit and rhubarb.

To be more precise, bacterial spores are much more easily destroyed in fruits at pH 3.0 than in vegetables at pH 5.0 to 6.0. These spores do not germinate and grow at pH values much below 4.5. They are not capable of causing any harm even if they are not destroyed in canned fruits. Thus, pH 4.5 is regarded as a dividing line between the acid and the nonacid foods. A canned product, having pH value of less than 4.5 can be processed in boiling water at 100°C, but a product with a pH value above 4.5 requires processing under pressure of 0.70 to 1.05 kg/sq. cm.

3.4. Effect of Pressure

When a can is closed, the pressure inside it is equal to that of the atmosphere, irrespective of the temperature of closing. Any increase in the temperature, after the closing of the can, develops pressure inside the can. This internal pressure depends upon several factors of which the following are important:

- Pressure due to water vapour inside the can
- Pressure due to expansion of any enclosed air
- Pressure due to the expansion of the contents of the can
- Pressure due to liberation and expansion of gases inside the vegetables
- Expansion of the can by the doming of the ends as well as compensating effect of the expansible rings on the lid.

Among these, water vapour produces the greatest effect. Water vapour has a pressure of 0.2 kg/sq cm at 60°C and 1.75 kg/sq.cm at 115°C. This indicates that very high pressures develop at higher temperatures.

The pressure inside the can during processing can kill thermophilic and mesophillic putrefactive anaerobes, including *Clostridium botulinum*.

- **Medium acid–pH 5.0–4.5**: Meat and vegetables mixtures, spaghetti, soups and sauces whose incredients produce a medium acid product, fall under this category. The thermophilic anaerobes (not producing hydrogen sulphide) that cause flat sours, are of increasing importance in this group.

- **Acid–pH 4.5–3.7:** This group includes tomatoes, pears, figs, pineapple nectarines and other fruits subject to spoilage regulated by following points:
  - Internal pressure in the can is equal to the vapour pressure at the processing temperature minus the vapour pressure at the closing temperature.
  - The nearer is the closing temperature to the processing temperature, the less would be the internal pressure during processing.
  - Small cans withstand greater internal pressure than the large ones.
  - To avoid bucking, cans larger than No. A21/2 should be coded under pressure.
  - To get a good vacuum, the closing temperature of the can should not be less than 71°C.

Cans closed at 71°C and processed at 115°C show a maximum pressure above that of the atmosphere, i.e., 1.40 to 1.47 kg/sq cm. These pressures are close to the theoretical value for water vapour alone.

4. COOLING

After processing, the cans are cooled rapidly to stop the cooking process to prevent stack-burning, especially in cans at the center of large stacks where they may remain hot for several hours. Prolonged heating results in an inferior and uneven pack. In extreme cases, fruits like peaches and pears become dark in colour, whereas, the tomatoes turn brownish and become bitter in taste, while peas become mushy with a cooked taste. In case of many of the vegetables, prolonged heating due to improper cooling may result in the development of flat sour by the heat resistant spore-forming bacteria, which survive ordinary commercial processing. In this case the cans do not show any bulging but the contents will be sour.

Cooling of cans is done by:

- Immersing or passing the hot cans in tanks containing cold water;
- Spraying with jets of cold water;
- Turning in cold water into the pressure cooker in the case of canned vegetables; or
- Exposing the cans to air in small lots when water supply is scarce.

The cooling equipment varies with the method as well as the output. Various types of equipment, from simple open galvanized iron tanks to the complicated automatic coolers are available. A simple cooling tank used in India is an ordinary galvanized iron tank fitted with water connections for constant flow of water. The cans are placed in crates for cooling in the tank, where the cold water is continuously circulated. In very hot places, it may even be necessary to cool the water by adding ice. Simple cement cisterns, with overhead traveling crane for moving crates of cans form the retorts to the cooling tank, have been found highly satisfactory in medium size canning industry.

5. TESTING FOR DEFECTS

In a canning unit, batches of finished cans should be finally tested for any leakage or imperfect seals. A simple way of doing this is to tap the can with a short steel rod. A clear ringing sound indicates a perfect seal, while a dull and hollow sound implies a leaky or imperfectly sealed can. This method is very rapid and an experienced workman seldom fails to detect a defective seal.

The other method commonly used for testing the vacuum inside the cans, is by means of a simple vacuum gauge fitted with a sharp point, which is protected with a thick rubber gasket. The sharp point is pierced into the lid of the can and a short twist given so that the rubber gasket presses against the puncture made in the air-tight seal. The vacuum in the can is directly read on the dial of the gauge. Faulty cans do not show any vacuum. The vacuum can also be determined with a vacuum tester without puncturing the can and losing its contents. This is a non-destructive type of vacuum testing and is based on the measurement of the concavity of the can lid which is proportional to the degree of vacuum inside the can. A battery-operated electrical circuit with a ringing bell facilitates the measurement of concavity and hence the vacuum inside. This device, first developed by the C.S.I.R.O. Australia at their Food Preservation and Transport Division, has since been modified and patented by the Central Food Technological Research Institute (CFTRI), Mysore. A Flip Tester can also be employed to determine the vacuum in the can by placing the can in a glass chamber and evacuating it. The can lid will give a flip sound when the vacuum outside becomes slightly more than that inside the can. This tester can be used for non-destructive testing of canned products for internal vacuum. It can be operated with ease either on the mains or with an ordinary dry cell. It is portable and handy.

6. LABELING, STORING AND PACKING

The outer surface of the cans should be perfectly dry, as even small traces of moisture are likely to induce rusting. When canning is done near the sea or in damp and moist regions, it is advisable to lacquer the can on the outside to prevent rusting during storage. The cans are then labeled by hand or by machine and packed in strong wooden cases or corrugated cardboard cartons, in standard packing. Till they are dispatched, they should be stored in a cool and dry place. In a hot country like India, where the atmospheric temperatures are high during several months in a year, basement stores are useful, specially during summer months, as the temperature in these is lower by about 6 to 8°C, compared to the outside temperature. Storage of cans at high temperature should be avoided, as it shortens the shelf-life of the product and often leads to the formation of hydrogen swells and perforations during extended storage.

Besides processing parameters, for the successful operation of a fruits and vegetables processing plant on commercial scale, it is important to take into consideration the other factors also like investments, location of processing unit, water & electricity supply, skilled and non-skilled manpower.

FURTHER READING

1. Salunkhe, D.K. and Kadam, S.S. Handbook of Fruit Science & Technology. Marcel & Dekkar, New York. (1995).
2. Sandhu, K.S. and Bakshi, A.K. Commercial canning of fruits and vegetables. In: Post harvest technology of fruits & vegetables (Eds. L.R. Verma and V.K. Joshi), Indus Publishing Co., New Delhi (2000) 637–680.
3. Srivastava, R.S. and Kumar, S. Fruit and Vegetable Preservation—Principles and Practices. International Book Distributing Co., Lucknow. (1998).
4. Thompson, A.K. Post harvest technology. Blackwell Science Ltd. London. (1996).
5. Woodroof, J.G. and Luh, H.S. Commercial Fruit Processing. 2nd edn. AVI Pub. Coo. Westport, CT. (1986).

Emerging Biotechnologies in Dairy and Meat Processing

B.N. MATHUR, R.K. MALIK
and
L. SABIKHI

National Dairy Research Institute
Karnal 132 001 (Haryana)

THE DISCIPLINE OF DAIRY and Meat Processing in the twenty-first century will require newer ways of capturing the application of microbial genetics, whether it is through *in vivo* or *in vitro* engineering, and will revolutionize our notions of food, from production to processing. In the words of Douglas McCormick, the editor of Bio/Technology,

> "By most standards, all of this biotechnology is still brand new. Yet biotechnology is the product of intellectual earthquakes that change things so completely that it is difficult, after the event, to remember that once the landscape was different. It is easy to forget that the change is recent and more change is just around the corner".

The report of the research committee of the Institute of Food Technologists states that, of the six areas of research priorities, the third highest belongs to food biotechnology. Various records, whether published or practised would indicate that through the application of biotechnology in the food industry, "an invisible revolution" is taking place.

1. STARTER CULTURES

Starter cultures comprising lactic acid bacteria have an ancient history in human cultural traditions. Their application makes possible the manufacture of thousands of fermented foods especially when used in mixed cultures with other types of bacteria. They are an essential component of all fermented dairy foods including cheese, yoghurt, sour cream and lactic butter. Research progress on the genetics of these bacteria has been substantial. The potentials for the application of dramatic discoveries made about the genetics of lactic acid bacteria are limitless. Good quality starter cultures are the basis of preparing good quality fermented milks. The commercial value of the fermented product is, therefore, mainly dependent on the performance of the starter cultures. Plasmid biology of lactic acid bacteria have opened new vistas for exploring possibilities for using recombinant DNA(r-DNA) technology and genetic engineering to improve the commercial value of these organisms. Genetic manipulation of these cultures with respect to commercially important traits has helped the dairy and meat industry in efficient production of high quality fermented products with higher consumer acceptability.

1.1 Flavour Production

In addition to extending the shelf life of perishable raw materials, food fermentation results in the development of products with excellent flavour and aroma characteristics. Although lactic acid bacteria (LAB) undoubtedly contribute to the ripening process, the specific mechanisms involved are not understood. Clearly proteolytic activity is of major significance. The component involved in mediating proteolysis have been elucidated. Molecular biologist is now in a position to deliberately manipulate the proteolytic system of LAB and to determine the consequential effect on flavour and body development during ripening. Genetic and protein engineering strategies have been applied to generate enzymes with new activities. Peptidase activity, the second major component involved in starter-mediated proteolysis is likely to have a more direct effect on cheese ripening. The generation of starter cultures expressing proteinases and peptidases with relative activities significantly different from those expressed by the wild type of strains will soon be an achievable goal. This will help point out those activities more directly involved in flavour development, and such strains may also help to develop novel flavour combinations that could lead to an expanded range of fermented foods.

The technology and knowledge that allow a thorough understanding of the role of proteolysis in flavour development in classical ripened fermented foods can also be exploited to generate flavour compounds that can be used as ingredients in a range of other food products. LAB also generate a range of other flavour products, including diacetyl, which gives a buttery aroma and is a highly desirable product in many foods. Strategies involving metabolic engineering, whereby metabolic pathways are deliberately manipulated to overproduce specific products, are currently being developed to generate cultures that elaborate elevated levels of diacetyl. Genetic approach offers the advantage of allowing overproduction to be deliberately controlled at levels appropriate to the process and the product.

1. 2 Exopolymer Production

Texture of fermented dairy products is also important in flavour perception. Amelioration in the body and texture of fermented foods has been attributed amongst other factors to polysaccharide production by the selected strains of LAB. The use of EPS predicting lactic starter cultures is of interest in the dairy industry for improvement of body and texture. There are variants of LAB, which produce exopolymers. These will change the mouthfeel of the product and encourage flavour volatiles to remain in the mouth for a longer period. It has been shown that the ropy phenotype is linked to plasmid DNA. Since the number of naturally ropy strains is limited, conjugal transfer of the mucoid plasmid to other strains could be used to construct additional ropy cultures useful in producing a ropy texture in a variety of fermented milk products.

1. 3 Bacteriophage Resistance

Food fermentations rely on actively growing lactic bacteria that either are added as starter cultures or which grow spontaneously in the food matrix. The fermentation capability of lactic acid bacteria can be severely inhibited by a nonviral inhibitors, but, the major commercial problem results from bacteriophage infections. In the area of food fermentations, the permanent threat of phage infection is particularly manifested in the dairy field. Due to rapid progress made in recent years in the understanding of both the phage genetics and the phage resistance mechanism of the bacterial starters, a variety of alternatives have become available. This has led us from conventional phage control to modern approaches. Since plasmid-encoded phage-insensitivity systems were first discovered in lactococci, starter cultures have become candidates for improving phage insensitivity by genetic techniques. The phage-resistant transconjugants also retain their "food grade" status since no r-DNA techniques are employed in their construction. This is an important consideration since it allows their use in the food industry.

2. ENZYMES IN DAIRY PROCESSING—RECOMBINANT CHYMOSIN

Cheese industry has been the major beneficiary of the biotechnological advances made in the last decade. In the early 1980s, r-DNA techniques were developed which made it possible to transfer DNA from one species to another. This revolutionary new technology and the growing market need for a standardized chymosin of high quality with a constant and guaranteed supply and stable price combined to lead to the eventual solution of the problem in the dairy industry and satisfy a market worth 65 million dollars of rennet per year. The chymosin story has been a primary example of how genetic modification can be used to bring benefits to the food processing industry and ultimately to the consumer. The DNA of calf chymosin has been successfully cloned into yeast, *(Kluyveromyces lactis),* bacteria *(Escherichia coli)* and moulds *(Aspergillus niger)* to give at least three commercial products which are now used widely in cheese manufacture. All three recombinant chymosins currently available in the market are produced by standard large-scale fermentation methods and extracted by down-stream processing. Technological performance of recombinant chymosins has been excellent since the cheeses produced using recombinant chymosin are essentially indistinguishable from traditional cheese.

3. CHEESE RIPENING—ACCELERATION OF THE PROCESS

Cheese ripening is a complex process of concerted biochemical changes, during which a bland curd is converted into a mature cheese having the flavour, texture and aroma characteristics of the intended variety. The gradual breakdown of carbohydrates, lipids and proteins during ripening is mediated by several agents including residual coagulant, starter bacteria and their enzymes, non-starter bacteria and their enzymes, indigenous milk enzymes, especially proteinases and secondary inocula. Ripening is an expensive and time-consuming process, depending on the variety. Owing to the cost of ripening cheese, there are obvious economic advantages to be gained by accelerating the process.

Greater control of ripening may also be gained by manipulating the process whereby end-product quality may be predicted with greater certainty. With the advent of biotechnological processes, acceleration of cheese ripening is, therefore, of benefit to the producer from both the economic and technological point of view, provided, of course, that the final product has the same flavour profile and rheological attributes as conventional cheese.

3. 1 Use of modified starter cultures

A number of approaches have been adopted to augment the contribution of modified/attenuated cultures to accelerate ripening, i.e., the use of (a) lysozyme-treated starters, (b) heat or freeze-shocked cells and (c) mutant cultures.

Modified starter cultures, with attenuated acid-producing abilities, are added with the normal starter culture during cheese manufacture and contribute to proteolysis during ripening. Cheddar cheese produced using defined-strain starters, while having a clean consistent flavour, does not satisfy the market for extra-mature cheese. Indeed, it now appears that the focus of accelerated ripening may switch from the use of added proteinase/lipase to the selection of starter strains with enhanced autolytic properties and increased peptidase activity.

3. 2 Mutant starter cultures

Lactose negative (lac-) mutants starter strains have also been used to provide 'packets' of uniformly distributed proteinase/peptidases, enhancing the production of peptides and free amino acids. These cells do not interfere with acid production during manufacture.

Modified starter cultures with enhanced complements of proteinase and/or peptidases which could be released early and evenly distributed in the curd would be an ideal method of accelerating cheese ripening and recent cloning of the genes for a dipeptidyl aminopeptidase and phage lysin would indicate that research is being directed to this end. Starter culture biotechnology offers the advantage of uniform distribution of increased amounts of proteolytic activity in cheese. Provided that the technique used to produce these enhanced starter qualities are not prohibited, a more efficient means of accelerating ripening will be achieved.

4. BIOPRESERVATION

Preservation of dairy and meat products in a sound and safe condition has ever remained a challenging subject for the mankind. The changing consumer requirements for fresh highly nutritious foods require a less processing treatment leaving many bacteria unaffected which may eventually grow and spoil these foods. On the other hand, the consumers of the present day are constantly looking for foods without any chemical preservatives, thus forcing the food scientists to search for natural antimicrobial compounds for preservation of foods. The antimicrobial metabolites of LAB are unique with respect to their role in the biopreservation of foods.

The diverse group of LAB involved in various food fermentations alter the sensory characteristics of raw materials, often resulting in dairy and meat products with increased nutritive and economic values. In addition to lactic acid, which is the major preservative in fermented foods, many LAB produce other compounds also, that contribute to antibiosis. In recent years, there has been considerable interest in the potential application of bacteriocins as food preservatives. This stems from the fact that they are natural compounds produced by food grade LAB. The potential for the use of bacteriocins of LAB in the meat and dairy industry has spurred research in this area for natural preservation of foods. Bacteriocins have been envisioned as effective means of ensuring food safety by inhibiting food-borne pathogens. Significant numbers of bacteriocins produced by LAB have been isolated and characterized. These bacteriocins form heterogeneous group of antibacterial proteins. Isolation of the antilisterial bacteriocinogenic strains has greatly helped in the biosafety of meat and meat products.

The recent success in the molecular analysis of the pronisin structural gene has given an impetus to genetic and protein engineering technologists; to modify the molecule to increase its stability and broaden its activity. Furthermore, the availability of the cloned genes for Nisin and other bacteriocins paves the way for the construction of strains that produce greatly increased levels of these compounds.

Bacteriocins of lactic acid bacteria do have a role to play in the control of undesirable bacterial growth and perhaps also in manipulating the properties of fermented products. The genetic analysis of bacteriocin operons, regulation and the ability to overproduce the inhibitor will certainly have impact on the cost effectiveness. Protein engineering such as that already initiated with nisin, promises much in the way of additional inhibitors with altered target ranges, stability and applications.

5. BIOPROTECTIVE PROTEINS OF MILK—LACTOFERRIN AND LACTOPEROXYDASE

Milk proteins are widely known and used as ingredients for their functional and nutritional properties. The major protein fractions in the milk that contribute to these properties have been well characterized. Milk also contains minor protein fractions that have gained interest because of their physiological functions. Among these proteins are lactoferrin and lactoperoxydase. Both proteins are associated with the whey fraction, which accounts for about 20% of the protein in the milk. About 3% of these whey proteins are represented by lactoferrin, whereas, the quantity of lactoperoxydase is smaller, about 0.4%. Yet it has been possible to isolate these minor milk proteins with high purity on industrial scale. The commercial availability of both proteins allows the development of applications both for food and non-food use. Based on the physiological/functional properties attributed to lactoferrin, a number of interesting and innovative, possibilities for the application of this bioactive protein from milk can be considered (Table 1).

Table 1: Envisaged application field of Lactoferrin

Functionality	*Infant formulae*	*Health foods*	*Medical/ Clinical*	*Veterinary & Feed*	*Cosmetics*
Iron carrier	*	*	*	*	
Anti inflammatory			*	*	*
Promotion of intestinal flora	*	*	*	*	
Antibacterial/ preservative			*	*	*
Cell growth	*	*	*	*	

Lactoperoxydase is the most prominent enzyme in bovine milk, where it is found in concentration around 30 mg/1. A wide range of bacterial species is affected by the LP system. Gram negative organisms may be inhibited or killed. Gram positive bacteria are more resistant and are generally only inhibited in their growth. Lactoperoxydase requires co-factors for anti-bacterial activity. Hydrogen peroxide and thiocyanate with either sodium or potassium serve this purpose. All these components together are referred to as the LP-system (Table 2).

Table 2: Envisaged application fields of LP system

Functionality	*Food Products*	*Feed & Veterinarian Products*
Preservative	— Liquid milk products	— Milk replacers
Anti- inflammatory	— Cheese products	— Anti-diarrhoea preparation
Promotion of intestinal flora	— Meat, fish and poultry products	
	— Functional foods	— Anti-mastitis preparation

6. BIOLOGICALLY ACTIVE PEPTIDES

Biologically active peptides in foods denote 'extra-nutritional' constituents naturally present in small quantities in food. They are released by enzymatic digestion of the food proteins, either *in vivo* or *in vitro.* Milk proteins also contribute to human health in the same manner as other food proteins. The *in vivo* production of peptides from milk proteins means the action of either the enzymes secreted by the gastro-intestinal tract (GIT) or those released by the intestinal microflora. The *in vitro* processes are the consequences of fermentations in food processing activities. Biologically active peptides originating from milk proteins have been isolated, purified and identified during peptide sequence studies. Such peptides with characteristic amino acid composition and sequences have been reported to exhibit opioid, antihypertensive and immunomodulating activity and modify nutrient absorption besides performing many other miscellaneous functions.

Some peptides which have opioid characters and affinity for opiate receptors have been isolated from hyrolysates of casein. There are two types of opioid peptides, the agonists which exhibit opiate-like traits and the antagonists which have anti-opiate-like properties. The latter reverse the effect of the former. Opioid peptides have been isolated from enzymatic digests of all the major components of casein.

Certain other peptides derived from α-s1 and β-Cn exhibit antihypertensive activity. The mode of action of peptides in regulating blood pressure is expressed in yet another manner, by their antithrombotic activity. A portion of the glycomacropeptide of κ-Cn inhibits blood platelet aggregation. It also combines with the receptor sites, thus preventing the binding of fibrinogen with platelets, hence expressing its antithrombotic activity.

Peptides have exhibited their physiological role of immunomodulation *in vitro* and *in vivo.* Some phagocytosis-stimulating and hypoallergenic peptides have also been isolated from casein hydroly sates. Bio-active peptides derived from the tryptic hydrolysates of casein, known as caseinophosphopeptides (CPP) possess physicochemical properties that enable the chelation of various bi- and trivalent minerals to be carried out, thereby enhancing mineral solubility in the lower small intestine. It is an established fact that calcium is solubilised in acidic environments and also that calcium as present in milk is highly absorbable in the intestine, although the pH of the intestinal contents is near neutral or slightly alkaline. It has been suggested that moderate and exchangeable binding of calcium to CPP is responsible for the high absorbability of calcium from milk.

Peptides that stimulate DNA synthesis were found in tryptic hydrolysates of P-Cn. Physiologically active peptides also have the ability to promote the growth of bifidobacteria in the GIT, thus increasing the number of these beneficial organisms in the intestines. Lacoferricin, a peptide derived from milk protein is said to have an antibacterial effect against *Helicobacter pylori,* a known cause of stomach and duodenal ulcers. Consumption of cultured milk products have been associated with reduced incidence of colon cancer, which is attributed to some casein-derived peptides.

7. PROBIOTICS

Diet, stress and modern medical practices have been implicated as factors capable of exerting an influence on human health and nutrition. As a result of these and other factors, the incidence of illnesses which may be induced by deficient or compromised microflora e.g. gastrointestinal tract infections; constipation, irritable bowel syndrome (IBS); inflammatory bowel disease (IBD)-Crohn's disease and ulcerative colitis; food allergies; antibiotic-induced diarrhoea; cardiovascular disease and certain cancers is increasing. Probiotics have been defined as live microbial food supplements which beneficially affect the host by improving the intestinal microbial balance or more broadly, as 'living microorganisms, which upon ingestion in certain numbers, exert health affects beyond inherent basic nutrition. Combination of various microorganisms, particularly species of *Lactobacillus* and *Streptococcus* have traditionally been used in fermented dairy products to promote human health through the influences they may exert on the microbial ecology of the host, lactose intolerance, incidence of diarrhoea, mucosal immune response, levels of blood cholesterol and cancer. Indeed, a number of probiotic bacteria are now being successfully exploited commercially.

In the development of probiotic dairy foods intended for human consumption, strains of lactic acid bacteria such as *Lactobacillus* bifidobacteria. and *Enterococcus* have been most commonly used. This is primarily due to the perception that they are desirable members of the intestinal microflora.

Over the past 5 years, the probiotic field has exploded with a number of new cultures each purported to elicit a variety of benefits. Lists of functional characteristics and benefits, *in vivo,* are now common to any presentation on probiotics. Industrial interest in developing probiotic functional foods is thriving, driven largely by the market potential for foods that target general health and well being. In recent years, advances in microbiology and molecular biology have stabilized a number of fronts that have been historically problematic for probiotics. First, the availability of molecular tools to properly identify probiotic species and individual strains has virtually eliminated confusion over strain identity and ancestry. The accumulating information on ribosomal RNA (r-RNA) sequences provides a gracing resource for comparative identification of probiotic cultures, both established candidates and potentially new candidates.

7.1 Selection of functional probiotics

The perceived desirable traits for selection of functional probiotics are many. The selection criteria that fall into four basic categories are appropriateness, technological suitability, competitiveness, performance and functionality. Understanding the mechanisms as to how these criteria impact *in vivo* functionality will present one of the major scientific challenges for probiotics in the coming decade. Genetic content and controls underlie each mechanism and dictate the performance of probiotic cultures across their expected roles. Therefore, this field is perfectly poised to exploit the recent progress in sequencing capacity and functional genomes towards the investigation of these bacteria and their probiotic capabilities. There is now a myriad of possible probiotic strains, coupled with a diverse set of phenotypes, which are being increasingly linked to a variety of benefits. It has become clear that defining and screening important genetic traits, which confer functional probiotic activities offers considerable promise to attack the insurmountable task of selecting superior strains and building combinations that can elicit unique or multiple effects.

Currently many opportunities are emerging for enhancing existing traits and developing novel properties by genetic modification of probiotic cultures. Targets for genetic modifications and improvement include: immunostimulation, antimicrobials and bacteriocins; vitamin synthesis and production, adhesion and colonisation determinants and metabolic engineering to alter products (e.g. polysaccharides; organic acids).

8. LACTIC ACID BACTERIA IN MEAT FERMENTATION

Meat is an important part of our diet and its storage requires special measures since it is highly sensitive to microbial spoilage. Traditional methods of preservation of meat are drying, salting and fermentation. The main fermented products consumed now-a-days in the western world are fermented sausages. Fermented sausages are cured meat products that are shelf-stable and are commonly consumed without application of any heating process.

It is remarkable that in the traditional fermentation process, bacteria, yeasts and fungi contribute together and specifically to the quality of the fermented sausages, although lactic acid bacteria (LAB) would contribute to all aims of the ripening process only by themselves. In modern technology, LAB can indeed control this process as sole agents of fermentation. A number of lactic starter cultures have been developed biotechnologically to be used in the ripening of meat. Different from starters in the dairy field, the resistance to phage attack has not been considered an important property of lactic starters for meat fermentation.

The study of the effect of LAB starter cultures has revealed that their application generally ensures that the aim of the ripening process is attained more safely than it was possible in the traditional spontaneous fermentations. Apparently strains of *L. sake* and *L. curvatus* are good candidates for further research employing the modern genetic tools. These bacteria may either be combined in multiple-strain lactic starters or may receive enhanced and even new enzyme activites with the aid of genetic modification. The recent molecular events in the bactericinogenic strains of pediococci may lead to the enhanced safety of the sausages. It has been shown that the methods developed for genetic modification of lactococci or *L. casei* can be adopted to meat starters. It appears even feasible to employ this potential for intrageneric gene transfer. The resulting LAB should combine the high competitiveness with an enlarged physiological activity for the production of various types of fermented sausages.

Recent physiological treatises in the nutritional and physiological research ascribe several health-related attributes to the microbial derivatives of dairy cultures. Several physiologically significant biochemicals are produced during the fermentation process that influence the cholesterol metabolism, and depress the proliferation of tumour cells. Microbial cells ingested through fermented milk products influence the gut microfiora and ecological system prevailing in the gut. Biochemical activity of the lactobacilli and bifidobacteria implanted in the gut in turn re-enforces the host resistance to various enteric infections, permits enhanced lactase activity to facilitate digestion of milk, reduces allergies of milk proteins, enhanced vitamin production and helps in detoxification of liver in chronic diseases. Excellent potential is indicated for biotechnological applications for developing food systems, which could potentially reduce the role of medicines in human health.

While beneficial aspects of cultured milk products are well documented, fermentation process cannot be applied indiscriminately to various kinds of dairy foods, because of the modification in the resulting taste and physico-chemical characteristics. Biotechnological developments have led to the production of range of dairy foods, where cultured cells could be incorporated through suitably prepared microbial cell concentrates, thus bypassing the fermentation process entirely.

Recent research findings have demonstrated technical feasibility of implanting desirable lactobacilli and bifidobacteria in the intestinal tract through oral ingestion of such biotechnologically modified foods. The need of the hour is to develop advanced infant foods with antimicrobial and protective factors of milk such as lysozyme, lactoperoxydase, lactoferrin and vitamin binding proteins.

The potential for use of lactoferrin as a therapeutic modality for dampening the hyperplasia associated with leukaemia has been demonstrated. Production of desired flavouring compounds in fermented dairy and meat products through genetically modified starter cultures may bring a revolutionary change in the palatability of these foods. Biotechnological developments in the biopreservation of these valuable foods are proving to be a boon in their biosafety. Scope is also indicated for the application of biotechnology for the development of range of dairy foods, which would meet the special health and nutritional requirements of athietes and during geriatric period.

FURTHER READING

1. Balasubramanyam, B.V., Kularni, B.V., Rao, K.J. and Ghosh, B.C. *Newer Concepts in milk processing,* **21**, (2002), 42.
2. Fox, P.F. Lactose: enzymatic modifications in advanced dairy chemistry. Vol. 3, Chapman & Hall Publishers, London. (1997). 77–125.
3. Gandhi, D.N. Fermented milk products. NDRI Publication No. 261, (1993), 1–11.
4. Gandhi, D.N. Fermented Dairy Products and their role in controlling food borne diseases. In: Food Processing: Biotechnological Applications (Eds. S.S. Marwaha and J.K. Arora), AsiaTech Publishers Inc., New Delhi. (2000), 209–220.
5. Waghela, M.N. and Kilara, A. Nutritional and therapeutic aspects of indigenous and related western fermented milk products—A review. *Indian J. Dairy Sci.*, **45** (1992) 393.

Possibilities of Use of Biotechnology in Dairy Industry

M. MUNJAL *and* R.K. SHARMA

Milkfed Composite Milk Plant
Mohali 160 059 (Punjab)

A COUPLE OF YEARS BACK, Milkfed started marketing its products under its brand name of VERKA and started a low key advertisement campaign. One of the pamphlets used in that campaign mentioned that "It all started with pure and whole-some Milk". This line was followed by details of products and finally a single liner "Simply because one can do wonders with Milk. Verka is making it happen".

Taking a cue from opening and closing, sentence of this advertisement campaign, idea struck to us that if we can adopt new technology, processing parameters and new ideas in manufacturing of new products from milk, why cannot Dairy Industry do and think something about application of bio-technology to tailor even the milk for meeting different requirements specific to different type of products and again different products specifically designed to meet the requirement of specific segment of customers. By doing so, a number of problems like lactose allergy to certain customers and changes in protein which may be less susceptible to deteriorating action of time and temperature between production and marketing can be solved.

The area, where application of biotechnology can be helpful for Dairy Industry is development of microbes which may be helpful in carrying-out such fermentations which are synergetic to human metabolism and also helpful in the following:

- Enrichment of diet by production of wide spectrum of flavours, aromas and by enhancing the body texture of food.
- Enhancement of shelf life of dairy products by production of such bio-preservatives as lactic acid, acetic acid and other similar products.
- Bioenrichment of food substrate with protein, essential ammo acids, fatty acids and vitamins.
- Removal by decompositions or otherwise, of toxic outputs of microbes which are harmful or create negative action on body metabolism.

Exploration, development and use of probiotics is another feasibility area where biotechnology application can be helpful to dairy industry. Probiotic bacteria are such bacteria which while growing, specifically in the human intestine, produce such mid or end products which carry out the following:

- Stimulate natural activities of body cells and consequently strengthen the immune system.
- Provide protection against infections.
- Breakdown carcinogenic substances.
- Protect against pathogenic bacilli and provide relief from constipation and diarrhoeal dispositions.
- Lower down cholesterol level in blood.
- Stop the mutagenicity of intestinal contents and reduce the chances of intestinal tumor.
- Help in metabolizing lactose and thus controlling the lactose intolerance.
- Improvement in assimilation of calcium in skeleton system
- Help in alleviating the chances of peptic ulcer.
- Help in boosting the synthesis of vitamins.

Biotechnology can be of help to dairy industry in development of these probiotic microbes which can be incorporated and used for such fermented milk products like curd, yoghurt and it has been found that these are being used in manufacturing of probiotic Cheese also. Milkfed Milk Plant Mohali has started working on probiotic cheese and curd. However, lot of R&D is required to access the actual roll of probiotic organism in Cheese.

There are approximately 200 recognized physical disorders which have been found to be food borne and most of the time, these dispositions may be related to rogue activities carried-out by microbes which include *Campylobacter jejuni*, *Escherchia coli*, *Listeria monocytogenes*, *Shigella dysenteriae*, *Staphylococcus aureus* and *Yersinia entrocolitica*.

Though India is numero uno in milk production, but, unfortunately we are lagging in the quality of milk produced. It is here where latest development in biotechnology can help dairy industry particularly in improvement of dairy cattle through genetic engineering.

Embryos produced by *in vitro* fertilization and placed at predetermined times into the uteri of estrous-cycled animals can result in higher conception rates than are now obtained by artificial insemination. This will accelerate genetic gains. Monoclonal antibodies used as diagnostic agents will greatly reduce the cost, time, and labour required to maintain animal health. Bovine somatotropin produced with recombinant DNA (r-DNA) technologies has the potential to greatly enhance milk production per cow.

Although bST can boost milk yield per cow significantly - doing in 1 year what it would take 10 to 20 years to achieve with current reproductive technologies. However, concerns have been raised about its safety for human and animals and .about the economic consequences of its use for the industry. The Food & Drug Administration (FDA), U.S.A., has determined that sufficient scientific information exists to indicate that the milk from bST supplemented cows is safe for human consumption.

Biotechnology related tests including DNA probes, Electro chemo-illuminescence technologies, ELISA (Enzyme Linked Immulo Sorbant Assay), Lux operon system and Polymerase chain reaction (PCRs) are already available. But, even these applications provide results in 18 to 24 h which is still a long period to keep certain milk products like processed milk away from customer. Hence, the need of day is to develop these tests further to shorten the time gap because milk in its most consumable/consumed form i.e. liquid milk except UHT milk, has to be delivered or should be delivered within 18-10 h of processing. Even milk meant for U.H.T. requires test results within 2-3 h so that decision can be taken to process the milk in U.H.T. or not. It has assumed more significance keeping in view the application of new regime of codex which will require food processors to identify and apply HACCP. Therefore, need for the quick test to declare the food safe from physical, microbiological and chemical point of view has assumed more priority.

Again while India accounts for 20% of population of milch animals, it is unfortunate that only 40 million out of 272 million milch animals are producing milk. If only this number could be doubled to 80 million, India can be a super power as far as milk production is concerned. Biotechnology can be of great help in improving the genetic stock by application of such techniques like DNA Finger Printing for verification of parentage and genotyping screened for milk protein locile. To quote Dr. Dbenezer "Self sufficiency in milk production could be achieved only if biotechnology proved to be useful to common man and the society". Already it has been reported that National Diary Development Board (NDDB), India has undertaken caryotyping of bulls and bull calves for carrying out further tests on this aspect of application of biotechnology for breed improvement.

On milk production side, the application of biotechnology has assumed greater importance, since the discovery of principles of cell fusion and also development of immortalized hybrid cells for production of monoclonal antibodies. These findings have opened new vistas for application of biotechnology for breed improvement and health-care of milch animals.

On animal health-care side, biotechnology can help in the field epidemiology, diagnostics and treatment/control of communicable diseases which are rampant in hinterland of India. Biotechnology based diagnostic techniques like ELISA, PCR and NAM can be used for detection of pathogens causing diseases like FMD. PCR has also been reported to be quite effective in diagnoses of *Rhinotritra cheitis*, Blue Tongue, Rinderpest and other diseases.

Other biotechnology based techniques, MAb profiling and Nucleic acid sequencing have also been found to be useful in tracking origin of viruses and strains of pathogens and thus can be helpful in understanding the diseases and entailing their effective control. Development of better and biosafe vaccines and application of transgenics for breed improvement is another important area of application of biotechnology.

Recently an environmental pact has been hammered out between 140 countries to regulate trade in genetically modified organisms used in food production. The protocol says that decision on import would have to be based on sound science. The bio-safety protocol is a set of legally binding rules for the bio-safety or the safety considerations associated with trade in biotechnology products. The pact also requires shipments of genetically modified products to have labels saying they 'may contain' genetically modified organisms. The labels can specify that the products are not intended for intentional introduction into the environment. It is apprehended that developing countries will be the losers as they are most likely to become the dumping ground of the mountains of GM food produced in the US, Canada and Australia. The experts have warned that developing countries are most likely to bear the burnt of runaway genetic pollution from foreign genes.

It can be concluded that biotechnology advances are not used for ends which by accident or design may lead to biological and psychological catastrophe. Already cloning has raised ethical brouhaha and such ruckuses will be a plenty if biotechnological research and applications are guided and used for selfish and ulterior motives. The need of hour is to address various issues involved such as labelling, legislature, import restrictions, regulatory authority and health & environmental problems or disaster, besides ethical issues involved such as "Should life become a commercial property though patenting" and "Who advocates for nature" and so on.

FURTHER READING

Hamid, B. Ghoddusi, Bifidobacterial in Dairy Products: Novel Aspects, *Indian Dairyman*, 7 (1999) 51.

Niyaz Ahmed, Sunita Grover, Pattnaik, P. and Batish, V.K. Recent Advancements in Biotech Diagnostics for Rapid Detection of Food-borne Pathogens, *Indian Dairyman*, **12** (1997) 49.

Impact of Biotechnology on Dairying

V.K. BATISH, S. GROVER *and* B.N. MATHUR

Molecular Biology Unit
Dairy Microbiology Division
National Dairy Research Institute
Karnal 132 001

BIOTECHNOLOGY HAS RECENTLY emerged as the most powerful and sought after technology having endless possibilities in almost all areas affecting human life. Dairying is yet another area where Biotechnology can make its impact in a spectacular way. Recent developments in Biotechnology have opened up new and exciting possibilities in Dairying and Animal Husbandry. With the emergence of India as the major milk producing country in the world, the relevance of Biotechnology to Indian Dairy industry has also increased considerably in the new millennium since, the amalgmation of white revolution in the country with the so called 'Biorevolution' through Biotechnology. It could be extremely valuable and rewarding for a developing country like India by uplifting the economical status of our people involved in this sector. Biotechnology can play a significant role in raising the status of dairy sector both in terms of milk production and processing. Dairy industry in particular can greatly benefit through the application and intervention of powerful magical tools of this technology such as r-DNA technology, Genetic Engineering, Gene cloning, Hybridoma Technology and Transgenics, PCR technology and Bioprocess Engineering to develop commercial products and processes that use living systems to provide high quality nutritious, clean, wholesome and healthful dairy foods to the people of our country.

1. POTENTIAL APPLICATIONS

Biotechnology has already made significant contributions in dairy industry. Some of the potential applications and target areas where Biotechnology has already made its impact along with future prospects are given below:

1.1 Dairy Production

- Recombinant bovine/buffalo somatotropin - rbst
- Recombinant vaccines
- Rumen manipulation
- DNA finger printing and RFLP
- PCR based diagnostics (animal health care)

- Embryo transfer technology
- Animal cloning
- Gene pharming and transgenics

1.2 Dairy Processing

- Designing milk through genetic engineering
- Genetically modified organisms (GMOs)
  - Starter cultures
  - Genetically modified foods
- Food grade biopreservatives
- Recombinant dairy enzymes/proteins
- Accelerated cheese ripening
- Probiotics
- Functional foods and neutraceuticals
- Dairy waste management and pollution control
- Gene probes and PCR based pathogen detection
- Membrane technology

Out of these envisaged areas, some have already been realized while others are still in the developmental stages and will take some time before reaching the end users. Some of these promising areas and their potential application in dairy industry are as below:

2. RECOMBINANT BOVINE SOMATOTROPIN (rbst)

The first FDA approved application of biotechnology to animal production was the use of recombinant *bst* in dairy cows. The *bst*, a protein hormone found naturally in cows is necessary for milk production. It can be isolated from cows/buffaloes, inserted into bacterial cells and produced in large quantities by those cells. The r*bst* can then be recovered, purified and injected into dairy cows/buffaloes. When *bst* is administered to cows/buffaloes under ideal management conditions, milk production has been shown to increase by 10 to 25%. After exhaustive experimentations and amid controversy, the FDA approved the use of *bst* in dairy cows in Nov., 1993. FDA based its ruling on findings that a) *bst* is species specific for cows b) *bst* is a protein that is digested in the intestinal tract of human beings and cows c) milk contains *bst* naturally and supplementation does not increase the amount of *bst* to levels outside the normal range d) *bst* supplementation does not change milk composition and e) *bst* has not been found to cause growth promoting activity in a variety of species. Although, results of one year following study of cow health issues have produced no new animal health concerns, the use of *bst* in dairy cows remains a political and social issue.

Other uses of biotechnology in animal production include development of vaccines to protect animals from diseases, production of several calves from one embryo (cloning), artificial insemination, improvement of growth rate or feed efficiency and rapid disease diagonstics. In addition, 'Transgenic animals' or animals with altered genetic make ups could be produced through biotechnological approaches.. Out of these, embryo transfer technology and transgenic animals merit special consideration since these are considered hot areas for commercial exploitation.

3. EMBRYO TRANSFER TECHNOLOGY

In livestock species, reproduction and production are two faces of the same coin. Events in reproductive processes of fertilization, gestation and parturition are invariably followed by lactation that essentially is the productive phase of this phenomenon. Hence, any intervention to augment the reproductive efficiency of the farm animals is bound to enhance the productivity of milk animals in one way or the other. There are certain inherent attributes specific for livestock species such as longer gestation period and generation intervals that limit the reproductive capability of animals. Through normal breeding, one can get one calf every 12-14 months and upto 6-7 calves in whole life time. With the help of Embryo Biotechnology, calf crop from the given elite female can be multiplied further. Three to four calves can be produced in a year and over 25 in the life time of the animal. In exceptional cases, more than 10 calves have also been produced from a single cow. This technology offers an opportunity to multiply the elite cows faster and make their contribution more towards the gene pool. The technology of multiple ovulation and embryo transfer (MOET) is drawing the attention of the breeders world over to develop newer breeding strategies for genetic improvement of livestock. In the Indian context, this technology has been fairly standardized for use in both cattle and buffaloes and has resulted in the birth of more than 1000 calves. NDRI scientists were able to produce first buffalo calf in the world called "Pratham" through embryo transfer technology.

4. ANIMAL CLONING

Birth of 'Dolly' first ever cloned sheep in the world that made history and has generated lot of interest in this technology. Transfer of nuclei from reproductive/non-reproductive tissues into enucleated oocytes followed by their reprogramming leads to development of complete embryos. Such embryos carry the exact replica of the genetic material of the donor animal. Animal cloning has now been perfected and has resulted in successful birth of several calves and has the potential of being used in humans, as well. However, many ethical issues have been raised about the appropriateness of this technology in human beings. Nevertheless, in livestock species, this technique has tremendous potential in multiplying the valuable and elite animals in large numbers.

5. TRANSGENIC ANIMALS

Recent advances in genetic engineering, r-DNA and embryo cloning have made it feasible to clone the protein encoding genes from any organism. These recombinant genes can be expressed in a variety of systems ranging from simple microbial systems such as bacteria, yeasts and cultured mammalian cells to complex eukaryotic systems such as plants and animals. An animal with genetic composition that has been altered by the addition of foreign (exogenous) DNA is said to be transgenic. The DNA that is introduced is called a transgene and the overall process is called transgenic technology or transgenesis. The lot of euphoria generated after the birth of Dolly prompted several groups in the world to explore the possibility of using animals as bioreactors for the production of commercially important proteins. Transgenic animals are useful tools for the production of human proteins of therapeutic, pharmaceutical or nutritional value in large quantities for commercial application. Many useful proteins such as human growth hormone protein C, alpha antitrypsin, serum albumin, tissue plasma activator, lactoferrin, alpha-lactablumin, human interleukin2, human CD4 protein, clotting factor IX and urokinase have been expressed using mammary gland specific promoters in milk of mice, rabbits, pigs, sheep, goats and other dairy animals.. Transgenic technology involves the transfer of a genetically engineered segment of DNA into genetic material of another species which allows the targeted expression of important human therapeutic and nutritional supplements in the milk of lactating female livestock. This can be achieved primarily by following two strategies i.e. by micro-injection technology or nuclear transfer i.e. cloning. The transgene offspring may express the target protein in mg per litre quantities, most frequently as soluble whey protein. Mammalian epithelial cells have the capability to carry out complex protein synthesis with a variety of post-translational modifications and proper protein folding. Co-expression of modifying enzymes in the epithelial cell golgi apparatus may allow the heterologous proteins to be engineered appropriately to confer specific therapeutic characteristics. The milk of transgenic livestock presents an excellent starting material from which human pharmaceutically active therapeutic proteins can be purified using established technologies of the dairy and pharmaceutical industries.

6. MAMMARY GLAND AS A BIOREACTOR

Because of unique design and specialized function of the mammary gland, it is considered as an ideal 'Bioreactor' for the production of recombinant proteins. With a cell density upto one thousand times that of a mammalian cell culture bioreactor, it is capable of producing tens of grams of recombinant protein per day. As per an estimate, one day's output from one goat, typically 2-4 liters of milk containing approximately 10 or more grams of recombinant protein can provide as much raw protein as harvested from a one liter cell bioreactor on a daily basis.

The ultimate goal of transgenically produced biopharmaceutical protein is the efficient production and recovery of commercially important protein from milk of transgenic animals. The expression of these proteins in milk is specifically intended because milk which serves as a source of raw material is safe, abundant, renewable, easily obtainable and socially acceptable as food. It is especially attractive for proteins that are required in large volumes because of high unit dosage requirements, multiple chronic administration, large patient population for industrial applications. Moreover, transgenic animals can express biochemically complex proteins in milk in an active form that can not be produced economically in cell culture. The significantly lower capital and operating costs for raw material production makes transgenically produced proteins more cost effective than other production systems.

Some potential benefits of production of pharmaceutical proteins in milk of transgenic dairy animals are:

- Efficient expression of the recombinant proteins in milk
- Stable transgene integration into host animals cell's genome/chromosome
- Reliable transmission of the transgene to the progeny
- Protein recovery with high yield and purity
- Retention of functional protein activity
- Appropriate protein structure stability and bioactivity

7. GENETIC ENGINEERING OF MILK

Milk is a complex biological fluid that is a fundamental dietary constituent of our population. Often referred to as the nearly perfect food, milk and milk synthesis by dairy ruminants have been the subjects of intense investigations to improve milk production or quality. With the recent advancements in biotechnological techniques such as genetic engineering and r-DNA technology, it is now possible to alter milk composition at will to add variety to dairy products produced after processing such altered milks. Molecular genetics has a valid place among other proven techniques such as nutritional management and quantitative genetics for the improvement of milk quality and quantity. A variety of ways have been suggested to modify milks to improve its nutritional quality and to improve the efficiency of manufacturing of milk products such as cheese, ice-cream and yoghurt etc. Our growing understanding of the lactation process and continual innovations in dairy processing has presented interesting opportunities for genetic manipulations that are not possible through traditional nutritional or classical genetic methodologies.

8. MANIPULATION OF MILK PROTEINS-CASEINS

Caseins which include alpha, beta and kappa caseins are the major proteins in milk representing 78 % of the proteins. Normal dairy cattle have a single copy of alpha -CN, beta -CN and kappa-CN within each cell. Introduction of one extra copy of casein in the bovine genome can have a dramatic impact on dairy industry. An increase of 20% in the content of alpha S1-CN of milk would result in increase of over US$ 190 million/year for the dairy industry. Increased beta-CN content in milk reduces rennet clotting time and increases the extent of syneresis. Therefore, higher protein content obtained by over expressing additional copies of endogenous bovine casein genes would be of significant advantage to the dairy industry.

The cloning of milk protein genes allows for their genetic modification in order to improve the nutritional quality or functional properties of milk. More economical cheese products could result from a more efficient cleavage of a kappa-CN by chymosin. Storage time for cheese maturation may be reduced by altering the chymosin cleavage sites of alpha S1-CN to more efficiently hydrolyzable peptide linkages. Other manipulations include alteration of physical properties particularly thermostability of casein. Thermostability can be achieved by increasing the expression of kappa -casein genes and masking the expression of beta-lactoglobulin in the mammary gland. Because caseins are relatively low in sulphur containing amino-acids, their nutritional value could be improved by increasing their methionine content.

Amino-acid sequence of alphaS1 casein can be manipulated by directed mutagenesis to make it more suitable for the reaction catalyzed by enzyme casein kinase which causes phosphorylation of casein and this makes them highly thermostable. Amphiphilicity of the caseins can also be augmented by gene technology. This property is associated with the surface activities of the casein that govern their emulsification and foaming properties. This can be extremely helpful in making ice-creams, softies, cadies and whipped toppings.

9. ALPHA-LACTALBUMIN (ALPHA-LA)

The second most abundant whey protein in human milk, alpha-LA is a minor component in bovine milk. Because alpha-LA is a small protein of 123 amino-acids, it is amenable to genetic engineering by site directed mutagenesis to modify its amino-acid composition. This technique has been used to modify alpha-LA and make it consumable by patients with phenylketonuria who lack the enzyme that metabolizes phenylalanine and thus require low phenylalanine diets. The three dimensional structure of alpha-LA is known. Therefore, the sensitivity of four phenylalanine residues in alpha-LA can be engineered without disrupting the structure of the molecule. The modified gene is then microinjected into the pronucleus of a fertilized egg and the protein is produced in milk and can be used as a supplement for an improved diet for phenyl ketonuria patients.

10. REMOVING COMPONENTS OF MILK BY GENETIC MANIPULATION

The essential availability of bovine ES cells should allow undesirable components of the milk to be removed by disrupting the corresponding gene in the bovine genome using homologous recombination.

10.1 Lactose

Lactose intolerance is a serious problem in 70% of the world population due to deficiency in the intestinal lactase (β–galactosidase) needed to hydrolyze the lactose in milk to its constituent monosaccharides resulting into gastrointestinal upset. Genetic manipulation can now be explored to reduce lactose content in milk by either removal of alpha-LA by ES cells and gene "knock out" methodologies or by introducing a lactase enzyme (β-galactosidase) into milk via mammary gland specific expression. An alternative approach for suppressing the alpha-LA gene expression is the use of antisense and ribozyme sequences. It is imperative, however, that such studies are conducted in ruminant tissues in order to delineate the actual reduction in lactose and the economic benefits.

10.2 β-Lactoglobulin (β-LG)

Whey proteins have nutritional benefit in fluid milk market, but, represent a less valuable milk component for cheese industry. β-LG is the most abundant (upto 50%) whey protein . Human milk is devoid of β-LG which is considered to be the main allergen in bovine milk. Not required for the lactation process, β-LG has no known function in the process of milk secretion. Further upon heat treatment, β-LG forms gel aggregates with other β-LG molecules because of the exposure of a thiol group. This thiol group interacts with a disulphide group of κ-CN and interferes with chymosin mediated hydrolysis which is important for curd formation and cheese production. It is possible now to manipulate the level of β-LG at genetic level. However, because β-LG is also the major source of cystein in milk, removal of β-LG may result in a milk of inferior nutritional value. Hence, it would be desirable to delete β-LG coding sequences and neutralize the endogenous regulatory elements to drive expression of other proteins i.e. casein and lactoferrin.

11. FAT CONTENT

It is now possible to manipulate even the fat content in milk by disrupting the fat synthesis in the mammary gland. By using this strategy, milk with 2% fat content (40% reduction) might be achievable. It has been suggested that *de novo* fat synthesis in mammary glands might be reduced by blocking the expression of acetylCoA- Caroxylase gene through stem cell (knock out), antisense or ribozyme technology. Natural milk with reduced fat content may be extremely beneficial for patients with heart ailments.

12. HUMANIZATION OF BOVINE MILK

Breast milk is nature's perfect food for human infants providing them all aspects of nutrition and protection against infections. However, a considerable number of infants are fed formulae based on bovine/buffalo milk. The composition of these infant formulations can be improved if the protein contain therein more closely resemble those of human milk. It is now possible to add human lactoferrin or lysozyme to bovine milk by genetic engineering to produce new functional foods. The shelf life of such products is also expected to be very high due to antimicrobial activity of these proteins.

The goal of the dairy industry has been to create an efficient, healthy cow/buffalo that can serve all the needs of the industry. Genetic engineering offers the opportunity for a paradigm shift, a reshaping of the industry from the producers to the processing plants. Dairy producers have the opportunity to choose to produce high protein milk, milk destined for cheese manufacture that has accelerated curd clotting time, milk containing neutraceuticals, orallly administered biologicals that provide health benefits or a replacement for infant formulae. Such a scenario would be a radical change for the dairy industry. There is now little doubt that the products of genetic engineering will become a part of the dairy industry in the new millennium.

13. GENETIC MANIPULATION OF STARTER CULTURES

Good quality starter cultures (Lactic acid bacteria, LAB) are the pre-requisite for preparing good quality fermented foods. The commercial value of the fermented foods is, therefore, mainly dependent upon the performance of the starter cultures. One of the most notable features of lactic starter cultures is that they generally harbour large complements of plasmids. It is very significant that many commercially important traits of these bacteria are in fact plasmid encoded. These include lactose and citrate utilization, phage resistance, proteinase production and bacteriocin production/immunity. The location of such important genetic information of considerable industrial value on the plasmids could be detrimental to the host as these extrachromosomal elements may be lost from the cell population on extended cell culture especially in absence of any population on extended cell culture especially in absence of any selective pressure to ensure their continued maintenance. However, it has proved to be a blessing in disguise to the researchers since the plasmid encoded genes are more accessible for manipulation and more easily analyzed than the genes located on the chromosomes.

Recent advances in gene transfer and cloning technologies provided numerous opportunities for the application of genetic approaches to strain development programme appropriate for their application in dairy industry. In addition, significant advances have already been made in the analysis of lactococcal and lactobacillus gene expression. Genetic manipulation of these bacteria has now been developed to the point where it is now possible to clone and express homologous and heterologous genes. Cloning vectors have also been constructed to isolate regulatory signals such as promoters, ribosome blinding sites and terminators. This information could be extremely valuable in allowing the expression of foreign gene in lactic acid bacteria and enhancing the expression of desired phenotypic properties. The development of plasmid integration vectors for lactic acid bacteria offers the possibility of introducing and stabilizing genes of interest by incorporating them into the host chromosomes. The strategy could be extremely useful to tackle the problem of instability in the lactic cultures as a result of frequent loss of plasmid during their propagation. Plasmid integration vectors can also be used to explore and help characterize chromosomal DNA from lactic acid bacteria. Indeed, proteinase and phage resistant determinants have been successfully integrated and stabilized in lactococcal chromosomes where they are amenable to substantial amplification.

14. FOOD GRADE BIOPRESERVATIVES

Another potential area for commercial exploitation of lactic acid bacteria through their genetic engineering is the biopreservative attributes of these organisms. Appropriate food-grade biopreservatives can be developed to control undesirable bacteria and molds in dairy foods by following these strategies for direct application in dairy industry.

14.1 Effective against bacteria

Suppression of spoilage and food-borne pathogens by lactic acid bacteria could be extremely beneficial to human health and dairy industry as these attributes can considerably improve the shelf life and safety of fermented foods. The antimicrobial activity in lactic acid bacteria has been ascribed to a number of factors and the extent of their antibacterial spectrum varies from culture to culture and strain to strain. Among these factors, the bacteriocins appear to be the most promising candidate for exploring the preservative potentials of these bacteria on a large scale. This is precisely due to their proteinaceous nature since the production of these polypeptides could be conveniently manipulated at the molecular level. The application of such food preservatives in dairy industry to combat undesirable food spoilage and pathogenic organisms could be more relevant to countries like India where the health of consumers is always at risk due to poor hygiene, tropical temperate conditions and inadequate refrigeration facilities. Biotechnological techniques can now be applied to develop strains of lactic acid bacteria capable of enhanced production of these natural food grade preservative and also to combine within a single strain the ability to produce a number of such bacteriocins to extend their antibacterial spectrum. Much of the work being applied to bacteriocins is often based on approaches and protocols originally developed for the analysis of Nisin which has already been given the GRAS status. In this case, the structural genes have been sequenced, protein engineering has been employed to generate the variants of the molecules and engineered strains which can be used to study expression of the compound available.

14.2 Effective against moulds

Moulds contamination in dairy foods and feeds is a serious problem in third world countries like India where tropical humid conditions and unhygienic environment are extremely favourable for the growth and proliferation of these types of organisms. Apart from causing serious spoilage problems in these commodities, many of the incriminating moulds could also pose serious health risks to the consumers on account of their ability to produce extremely toxic and carcinogenic metabolites such as aflatoxins. Although, mould contamination in foods can be effectively controlled by the application of some chemical additives, there are serious concerns regarding the use of these unnatural additives in food chain. Exploring the antifungal attributes of lactic acid bacteria could be a very useful strategy to tackle the mould problem in dairy foods. In this regard, some work has been initiated at NDRI, Karnal, India. A very promising lactic culture characterized as *Lactococcus lactis* subsp. *lactis* CHD 28.3 exhibiting a very strong antifungal activity against a variety of moulds has been isolated and studied extensively. Because of its strong broad spectrum antifungal activity, this culture could be a very attractive candidate for exploring its in-built biopreservative potentials on commercial scale. The nature of the antifungal substance (AFS) produced by the test culture was determined to be proteinaceous which was partially purified and characterized. The production of AFS in the test culture has been found to be encoded on chromosomal DNA rather than a plasmid. A partial nucleotide sequence of open reading frame (ORF) for the AFS has been determined. The nucleotide sequence analysis strongly suggests the involvement of a transposon like element in the production of AFS in the test culture. Efforts are now being made to use a strong nisin promoter for over expression of AFS gene for its commercial application in the development of a food grade bio-preservative.

15. FOOD GRADE VECTOR SYSTEM

Application of genetically engineered lactic acid bacteria constructed with the help of vector systems having antibiotic/drug resistance markers based on *E. coli* in foods could raise serious public concerns. Hence, there is a need to develop food grade vector system where only naturally occurring genes such as bacteriocin production/immunity or lactose/sucrose utilising genes from LAB are incorporated into the food-grade vectors as a marker for the selection of the desired recombinants. The use of starter cultures engineered with such food-grade vectors for different desirable characteristics in dairy foods will not raise any controversy and dairy industry will have no hesitation to use such genetically engineered organisms in the manufacture of high quality fermented foods.

16. NICE SYSTEM FOR OVER EXPRESSION OF PROTEINS

Controlled over production of proteins by industrially important micro-organisms is highly desirable. Lactic acid bacteria are ideally suited for use as cell factories since they are food-grade, genetically accessible and easy to grow on a low cost substrate. Until now several systems based on broad host range vectors and constitutive promoters have been developed for different LAB. However, a good food grade system was lacking that enable high and controllable expression of target genes. Scientists working at NIZO, Netherlands have now developed a complete nisin inducible expression system that can be applied in several lactic acid bacteria and also in other gram positive organisms. Some advantages of NICE system include food grade components i.e. plasmids, strains and inducer, all components are tailor made and ready to use, high expression levels (upto 60% of total cellular protein), low costs, versatile and flexible, expression tightly controlled enabling production of lethal proteins, expression levels modulated by varying nisin concentrations and *in situ* induction is possible in nisin producing lactococci. This system works efficiently for expression of even heterologous proteins in LAB, and hence, can have lot of commercial value in dairy industry.

17. DAIRY ENZYMES

Advances in biotechnology have also made a strong impact on production of several enzymes and proteins used in dairy industry in the processing of milk for the manufacture of some fermented products. Many of the enzymes are by and large of microbial origin. However, their production levels are not very high in the producer organisms and thus their production is not cost effective for commercial exploitation. Food processing has benefited from biotechnologically produced enzymes such as chymosin (rennet), proteases, lipases, α–amylase and lactase which can find lot of application in dairy industry. Many of these enzymes like α-amylase, chymosin produced by genetically modified organisms have been granted GRAS status by FDA- thereby allowing their use in place of conventional sources of these enzymes in the dairy industry. Genetically engineered enzymes are easier to produce than enzymes isolated from original sources and are favoured over chemically synthesized substances because they do not produce by-products or off flavours in foods.

18. RECOMBINANT CHYMOSIN

Bovine chymosin traditionally known as calf rennet has been extensively used as milk coagulant during the manufacture of a variety of cheese all over the world. India's cheese industry is also expanding rapidly and hence the demand for calf rennet or its substitutes is also showing a phenomenal increase. However, in India, the cheese made with the help of calf rennet is not acceptable to a large sector of the consumers due to religious sentiments. This has stimulated lot of interest in search for rennet from alternate sources. Although, application of milk clotting enzymes from microbial sources have been in vogue in cheese manufacture, the quality and quanity of cheese made with such enzymes does not match with that made from calf rennet. In this context, buffalo rennet could be an excellent alternative to bovine calf rennet on account of its inherent compatibility with buffalo milk for producing high quality cheese. With the rapid expansion of fast foods entering into the Indian market, the demand for specialized cheeses such as Mozzarella cheese for Pizza etc. has increased tremendously. With this, demand for chymosin has also increased several folds. Hence, for large scale production of buffalo calf rennet, application of rDNA technology and genetic engineering techniques could be extremely valuable. By cloning the genes of interest from buffalo calves in *E. coli* and other hosts, it is now possible to produce recombinant buffalo chymosin for commercial use in cheese industry. Scientists at NDRI, Karnal, India, have already cloned and expressed buffalo chymosin in *E. coli* and attempts are now being made to express in a yeast system for commercial application in cheese industry.

Similarly other dairy enzymes like proteinases lipases, β-galactosidase etc. can also be produced through rDNA technology. Proteases and lipases can be used in cheese industry for flavour development whereas β-galactosidase can find application in the production of lactose free milk intended for lactose intolerant patients.

19. PROBIOTICS

There is currently a growing interest in certain lactic acid bacteria that have been suggested and shown to provide health benefits when consumed as food supplements or as food components. Such type of LAB have been invariably referred to as "Probiotics". Probiotics are defined as living microorganisms which upon ingestion in certain numbers exert health benefits beyond inherent basic nutrition. Probiotics may be consumed either as a food component or as a non-food preparations. Among the LAB, *Lactobacillus* species such as *L. acidophilus, L. casei* and *L. bulgaricus* and *Bifidobacterium bifidum* have tremendous probiotic potentials by virtue of which these bacteria can establish in the gastro-intestinal tract of man and animals. After implanting in the GI tract, these organisms do not allow potential pathogens of enteric disorders to inhabit the intestine and hence can play an important role in controlling the enteric disorders. However, this attribute of considerable medical significance is limited to only a few strains of LAB thereby leaving ample scope to manipulate this attribute at molecular level to exploit the full potentials. Apart from the colonization attributes, these probiotic organisms can also exhibit strong anticarcinogenic and anticholesterolemic activities as well and hence can play a vital role in the alleviation of cancer and blood pressure in human beings. Recently, these probiotic organisms are attracting lot of attention as food grade biotherapeutics and have led to the development of a new concept of functional foods. These functional foods in which viable probiotic organisms are incorporated can find lot of medical applications as a special diet for patients. A number of companies are now venturing into the development of a variety of probiotic formulations. One such company M/s. Probiotics International is the leading manufacturer and supplier of nutritional health care products which include the natural live microbial supplements that are known to be beneficial in the treatment of a variety of digestive disorders and diseases associated with stress. A few of these products are listed below and commercially available.

- Protexin-Natural Care Tablets (Maintains healthy nervous and immune system)
- Protexin-Natural Care Powder (Maintains healthy nervous and immune system)
- Protexin-Vital (Adults) (Maintains healthy intestinal flora in adults)
- Protexin-Vital (Child) (Maintains natural healthy flora in children)
- Protexin-Travel (Diarrohea and sickness during travel)
- Protexin-Relief (Relieves constipation in elderly)
- Protexin-Lacta (Lactose intolerance)
- Protexin-Restore (Restores intestinal flora in children)

Some of the future targets aimed in this direction include identification of the genetic determinants for probiotic function and regulatory signals to improve the colonization ability of these organisms in different hosts, inter and intraspecies conjugal transfer of this character and understanding the exact mode of action of probiosis.

20. FUNCTIONAL FOODS/NEUTRACEUTICALS

Functional foods produced with the help of genetic engineering methods are regarded as a way to obtain public acceptance for this technology. Functional foods are supposed to have an added value for consumers. The added value may be given either due to health improvements and or minimised risks to develop certain diseases. Internationally apart from functional foods, the terms neutraceuticals, designer foods, healthy foods or pharma-foods are used as synonyms for functional foods. So far only Japan has a legally binding definition for functional foods. Functional foods are primarily targeted at the metabolism of macronutrients, the defence against reactive oxidants, cardiovascular system and physiology of the gut.

Based on the envisaged targets, functional foods are specifically designed to control a number of chronic diseases such as obesity, diabetes, hypertension, other cardiovascular diseases, cancer and enteric disorders. Functional foods are designed and produced by using different approaches i.e. by adding either the probiotic organisms, prebiotics in the food formulations or other ingredients such as anti-oxidants, phytochemicals, fat substitutes, N-3 fatty acids and bioactive peptides. Out of these, pro and prebiotics and bioactive peptides are more popular.

20.1 Bioactive Peptides

Research on bioactive peptides mainly deals with milk proteins such as caseins which are regarded as precursors of bioactive peptides, lactalbumin and lactoferrin. The former is supposed to have anti-carcinogenic effects and latter has antimicrobial, anti-oxidants and anticarcinogenic effects. Furthermore, bioactive peptides are said to have regulating and hormonal effects. However, these effects are mainly concluded from *in vitro* experiments or feeding trials. Their effects on humans are still to be examined.

20.2 Market for Functional Foods

The world market for functional foods has already reached a staggering figure between 10 billion US$ and 22 billion US$. It is mainly dominated by non-alcoholic drinks and dairy products. In Europe, functional foods account for 1.4 to 1.7 billion US$ and dairy products account for 65%. In Germany, probiotic yoghurts have a market share of 15%. Germany is the biggest customer for functional foods followed by France and Great Britain. Apart from big food companies, pharmaceuticals and biotech companies are increasingly trying to break into functional food market. In India too, there is bigger future for these functional foods.

21. LACTIC ACID BACTERIA AS ORAL VACCINES

Lactic acid bacteria possess a number of properties which make them attractive candidates for oral vaccination purposes. These bacteria have been used for centuries in the fermentation and preservation of foods and are considered to be 'safe' organisms with a GRAS status. Apart from this, LAB also have adjuvant properties and low intrinsic immunogenicity. A particularly attractive feature of lactobacilli is their ability to colonize certain regions of mucosa which permits in principle the induction of a local immune response. An additional benefit of the use of lactobacilli is that some strains are considered to show health promoting activities for humans and animals. Many of these lactobacilli also produce surface proteins which have been very well characteized. By applying genetic engineering techniques on these organisms, it is now possible to express antigens from potential pathogens such as malarial parasite, *Vibrio cholerae* and Hepatitis B virus either on their surface or extracellularly or intracellularly which eventually can elicit antibody production in the consumers fed with such genetically modified cultures to protect against the targeted pathogen. The elucidation and characterization at DNA level of bacterial surface proteins involved in adherence to mucosal tissues has provided new and challenging opportunities which will allow us to better understand the interaction between intestinal lactobacilli and mucosal tissues. The recent developments of molecular biology techniques that permit construction of LAB presenting foreign antigens in different ways to the immune system offers excellent opportunities to further explore the potentials of these food bacteria as vaccine vehicles.

22. ACCELERATED CHEESE RIPENING

Many cheese varieties require long ripening periods at low temperature for characteristic flavour and texture development. This process significantly increase the cost of the product. Controlled use of biotechnological products like genetically engineered proteolytic and lipolytic enzymes can accelerate cheese ripening. Modified/genetically tailored microorganisms and enzymes are now being used for enhancing flavour production in cheese. Enzyme addition is now one of the few preferred methods of accelerated ripening of cheese. The enzymatic reactions are specific and hence undesirable side effects caused by live microorganisms are avoided in the cheese. Enzymes may be immobilized or encapsulated for long term action on the production for quick action and homogenous distribution in the product. The molecular analysis of the lactococcal proteolytic system has created the exciting prospect of being able to use genetic strategies to manipulate these hosts to produce strains with new flavour characteristics. It is also likely that these approaches will allow cheeses to be ripened in a more controlled fashion and an accelerated rate, if desired. This will result in the production of fermented foods with new attractive properties as well as creating economic savings for the producers.

23. REGULATORY CONSIDERATIONS

One of the major concerns in the public mind is whether food prepared with genetically engineered lactic starter cultures or recombinant dairy enzymes or proteins are safe for consumption. To be used in food systems, genetically engineered dairy starter cultures will need to be approved for use by appropriate regulatory agencies. The most important issue which needs immediate attention is the use of artificial drug resistance markers in the construction of genetically engineered organisms since the foods prepared with such organisms carrying drug resistant genes could pose serious health risks to the consumers. This necessitates the need for replacing drug resistant marker genes with some natural food-grade markers derived from lactic acid bacteria themselves. In this context, genes encoding bacteriocin production and immunity could be extremely valuable in the construction of food-grade vector systems which can easily replace the conventional *E. coli* based vectors. The application of engineered starters with such food grade vectors in food processing should not pose any problem in getting the approval of regulatory agencies.

24. RAPID DETECTION OF FOOD PATHOGENS

Although there has been dramatic increase in milk production in the country and more surplus milk is now available to dairy industry for processing, the microbiological quality of such milk and processed products has to be of acceptable quality. The safety of milk and milk products from public health point of view can be ensured by rigorous monitoring of dairy foods especially for high risk food pathogens such as *E. coli* 0157 : H7, *Listeria monocytogenes*, *Salmonella*, *Shigella* etc. which gain access into these foods due to mishandling during production and processing. Since, the conventional methods currently used in quality assurance labs for the detection of these potential pathogens are extremely laborious, lengthy and tedious, they do not serve any useful purpose to dairy industry as by the time results are available, the concerned food lot is already sold in the market and hence no follow up action is possible to avoid possible outbreak of diseases associated with such food pathogens. Hence, there is a need to develop simple and new innovative techniques for rapid detection of pathogens in dairy foods.

24.1 Gene Probes

One such approach which is in vogue these days in most of the developed countries is the use of molecular techniques for detection of pathogens. Gene probes and immunological assays using poly or monoclonal antibodies, targeted against the specific pathogen are highly specific and sensitive and have turned out to be extremely valuable tool for the reliable detection of pathogens in foods even at low levels. However, application of gene probe requires highly sophisticated procedure, adequate infrastructure and expertise in terms of skilled manpower. Nevertheless, a number of companies are entering into the development of kits based on gene probes and immunological principles for rapid detection of targeted pathogen in the foods. Many such kits are commercially available now.

24.2 PCR Based Detection of Food Pathogens

Another extremely versatile, reliable, sensitive and practical method for rapid detection of food pathogens directly from foods without following lengthy culturing steps is polymerase chain reaction (PCR). Application of such methods in Food quality assurance and public health labs can go a long way in protecting the health of the public. The crucial feature of sample preparation in the direct analysis of food without cultural enrichment is the isolation of target DNA with high reproducibility and efficiency. Construction of the actual PCR assay requires nucleotide sequence information and knowledge about the chosen target gene to guarantee maximum specificity and sensitivity. By using different formats of PCR, it is now possible to simultaneously detect more than one food pathogen in the food. PCR based detection of food pathogens has completely revolutionized the area of diagnostics as the results are available in less than 24 h to enable the food processing units to take appropriate follow up action in case a pathogen is detected in a food to avoid any risk to the consumers.

Molecular biology group at NDRI, Karnal has recently developed a multiplex PCR assay for rapid detection of *E. coli* 0157:H7 in milk and milk products. The entire test takes just 8 h to unequivocally prove the presence of *E. coli* 0157:H7 in the given food. Attempts are now being made to extend this technique to other high risk food pathogens like *Listeria monocytogenes, Salmonella* and *Shigella*,etc

From the foregoing test, it can be concluded that biotechnology is bound to play a significant role in reshaping the dairy scenario in the country. By judiciously applying biotechnological approaches, Indian dairy industry can make a niche and add new dimensions to our economy in the world market by producing high quality dairy products, which are not only nutritious, clean and wholesome, but, also safe for local consumption as well as for export.

FURTHER READING

Banerjee, A. Challenges and opportunities in Dairy Sector for 2020. *Indian Dairyman*. **50**: (1998). 17.

Batish, V.K, Roy, U.; Lal, R. and Grover, S. Antifungal attributes of lactic acid bacteria—A review. *Critical Review in Biotechnology*. **17**: (1997). 209.

Grover, S. and Batish, V.K. PCR based detection of high risk pathogens in dairy products. *Indian Dairyman*. **53**: (2001). 41.

Gurdon, J.B. and Colman, A. The future of cloning. *Nature*. **402**: (1999) 743.

Janne, J., Alhonen, L., Hyttinen, J., Peura, T., Tolvanen, M. and Korhonen, V. Transgenic Bioreactors. *Biotech. Ann. Rev.* **4**: (1999) 55.

McBean, L.D. Functional foods—An overview. *Dairy Council Digest*. **70**: (1999) 31.

Mohanty, A.K., Mukhopadhyay, U.K., Grover, S. and Batish, V.K. Bovine Chymosin: Production by rDNA technology and its application in cheese manufacture. *Biotechnology Advances*. **1**: (1999) 1.

Niyaz Ahmed, Grover, S., Pattnaik, P. and Batish, V.K. Genetic engineering of milk proteins for better manufacturing properties. *Indian Dairyman*. **50**: (1998) 45.

Pouwels, P. H., Leer, R.J. and Shaw, M. Lactic acid bacteria as antigen delivery vehicles for oral immunization purposes. *Int. J. Food Microbiol*. **41**: (1998) 155.

Sanders, M.E. Probiotics. *Food Technology*. **53**: (1999) 67.

Functional and Nutritional Components of Fish

A.S. BAWA
Department of Food Science and Technology
Guru Nanak Dev University
Amritsar 143005, Punjab

SEA FOODS HAVE BEEN A GOOD source of animal protein and lipid since early civilization. No doubt sea food proteins have a well balanced amino acid composition, even then many species of fish remain under utilized because of their being small in size, possessing high bone content, unappealing shape and look or fatty nature. Such raw materials can be utilized for novel preparations such as protein concentrates/dispersions and hydrolysates. Sea food lipids are rich in omega-3 poly unsaturated fatty acids (PUFA). The interest in sea food lipids has increased since the observation of low incidence of cardiovascular diseases (CVD) in Eskimos consuming fish as well as seal meat, in spite of high dietary fat intake.

Traditionally food products have been developed for taste, appearance, value and convenience for the consumer. The development of products for health benefits is relatively new trend and is based on the acceptance of the role of diet in disease prevention and treatment. The terms 'Functional Food' and 'Nutraceutical' have been used for promotion of such products. A functional food is similar in appearance to conventional foods, is consumed as part of diet, and has demonstrated physiological benefits and/or reduces the risk of chronic diseases. A nutraceutical is a product produced in the form of pills, powders and other medicinal forms not generally associated with food and demonstrated to have a physiological benefit and provides protection against chronic diseases. These terms have been used simultaneously as well as separately.

The omega-3 PUFA lower serum triacylglycerol and cholestrol, are essential for growth and development, and play a role in prevention and treatment of hypertension, arthritis and, other inflammatory and autoimmune disorders and cancer.

1. MARINE LIPIDS

Fish oils consist of saturated, monounsaturated and poly unsaturated fatty acids (PUFA). There are two classes of PUFA namely omega-3 and omega-6, different from each other w.r.t. location of double bond from the terminal methyl group of the fatty acid molecule. Unlike saturated and monounsaturated fatty acids which can be synthesized by all mammals including humans, PUFA can not easily be synthesized in the body, but, must be provided through diet. Omega-3 descend from linolenic acid while omega-6 descend from linoleic acid. Lipids from marine species contain long chain PUFA i.e. eicosapentaenoic acid (EPA; C 20:5,W3) docosahexaenoic acid (DHA; C 22:6W3) and to a lesser extent docosapentaenoic acid (DPA; C 22:5 W3). These are formed in unicellular and multicellular sea algae and pass through the food web to become incorporated into the body of fish. The high content of W3 fatty acids in marine lipids is a consequence of cold temperature adaptation in which it remains liquid and does not crystallize.

Consumption of marine oil decreases the plasma lipids by reducing the synthesis of fatty acids and low density lipoproteins (LDL). Long chain omega-3 PUFA have a direct effect on heart muscle, increase blood flow, decrease arrhythmias, improve arterial compliance, and reduce several cellular processes that comprise heart function. Marine oils also retard artherogenesis through their effect on platelet function, platelet endothelial interactions and inflammatory response of long chain PUFA, which produce a range of regulatory and biologically active substances collectively known as eicosanoids. In general, compounds derived from omega-3 are less powerful than those derived from omega-6 PUFA. Consumption of high amounts of dietary marine oils, produce less powerful compounds .Both arachidonic acid and EPA are precursors of eicosannoids such as prostaglandins, thrombpxanes, and leucotrience, which are oxygenated derivatives of C 20 fatty acids. Eicosanoids from the two acids differ in structure and function Physiological effects of omega-3 fatty acids are in the heart and blood circulation, immune response and cancer. The first group includes prevention and treatment of artherosclerosis, thrombosis, hypertryacyl glycidemira and high blood pressure. The second area relates to treatment of asthma, arthritis, mignrine headache, psoriasis and nephritis. The third category involves cancer of breast, prostate and colon.

Since eicosanoids are derived from PUFA in the diet so the supply of dietary PUFA affects their production. Omega-3 fatty acids are essential as a component of brain, retina, testes and sperm. DHA is linked to proper tissue function and needs to be supplied in sufficient amounts during tissue development. DHA supplementation is also necessary during pregnancy and lactation to meet the high fetal requirements. The infant formulas should be supplemented with DHA to provide as much DHA as supplied by the breast milk. British nutrition foundation task force has recommended that 5% of total daily energy supply should be from omega-3 fatty acids.

2. PROCESSING OF MARINE OILS

It includes heat processing or rendering to release the oil, degumming , alkali refining ,blanching, and deodorization. The quality of marine oil is less uniform than crude vegetable oils and it must be processed at low temperature under nitrogen.

2.1 Oxidation of Marine Oils

Oils rich in PUFA are prone to oxidation and produce off flavours and off odours. Therefore inhibition of oxidation is a major criterion in marine oils during processing, storage and incorporation into foods.

Peroxides are the primary products of oxidation, but, are unstable and degrade further to malonaldehyde and 4-hydroxynonenaI which are toxic. These secondary oxidation products cause off-flavours and off-odours in stored sea foods and marine oils. In order to get maximum benefit from omega-3 oils, their oxidation must be prevented.

The marine oils can be stabilized by addition of autooxidants, microencapsulation and hydrogenation. Hydrogenation negates the health benefits of PUFA. Synthetic autooxidants like Butylated hydroxy anisole (BHA), Butylated hydroxy toluene (BHT), Propyl gallate (PG) and Tert-butyl hydro-quinone (TBHQ) as well as synthetic α-tocopherol may be incorporated to control oxidative deterioration. The effectiveness of antioxidants other than THBQ in prevention of oxidation in marine oils is limited. They are also regarded as carcinogenic when used at high concentrations. BHA has been removed from GRAS. THBQ has not been aproved in Europe, so natural anti-oxidants may be used since they have been traditionally used. They may augment body antioxidant defence systems for combating cancer and degenerative diseases of ageing. Green tea extract has been found to be effective means of controlling oxidation and may exert beneficial health effects in the prevention of cancer. Thus inclusion of green tea extracts in marine oils may provide products that are effective in the control and possible treatment of both CVD and cancer. They have to be completely dechlorophillized for incorporation into marine oils. The dechlorophillized green tea extracts also give similar inhibition as catechnis in oxidation of marine oils.

3. MICRO ENCAPSULATION

Oils rich in PUFA may be stabilized by their inclusion in microcapsules. In this, oil is trapped in a wall material for protection against moisture, oxygen and light. The carbohydrates as such or modified, as well as proteins may be used for encapsulation e.g. sardine oil in egg powder, gelatin and caseinate in squid oil. β cyclodextrins, maltrin, and corn syrup have also been used for encapsulation.

4. OMEGA-3 CONCENTRATES

Omega-3 concentrates from marine oils may be prepared as natural triacylglycerols or modified triacyl-glycerols or as free fatty acids. Winterization or crystallisation may be used to prepare concentrates with a total of EPA, DHA and SPA content of upto 30%. Production of higher concentration of fatty acids is difficult. However omega-3 concentrates may be easily prepared by transforming into FFA.

Urea complexation allows handling of large quantities of material. In this process saturated and longer chain unsaturated FA are complexed with urea as inclusion compounds while PUFA remains in the mixture which can be separated. In addition, lipase assisted hydrolysis, alcoholysis and acidolysis may allow selective concentration of omega-3 fatty acids. The total content of omega-3 fatty acids from enzymatic process is 54% while for urea, complexation, it is 88%.

5. SEA FOOD PROTEINS

Sea food proteins possess excellent amino acid scores and digestibility characteristics. These constitute 11–27% of sea foods and consist of sarcoplasmic, myofibrillar and stroma-type. Dark muscles of fish generally contain more NPN than their light counterparts. The sarcoplasmic proteins account for approx 30% of the total muscle protein. The dark muscles contain less sarcoplasmic proteins than white muscles. The sarcoplasmic enzymes (glycolytic, hydrolytic) cause spoilage in fish after death.

The myofibrillar proteins undergo, changes during rigor mortis, resolution of rigor and long term frozen storage, affecting the texture of fish and gel forming ability of fish . Stroma consists of collagen and elastin (0.2–2.2%). No doubt collagen accounts for toughness, but, no such problem is encountered in fish. NPN contributes to the taste of sea foods.

6. NOVEL FUNCTIONAL PROTEIN PREPARATION

Surimi is a mechanically deboned fish flesh which has been washed with water or dilute salt solutions at 5–10°C. Removal of NPN water soluble sarcoplasmic proteins, haem proteins and enzymes gives colourless and tasteless products. The yield is 50–60%. The surimi can be used for value added products. The surimi quality and yield are affected by storage technique, raw material quality and various processing factors.

7. LOW VISCOSITY THERMO STABLE PROTEIN DISPERSION

The protein from low cost fish are not fully utilized because of lack of solubility of myofibrillar proteins in water and their sensitivity to denaturation. The preparation of thermostable water dispersions of fish proteins involve washing of comminuted muscle tissue to remove water soluble components and lipids. Conformational changes in muscle proteins enhance the entrapment of water in gel matrices. After this, the solids are further washed with salt and sodium bicarbonate solutions followed by a subsequent wash with cold water. In some cases, after this it is necessary to lower pH to 3.0–3.5 to enhance gelation. The pH adjusted dispersion samples are heated to 50°C.

Washed meat has higher moisture content and lower lipid content than light colour and bland taste. The washings increase the mass of proteins by 20% due to hydration. The aqueous dispersions of washed meats are highly viscous and their viscosity is temperature and concentration dependent. The viscosity decreases at increased temperatures which, however, is mostly recovered on cooling. For capelin dispersions, heating to 80–100°C results in permanent loss of viscosity. In herring, mackerel and shark dispersions, loss of viscosity is achieved only when small amount of acetic acid is added before heating. The dispersions with reduced viscosity are thermally stable. The proteins are stable when heated to 100°C. for 30 min and even moderate concentration of salt in the solution does not influence the solubility characteristics of such dispersions. Stable proteins are useful to study structure-function relationships and examine the possibility of their use in biological applications. These dispersions can be used as protein supplement in extrusion cooking of cereal based products and development of a functional powder by suitable dehydration of dispersions.

The fish protein concentrates (powders) have 85% protein. The amino acid composition of products is similar to that of proteins present in the starting muscle tissue. However, individual amino acid contents are lower than that in the muscles. Thus the dispersions and powders so prepared have nutritional value equal to their original fish proteins, but, have bland taste. In addition, most of the haemproteins are removed during washing giving a milky white colour. These can be used for fortification of cereal-based products.

8. PROTEIN HYDROLYSIS

The comminuted fish muscles are suspended in water and enzymes are added to the slurry. The reaction takes place for two hours to one week. After separation of solids, aqueous layer is clarified, the pH adjusted and material dehydrated. The type of enzyme employed has a marked effect on the composition as well as characteristics of the product. The enzymes used are alcalase, papain, neutrase. Alcalase gives high yields and is less costly so can be commercially exploited. The protein hydrolysates generally have low lipid content, are more stable to oxidative deterioration and have ivory white colour. So they can be used in fabricated foods where colour addition is desirable. The amino acid profile remains unchanged. Sensitive amino acids like tryptophan, methionine may be slightly affected. Content of free amino acids increases from 123 mg/l00g to 2944 mg/ l00g. So these hydrolyzates can be useful in thermal generation of aroma, when combined with sugars and glucose. They are excellent fat absorption, moisture retention, emulsification properties and whippability. They are excellent solubility in the pH range of 2.0 to 10.4. Biocapelin inhibits oxidation of meat lipids due to chelating effect of hydrolyzed molecules towards metal ions, released from haem proteins during processing of meat. Some amino acids like tyrosine may act as free radical scavengers because of their phenolic nature.

Novel lipids and protein products may be prepared from underutilized fish species and their processing discards. These products may be used as pharmaceutical, food and speciality commodities. The knowledge about health effects of sea food lipids and the nature of their proteins would result in change in dietary habits of the consumers.

FURTHER READING

1. Ahmed, J., Ramesh, B.S. and Mahendrakar, N.S. Changes in Microbial Population during fermentation of tropical fresh water fish viscera. *J. Appl. Bacteriol.* **80**: (1996) 153.
2. Bawa, A.S. and Jayathilakan, K. Fresh water fish processing—A Review. *Ind. Food Industry*, **21**: (2002) 34.
3. Ignatius, O.A. Biological evaluation of the effect of traditional processing methods on protein quality of fishes. *J. Food Sci. Technology* **31**: (1994) 409.

4. Jeyasekaran, G., Shakila, R.J. and Sundararaj, V. Opportunities in fish processing. *Ind. Food Industry*, **21**: (2002) 40.
5. Yoon, H.D., Lee, D.S., Kim, T.J., Jang, Y.S. and Suh, S.B. Processing conditions and nutritional quality of fresh water fish extractives. *Bull Natl. Fish Res. Dev. Inst. Korea*. **20**: (1997) 55.

Starting a Biotech Venture

S.R. Nair

Biotech Consortium India Limited
Anuvrat Bhawan, Deen Dayal Upadhaya Marg
New Delhi-110002

Biotechnology is a frontier area in modern science and technology. It deals with the practical applications of living organisms and products derived from them. The reach of biotechnology now extends to major sectors of economy such as agriculture, drugs and pharmaceuticals, industrial processes and pollution control and abatement. Traditional biotechnology has been in use for the production of alcohol and various food products using conventional fermentation methods. Modern biotechnology refers to application of genetic engineering or recombinant DNA (r-DNA) technology.

Substantial growth in the commercial applications of biotechnology started in the 1970s. The complexion of biotechnology changed forever by the developments in r-DNA technology. The first r-DNA product, human insulin was introduced in 1982. Since then, new products were introduced almost every year. So far, over 30 such products have been commercialised and nearly 800 products are in trial production stage. Biotechnology industry is now being recognised as one of the greatest entrepreneurial options for investment around the world.

In comparison to the developed world, the Indian biotech industry is still in its infancy. The majority of our biotech industries are into production of traditional products like alcohol, antibiotics, vaccines and low technology intensive products such as biopesticides, biofertilizers, mushrooms, tissue culture and floriculture. Commercialisation of the first indigenously developed r-DNA product in India, the hepatitis-B vaccine of Shantha Biotechnic, Hyderabad in 1997 has proven the indigenous capabilities in modern biotechnology and has given a boost to the Indian industries planning to enter this field. Considerable research and development work has been taken up during the last 15 years, mainly under funding from the Department of Biotechnology (DBT), Government of India, which has helped to expand the technology base in biotechnology in the country and has thrown up many promising commercial leads.

The development of biotech products still remains a high-risk, but potentially highly rewarding exercise. To cite an example, it may take upto 12 years for any bio-pharmaceutical substance to reach the pharmacy shelves from the time of its conception in the laboratories. The American Pharmaceutical Association estimates the cost of developing a single new pharmaceutical drug to be in the order of

$200-$250 million. Against this, , when the product is successful, the ultimate financial rewards can be considerable and well worth the effort.

Long-term commercial viability is obviously essential to the success of any biotechnological venture. Apart from a strong technology base, factors such as good management, growing market potential, safety issues etc. also play an important role in the success of a venture. Many instances may be quoted where products or processes were resounding technical successes, but dismal commercial failures. Factors leading to failure include adverse customer perception of product safety or efficacy, the appearance on the market of an alternative, cheaper or more effective competing product, exchange rate fluctuations (affecting export markets), government price control or other regulatory measures.

1. STARTING A BIOTECH VENTURE

Following are some of the important steps in starting a biotech venture:

1.1 Selection of the product and technology source

The first and foremost step is the selection of the product. This should be undertaken carefully by evaluating various factors such as market demand, target users, production technology, raw material availability, profit margin and comparison with the competing products.

Once the product has been identified, the promoter needs to evaluate the available technologies that can be adopted for commercialisation. Independent expert opinion and demonstration should always be sought before finalising selection of the technology.

While identifying the source of technology, it is important to critically evaluate the IPR ownership and related issues at the initial stage itself. One should keep in mind not only the existing rules in this respect, but, also expected changes in the near future which can affect the future prospects of the venture

1.2 Management team

A sound management team with the skills of the members complementing one another is an important key to a successful venture. A good management team can create very successful companies even based on ordinary products. Therefore, it is necessary to form a multidisciplinary management team, which should have at least three experts, i.e. a technocrat, a finance executive and a marketing manager. It has been observed in India as well abroad that projects of small biotech ventures could not take off wherever a technocrat took up all the responsibilities of production, marketing and financial management upon himself with limited support from outside. This is because technology is only one of the inputs for a successful commercial venture. On the other hand, a multi skilled team, in which the members stay together long enough and supplement each other's skills had generally led to successful ventures.

1.3 Business plan

Preparation of a business plan is the next important step. The business plan should be comprehensive, specific and clear. It should begin with a brief summary i.e. executive summary covering highlights of business, markets and financial projection. The major heads, which would be covered in the plan, are products, markets and marketing, manufacturing and operations, management and financial analysis. In case of biotech ventures, the promoters should properly identify the risks inherent in the venture and see that they are minimised to acceptable levels. Since most of biotech products are innovative in nature, a detailed marketing strategy should be presented in the business plan covering market position, pricing policy, promotion and advertising etc.

1.4 Regulatory approvals

Since modern biotech products using genetic engineering can have far reaching effects on life forms, certain rules and regulations have been formulated by the government for the safe and regulated introduction of such products in the country. Any institution initiating work in this area should first take the approval from Review Committee of Genetic Manipulation (RCGM) or Genetic Engineering Approval Centre (GEAC), functioning under DBT and Ministry of Environment and Forests (MOEF) respectively. Secondly, if the product belongs to category of drugs and pharmaceuticals, approvals need to be taken from Drug Controller General of India (DCGI) as well. All recombinant products for therapeutic use are considered as new products as per the current Indian Drugs Act and the drug rules, and import as well as local production of the these products require the permission of the drug control authorities.

The promoter should submit the applications for approval at the earliest, based on the nature of the product. Applications to any other authority such as Department of Industry, Department of Agriculture etc. as per the rules may also be submitted for approval by the entrepreneurs.

1.5 Financing of the venture

An entrepreneur, in consultation with his financial advisors, should research carefully on the options available from potential investors for financing the venture. Care should be taken to avoid unnecessary circulation of the plan to a large number of financial institutions. In India the leading financial institutions such as IDBI, ICICI and IFCI have promoted venture capital divisions or companies which may be approached by the entrepreneurs apart from conventional agencies such as the state IDCs, SFCs and banks. Sound technology, market, competence of the management team and the financials are the major considerations which favour an investment decision by a funding agency.

1.6 Role of venture capital

For new entrepreneurs having innovative ideas for commercialization, the role of venture capital as a source of funding cannot be overemphasized. While in USA and Europe, where the venture capital industry is well developed, small amounts of money are available to entrepreneurs from 'angels' to try out novel ideas without any obligation of returning the money if the idea fails. Such liberal facilities are not yet available in India. However, the venture capital industry is growing fast under favourable guidelines and incentives from the government for venture investment. Sectoral funds for high technology, fast growth areas are also being set up, managed by people highly knowledgeable in the respective fields. All these will make access to venture funds easier for a new entrepreneur.

The major advantages for a new company obtaining venture capital assistance are:

- As most of the assistance comes in the form of equity or equity like instruments, the initial burden on the company for payment of interest or repayment of loans is minimal
- The venture capitalist is a partner in the business, sharing his expertise as well as money which will greatly improve the prospects of success of the enterprise.

However, venture capitalists are extremely selective and will invest only if they are convinced of the soundness of the project, especially of the management team and the potential for much above average returns in the medium term. Once the company makes profits and grows, the pay out to the venture capital can be several times more than the normal outgo to financial institutions in a conventional industry.

2. PROBLEMS FACED BY BIOTECH VENTURES

2.1 Lack of awareness

The possibilities offered by biotechnology are little known among the users in our country irrespective of the several applications already proven in various areas such as agriculture, health care, environment or process industry. This lack of awareness results in restricted markets and more expenses towards popularising the product.

2.2 Adverse customer perception of safety

It has been observed that there is undue fear in the minds of people about the risks in using the modern biotech products, inspite of a fairly satisfactory regulatory framework in the country to evaluate the risks before the release of these products. This plays a negative role in successful marketing of the product.

2.3 Non-availability of technology

In spite of extensive research in various areas of biotechnology, there are not many proven commercial leads. Only limited data are available on to various aspects such as process economics, quality control etc., which leads to lack of interest from prospective investors. A dedicated consultancy agency can help in providing the linkages in biotechnology commercialization in such cases.

Biotechnology being a highly skill based technology, leading foreign companies are not too keen, to transfer their technologies or set up production units at different global locations for fear of exposure. Therefore both technology purchase as well as joint ventures are difficult options for the Indian industry. Only less intensive technologies have been traded globally and recently some of the established technologies are shifting to multi locations primarily because their patents have either expired or are going to expire in the short run.

2.4 Delay in regulatory approvals

Procedures for getting approvals for commercial exploitation of biotech products are long and time consuming acting as a limiting factor in biotech ventures.

2.5 Lack of funds for new venture

Financing for ventures based on new and unproven technologies is extremely difficult. Even though various venture funds are expected to finance such projects, a perception of risk has led to limited investment in this area. Setting up of a sector specific fund for biotechnology could help in mitigating this problem.

Inspite of the above constraints, biotechnology holds great potential in India and is expected to attract considerable investment in future.

For further details the contact address is

Managing Director
Biotech Consortium India Ltd.,
Anuvrat Bhawan, 5th Floor,
210, Deen Dayal Upadhaya Marg
New Delhi - 110002
Ph. (011) 23219064–67
Fax (011) 23219063
e-mail: bcilegiasdl01.vsnl.net.in

NRDC—Technology Transfer Organisation: Activities and Experience

A.B. ILANGO

National Research Development Corporation
Zamroodpur Community Centre
Kailash Colony Extension
New Delhi 110 048

INDUSTRIAL PROGRESS OF A COUNTRY is measured in terms of its Scientific and Technological advancements. There is no doubt that the advances made by India in the sphere of Science and Technology (S&T) are definitely impressive compared to other developing countries. India adopted a planned approach to achieve self-reliance in industry through the application of science and technology.

It was made amply clear in the plan document itself that building competency and expertise in S&T is a pre-requisite for industrial growth. A chain of national laboratories under Council for Scientific and Industrial Research (CSIR) network were set up in the country to give impetus to industrial research and also laboratories under Defence Research and Development Organisation (DRDO), Indian Council for Medical Research (ICMR) and Indian Council of Agricultural Research (ICAR) were set up to keep pace with the development in the field of defence, medical and agricultural related sectors of industry. A key role was assigned to National Research Development Corporation (NRDC) for developing and exploiting indigenous know-how, inventions, patents and processes emanating from national laboratories, academic and other research organizations. During the last four decades NRDC has built up bridges with Research & Development (R&D) institutions and industry and acquired over 2500 technologies for commercial exploitation, out of which half of them have gone into production with an annual turnover in the order of rupees 5000 million. It has helped to establish over 4200 small and medium scale industries. Research and development in India has led to self-sufficiency in various fields. India now has entered the R&D area for sophisticated and new technologies and developing technologies in High-Tech areas which have been the preserve of developed nations earlier.

1. TECHNOLOGY TRANSFER THROUGH NRDC

Established in 1953, NRDC is playing a vital role in building up a strong R&D base in the country. By virtue of its close linkages with the R&D institutions on one hand and the industry on the other, NRDC has brought to fruition many innovations and had several technological breakthroughs that have had a significant impact on the overall economic and technological development of the country. To make industry self-reliant in indigenous technology, NRDC has launched following programmes.

1.1 Development of Technology

National Research Development Corporation is committed to develop and promote inventions, new ideas and processes and convert them into innovations. Towards this end, NRDC sponsors the developmental projects for proving indigenous technologies. At present, NRDC has following methods of funding its technology development schemes:

- By providing technology development loans to industry.
- By equity participation in ventures based on NRDC technologies.
- By joint development projects with industry/R&D institutions

A few specific examples of successful cases of technology transfer where Indian Industry made significant break-through towards attaining self-reliance are:

1.1.1 Baby food from buffalo milk (AMUL)

Till 1960 India was meeting its entire requirement of infant food by importing brands such as Glaxo, Lactogen and Dumex. This involved drain of foreign exchange reserves and during period of crisis the supplies became erratic resulting in shortages and soaring price. It was important, therefore, to develop indigenous technology to manufacture infant food. Central Food Technological Research Institute (CFTRI), Mysore carried out R&D work on buffalo milk and established that buffalo milk could be processed into an excellent baby food. This technology could be commercialised, because, NRDC came forward to put up a pilot plant for upscaling the technology and demonstrating its commercial viability. Two plants based on the above technology having a total production capacity of about 4000 tonnes per annum are in operation in the country.

1.1.2 Swaraj tractor

CMERI, Durgapur developed know-how for the manufacture of a 100% indigenous tractor specifically designed for Indian conditions. This tractor technology was taken for commercial production by a company, M/S. Punjab Tractors Ltd., SAS Nagar, Punjab. Under stiff competition from multi-nationals, this technology has even penetrated foreign market and contributed significantly to the green revolution in India. Company has so far produced more than 1 lac tractors and established a good reputation in Agro-machinery field.

1.1.3 Rice husk particle board

To give boost to agro-based industries and converting waste into wealth, this process developed by Indian Plywood Industries Research Institute (IPIRI), Bangalore was selected by the Corporation for technology promotion, both on account of its commercial production and the positive effect it could have on saving scarce forest resource. NRDC provided soft loan for technology development to M/S. Padmavathy Panel Boards Pvt. Ltd. (PPBL), Bangalore for setting up commercial plant having a capacity of 600 tonnes per annum and arranged soft loan from Venture Capital Fund (VCF) of IDBI for the above company. The party is already in production and technology has been so far licensed to 9 parties in different parts of the country. A turnkey plant was also set up in Malaysia in 1996. The boards produced by these firms have shown excellent market acceptance and have gained popularity due to their aesthetics, better mechanical properties and lower prices as compared to wood. It is thus poised to make a significant impact on the economy in the coming years.

1.1.4 Spice oleoresin

The use of spice oils and oleoresins from agricultural products like ginger, pepper, cardamom, turmeric and chilies have now become common, both in hotels and in domestic kitchens. Oleoresins find extensive use as flavouring materials in canned meat, sauces, soft drinks, pharmaceuticals preparations, perfumes, soaps, tobacco, confectionery and bakery products. The technology for extraction of oleoresin and oils from spices has been developed by CFTRI, Mysore.

The significant feature of this technology is its suitability for small-scale production, helping in the utilisation of low priced raw materials in the production of value added items. As such, this technology is most relevant to predominantly agro-based economies and is applicable to a variety of spices like ginger, cardamom, pepper, turmeric and chilies. The technology has been successfully used on a commercial scale by a number of entrepreneurs in India and also in some other countries with similar culinary tastes such as Malaysia and Nepal.

1.1.5 Invert sugar

Carbohydrate based sweeteners have emerged on the scene to keep pace with the advances in food technology. These sweeteners have tremendous application in the food and pharmaceutical industries. The most important among them is Invert Sugar, an equi-molecular mixture of glucose and fructose. Invert sugar is a valuable sweetener in food and pharmaceutical applications because of its functionally more desirable properties. The Invert sugar can be used in:

- **Biscuits:** For caramalisation, enhanced flavour and better texture.
- **Bread, Cakes and Pastries:** For better crust colour, softer crumb and faster yeast activation.
- **Fruit Processing:** For better shelf life due to good humectant properties. Provides better taste profile and enhances flavour.
- **Honey:** For bee feeding and blending with honey. Restricted bacterial activity. Good and chemically purer replacement of Honey.
- **Squashes, Lemonades and Instant Energy Drinks:** Glucose and fructose provide instant energy and better taste.

- **Pharmaceuticals:** Ready to use for cough syrups and glucose and fructose based intravenous fluids.

The Food Technology and Enzyme Engineering Division of the Bhabha Atomic Research Centre (BARC) has developed the process using immobilized technique for enzymatic production of concentrated invert sugar from cane sugar for the first time in the world. Sophisticated biotechnological processes ensure that the invert sugar is entirely safe for human consumption.

At present, Dhampur Invertose Ltd, Dhampur, Uttar Pradesh and Invert Sugar India Pvt. Ltd, Gwalior are manufacturing the invert sugar and other parties are in the process of setting up plants. The process has been patented in India.

1.1.6 *Fly ash bricks*

Based on the process for the manufacture of fly ash bricks developed by CFRI, Dhanbad and licensed by the Corporation, a pilot plant with a production capacity of 20,000 bricks/day has been set up by M/s. Jagatdhatri Brick Industries (JBI), Barrackpore, West Bengal. Keeping in view its techno-economics, Corporation provided a developmental loan to M/S. Jagatdhatri Brick Industries to upscale the technology. The party is successfully producing the product and the technologies has been licensed to other parties also.

The Corporation thus hopes to promote this technology on a large scale in the coming years and thus utilise waste material for building industry.

1.1.7 *Spirulina algae*

It is a high quality food supplement containing vitamin B_1, B_2, B_6, C, E, besides protein etc. It has tremendous potential use in foods, cosmetics and health care applications. The Corporation recognised the importance of making available spirulina algae on large scale and joined hands with MCRC, Madras to upscale the technology developed. After successful demonstration and field trial this important know-how has been licensed to M/S. New Ambadi Estates Pvt. Ltd., Madras. The firm has set up a plant having capacity of 100 tonnes per annum. Tamil Nadu Govt. is experimenting with this product for feeding children in their Mid Day Meal programme to remove the deficiency of vitamins/proteins etc. among school going children. A work plan has also been prepared by the Corporation to promote this technology on National Level.

1.1.8 *Blood bags*

M/s. Peninsula Polymers Ltd., Trivandrum, the country's first company to produce blood bags based on technology developed by Sree Chitra Tirunal Institute for Medical Sciences and Technology has established the commercial viability of this life saving product and besides the national market has penetrated the foreign market. The Corporation had put in equity funds as a measure of risk investment for this pioneering technology. The product has been widely accepted in developed countries and patented worldwide. The Corporation has licensed this technology to M/S. Hindustan Latex Ltd., Trivandrum.

1.1.9 Artificial heart valve

National Research Development Corporation recently crossed a major milestone in the development and commercialisation of indigenous technology when it licensed to M/s. TTK Pharma Ltd., Madras, the technology for the manufacture of Artificial Heart Valves. The design and development of the valve had been underway with funding from NRDC at the Sree Chitra Tirunal Institute of Medical Sciences and Technology, Trivandrum since 1985. Sree Chitra valve will be a boon to the estimated one million persons in the country assessed to be suffering from rheumatic heart valve replacement early in their lives. The fact that the Sree Chitra heart valve is costing 'considerably' less than the imported after commercially manufactured and sold by TTK Pharma Ltd., from mid 1993. It is a major break-through by Indian Science and Technology in the service of common man.

2. BIOTECHNOLOGY

Biotechnology, the Science of future, holds unlimited opportunities for food production and combating hunger as it does for other facet of life, such as the development of therapeutics and diagnostics in combating human and animal diseases. In the area of primary food production great strides have been made in the production of high yielding pest and pathogen free plants and animals via tissue culture and other techniques. Wide opportunities have been opened in the field of bio-fertilizers and biological control of agricultural pests. In addition, the management of agro-waste through harnessing enzymatic capabilities of various microorganisms to convert waste into useful products such as fuels, chemicals solvents, antibiotics, soil conditioners and microbial bio-mass is leaving an impact on our daily life.

Rural based industries involved in food fermentation, setting of fibrous plants, some kinds of leather tanning, farm manure ink fermentation, mushroom cultivation can get benefited from new development in the areas of biotechnology.

A striking feature over the last 50 years of Indian economic development has been impetus given to industrialisation. The process of industrialisation was launched, as a conscious and deliberate policy; large investments have been made in building up infrastructure facilities over a wide spectrum of industries. Industrial production has made strides in terms of variety, quantity and quality. There has been a substantial diversification of the industrial base during this period with the consequent ability to produce a very broad range of industrial goods. Indigenous capabilities have been now established upto the point of self-sufficiency even in critical areas, so that further expansion in various sectors such as mining, irrigation, power, chemicals, transport and communication can be based primarily on indigenous equipment. The process of industrialisation has also fostered entrepreneurship and development of wide variety of technical, managerial and operative skills. Today India is in a position to provide consultancy services as well as managers, technicians and skilled workers for setting up of industrial projects abroad. Indian industry has been competing in the international technology market by improving indigenous technology and assimilating that from abroad and utilizing modern managerial skills.

Technology is fast changing and no country can be self-sufficient. Even Japan with enormous advances in various industrial fields is a net importer of technology. The contribution made by NRDC may be viewed in this global context. NRDC has made a significant contribution to develop and to promote indigenous technology to build self-reliance in Indian industry.

For further details the contact address is:

Managing Director
NRDC, 20–22, Zamroodpur Community Centre
Kailash Colony Extension
New Delhi-110048
Ph. (011) 26417942, 26417821
Fax (011) 26231872, 26478010
e-mail: nrdc@x400.nicgw.nic.com.in
nrdc@nda.vsnl.net.in
website: www.nrdcindia

Small Industries Development Bank of India (SIDBI): Range of Services

P.E.S. VIDYASAGAR

Small Industries Development Bank of India (SIDBI), Chandigarh 160 017

SMALL INDUSTRIES DEVELOPMENT BANK OF INDIA (SIDBI) set up under a special Statute of Parliament is the principal institution in the country for the promotion, financing and development of industries in the small scale sector. SIDBI became operational in April 1990. The Banker (May 1999 issue), London has ranked SIDBI 23rd in terms of assets and 24th in terms of Capital among the top 50 Development Banks of the world. It has also won Asian Banking Award 1999 for its Micro Credit initiatives. SIDBI has been serving small scale sector which is a vibrant and dynamic sub-sector of India's industrial economy. The ceiling on investment in plant and machinery of SSIs stands at rupees 100 lakh effective from December 24, 1999.

The bank operates a number of schemes tailored for the benefit of small scale units. While some of its schemes of assistance are routed through the State-Level financial corporations and the banking system in the country. The other schemes are directly implemented by SIDBI. While SIDBI refinances loans given by Primary Lending Institutions (PLIs) to new/ existing SSI projects for setting up of manufacturing facilities/ for modernisation and technological upgradation of existing units to enhance their competitiveness. It also provides financial supports to banks/co-operative banks in respect of outstanding portfolio to SSI sector. A few innovative projects which SIDBI has initiated include:

- Marketing development projects which expand the domestic and international marketability of SSI products;
- Modernisation-cum-Technology upgradation projects of existing well run SSI units and ancillaries
- Financial support to infrastructural development agencies for developing modern infrastructure for SSIs
- Support to leasing and hire purchase companies for offering leasing and hire purchase facilities to SSI units
- Financial support to SSI units to enable them to acquire ISO 9000 certification/credit rating from accredited agencies,
- Pre-shipment/ post-shipment credit in foreign currency to SSI units.

SIDBI has been made as nodal agency for recently announced Technology Upgradation Fund for textile industry. Discounting of supply bills by SIDBI has proved to be a great boon to SSI suppliers of large companies in augmenting their working capital. Its bills discounting facilities for acquisition of equipments/sale of equipments by SSI units on deferred payment basis has also given great soccour to SSI units. SIDBI has sanctioned around rupees 45,000 crore and disbursed around rupees 33,000 crore to the SSI sector in its nine years of operations.

The`development and support services of SIDBI include loans, grants and corpus support to NGOs, technology institutions, management institutions etc. to act as implementing agencies of the programmes of the bank which mainly include Skill and Technology Upgradation Programmes (STUP), Small Industries Management Programme (SIMAP) and various Entrepreneurship Development Programmes (EDPs). It has started a full-fledged foundation for micro credit to foster micro credit initiative.

In view of the importance accorded by the Government of India to the Information Technology sector, SIDBI took the initiative in setting up State level dedicated Venture Capital (VC) Funds for software and IT industry. SIDBI has been successful in motivating various State Governments/State Government agencies to set up VC funds in which 50% of the corpus was contributed by SIDBI. During 1998-99, sanctions of VC assistance of rupees 86.3 crore rupees was 45.9% higher than the sanctions of rupees 59.15 crore rupees during 1997-98 and disbursements by 53% at rupees 15.02 crore.

1. VENTURE CAPITAL SCHEME

1.1 Objective

The objective of the scheme is to provide a window within SIDBI to entrepreneurs with ventures which have special characteristics to be innovative, but, at the same time may not qualify for assistance through the conventional route of term financing. Projects involving new and untried processes and technologies which have scope for commercial application with characteristics of high risk and high return are one example of the type of projects which SIDBI looks for extension of support out of its VC Fund. Computer software development is another example. SIDBI proposes to keep the objective wide and flexible to include all deserving projects for examination.

1. 2 Eligibility of Ventures

To be eligible for assistance, the intending units should be set up as private or public limited companies. Existing companies also will be eligible to propose projects for assistance. Their track record would be an important factor in deciding their eligibility for assistance.

1. 3 Nature and extent of Assistance

There are no maximum/ minimum limits. The assistance will be decided on case to case basis. It could be equity investment/ conditional loan/ convertible instruments.

1.4 Venture Selection

Projects will be evaluated by a team of experts possessing specialisation relevant to the projects concerned. Recommendation of the expert group will be kept in view while considering SIDBI's assistance to the project. Projects assisted under the scheme should satisfy the following broad parameters.

1. 4. 1 Management, market and competitive advantages

A sound management team comprising professionals endowed with entrepreneurial traits and skills, integrity and ability to innovate with a total commitment to the venture. SIDBI looks for proposals which address a rapidly growing market opportunity and have long-term competitive advantage in terms of quality, price of product and growth potential. They should further demonstrate a potential for above average profitability leading to attractive returns on investment.

1.5 Procedure

Intending entrepreneurs/ promoters may initiate the process by furnishing certain basic particulars of their proposed venture to Venture Capital Division at SIDBI's Bombay office which will take a preliminary view of the proposal.

A good business plan will help its assessment bank. A plan should be brief and include the following:

- Objective of the project;
- Resume and reference on promoters and key management personnel;
- A note on performance of the company during the preceding 3 years (where applicable) covering financial performance, products manufactured, market shares and competitive strengths and possible threats;
- Present status/stage of the project;
- Speciality of the process/technology and sources;
- Innovative content, nature and advantages of the proposed product/ service;
- Market assessment and strategy - including details of substitute products, company's strength vis-a-vis competitors;
- Cost of venture and pattern of financing envisaged;
- Manufacturing, marketing and distribution plans;
- Financial projections;
- Implementation schedule;
- Manpower requirements and
- Analysis of risks.

On receipt of the above information, Venture Capital Division of SIDBI, Mumbai will take a preliminary view of the proposal and fix up discussions with the promoters in Mumbai. Interested entrepreneurs/ promoters may also have a discussion with any of the offices of SIDBI.

For enquiries/clarifications contact address is:

Chief General Manager
Small Industries Development Bank of India,
SCO 145-146, Sector - 17-C,
Chandigarh
Tel. No.: 709287-88/ 701634
Fax No.: 701633

Financing of Hi-tech Projects in Agriculture Sector

P. SINGH

Punjab National Bank
(High Tech Agriicultural Finance Branch)
12, Gole Market, Model Town
Ludhiana 141 001 (Punjab)

WITH ADVANCEMENT IN SCIENCE AND Technology as well as its application in the field of Agriculture and allied activities on commercial lines, has opened new avenues for investments in agricultural activities. The agricultural scenario has further brightened with liberalization in the national economy and setting up of World Trade Organisation (WTO) promising a boost in international economy.

Agriculture is fast gaining the status of industry. It has opened immense potential for us to undertake multi-profile environment controlled high-tech agricultural projects so as to make a dent in production and export of quality farm products, preferably processed and value added.

FINANCING SCHEME OF PUNJAB NATIONAL BANK (PNB)

1. Eligibility

Progressive Enterpreneurs/Corporate houses/Business firms/Company having adequate exposure/knowledge of latest technology in the field of agricultural and allied activities with a stable financial position and good banking habits.

2. Scope and Coverage

The bank will undertake technically feasible and economically viable projects for finance involving the application of latest technology in the following fields.

2.1 Agriculture crops

Projects relating to adoption of genetic techniques i.e. building resistance to virus, insects, pests & draughts and exploiting hybrid vigour. Projects of tissue culture techniques, production of biofertilizers and bioinsecticides also qualify for funding.

2.2 Forestry

Clonal multiplications for Bamboos, Eucalyptus and Salvadora tree species.

2.3 Animal husbandry/Dairy

Production of pure line/cross poultry breeding, deep freezing cattle semen ,embryo transfer technology.

2.4 Fishery

Activities relating to fresh and brackish aquaculture and processing units including individual quick freezing (IQF) etc.

2.5 Mushroom

Spawn/mushroom cultivation and processing, canning and preservation.

2.6 Floriculture

Projects relating to floriculture for production and export of cut flowers etc.

3. Quantum of loan

The bank will finance 75% of the project cost of technically feasible and economically viable projects duly appraised by concerned agencies like NABARD in case of big projects. NABARD guidelines will be followed for deciding the loan amount.

4. Margin

25% of the project costs.

5. Security

- Hypothecation of movable assets/primary security created out of bank finance.
- Mortgage or charge over agricultural land valued 200% of the loan amount or equitable mortgage of immovable properties in the name of borrower/guarantor value (200%) of the loan amount

6. Interest

The interest will be charged as pier RBI guidelines for loan upto rupees 2 lacs as well as bank's guidelines for loans above rupees 2 lacs. The present rate of interest is 13% + int tax upto rupees 2 lacs and 16.75% + int tax for loans above rupees 2 lacs for 'A' category borrowers.

7. Moratorium period

Depending on the activity to be undertaken as stipulated by NABARD.

8. Repayment

The term loan will be repaid in 7 to 9 y including moratorium period in yearly/half yearly/quarterly/monthly instalments along with interest.

9. Insurance

Comprehensive insurance of the movable/immovable assets as per bank's requirement.

For further details and banking facilities in the entire State of Punjab, contact address is

Chief Manager
Punjab National Bank,
Hi-tech Agricultural Finance
Gole Market, Model Town,
Ludhiana-141 001 (Punjab)
Phone: 425747.

Precision Farming Using Geographical Information System

G. SINGH

Canadian Institute of Computer Science & Technology
C-2, Phase-I, Industrial Area
Mohali 160 055 (Punjab)

IN JUST ONE GENERATION, our world population will double. The task of supplying the needs of an ever-increasing population will be staggering. Chief among the demands of a growing population will be to maintain an adequate food and fibre supply. This is what the future entails, but, even without the impact of the future population boom, farmers face many challenges to produce enough food and fibre to meet the needs of today. By the year 2050 the world's population will total approximately 10 billion people with India being the most populous of all. Consequently the challenge is to be able to feed two people in 2050 where today there is only one.

Feeding these additional billions poses a dilemma of immense proportions. There are just two ways to increase food production-put more land under cultivation or increase yields. Because land is needed to house a growing population and to save the earth's wildlife and natural resources, focusing our resources on increasing crop yields is the logical alternative.

Expanding our food and fibre supply by increasing yields will require great creativity, enormous intellectual effort, unprecedented cooperation between public and private research institutions and the different sectors that contribute to food and fibre production. Biotechnology and Information Technology will play a crucial role in expanding agricultural productivity in the 21st century.

1. PRECISION FARMING

"Precision Farming" is a current buzzword among agricultural circles. The term "precision farming" means carefully tailoring soil and crop management to fit the different conditions found in each field. Precision farming is sometimes called "prescription farming", "site specific farming" or "variable rate technology." It has caused a focus on the use of three technologies that are very central to the LARS programs-remote sensing, geographic information systems (GIS) and global positioning systems (GPS). Some people incorrectly use the term "GPS" to imply precision farming. We have literally taken "agriculture into the space age." Farmers have services available that involve satellites collecting data, transmitting locational information, or providing data from a variety of sources. Farmers can analyze this satellite information or they can rely on companies to do this service for them for a fee.

Precision farming is a farming system concept that involves the development and adoption of knowledge-based technical management systems with the main goal of optimizing profit. This management system will enable micro-management concepts, that is, the ability to appropriately manage every field operation at each location in the field, if it is technically and economically advantageous to manage at that level. The system will likely include the ability to vary or tailor the rate of application of all inputs such as tillage, seeds, weed, insect and disease control, cultivation and irrigation.

It is possible to implement precision farming at many different levels. In its most extensive form, it will include precise micro-management of every step of the farming process. It is expected that the advisability of micro-management depend upon many factors, such as soil type, crop, seasonal weather, and other factors. For example, in a dry year, it may be possible to control insects by spraying only small areas where the insects are known to exist; in a wet year, it may be advisable to uniformly spray the whole field. Technically, one important aspect of the development of precision farming concepts is the development of the hardware and software necessary to vary the rate of the application of agricultural inputs. A number of research projects have been conducted in this area, and several companies have been developing variable rate application equipment in recent years.

Although the rationale behind precision farming is simple enough-to control the spread of fertiliser to an accuracy of a few metres-the actual process is highly complex and involves the collection, analysis and dissemination of data on a massive scale.

The internet and mobile wireless communications technologies offer obvious advantages in facilitating this data-intensive process, and researchers have collaborated to develop an integrated software solution geared to precision farming.

2. ROLE OF GIS, GPS AND REMOTE SENSING

Some farmers have already received benefits of satellite remote sensing data. Satellite images have been used to distinguish crop species and locate stress conditions. The cost of obtaining the images has been the biggest deterrent in regular use of such data. The remote sensing part of precision farming will need to be proven.

More recently farmers have gained access to site-specific technology through Global Positioning Systems (GPS). GPS makes use of a series of military satellites that identify the location of farm equipment within a meter of an actual site in the field.

The value of knowing a precise location within inches is that

- locations of soil samples and the laboratory results can be compared to a soil map,
- fertilizer and pesticides can be prescribed to fit soil properties (clay and organic matter content) and soil conditions (relief and drainage),
- tillage adjustments can be máde as one finds various conditions across the field,
- one can monitor and record yield data as one goes across the field.

The real value for the farmer is that he can adjust seeding rates, plan more accurate crop protection programs, perform more timely tillage and know the yield variation within a field.

Precision farming using the above technologies can enhance the overall cost effectiveness of crop production in different areas:

2.1 Seeding

Hybrid seeds perform best when placed at spacing that allows the plants to obtain such benefits as maximum sunlight and moisture. This is best accomplished by varying the seeding rate according to the soil conditions such as texture, organic matter and available soil moisture. One would plant fewer seeds in sandy soil as compared to silt loam soils because of less available moisture. The lower seed population usually has larger heads (ears) of harvested seeds providing for a maximum yield. Since soils vary even across an individual farm field, the ability to change seeding rates as one goes across the field allows the farmer to maximize this seeding rate according to 'the soil conditions. A computerized soil map of a specific field on a computer fitted on the tractor along with a GPS can tell farmers where they are in the field allowing the opportunity to adjust this seeding rate as they go across their fields.

2.2 Crop Protection

The application of chemicals and fertilizers in proper proportions are of environmental and economic concern to the farmers. Environmental regulations are calling for the discontinuance of certain pesticide applications within 100 feet of a stream or water body or well or within 60 feet of an intermittent stream. Using a GPS along with a digital drainage map, the farmer is able to apply these pesticides in a safer manner. In fact, the spraying equipment can be pre-programmed to automatically turn off when it reaches the distance limitation or zone of the drainage feature. Additionally, farmers can pre-program the rate of pesticide or/fertilizer to be applied so that only the amount needed determined by the soil condition is applied varying this rate from one area of the field to another. This saves money and allows for safer use of these materials.

2.3 Tillage

The ability to vary the depth of tillage along with soil conditions is very important to proper seedbed preparation, control of weeds and fuel consumption and therefore cost to the farmer. Most farmers are using conservation tillage, which means leaving residues on the soil surface for erosion control. The use of GPS in making equipment adjustments as one goes across the different soil types would mean higher yields and safer production at lower cost. This part of precision farming is in its infancy. The equipment companies will be announcing tillage equipment with GPS and selected controls tailored to precision farming in the near future.

2.4 Harvesting

The proof in the use of variable rate technology (adjusting seed, pesticide, fertilizer and tillage) as one goes across the field is in knowing the precise yields. Combines and other harvesting equipment can be equipped with weighing devices that are coupled to a GPS. One literally measures yield on the go. With appropriate software, a yield map is produced showing the yield variation throughout the field. This allows farmers to inspect the precise location of the highest and the lowest yielding areas of the field and determines what caused the yield difference. It allows one to program cost and yield to determine the most profitable practices and rates that apply to each field location. Thus, the use of yield monitors is a good place to start if one wants to get started in precision farming. Yield data from the same field over 3+ years would define the weak spots in the field and narrow down the probability of what is causing a log yield.

GPS has already made a big impact on precision farming. Several companies are starting to market GIS record keeping systems so farmers can record all of the field operations such as planting, spraying, cultivation and harvest (along with specific information such as type of equipment used, rates, weather information, time of day performed, etc.). Additionally the farmers are able to record observations through the growing season such as weed growth, unusual plant stress or coloring and growth conditions. Data collected by the GPS operations can be automatically recorded with the GIS program. Remotely sensed data can be analyzed and added to the GIS using soil maps, digital terrain and field operations information as ground truth. This can be used to guide further field operations like spraying, fertilizing and irrigating plus it is part of the permanent record. Precision farming will make a strong impact on the way farmers manage their farm operations in the future. A steady growth of the remote sensing, GIS and GPS technologies will be seen as a result of this method of "farming by the inch."

3. PRECISION FARMING IN ASIA

The growing food demands due to ever-rising human populations forced Asian farmers to adopt resource-intensive and unsustainable practices that increased both economic and environmental costs. Asian farming systems, therefore, present both obstacles and opportunities for adoption of precision agriculture.

Tailoring soil and crop management to match varying conditions (soil texture, moisture and nutrient status, pest distribution, etc.) within a field is not entirely new to Asian farmers. The growers traditionally noted yield variability in both space and time, and changed farm practices (e.g., seeding rates) depending on site conditions to optimize soil resources and external inputs. This was possible because most Asian farms were relatively small and farmers were familiar with spatial and temporal variation. However, precision farming in terms of using technologies such as GPS, GIS, remote sensing, yield monitors, guidance systems for variable rate application, etc. to manage within-field variation is still in its infancy in Asia.

Ever since the introduction of Green Revolution technologies in 1960s, Asian farmers have tended to treat all fields within a region uniformly resulting in application of inputs to areas not needing them or where the crops cannot make full use of them. It is now well known that such practices are not always in harmony with the environment. In addition, demographic and market changes in industrial Asian countries such as Japan, Taiwan and Korea, are compelling farmers to make decisions that are both economically and environmentally sound. The effective decision making, however, depends on quick and accurate analysis of agronomic data. In this context, precision farming technologies are widely known to assist growers in making informed decisions.

Two basic steps in precision farming include the measurement of variability in a field and the management of such variation by optimizing input application, tillage, cultivar's selection, etc. The emphasis on measurement, however, seems to vary in different regions. North American researchers have so far emphasized measuring variation in soil fertility and developing sensors, while the European researchers focused on producing and characterizing yield maps. While it is important to measure variation as precisely as possible, it is not clear how far collection of precise data *per se* is effective, for it is not always easy to determine the cause-effect relationships impacting on yield and profitability. This is especially relevant in Asia, where complex cropping systems characterized by high spatial and temporal variation are common. In such cases, it is perhaps more important to simultaneously address the known causes of yield variation while data collection on field variability is in progress. In order to know the relevance of precision farming in Asia, it is essential to examine the current agricultural status.

4. CURRENT STATUS OF ASIAN AGRICULTURE

Agronomic research in Asia was so far mainly targeted to enhance productivity at the farm level in the shortest possible time. As a result, very little attention was paid to monitoring the after-effects of adopted technologies on the quality of the resource base. Precision farming technologies are likely to be of great benefit in addressing this issue. However, as Asian countries and regions are at various stages of technological and economic development, there is considerable diversity in yields and agronomic practices both within and between nations. For example, in northwest India, where intensive farming is common, productivity is similar to that in developed countries. In contrast, the yields in Eastern India are very low. In countries such as Japan, where farm mechanization is more common than in others, grain yields are high, but, input efficiency is low. The efficiency in farming is thus interpreted variously in different regions, but, the need for improving farm efficiency and creating systems that are less wasteful and less damaging to the environment than the existing ones, is recognized everywhere.

The enhanced productivity that characterized the green revolution era in developing countries of Asia resulted from a combination of factors including the introduction of high yielding varieties, the expansion of irrigated area, and the intensive use of inputs including fertilizers and pesticides. The increase in input use over time has been especially high in Asian countries. For example, in India, chemical fertilizer use rose from a mere 69,000 tons in 1950 to 13.6 million tons in 1995, with many farmers applying high doses of about 400-500 kg nutrients/ha/year. Likewise, the use of pesticides (active ingredient) increased from 2,350 tons in 1950 to 63,270 tons in 1995. Further, nearly 78 out of 124 million ha. of area under food grains were covered by modern varieties in 1995. Despite this intensive input use, growth rates have been dropping recently causing great concern. Many farmers realize they need to continue raising the input levels to maintain past yields due to a downward shift in response functions across a wide range of input application rates. The occurrence of deficiencies of secondary and micro- nutrients has become common and farmers now apply P, K, S, Zn, B, Fe and Mn to alleviate such deficiencies. The intensive cropping systems led to the lowering of groundwater levels to as much as 15 to 20'm with the rate of decline continuing at 0.5-1 m per year. For example, in Punjab and Haryana, the net annual draft exceeded utilizable groundwater resources in as many as 53 and 42% of blocks respectively. The development of insecticide resistance in pests such as cotton bollworm and herbicide resistance in weeds such as *Phalaris minor,* are forcing growers to use high doses of chemicals without the benefit of effective control. It is, therefore, realized that future gains in input efficiency will have to come from technologies such as precision farming.

In industrial Asian countries such as Japan, Taiwan and Korea, a fierce price competition following globalization of markets and demographic changes in farm population are putting extra pressure on agricultural competitiveness. The changes include aging farm populations, a decline in number of full time farmers, and rising costs due to labour shortage. In Japan, up to 58.4% of farmers are 60 years or older. This means that at least one-third of all farmers would retire from farming in a few years. Between 1960 and 1995, the number of farm households dropped from 6 to 3.4 million, and farming population from 34.4 to 15 million. The number of part time farmers also declined by as much as 330,000 households. The numbers decreased by more than 10% in all prefectures from 1990 to 1995, the largest drop ever for any five-year period. In Hokkaido prefecture, for example, the number of farm households declined by about 50,000 over a span of 20 years, and the population engaged in own farming by about 6,000 people per year. The positive side of this bleak picture is that farming households with 5 ha or more of farmland in Japan increased from 2,000 in 1960 to 36,000 in 1995.

Input application in-farms of industrial Asian countries has been high for quite a long time, with the result that accumulation of excess nutrients has become a serious problem in some watersheds. Farmers usually apply more than endorsed doses of fertilizers and pesticides because of their relatively low cost and high input subsidies. Soil type is another factor influencing excessive use of fertilizers in some regions. For example, excessive P fertilization is common in Tokachi region of Hokkaido, where Andosols characterized by low bulk density and high P retention, occupy nearly 50% of cultivated lands.

It is important to bear in mind the two major changes occurring all over Asia. First, increased urbanization and migration of people from rural to urban areas is causing a severe labour shortage, which in turn is contributing to an indiscriminate use of chemicals. For example, the direct seeding of rice is gaining appeal in Malaysia, Indonesia and the Mekong Delta, despite major weed problems associated with this system. Herbicide use in these regions, therefore, increased substantially.

Secondly, a rapid shift in area out of grains to vegetables and fruits is leading to a shift in resource allocation from land-extensive activities to labour, capital and technology-intensive ones. In view of such changes, Asian growers are likely to use precision farming technologies to remain globally competitive.

5. FARM INFORMATION TECHNOLOGIES IN ASIA

Precision farming obviously requires some degree of competence in the use of software and hardware on the part of growers and/or crop consultants. Indeed the success of precision farming largely depends on creation of management systems, which will involve some combination of computerized decision support systems and the wisdom of farmers. Growers will adopt information technologies only if they are reliable and easy to use, offer some competitive advantage and can be introduced into farming without too much difficulty or expense.

Three levels of technology adoption have been in precision farming, where the first level represents conventional practice and the third level has fully supported variable application rate capability.

The use of GPS in farming is limited yet, but, it is fair to expect widespread use of GPS in future. Recently, a GPS-based crop duster (Precision GPS Helicopter), which can spray an area as small as 4 × 4 m, is attracting great attention. Some progressive farmers are now beginning to use GPS for recording observations such as weed growth, unusual plant stress, colouring and growth conditions, which can then be mapped with a GIS program.

Application of remote sensing for regional or landscape studies is common in Asia. Many countries, using an area frame sample, have begun to integrate GIS and remote sensing in their crop inventory statistics programs and as the basis for crop forecast. In Japan, corporations like Hitachi and Mitsubishi are involved in the launch of satellites in collaboration with the US companies and are attempting to supply data for agricultural use in Asian countries. Remote sensing has also been used in conjunction with GIS for monitoring changes in crop conditions. The use of remote sensing to estimate soil and plant characteristics is also common.

6. OBSTACLES TO ADOPTION OF PRECISION FARMING

There are many obstacles to adoption of precision farming in Asia. Some are common to those in other regions, but, the others are specific to Asian conditions.

6.1. Culture and Perceptions of the Users

It is widely felt that Asian farmers are resistant to new technologies and do not accept them unless they are fully convinced of benefits. Many growers are also reluctant to share data concerning their fields, as they do not want the information exploited too early by the commodity traders. The planners, on the other hand, believe that running costs in using precision technologies are high. While many growers understand that the intent of precision farming is to manage input needs to achieve maximum economic yield for each area, they are skeptical of how the data is transformed into useful information for farm level decisions. Most farmers are over 60 years of age and/or part-time, and are reluctant to use computers. Further, they are unaware of agro-environmental problems such as groundwater pollution caused by ineffective fertilization methods. Therefore, the environmental benefits from precision farming need to be popularized.

6.2. Small Farm Size

The meaning of small farms in Asia is different than in the West. Family farms range from 80 to 120 ha in the US, but they are only from 0.1 to 8 ha in Asia. Land fragmentation and sub-division are common in Asia and slow progress in consolidation is a major obstacle to adoption of precision farming. If typical field size is about 1 ha, it is perhaps necessary to get imagery acquired by a sensor with greater than 1 m resolution. Many studies in the West show that farm size is positively related to the adoption of new technologies including soil conservation integrated pest management and modern irrigation. This may be due to the better ability of big farms to hire crop consultants, easier access to credit and information, and increased contacts with representatives of extension agencies and agribusiness.

6.3. Lack of Success Stories

There have been no success stories of precision farming to date under Asian conditions. In order for farmers to adopt the technologies with confidence, they need to see the value of adoption.

6.4. Heterogeneity of Cropping Systems and Market Imperfections

Farm units in Asia are highly heterogeneous and the costs and benefits of adoption vary across widely. Further, market imperfections such as distortions in input and output markets, institutional hurdles to efficient input application, prices that do not reflect their true scarcity values, and absence of penalties for residual generation lessen the potential benefit of adoption. Because of high subsidies, the cost of fertilizers and pesticides is so low in Japan that saving on inputs cannot be a major incentive to adopt precision farming.

6.5. Land Ownership, Infrastructure and Institutional Constraints

In developing countries of Asia, policy and bureaucratic constraints slow down the process of timely information transfer. Further, poor coordination among the private sector, research and administration agencies makes the adoption of these technologies slower. The slow diffusion can also be due to the small number of input dealers offering such services to growers.

6.6. Lack of Local Technical Expertise

Adoption of precision farming requires a relatively advanced and computer literate workforce. But, there is a severe shortage of skilled manpower. Moreover, the learning costs of adoption of precision farming are influenced by variations in personal ability and age of farmers, motivation, quality and number of years of education, and exposure to extension services. Because Asian farmers lack enough exposure to data management technologies, they have to rely on farm consultants. It is, therefore, unlikely that precision farming will be adopted soon if governments and private associations do not make additional efforts.

6.7. Knowledge and Technological Gaps

Inadequate knowledge of the user's need and poor understanding of researchers to tackle major grower's concern are important constraints. For instance, very few crop growth models are directly applicable to the concurrent management of the varied growing conditions in Asia. Studies to improve capacity for quantitative modeling of agricultural systems are, therefore, necessary. Other constraints include gaps in communication between computer users and data suppliers, and between growers and the precision farming industry. Accurate analysis of farm data is another weak link in the spread of these technologies.

6.8. Data Availability, Quality and Costs

Difficulties in obtaining reliable data in a timely and cost effective way are limiting the adoption of precision farming in Asia. If data are made available for an area of single agricultural association or a municipality, people may receive the idea of precision farming more enthusiastically. High price of imagery has also been the biggest obstacle in routine use of such data.

7. OPPORTUNITIES FOR PRECISION FARMING IN ASIA

Despite many obstacles listed above, business opportunities for precision farming technologies including GIS, GPS, remote sensing, and yield monitor systems are immense in Asia. The scope for founding new hardware, software and consulting industries related to precision agriculture is gradually widening. Within each Asian country, however, some regions are in a better position than others to adopt precision farming. Punjab and Haryana states in India, where farm mechanization is more common than in other states, may be the first to adopt precision farming on a large scale.

Recently, the governments of certain Asian countries initiated special efforts to promote precision farming.

Precision farming is useful in many situations in Asia but only a few specific cases are discussed in detail. Rice, wheat, sugarbeet, onion, potato, and cotton among the field crops, and apple, grape, tea, coffee and oil palm among horticultural crops are perhaps the most relevant. Some have a very high value per acre, making excellent cases for site-specific management. For all these crops, yield mapping is the first step to determine the precise locations of the highest and lowest yield areas of the field, and to analyze the factors causing yield variation.

Further, the preparation of product quality maps for these crops is as important as yield maps. In India, a few researchers in the private sector initiated studies on precision agriculture in high value crops like cotton, coffee and tea. In cotton, remote sensing coupled with GIS can assist in improved precision of insect pest management and harvesting. In Sri Lanka, researchers at the Tea Research Institute are examining precision management of soil organic carbon.

So far as dairy farming in Asia is concerned, precision farming techniques can help in improving efficiency of methods, timing, and rate of application of animal wastes leading to high application efficiency and low environmental pollution while considering soil and climatic conditions. For instance, factors determining the risk of nitrate leaching, release of N_2O through denitrification and contamination of surface and ground waters by runoff can be mapped and analyzed. Likewise, poorly managed areas in grasslands can be identified and the optimum period for cutting on a plot basis determined.

Nutrient stress management is another area where precision farming can help Asian farmers. In semi-arid and arid tropics, precision technologies can help growers in scheduling irrigation more profitably by varying the timing, amounts and placement of water. For example, drip irrigation, coupled with information from remotely sensed stress conditions (e.g., canopy temperature), can increase the effective use of applied water from 60 to 95%, thereby reducing runoff from 23 to 1%, and deep percolation from 18 to 4%.

Pests and diseases cause huge losses to crops in Asian countries.. If remote sensing can help in detecting small problem areas caused by pathogens, timing of applications of fungicides can be optimized. Likewise, airborne video data and GIS have been shown to effectively detect and map blackfly infestations in citrus orchards, making it possible to achieve precision in pest control. Perennial weeds, which are usually position-specific and grow in concentrated areas, are also a major problem in Asian countries. Remote sensing combined with GIS and GPS can help in site-specific weed management.

8. STRATEGIES FOR ADOPTION OF PRECISION FARMING IN ASIA

Precision farming is still only a concept in Asia and strategic support from the public and private sectors is essential to promote its rapid adoption. Successful adoption, however, comprises at least 3 phases including exploration, analysis and' execution. Data on crop yield, soil variables, weather and other characteristics are collected and mapped in the exploratory stage, which is important for increasing the awareness among farmers of long term benefits. The approaches to data collection and mapping must, therefore, reflect local needs and resources. For instance, soil sampling on a grid basis can provide the most accurate guide for variable rate application, but it is prohibitively expensive for most farmers in Asia. In such cases, sampling can make use of other information such as past field history, yield maps, and remote sensing as guides to reduce the number of soil sampling sites. Additional research is also necessary on various criteria for indicating spatial patterns in crop yield and fertility.

In the analysis stage, factors limiting the potential yield in various areas within a field and their interrelationships are examined using GIS-based statistical modeling.

Variable application of inputs or cultural operations comprises the execution phase. In most developing countries of Asia, however, it is not always necessary and/or possible to use variable rate applicators. Efforts must, therefore, initially focus on limiting indiscriminate use of inputs in conventional methods. Once the economic and environmental benefits are known widely, variable rate technology would be rapidly implemented at least in high value crops.

To spur adoption of precision farming methods in Asia, pilot demonstration projects must be conducted at various growers' locations by involving farmers in all stages of the project. The pilot projects must attempt to answer the grower's needs and emphasize the operational implementation of technology and complete analysis of the costs and savings involved. Documentation of pilot projects would help in examining the operational weaknesses and identification of remedial measures. The projects can be used to -train innovative farmers and early adopters, expose the neighboring non-participating farmers to the new technologies, and show the usefulness of the technology for short- and long-term management. Based on these projects, the extension agencies can gather and convey success stories to the public in other regions.

The role of agricultural input suppliers, extension advisors, and consultants in the spread of these technologies is vital. For instance, public agencies should consider supplying free data such as remotely sensed imagery to the universities and research institutes involved in precision farming research. Also, professional societies of agronomy, agricultural informatics, and engineering must provide training guidance in the use of technologies. The involvement of interdisciplinary teams is essential in this. Small farm size will not be a major constraint, if the technologies are available through consulting, custom and rental services. For instance, by renting the equipment, manufacturers can enable early adapters to avoid the risks associated with purchasing costly machinery.

The role of agricultural cooperatives is important in dissemination of precision farming technologies to small farmers. If precision farming is considered a series of discrete services: map generation, targeted scouting, etc., it is possible to fit these services within the structure of a progressive agricultural cooperative in each Asian country. For instance, satellite imagery can be acquired for the entire region of a primary society and the costs may be distributed to individual farmers. In addition, agricultural associations can assign some funds to increase farmer awareness of the value of precision farming.

Changes in agricultural policies are also necessary to promote the adoption of precision farming. There are basically two policy approaches: regulatory policies, and market-based policies. The former refer to environmental regulations on the use of farm inputs, and the latter refer to taxes and financial incentives aimed at encouraging growers to efficiently utilize farm inputs. In most Asian countries, the lack of penalties for pollutant generation has partly contributed to an excessive use of inputs. Subsidies on inputs and outputs, and mechanisms that prevent the price system from rationing limited resources are also common. The latter include state-guaranteed crop prices, tariffs, import quotas, export subsidies, etc. Inputs such as water and fossil fuels are usually sold at prices that are well below the real resource cost of their use, which consists not only production costs, but, also includes scarcity value and costs of pollution. In such cases, the formulation of policies that reflect the real scarcity value of natural resources and penalize pollution, and policies such as green payments for farmers adopting techniques that would lower environmental costs can promote the adoption of precision farming technologies.

In most Asian countries, the pollution effects of agriculture have been largely ignored so far because of inability to effectively monitor such effects. The advent of precision farming, and the computerization of input and output flows, will now enable such monitoring.

The ability to finance a creative information venture in agriculture will affect the speed of diffusion of precision farming technologies. Commercial banks, as well as other sources of funding, have to be educated regarding the potential of precision farming. In many Asian countries, it may be worthwhile to develop programs of subsidized credit to enable R&D activities on precision farming.

No doubt the precision farming in Asia is in its infancy, but, there are numerous opportunities for adoption. It is believed that the progressive Asian farmers, with guidance from the public and private sectors, and agricultural associations, will adopt it in a limited scale as the technology shows potential for raising yields and economic returns on fields with significant variability, and for minimizing environmental degradation. Although it is recognized that agriculture is a major polluter of the environment in many Asian countries, farmers will not adopt precision farming unless it brings in more or at least similar profit as compared to traditional practice. The support from governments and the private sector during the initial stages of adoption is, therefore, vital. It must be remembered that not all elements of precision farming are relevant for each and every farm. For instance, introduction of variable rate applicators is not always necessary or the most appropriate level of spatial management in Asian farms. Likewise, not all farms are suitable to implement precision farming. Some growers are likely to adopt it partially, adopting certain elements, but not others. Precision farming cannot be convincing if only environmental benefits are emphasized. On the other hand, its adoption would be improved if it can be shown to reduce the risk. We must be cautious, however, in not overselling the technologies without providing adequate product support. The adoption of precision farming also depends on product reliability, the support provided by manufacturers and the ability to show the benefits. Effective coordination among the public and private sectors and growers is, therefore, essential for implementing new strategies to achieve success.

To conclude

"Technology made large populations possible; large populations now make technology indispensable"

Joseph Wood Krutch
US Naturalist

FURTHER READINGS

Branden, J.B., Netsusil, N.R. and Kosobud, R.F. Incentive-based non-point source pollution abatement in a re-authorized clean water act. *Water Resources Bull.* **30**: (1994) 781.

Byerlee, D. Modern varieties, productivity, and sustainability: Recent experience and emerging challenges. CIMMYT, Mexico D.F., Mexico. (1994).

Changyao, W. Remote sensing monitoring on land use change in China, p. 20-26. In: Vegetation Monitoring. Proc. Int. Symp. 29-31 Aug. 1995. Ctr. Environ. Remote Sens., Chiba Univ. Chiba, Japan. (1995).

Kar, T. and Misra, A.K. Internet Applications in Food Industry, *Ind. Food Industry*, **18**, (1999) 218.

Larscheid, G. and Blackmore, B.S. Interactions between farm managers and information systems with respect to yield mapping. In: Precision Agriculture. (Eds. P.C. Robert *et al.*) Proc. Third Int. Conf. Minneapolis, MN. June 23-26, 1996. ASA, CSSA, SSSA, Madison, WI. (1996) 1153-1163.

Maji, A.K., Sidhu, G.S. Kandpal, B.K. Pande, S. Velayudham, M. and Ahmed, M.I. Evaluation and Management of Rice-wheat cropping systems in Indo-Gangetic Plains using GIS. In: Use of GIS in analysis of cropping systems, Training Workshop, Patancheru. 20-29 Aug. 1997. ICRISAT, Patancheru, India. (1997).

Okano, C., Nishimune, A. Fukuhara, M. Okamoto, K. and Hayasaka, M. Sugar beet nutritional diagnosis using remote sensing, In: Plant Nutrition for sustainable food production and environment (Eds. T. Ando *et al.*). Kluwer Acad. Publ., Tokyo, Japan. (1997) 381-382.

Sadler, E.J., Busscher, W.J. Bauer, P.J. and Karlen, D.J. Spatial scale requirements for precision farming: A case study in the Southeastern USA. *Agron. J.*, **90**: (1998) 191.

Section II

Research Papers

Plant Regeneration and Agrobacterium Mediated Gene Transfer Studies in Tomato

A. KANT *and* D.K. SRIVASTAVA
Department of Biotechnology
Dr. Y.S. Parmar University of Horticulture & Forestry
Solan 173 230, Himachal Pradesh

ABSTRACT

Plant regeneration and genetic transformation studies were carried out to standardize a protocol for high frequency plant regeneration and genetic transformation in tomato tissue. Plants were regenerated from the cotyledonary explant. The cotyledons were excised from the aseptically grown 10 d old seedlings of tomato. Shoot and root differentiation from the cotyledon segment was obtained after callusing on medium containing BAP and IAA in combination and separately. Root regeneration was obtained in the regenerated shoot, when transferred to the medium containing NAA. Genetic transformation studies were carried out using binary vector agrobacterial system containing *npt-II* and gus genes. After co-cultivation and selection, transformed callus was obtained which was able to grow on the selective medium containing kanamycin. The DNA transferred to the plant cell from the binary vector (T-DNA) included genes which encoded the enzyme neomycin phosphotransferase-II (NPT-II) and β-glucuronidase (GUS) and thus conferred resistance to kanamycin on the tomato cells. The putative transformed calli expressed β-glucuronidase activity.

INTRODUCTION

Tomato is an important vegetable crop of family Solanaceae grown all over the world. Tomato is a rich source of vitamins A, B & C, and minerals like calcium, iron, phosphorus besides it has been reported quite useful in controlling liver problems, indigestion, arthritis and urinary troubles. Being an economically important crop, the application of genetic engineering in tomato cultivation is of great value so as to obtain improved or desirable traits like insect resistance or disease resistance. Genetic transformation of the plants can be expected to prove a new possibility for increasing the genetic variability in crops. Of all the gene transfer methods available for plants, *Agrobacterium*-mediated transformation has been most widely used. This technology is now available for many crops[1-8]. However, before undertaking this approach, it is important to know whether the somatic cells of the cultivar (cv. Solan Vajr) are able to regenerate in such a way so as to give rise whole plantlets and to standardize the protocol of genetic transformation in tomato cells/tissues. *In vitro* regeneration through explant culture has an advantage over callus, protoplast and cell suspension cultures as cytological aberrations (somaclonal variations) are quite common in such cultures. Genetic manipulation of crop plants in *in vitro* is dependent on the reproducible regeneration of plants from cultured tissues. Different types of explant showing morphogenetic ability have been used for genetic transformation of tomato. A number of successful genetic transformation in tomato have been achieved.[9-19]

Recent biotechnological developments have enabled investigators to produce transgenic tomato plants. Genetic transformation studies were carried out in tomato using genetically engineered A*grobacterium tumefaciens* strain LBA 44O4 with neomycin phosphotransferase-II (npt-II) and β-glucuronidase (gus) genes. The permission to use genetically engineered A*grobacterium tumefaciens* with marker gene was obtained from the Director, Department of Biotechnology, Ministry of Science and Technology, Government of India, New Delhi and Institutional Biosafety Committee of our University as per the norms of "Recombinant DNA Safety Guidelines" DBT, Government of India, New Delhi.

In this paper, we report the regeneration of tomato plants from cotyledon explant and the use of A*grobacterium tumefaciens* LBA 4404 strain to transfer marker gene and reporter gene in tomato cells/tissues and expression of chimeric neomycin phosphotransferase-II (npt-II) and β-glucuronidase (gus) genes in the transformed calli of tomato cv. Solan Vajr using binary vector (pBI121) agrobacerial system.

MATERIALS AND METHODS

Plant Material

Certified seeds of tomato (*Lycopersicon esculentum* Mill. cv. Solan Vajr) were procured from the Department of Vegetable Crops, Dr Y.S. Parmar University of Horticulture and Forestry, Solan and grown in pots Ten days old seedlings were used as source of explant (cotyledons). The cotyledons were excised from the germinated seedlings and thoroughly washed with tap water and teepol. The explant was surface sterilized with 0.1% mercuric chloride for 2 min and then thoroughly washed with sterilized distilled water to remove the traces of mercuric chloride. The surface sterilized explant (cotyledons) was used for plant regeneration and genetic transformation studies.

Nutrient Media

Murashige and Skoog's (MS) medium supplemented with 100 mg/1 meso-inositol, 3% sucrose and gelled and with 0.8% agar-agar was used as basal medium[20]. Three different combination and concentration of benzyl amino purine-indole acetic acid and three different concentrations of naphthalene acetic acid were used in MS medium for shoot regeneration and root regeneration, respectively. The pH of medium was adjusted to 5.8 before adding agar-agar and sterilization. After mixing agar-agar, the medium was poured in culture vessels plugged and sterilized.

YMB medium was used for maintenance and preparation of fresh cultures of *Agrobacterium* strain. The pH of the medium was adjusted to 7.5 with 0.1N NaOH. Kanamycin, filter sterilizated using membrane filter of pore size 0.22 µm. was added into sterilized molten or liquid YMB medium. The *Agrobacterium* strain and information regarding its cultural conditions and gene construct were obtained from Dr D.P.S. Verma, Plant Biotechnology Centre, The Ohio State University, Columbus OH 43210, USA.

Plant Regeneration

Cotyledons were excised from ten days old seedlings and surface sterilized. The explants were first cut. into approximately 0.5–1.0 cm long sections and were cultured on MS medium supplemented with BAP and IAA. All the cultures were kept in a culture room at 26 ± 2°C with a 16 h photoperiod. The shoots regenerated from cotyledon explants were transferred to shoot elongation medium containing 1.25 mg/1 BAP. The elongated shoots were then transferred on HS medium supplemented with different concentrations of NAA for root regeneration. After 3-4 weeks, the rooted plantlets were transferred to small plastic pots containing a mixture of sand : soil (1:1) for hardening. The plantlets were kept covered with jam jars to maintain high relative humidity.

Plant Transformation

Agrobacterium strain and plasmid

Disarmed *Agrobacterium tumefaciens* LBA 4404 strain containing β-glucuronidase (gus) gene in binary vector pBI 121 system alongwith a kanamycin resistance gene (npt-II) for selection in both bacteria and plant was used for co-cultivation experiment to transfer gus and *npt-II* genes in tomato cells.

Maintenance of Agrobacterium strain

Agrobacterium strain was maintained by sub-culturing the bacterial colonies on the YMB medium containing 50mg/l kanamycin. After the proper growth of agrobacterial cells, the plates were kept at low temperature for storage.

Preparation of fresh culture of Agrobacterium

Fresh cultures of *Agrobacterium* strain were prepared by inoculating a small colony into the 10 ml of liquid YMB medium containing 50 mg/1 kanamycin (filter sterilized). These cultures were kept, overnight, at 28°C in orbital shaking incubator.

Pre-incubation of cotyledon explants

Cotyledon explants were cut into small pieces of 0.5 to 1.0 cm and inoculated on the shoot regeneration medium (on which high frequency shoot regeneration was obtained) for 24 h and then subsequently co-cultivated with *Agrobacterium* strain.

Co-cultivation of cotyledon explants

The fresh cultures of *Agrobacterium* were centrifuged for 10 min at 5000 rpm and supernatant was discarded to get a pellet. The pellet, was resuspended in MS liquid medium to get the concentration of 10^8 cells/ml. Cotyledon explants were immersed into bacterial suspension for 15–20 sec and then they were blotted dry on pre-sterilized filter paper and inoculated on the shoot regeneration medium for co-cultivation.

Preparation of selective regeneration medium

Kanamycin and cefotaxime were added into the shoot regeneration medium at concentration of 50 mg/1 and 500 mg/1, respectively after filter sterilization under aseptic conditions.

Transfer of cotyledon explants on selective medium after co-cultivation

After the co-cultivation period of 48 h, the cotyledon segments were transferred to the fresh selective shoot, regeneration medium containing kanamycin (50 mg/1) and cefotaxime (500 mg/1) for the selection of transformed cells and to inhibit the further growth of agrobacterial cells. All the cultures were kept in the culture room for further growth and differentiation. Cotyledon segments were subsequently subcultured to fresh selective shoot regeneration medium in order to check excessive agrobacterial growth.

β-Glucuronidase assays

Spectrophotometeric determination of β-glucuronidase activity in the plant tissues (non-transformed and transformed callus) was conducted[21]. Reaction mixture in total volume of 1 ml contained 1mM p-nitropheriyl β-D glucuronide and suitable volume of enzyme extract. The reaction mixture was incubated at 37°C for 60 min and was stopped by stop buffer. The absorption was measured at 415 nm using time point zero as blank.

RESULTS AND DISCUSSION

Plant Regeneration

Ten days old seedlings of tomato were used as source of explant (cotyledon) for regeneration and genetic transformation experiments. The cotyledon segments were cultured on MS medium supplemented with 2 mg/1 BAP and 0.5 mg/1 IAA, 2.5 mg/1 BAP + 0.25 mg/1 IAA and 1 mg/1 BAP + 0.5 mg/1 IAA. After one week of inoculation, the size of cotyledon explant had increased and a little callus initiation was seen at the cut ends of the explants. The colour of the explant had not changed. There was no change in the colour of the media. Shoot bud regeneration was observed on media containing different concentrations of BAP and IAA (Table 1). The shoot regeneration was observed along the edges of the explant after 20 d in culture. The percentage shoot regeneration was recorded and it was comparatively higher (53.33%) on the medium containing 2 mg/1 BAP + 0.50 mg/1 IAA (Table 1). The regenerated shoots did not elongate on the same medium and were transferred to MS medium containing lower level of cytokinine i.e. 1.25 mg/1 BAP for elongation. Dwivedi and his coworkers have also reported a prerequisite sequential exposure of explants to different level of morphogenetic stimuli[22]. Different concentrations of NAA were used in MS medium for root regeneration in *in vitro* developed shoots. Root regeneration was observed after 10–12 d and percentage of root regeneration was highest in medium containing 0.1 mg/1 of NAA (Table 2). The *in vitro* raised plantlets were then transferred to the pots containing a mixture of sand and soil (1:1) and acclimatized.

Table 1: Shoot regeneration from cotyledon explant of tomato

Medium	*Per cent shoot regeneration*
MS basal medium + 2 mg/1 BAP + 0.05 mg/1 IAA	53.33
MS basal medium +2.5 mg/1 BAP + 0.25 mg/1 IAA	46.67
MS basal medium + 1 mg/1 BAP + 0.50 mg/1 IAA	26.67

Table 2: Root regeneration in *in vitro* developed shoots of tomato

Medium	*Per cent root regeneration*
MS basal medium + 0.05 mg/1 NAA	80
MS basal medium + 0.10 mg/1 NAA	100
MS basal medium + 0.20 mg/1 NAA	75

In our plant regeneration and genetic transformation experiments, 10 d old seedlings were used as a source of cotyledon explant. The effect of age of donor seedlings on shoot induction from cotyledon or hypocotyl or leaf have been studied in different species[22-23] Of the various plant growth regulators tried in MS medium for shoot regeneration in tomato (cv. Solan Vajr) from cotyledon explants, the combination and concentration of BAP (2 mg/1) and IAA (0.50 mg/1) proved better than any other concentration. Different concentrations of NAA were used in MS medium for root regeneration in *in vitro* developed shoots. The percentage root regeneration was highest in the medium containing 0.10 mg/1 NAA. The concentration of different cytokinin and auxin in combination or alone required for shoot regeneration and root regeneration varies with the type of explant or cultivar used.[34–41]

Genetic Transformation Experiment

Cotyledons were excised from 10 d old seedlings of tomato, surface sterilized and used for genetic transformation studies. Cotyledon segments were pre-cultured on the shoot regeneration medium for 24 h. Pre-cultured explants were co-cultivated with *Agrobacterium tumefaciens* strain containing the binary vector pBI 121 alongwith the disarmed helper Ti-plasmid for 48 h and then transferred to the selective regeneration medium containing antibiotics. Cefotaxime and kanamycin were added into the medium to kill the agrobacterial cells and to select the transformed tomato cells. Callus initiation was seen after 20-25 d at the cut edges of the cotyledon segment, and also at the wound site where the tissue was damaged during inoculation .

Agrobacterium tumefaciens strain LBA 44O4 was infectious to tomato cells. After co-cultivation, the explants were able to grow on the selective medium, whereas, control explants were unable to grow on the selective medium. Transformed calli were selected on the medium containing 50 mg/1 kanamycin. Several kanamycin resistant calli were obtained and they were tested for β-glucuronidase enzyme assay.

Gene Expression in Transformed Calli

β-glucuronidase enzyme assay was carried out to confirm the transfer of gus gene from *Agrobacterium tumefaciens* into the genome of cells/tissue (calli) of tomato. The transfer of chimeric *gus* gene was confirmed with the presence of β-glucuronidase activity in transformed calli. Tomato calli were transformed with *Agrobacterium* binary vector containing transcriptional fusion of CaMV35S promoter with the coding region of gus. Several kanamycin resistant calli were obtained which showed β-glucuronidase activity. The enzyme activity was not observed in the non-transformed (control) calli of tomato (Table 3).

Table 3: β-glucuronidase activity (enzyme unit/g/min) in non-transformed (control) and transformed calli of tomato

Sample	*Enzyme activity (enzyme unit/g/min)*
Control callus (Non-transformed)	0.13
Transformed callus	175

0.001 change in O.D/min. = 1 enzyme unit of β-glucuronidase

Control of gene activity can be manifested at many levels. Analysis of mutationally altered genes in organisms accessible: to transformation techniques is greatly facilitated by the use of sensitive reporter enzymes. By using a reporter gene that encodes an enzyme activity not found in the organism being studied. The sensitivity with which chimeric gene activity can be measured is limited only by the properties of the reporter enzyme and the quality of the available assays for the enzyme. To date, a number of reporter genes have been used in studies of gene expression in higher plants[8]. The two most useful reporter genes have been the bacterial genes chloramphenicol acetyl transferase (CAT) and neomycin phosphotransferase-II (*npt*-II) which encodes enzymes with specificities not normally found in plant tissues[42-44]. However, both CAT and *npt*-II are relatively difficult, tedious and expensive to assay[45-46] Competing reactions catalysed by endogenous esterases, phosphatases, transferases and other enzymes also limit the sensitivity and make quantitations of CAT and *npt*-II by enzyme kinetics very difficult. Another reporter gene i.e. the firefly luciferase gene has been used as a marker in transgenic plants but the enzyme is labile and difficult to assay with accuracy[47,48].

β-glucuronidase is superior reporter gene system for plant transformation studies. Many plants assayed to date lack detectable β-glucuronidase activity providing a null background in which to assay chimeric gene expression. β-glucuronidase is easily, sensitively and cheaply assayed spectrophotometerically, fluorometerically in *in vitro* and can also be assayed histochemically to localize *gus* activity in plant cells and tissues. *gus* is very stable and tissue extracts continue to show high level of this activity after prolonged storage[21,49].

The genetically engineered *Agrobacterium tumefaciens* strain LBA 44O4 which we have used in the genetic trnasformation studies in tomato has two genes i.e. neomycin phosphotransferase-II (*npt-II*) and β-glucuronidase (gus) in a binary vector system under the control of NOS-promoter and 35S—promoter[49]. Transformed calli. were obtained on the selective medium containing 50 mg/l Kanamycin. The non-transformed (control) calli did not survive on the selective medium containing Kanamycin (50 mg/1). In this experiment, cloned β-glucuronidase gene in binary vector has been used successfully to monitor the expression of chimeric genes introduced into tomato cells via *Agrobacterium*-mediated gene transfer technique. The gene expression was reliably measured from a very small tissue (callus) of tomato.

Phenotypic and enzymatic data indicate that the chimeric β-glucuronidase gene was expressed in tomato cells (calli) using binary vector agrobacterial system. Tomato transformation technology in combination with advances in tomato plant regeneration (our present data on tomato plant regeneration) may eventually lead to the recovery of transgenic plants of tomato. These methodologies are simpler than the co-cultivation method in which plant protoplasts are transformed. The co-cultivation method of protoplast transformation is labour intensive and prone to contamination, whereas the methodologies presented here consist of a few simple steps that can be effectively performed in standard laboratories. Furthermore, somaclonal variation frequently results from protoplast culture, whereas organised tissue is known to remain more stable[50,51]. Finally, the markers would be of great value in transfer of agronomically important genes, since the presence of an easily identifiable marker linked to an agronomically desirable gene would permit efficient selection in tissue culture.

ACKNOWLEDGEMENTS

The authors are grateful to Professor D.P.S. Verma, Plant Biotechnology Centre, The Ohio State University, Columbus, USA for providing the genetically engineered *Agrobacterium* strain LBA 4404. We are also thankful to Dr. A.K.Nath for her help in biochemical estimation of the transformed callus. Financial support obtained from the ASPEE Agricultural and Research Development Foundation, Mumbai, in the form of research project to D.K. Srivastava and Anil Kant is also thankfully acknowledged.

REFERENCES

1. Gasser, C. S. and Fraley, R.T. *Science*, **244** (1989) 1299.
2. Day, A.G. and Lichtenstein, C.P. *Plant genetic transformation.* In: Plant Biotechnology (Pergamon Press, Oxford) (1992) 151.
3. Lindsey, K. *J Biotech* **26** (1992) 1.
4. Jain, S.H. *Current Science*, **64** (1993) 715.
5. Srivastava, D.K. Agrobacterium-mediated gene transfer in plants — A review. In: Trends in plant tissue culture and biotechnology. (1997) 17.
6. Srivastava, D.K. Biotechnological studies in some horticultural and forestry plants. In: Proceedings of Prof. A.R.Mehta, Feliciation Symposium—Down The Memory Lane of a Botanist. (1997), 55.
7. Srivastava, D.K. Plant genetic engineering in crop improvement. In: Proceedings of National Symposium on Biotechnology in Agriculture and Environment (1998) 21.
8. Singh, Z. and Sansavini, S. *Plant Breeding Reviews*, **16** (1998) 87.
9. McCormick, S., Niedermeyer, J., Barnason, A., Horsch, R. and Fraley, R. *Plant Cell Reports*, **5**(1986) 81.
10. Fillati, J.J., Kiser, J., Rose, R. and Comai, L. *Bio-Technology,* **5**(1987) 726.
11. Fischhoff, D.A., Bowdish, K.S., Perlak, F.J., Marsona, P.B., McCormick, S.M., Niedermeyer, J.G., Dean, D.A., Kusano, K., Mayer, E.J., Rocherter, D.E., Roger, S. G. and Fraley, R.T. *Bio-Technol*ogy, **5**(1987) 807.
12. Horsch, R.B., Fry, J.E., Hoffman, N.L., Wallroth, M., Eichholtz, D., Rogers, S.G. and Fraley, R.T., *Science*, **227**(1988) 1229.
13. Nelson, R.S., McCormick, S.M., Delannay, X., Dubi, P., Layton, J., Anderson, E.G., Kaniewska, M., Prakash, R.K., Horsch, R.B., Roger, S.G. Fraly, R.I. and Beachy, R.N., *International J Plant Sci.* **6** (1998) 403.
14. Hamilton, A.J., Lycett, S.W. and Griesson, D., *Nature*, **346** (1990) 284.
15. Norva, G.E., Kaliev, A.B., Andrinov, V.M. and Piruzyan, E.S., *Soviet Biotechnol.*, **2** (1991) 19.
16. Xue, B., Gonasalves, C., Provvindenti, R., Slighton, J.L., Fuchis, M. and Gonasalves, D., *Plant Disease*, **78** (1994) 1038.
17. Dong, C.Z., Jiang, C.X., Feng, L.X., Li, S.D., Gao, Z.H., Guo, J.Z. and Zhu, Dewei, *Acta Horticulture*, **402** (1995) 78.
18. Nagata, M., Mori, H., Tabei, Y., Sato, T., Hirai, M., Imseki, H., Nito, N., Looney, N.E., Nevins, D.J. and Halevy, A.H., *Acta Horticulture*, **394** (1995) 213.
19. Rudas, V.A., Piven, N.M., Rivkin, M.I., Vershinin, A.V. and Infante, D., *Tsitute Genetike,* **31** (1997) 17.
20. Murashige, T. and Skoog, F., *Physiologia Plantarum,* **15** (1962) 473.
21. Herman, L. and Depicker, A., EMBO course on Plant Molecular Biology (1987) 61.
22. Dwivedi, K., Srivastava, P., Verma, H.N. and Chaturvedi, H.C., *Indian J. Experimental Biol.* **28** (1993) 32.
23. Dai, C.X., Mertz, D. and Lambeth, V.N., Genetic manipulation in crop. *Newsletter*, **4** (1998) 26.
24. Kathal, R., Bhatnagar, S.P., and Bojwani, S.S., *Plant Cell Reports,* **7** (1968) 449.
25. Niedz, R.P., Smith, S.S., Dunbar, K.B., Stephans, C.T. and Murakishi, H.H., *Plant Cell Tissue and Organ Culture,* **14** (1989) 313.
26. Chi, G.L. and Pua, E.C., *Plant Science*, **6** (1987) 105.

27. Srivastava, D.K., Andrinov, V.M. and Piruzian, E.S., *Soviet J. Plant Physiol.* **35** (1988) 1243.
28. Srivastava, D.K., Andrinov, V.M. and Piruzian, E.S., *Plant Cell Reports,* **8** (1989) 300.
29. Srivastava, D.K., Andrinov, V.M. and Piruzian, E.S. In: Horticulture New Technologies and Applications. (Eds. R.L.M. Pierik and C. Prakash), Kluwer Academic Publisher. The Netherlands (1991) 127.
30. Eisner, G.I., Mar, Y.I. and Ya, S.M.F., Soviet Agricultural Sciences, **11** (1992) 15.
31. El-Farash, E.M., Abdalla, H.I., Taghian, A.S. and Ahmad, M.H., *Asso. J. Plant Sci*, **24** (1993) 3.
32. Hou, X.L., Li, X.L., Chun, S.C. and Cao, S.C., *Advances Horticulture* **1** (1994) 304.
33. Petrova, S. and Antonova, G., Cruciferae-Newsletter **18** (1996) 31.
34. Zelcer, A., Soferman, O. and Izhar, S., *Newsletter International Assoc., Plant Tissue Culture*, **40** (1983) 14.
35. Izadpanah, M. and Khosh, K.M., *Iran Agriculture Research Publication,* **8** (1989) 37.
36. Kageyama, K., Yabe, K. and Miyajima, Research Bulletin of the Aichi Ken Agricultural Research Centre, **20** (1988) 167.
37. Duzyaman, E., Tanrisever, A., Gunver, G., Cockshull, K.E., Tuzel, Y. and Gul, A., *Acta Horticulture*, **366** (1994) 235.
38. Davis, D.G., Breiland, K.A., Frear, D.S. and Secor, G.A., *Plant Growth Regulator Society of America Quarterly* **22** (1994) 115.
39. Lech, M., Miczynski, K. and Pindel, A., *Acta Societates Botanicorum Ploniae* **65** (1996) 53.
40. Lu, R.J., Hung, J.H., Sun, Y.F. and Zhou, R.M., *Acta Agriculturae Shanghai* **13** (1997) 16.
41. Plastira, V.A., Peridikaris, A. K., Actman, A. and Zir, M., *Acta Horticulture,* **447** (1997) 231.
42. Sevan, M.W., Flavell, R.V. and Chilton, M.D., *Nature,* **304** (1987) 184.
43. Fraley, R.T., Rogers, S.G., Horsch, R.B., Sanders, P.S., Flick,.J.S., Adams, S.P., Bittner, M. L., Brand, L.A., Fink, C.L., Fry, J.S., Gallupi, G.R., Goldberg, S.B., Hoffman, H.L. and Wood, S.C., *Proc. Natl. Acad. Sci.* (USA) **80** (1983) 4803.
44. Herrera-Estrella, L., Depicker, A., Van-Montagu, M. and Schell, J., *Nature*, **303** (1983) 209.
45. Gorman, C.M., Moffatt, L.F. and Howard, B.H., *Mol. Cell Biol.* **2** (1982) 1044.
46. Reiss, B., Sprengel, R., Will, H. and Schaller, H., *Gene* **30** (1984) 213,
47. Ow, D.W., Wood, K.V., DeLuca, M., Devid, J.R., Heliuski, D.R. and Howell, S.H., *Science,* **234** (1986) 856.
48. Deluca, M. and McElroy, W.D., *Method Enzymol.* **57** (1978) 3.
49. Jefferson, R.A., Kavanagh, T.A. and Bevan, M.W., *The EMBO Journal*, **6** (1987) 3901.
50. Shepard, J.F., Bindey, D. and Shahin, E., *Science*, 208 (1980) 17.
51. Horsch, R.B., Fry, J.E., Hoffman, N.L., Wallroth, M., Eichholtz, D., Rogers, S.G. and Fraley, R.T., *Science*, **227** (1985) 1229.

Plant Regeneration and Genetic Transformation Studies in Cabbage (*Brassica oleracea* L. var. *capitata*)

S. SHARMA *and* D.K. SRIVASTAVA

Department of Biotechnology
Dr. Y.S. Parmar University of Horticulture & Forestry
Nauni-Solan 173 230 H.P.

ABSTRACT

In order to establish a model system for introduction of foreign genes into cabbage (*Brassica oleracea* L. var. *capitata* cv. Pride of India) plants by *Agrobacterium*-mediated transformation, hypocotyls were used as an explant and bacterial neomycin phosphotransferase-II (npt-II) and β-glucuronidase (gus) genes were used as a selection marker and a reporter gene, respectively. Plantlets were regenerated from the hypocotyl explants. Hypocotyl segments obtained from 9 d old aseptically grown seedlings were co-cultivated with genetically engineered *Agrobacterium tumefaciens* LBA 4404 containing npt-II and gus genes. The putative transformed cells/tissues were identified by their growth on the selective medium containing kanamycin. The putative transformed calli obtained on the selective medium were GUS positive. Using this technique, chlorophyll deficient transgenic shoots of cabbage were recovered which were able to grow on the selective MS medium containing 50 mg/l kanamycin. A protocol for plant regeneration and genetic transformation in cabbage cells/tissue have been standardized.

INTRODUCTION

Cabbage is one of the most popular and widely grown cole crops in the world. The introduction of specific trait(s) into cultivated plants is especially desirable when species are incompatible with conventional methods. Hybrids obtained by embryo culture are limited to related species. Protoplast fusion poses problems such as the selection of fused cells and their subsequent regeneration[1]. Even if these barriers are overcome, undesirable genes may also be introduced from the donor parent, necessitating repeated backcrossing. Gene transfer systems using *Agrobacterium tumefaciens* and *Agrobacterium rhizogenes* have been explored[2,3]. The application of genetic engineering in cabbage cultivation is of great value to obtain new improved and desirable traits like insect and disease resistant varieties. Genetic transformation is thus expected to provide new possibilities to increase the genetic variability in crops. However, before undertaking this approach, it is important to know whether, the somatic cells of the cultivar (Pride of India) are able to regenerate in such a way, so as to give rise to whole plants and to standardize the protocol of genetic transformation in cabbage cells/tissues. *In vitro* regeneration through explant culture has an advantage over callus, protoplast and cell suspension cultures as cytological aberrations (somaclonal variations) are quite common in such cultures. Genetic manipulation of crop plants *in vitro* is dependent on the reproducible regeneration of plants from cultured tissues. A variety of explants showing morphogenetic ability have been used for *Agrobacterium*-mediated transformation of Brassica *oleracea*. A number of successful transformation in cabbage have been achieved[4-11].

The research described in this paper was initiated to establish a protocol for the regeneration of cabbage plants (cultivar: Pride of India, which is widely grown in Himachal Pradesh) from hypocotyl explant and genetic transformation and expression of neomycin phosphotransferase-II (npt-II) and β-glucuronidase (gus) genes in the transformed calli using the binary vector (pBI121) in *Agrobacterium tumefaciens* LBA 4404.

MATERIALS AND METHODS

Plant Material

Certified seeds of cabbage (*Brassica* oleracea L. var. capitata cv. Pride of India) were procured from the Department of Vegetable Crops, Dr. Y.S. Parmar University of Horticulture and Forestry, Solan and thoroughly washed with tap water and teepol and soaked for an hour. They were then surface sterilized with 0.1% mercuric chloride ($HgCl_2$) for 2 min and then rinsed 3-4 times with sterilized distilled water to remove the traces of $HgCl_2$. The seeds were then inoculated on MS half strength basal medium containing 0.5% sucrose for seed germination. Nine days old aseptically grown seedlings were used as a source of explant (hypocotyl) for plant regeneration and genetic transformation studies in cabbage.

Nutrient Medium

Murashige and Skoog's (MS) medium supplemented with 100 mg/l meso-inositol, 3% sucrose and gelled with 0.8% agar agar as the basal medium was used for the present investigation[12].

Two different concentrations and combinations of indole acetic acid-kinetin and napthalene acetic acid-benzyl aminopurine, were used in the MS medium for plant regeneration. The pH of the medium was adjusted to 5.8 before adding agar agar and sterilization. After mixing the agar, the medium was poured in culture vessels, plugged and sterilized.

YMB medium was used for the maintenance and preparation of fresh cultures of *Agrobacterium* strain. The pH of the medium was adjusted to 7.5 with 0.1 N NaOH. Kanamycin was added into the sterilized molten or liquid YMB medium by filter sterilization using membrane filter of pore size 0.22 μm. The *Agrobacterium* strain and informations regarding its cultural conditions and gene construct were obtained from Dr. D.P.S. Verma, Plant Biotechnology Centre, The Ohio State University, Columbus, USA.

Plant Regeneration

Hypocotyl explants were excised from nine days old aseptically grown seedlings. The explants were first cut into approximately 0.5–1.0 cm long sections and were cultured on MS medium supplemented with BAP-IAA and Kinetin-IAA. The regenerated shoots from the hypocotyl explants were transferred on MS medium supplemented with IBA for root regeneration. After 3-4 weeks, the rooted plantlets were transferred to small plastic pots containing mixture of sand:soil (1:1) for hardening. The plants were kept covered with jam jars to maintain high relative humidity. All the cultures were kept in a culture room at 26 ± 2°C with a 16 h photoperiod.

Plant Transformation

Agrobacterium strain and plasmid

Genetically engineered *Agrobacterium tumefaciens* LBA 4404 strain containing β-glucuronidase (gus) gene in binary vector (pBI121) system along with a kanamycin resistance gene i.e. neomycin phosphotransferase-II (npt-II) for selection in both bacteria and plant was used for co-cultivation experiment to transfer gus and npt-II genes in cabbage cells.

Maintenance of Agrobacterium strain

Agrobacterium strain was maintained by sub-culturing the bacterial colonies on the YMB medium containing 50 mg/1 kanamycin. After the proper growth of the agrobacterial cells, the plates were kept at low temperature for storage.

Preparation of fresh cultures of Agrobacterium

Overnight fresh cultures of *Agrobacterium* strain were prepared by inoculating a small colony into the 10 ml liquid YMB medium containing 50 mg/1 Kanamycin (filter sterilized). These cultures were kept overnight at 28°C in orbital shaking incubator.

Pre-incubation of hypocotyl explants

Hypocotyl explants were cut into small pieces of 0.5-1.0 cm and inoculated on the shoot regeneration medium (on which high frequency shoot regeneration was obtained) for 24 h and then subsequently co-cultivated with *Agrobacterium* strain.

Co-cultivation of hypocotyl explant

The fresh cultures of *Agrobacterium* strain were centrifuged for 10 min at 5000 rpm and the supernatant was discarded to get, a pellet. This pellet was resuspended in MS liquid medium to get the concentration of 108 agrobacterial cells/ml. This bacterial suspension (from fresh cultures of *Agrobacterium*) and pre-cultured hypocotyl explants on regeneration medium were used for co-cultivation experiment. Hypocotyl explants were infected with agrobacterial cells (bacterial suspension) for 15-20 sec and then they were blotted dry on pre-sterilized filter paper and inoculated on the shoot regeneration medium for co-cultivation.

Preparation of selective shoot regeneration medium

Kanamycin and cefotaxime were added into the shoot regeneration medium at concentration 50 mg/1 and 500 mg/1, respectively and sterilized by filter sterilization. The whole assembly of the filter unit and the syringe were sterilized before using and the antibiotics were added under aseptic conditions.

Transfer of hypocotyl explants on the selective shoot regeneration medium after co-cultivation

After the co-cultivation for 48 h, the explants were transferred to the fresh selective shoot regeneration medium containing kanamycin (50 mg/1) and cefotaxime (500 mg/1) for the selection of transformed cells and to inhibit further agrobacterial growth. All the cultures were kept in the culture room for further growth and differentiation. Explants were subsequently subcultured to fresh selective shoot regeneration medium in order to check excessive bacterial growth.

β-glucuronidase assay

Spectrophotometric determination of β-glucuronidase activity in the plant tissue (transformed callus and non-transformed callus) was conducted[13]. Reaction mixture in a total volume of 1 ml contained 1 mM p-nitrophenyl β-D-glucuronide and suitable volume of enzyme extract. The reactants were incubated at 37°C for 60 min and was stopped by stop buffer. The absorption was measured at 415 nm using the time point zero as blank.

RESULTS AND DISCUSSION

Plant Regeneration

Aseptically grown 9 d old seedlings of cabbage were used as a source of explant (hypocotyl) for regeneration and genetic transformation studies. The explants were cut into small pieces of 0.5–1.0 cm in size and then cultured on MS medium supplemented with 2 mg/1 BAP + 0.25 mg/1 IAA and 1.5 mg/1 Kn + 0.75 mg/1 IAA, respectively. The size of the hypocotyl explants increased after about one week in culture. Callus formation was seen around the cut ends of the explant. Shoot buds and shoots started originating from the tissue around the cut edges. Shoot bud initiation was observed after 15–20 d in culture. The average number of shoots formed per explant was recorded and it was comparatively higher on the medium containing 2 mg/1 BAP + 0.25 mg/1 IAA (Table 1). Shoot elongation occurred on the same medium.

Table 1: Shoot regeneration frequency of cabbage (*Brassica oleraceae* L. var. *capitata* cv. Pride of India) from hypocotyls explant

Medium composition	*Average No. of shoots/explant*
MS + 2 mg/1 BAP + 0.25 mg/1 IAA	6.00
MS + 1.5 mg/1 Kn + 0.75 mg/1 IAA	5.66

Elongated shoots (about 2-3 cm in length) were excised and cultured on MS medium supplemented with different concentration of IB A (Table 2). Root regeneration from *in vitro* developed shoots was observed after 7-8 d in culture. The number of shoots forming roots was comparatively higher on the MS medium containing 0.20 mg/1 IBA (Table 2). Cabbage plantlets were able to regenerate within 2 months. The young plantlets were transferred to pots containing mixture of sand and soil (1 : 1) and acclimatized.

Table 2: Root regeneration in *in vitro* developed shoots from hypocotyl explants

Medium composition	*No. of shoots forming roots(%)*
MS + 0.05 mg/1 IBA	81.88
MS + 0.20 mg/1 IBA	90.90

In our experiments 9 d old aseptically grown seedlings were used as a source of explant for plant regeneration and genetic transformation studies. The effect of age of donor seedlings on shoot induction from cotyledon, or hypocotyl or leaf have been studied in different plant species[14–20].

Of the various plant growth regulators tried in MS medium for shoot regeneration in *Brassica oleracea* var. *capitata* hypocotyl explants, combinations of BAP and IAA proved superior to any other growth regulator combinations. Different concentrations of IBA were supplemented in MS medium for root regeneration of *in vitro* developed shoots. Root regeneration was observed after 7-8 d in culture, the percentage root regeneration was highest in the medium containing 0.20 mg/1 IBA. The concentration of different cytokinins and auxins in combination or alone required for shoot regeneration and root regeneration varies with the type of explant or cultivar used[21–24].

Genetic Transformation

Hypocotyl explants were excised from aseptically grown 9 d old seedlings of cabbage and were used for the genetic transformation studies. Hypocotyl segments were pre-cultured on the shoot regeneration medium for 24 h, before carrying out the co-cultivation experiment. Pre-cultured explants were co-cultivated with *Agrobacterium tumefaciens* strain containing the binary vector pBI 121 along with the disarmed helper Ti-plasmid for 48 h and then transferred to the fresh selective shoot regeneration medium containing antibiotics. Cefotaxime (500 mg/1) and kanamycin (50 mg/1) were added into the medium to kill the agrobacterial cells and to select the transformed cabbage cells, respectively. Callus formation was seen after 15-20 d at the cut edges of the hypocotyl segments and also at the wound site where the tissue was damaged during inoculation. The non-transformed (control) tissue did not survive on the selective medium containing kanamycin. Hypocotyl segments with developing calli were transferred after 2-3 weeks to fresh selective medium. The developing calli from the hypocotyl segments were tested for β-glucuronidase enzyme assay.

Gene Expression in Transformed Callus

β-glucuronidase enzyme assay was carried out to confirm the transfer of the *gus* gene from *Agrobacterium tumefaciens* into the genome of cells/tissue of cabbage. The transfer of the gus gene was confirmed with the presence of β-glucuronidase activity. Chimeric gus gene was expressed in transformed calli of cabbage. Cabbage cells/tissue were transformed with *Agrobacterium* binary vector containing transcriptional fusion of CaMV35S promoter with the coding region of gus25. Several kanamycin resistant calli were obtained which showed β-glucuronidase activity. The enzyme activity was not observed in the non-transformed calli of cabbage (Table 3).

Table 3: β-glucuronidase activity (Enzyme unit/g/h) in non-transformed an transformed tissue of cabbage

Sample	*Enzyme Activity (enzyme unit/g/h)*
Control tissue (Non-transformed tissue)	0.13
Transformed callus	3.45

0.001 change in O.D./min = 1 enzyme unit of β-glucuronidase

The transformed calli which were cultured on the selective shoot regeneration medium gave rise to transgenic shoots after 4 weeks in culture. The transgenic shoots were chlorophyll deficient and were able to grow on the selective medium containing kanamycin.

The genetically engineered *Agrobacterium* strain, containing npt-II and gus genes has been successfully used to transfer the chimeric genes into cabbage cells. The gene expression was reliably measured from a very small callus of cabbage. Phenotypic and enzymatic data indicate that the chimeric *npt*-II and *gus* genes were expressed in cabbage cells (callus) using the binary vector agrobacterial system. During the genetic transformation studies, chlorophyll deficient shoots were obtained. Morphological variants and chlorophyll mutant shoots/plants were also obtained in *Brassica* species[26–28]. Changes at morphological level and chlorophyll deficient mutant are expected due to the insertion of foreign gene in plant genome. The genomic alterations include the events of chromosomal rearrangements and insertion mutagenesis. The events of genomic changes after the insertion of T-DNA in plant cells lead to the occurrence of morphological variation and formation of mutants. *Agrobacterium*-mediated transformation may result in the insertion of T-DNA in regulatory or coding regions of functional genes thus giving rise to insertional mutants.

A protocol for plant regeneration from hypocotyl explants and genetic transformation in cabbage cells have been standardized. Cabbage transformation technology in combination with advances in cabbage plant regeneration, may eventually lead to the recovery of transgenic cabbage plants.

ACKNOWLEDGEMENT

We are grateful to Prof. D.P.S. Verma, Plant Biotechnology Centre. The Ohio State University, Columbus, USA for providing us the *Agrobacterium* strain LB A4404. We are thankful to Dr. A.K. Nath for her help in biochemical estimation of the transformed tissue. The financial assistance obtained from the University to Shalini Sharma is also thankfully acknowledged.

REFERENCES

1. Horsch, R.B., Fry, J.E., Hoffman, N.L., Wallroth, M., Eichholtz, D., Rogers, S.G. and Fraley, R.T., *Science* **227** (1995) 1229.
2. Barton, K.A., Binns, A.N., Matzke, A.J.M. and Chiltan, M.D., Cell **32** (1983) 1033.
3. David, C. and Tempe, J. *Plant Cell Reports*, **7** (1988) 88.
4. Shahin, E.A. and Yashar, M., *Crucifer genetics workshop III Proceedings,* May 29-30 (1986) 41.
5. Christey, M.C. and Earle, E. D, *Cruciferae Newsletter* **13** (1988) 94.
6. He, Y.K., Gong, Z.H., Wang, F., and Wang, M., *Chinese J. Biotechnology* **7**: (1991) 382.
7. Bai, Y.Y., Mao, H.Z., Cao, X.L., Tang, T., Wu, D., Chen, D.D., Li, W.G., Fu, W.J., You, C.B. and Ding, Y., ed. *Current Plant Science* **15** (1992).
8. Berthomieu, P., Beclin, C., Chariot, F., Dare, C. and Jouanin, L., *Plant Science Limereck*. **96** (1994) 223.
9. Metz, T.D., Dixit, R. and Earle, E.D., *Plant Cell Reports* **15** (1995) 287.
10. Beclin, C., Chariot, F., Bolton, E., Jouanin, L. and Dore, C. Transgenic Research, **2**: (1998) 48.
11. Zhang, F.L., Takahata, Y. and Wattanabe, M. *Plant Cell Reports*, **19** (2000) 569.
12. Murashige, T. and Skoog, F., *Physiologia Plantarum* **15** (1962) 473.
13. Herman, L. and Depicker, A., *EMBO Course on Plant Molecular Biology* (1987) 61.
14. Kathal, R., Bhatnagar, S.P. and Bhojwani, S.S., *Plant Cell Reports* **7** (1988) 449.
15. Chi, G.L. and Pua, E.C. *Plant Science* **64** (1989) 243.
16. Srivastava, D.K., Andrinov, V.M. and Piruzian, E.S. *Soviet J. Plant Physiology* **35** (1988) 1243. (In Russian)
17. Srivastava, D.K., Andrinov, V.M., and Piruzian, E.S. *Plant Cell Report*, **8** (1989) 300.
18. Srivastava, D.K., Andrinov, V. M., and Piruzian, E.S. In: Horticulture-New Technologies and Applications. (Eds. J. Prakash and RLM Pierik). Kluwer Academic Publisher. *The Netherlands* (1991) 127.

19. Eisner, G.I., Mar Yakhina, I. and Ya Shengakin, M.F. *Soviet Agricultural Sciences* **11**: (1992) 15.
20. Petrova, S. and Antonova, G., *Cruciferae-Newsletter* **18** (1996) 31.
21. Koda, T., Ichi, T., Yamagislli, H. and Yoshikawa, H., *Agricultural and Biological Chemistry* **52**: (1998) 2387.
22. Cao, M.Q., Chariot, F. and Dore, C. *Academie-des-science-series* 3, **310**: (1990) 203.
23. Hou, X.L., Li, X.L., Chun, S.C. and Ceo, S.C. *Advances in Horticulture* **1** (1994) 304.
24. Zhang, S., Wen, F.J., Wei, Y.T. and Pu, L.H. *Cruciferae Newsletter* **19** (1997) 51.
25. Jefferson, R.A., Kananagh, T.A. and Beuan, M.W. *The EMBO Journal* **6** (1987) 3901.
26. Srivastava, V., Reddy, A.S. and Guha Mukherjee, S., *Plant Cell Reports*, **7** (1988) 504.
27. Mukhopadhyay, A., Arumugum, N., Nandakumar, P.B.A., Pradhan, A.K., Gupta, V. and Pental, D., *Plant Cell Reports*, **11** (1992) 506.
28. Dixit, S., Srivastava, D.K., and Sharma, D.R. In: Proceedings of National Symposium on Biotechnology in Agriculture and Environment (25-26 March) Eds. S.S. Marwaha, J.K. Arora and R.K. Grover, Abstract A.2 (1998).

Distribution of *Bacillus Thuringiensis* in Punjab Soil and their Toxicity Against *Chilo Partellus*

V. KHANNA, S.S. MARWAHA*,

and

H.S. DHALIWAL

Department of Genetics and Biotechnology
Punjab Agricultural University
Ludhiana
** Punjab State Council for Science and Technology*
Chandigarh

ABSTRACT

Ten crystal positive isolates *of Bacillus thuringiensis* were ascertained out of 125 isolates obtained from ten samples of soil. Though the crystal number varied between the isolates, all isolates showed the presence of bipyramidal crystals specific to the lepidopteran pathotype. Two of the local isolates showed high mortality against larvae of *Chilo partellus,* and their insecticidal potential ranged between 78%-95% with early stage larvae. A decline in susceptibility was observed with the increase in larval age.

INTRODUCTION

Many strains of the gram positive, sporulating bacterium *Bacillus thuringiensis* are entamopathogens of economic importance because of their toxicity to the larvae of insect pests. The toxicity of *B. thuringiensis* preparations is due to delta-endotoxins, a protein in the sporulating cell as a phase retractile inclusion. This attribute has led to the development of bioinsecticides based on *Bacillus thuringiensis* for control of certain insect species among the orders Lepidoptera, Diptera and Coleoptera[1]. There are more reports of *B. thuringiensis* isolates active against other insect orders Hymenoptera. Homoptera, Orthoptera and Mallophaga and against nematodes, mites and protozoa[2]. Therefore. the identification of novel *B. thuringiensis* strains and insecticidal genes provides a basis for development of new biopesticides. Systematic research is a routine matter in many laboratories to explore and develop their own natural biopesticide resources in crop protection, thereby ensuring no threat to themselves, the environment and the consumer.

MATERIALS AND METHODS

Ten soil samples were collected from different locations and *B. thuringiensis strains* were isolated by the modified method of earlier workers[3]. One gram of soil was suspended in 10 ml of Luria Bertani medium in 0.25M sodium acetate which selectively inhibits the germination of *B. thuringiensis* spores and incubated at 30°C for 4h under shake conditions. Small volumes of this suspension were then heat treated at 80°C for 5 min. Aliquotes of serial dilutions were then plated on T_3 medium (0.3% tryptone, 0.2% tryptose. 0.15% yeast extract 0.0005% $MnCl_2$, 0.05% sodium phosphate, 1.5% agar. pH 6.8) and incubated at 30°C + 2°C for 48h. Individual creamish colonies with a matte texture were selected at random and examined under Phase Contrast microscope for the presence of spores and crystals.

Insect toxicity assay

Acetone dried spore-crystal (SC) mixture[4] from ten local and two standard isolates of *B. thuringiensis* were suspended in CMC-Triton X 100 solution (1% CMC in 0.01% Triton X 100) to a final concentration of 1.0 mg/ ml. The above suspensions were used to treat maize stalks/leaves (6-8 cm in length / 15-20 g). The treated maize stalks were air dried for 30 min and incubated in battery jars with 50 newly hatched larvae of *Chilo partellus.* After covering with muslin cloth each jar was kept at ambient temperature of 30°C in dark. Maize stalks treated similarly but with CMC-Triton X solution only, served as control.

To determine the relative toxicity of potential Bt isolates viz-a-viz standard strain HD-1, the SC-suspension was diluted ten-fold to obtain final concentration ranging between 1.0-0.01 mg/ml and these were used to coat maize stalks/leaves.

The observations on larvae for development of toxicity symptoms were made 12h after their release and larval mortality recorded after 24-48h. To study the effect of larval age on % mortality by SC-mixtures, 72h old larvae were allowed to feed on maize leaves treated with varying concentrations of 5.0, 1.0 and 0.1 mg/ml of SC-mixtures.

RESULTS AND DISCUSSION

Of the ten soil samples collected eight tested positive for *B. thuringiensis* (Table 1). Out of the 125 randomly picked *B. thuringiensis* type colonies, 8% of the isolates were found to be crystal positive although the crystal number varied from isolate to isolate. Moreover, all showed the presence of bipyramidal lepidopteran pathotype specific crystals. Numerous studies have found that soil and grain dust are environments in which *B. thuringiensis* is abundant[5]. In a collection of samples surveyed from 80 countries for *B. thuringiensis*, grain dust and soil were reported to be the richest habitats[6]. The natural occurrence of *B. thuringiensis* in the environment is evident from the presence *of B. thuringiensis* in locations unlikely to have been exposed to commercial *B. thuringiensis* formulation sprays.

Table 1: Isolation of crystal forming Bt-isolates from soil

Location	*No. of colonies Examined**	*No. of crystal forming isolates***
Sanet	15	2
Ludhiana	20	1
Barewal	8	2
Threeke	10	—
Barnala	12	1
Haibowal	15	1
Gobindgarh	8	—
Khanna	10	1
Lalton	15	—
Sahnewal	12	2
Total	125	10

* Showing characteristics typical to Bt
** Phase contrast microscopic examination

Insecticidal activity of local *B. thuringiensis* isolates

In vitro toxicity of SC-mixtures from 10 local isolates when compared with two standard strains HD-1 and HD-73, against neonate larvae *of Chilo partellus*

Table 2: Relative toxicity of different concentrations of SC-suspension of B*t*-isolates on neonate *C. partellus*

Isolate/Strain No.	*Larval mortality (%)***		
	SC-Concentration (mg/ml)		
	1.0	*0.1*	*0.01*
B*t* var. kurstaki HD-1*	88	74	50
Bt var. kurstaki HD-73*	88	82	58
B*t*C-1	50	44	21
B*t*C-2	87	72	31
B*t*C-4	55	40	26
B*t*C-5	97	82	49
B*t*C-6	82	76	24
B*t*C-7	52	40	21
B*t*C-8	86	74	21
B*t*C-9	98	94	57
B*t*C-10	95	80	36
B*t*C-11	95	80	28

* Standard strains; ** Control treatment: 2% mortality

showed percent mortality ranging between 50-95% at SC-conc. of 1.0 mg/ml (Table 2). Four local isolates BtC-5, BtC-9, BtC-10 and BtC-11 showed more than 95% mortality. Of these isolates BtC5 and BtC9 showed 82% and 94% mortality even at 0.1 mg/ml SC-concentration. The affected larvae showed toxicity symptoms as weakening, sluggishness and stoppage of feeding within 12h as compared to larvae in control treatment which remained healthy and active with extensive damage to maize stalks. Many reports have indicated paralysis of mouth parts as the first effect of ingestion of Bt toxin leading to non-feeding in later stages and coupled with toxicity due to activated toxin in the mid-gut resulting in larval mortality[7].

Decrease in susceptibility towards chemical insecticides has been reported with increase in age or instar stage in many insect larvae. Similar observations have been recorded with a few insects exposed to Bt-based formulations [8]. The results in Table 3 & 4 revealed a comparatively low per cent mortality of 48-72h old larvae as compared to the corresponding values of neonate larvae (Table 2). It was also observed that higher dose concentration (5 mg/ml) of SC-mixture was required to achieve same level of mortality as was observed on newly hatched larvae. Toxicity symptoms and percent mortality were time dependent as was reflected in the difference in mortality rate at 24h and 48h of contact (Table 4).

Table 3: Relative toxicity of selected B*t*-isolates against *C.partellus* larvae (24 - 48h old)

	*Larval mortality (%)***		
	SC-concentration (mg/ml)		
Isolate No.	*1.0*	*0.1*	*0.01*
B*t* var. kurstaki HD-1*	82	66	31
B*t*C-5	88	76	39
B*t*C-9	88	78	42
B*t*C-10	76	65	28

*Standard strain; **Control treatment: 4% mortality

Table 4: Relative toxicity of selected *Bt* isolates against *C.partellus* larvae (72h old)

	*Larval mortality (%)***					
	SC-concentrtion (mg/ml)					
	5.0		*1.0*		*0.1*	
Isolate No.	*24h*	*48h*	*24h*	*48h*	*24h*	*48h*
B*t* var. kurstaki HD-1*	82	91	74	82	68	69
B*t*C5	88	94	76	82	70	72
B*t*C9	88	97	78	85	71	74

*Standard strain; **Control treatment: 4% mortality

The various effects induced by Bt toxicity like, inhibition of feeding, sluggishness, stunning of growth etc.are characteristics of delta-endotoxin protein[9]. The decline in larval mortality with increase in larval age suggest a need for effective management of the pest larvae at proper stage of their appearance. These observations can be useful for exploitation of local Bt-isolates for the control of *C. partellus*.

REFERENCES

1. Beegle, C.C. and Yamamoto, T. *Can. Entomol.,* **124** (1992) 587.
2. Feitelson, J.S., Payne, J. and Kim, L. *Biotechnology.* **10** (1992) 271.
3. Travers, R.S., Martin, P.A. and Reichelderfer, C.F., *Appl. Environ. Microbiol.,* **53** (1987) 1263.
4. Dulmage, H.T., Correa, J.A. and Martinez, A.J. *J. Invert. Pathol.*, **15** (1970) 15.
5. Meadows, M.P. (ed.). *Bacillus thuringiensis an environment biopesticide: theory and practice,* Wiley, New York, 1993.
6. Bernhard, K.P., Jarret, M., Meadows, M., Butt, J., Ellis, D.J., Roberts, G.M., Pauli, S., Rodgers, P. and Burges, H.D. *J. Invert. Pathol.,* **70** (1997) 59.
7. Aronson, A.I., Beckman, W. and Dunn. *Microbiol. Rev.,* **50** (1980) 1.
8. Miranpuri, G.S., Ubhi, M.S., King, M. and Khachatourians, G.G. *Proc. Natn. Symp. "Recent Adv. in Integr, Pest Management"* Punjab Agricultural University, Ludhiana, Oct. 12-15 (1992) 84.
9. Luthy, P. and Ebersold, H.R. *Pathogenesis of Invertebrate Microbial Diseases,* Allenheld Osmum, Totawa, N.J., (1981) 235.

Liquefying and Saccharifying Amylases *of Bacillus* sp. SN-1 and *Aspergillus oryzae* HS-3 and their Application in Hydrolysis of Starchy Biomass for Ethanol Production

S.K. SONI*, H. SINGH, P. VARMA
and
J.K. GUPTA

Department of Microbiology
Panjab University
Chandigarh 160014

ABSTRACT

Bacillus sp. SN-1 and *Aspergillus oryzae* HS-3, obtained after an extensive screening of natural habitats, proved to be suitable for producing very high titers of thermostable α-amylase and glucoamylases in solid state fermentation (SSF) of inexpensive and simple agricultural substrates. Rice bran moistened with 1.5 parts of distilled water and supplemented with 0.5% glucose, 0.5% soyabean meal and 0.01M $CaCl_2.2H_2O$ supported the highest production α-amylase (75,000 U/g of original rice bran) by *Bacillus* sp SN-1 after 48 h incubation at 37°C. On the other hand wheat bran moistened with 1.5 parts of distilled water and supplemented with 1.0% lactose, 1.0% soyabean meal and 1mM $CaCl_2.2H_2O$ induced the maximum fungal glucoamylase production (3460 U/g of original wheat bran) after 96 h incubation at 37°C.

The kinetic characteristics of the α-amylase revealed maximum activity at 60°C, pH 6.5 and was very stable at this temperature, retaining 94% of its original activity after 7 h. The glucoamylase showed its maximum activity at 50°C and a pH range of 6.0-7.0 with half life of 6 h. The bacterial and fungal amylase preparations exhibited very high liquefying and saccharifying abilities, respectively, when evaluated for the hydrolysis of 15% soluble starch gels. These worked very well in the hydrolysis of wheat mashes for alcohol fermentation and resulted in significant mashing and fermentation efficiencies.

INTRODUCTION

Production of industrial enzymes has shown a spectacular rise in the last three decades. Before 1965 the size of the enzyme market was only a few million dollars. With the introduction of detergent proteases, enzyme sales grew to about a hundred million dollars within a few years. The enzymes for making high fructose com syrup were introduced between 1972–74. This application became almost as large as the use of enzymes in detergents. From 1974 to 1986, no major new enzyme applications emerged. Still, the enzyme market grew at an average rate of 10-15%. In 1983, the estimated sales of all industrial enzymes world wide was US $400 million . The size of the enzyme market in 1989 was about $600 million. Of which 90% were thermostable enzymes used by the detergent and starch industries.[1] The estimated value of sales of all industrial enzymes in 1995 was $1 billion including some $100 million assessed for the 'inhouse' processing. By 2005, the whole world market for industrial enzymes is expected to be in the range of $1.7 and $2 billion.[2] There are now 12 major producers of enzymes and 400 other minor companies which take care of the demand of enzymes in the world. Around 60% of the world's total supply of industrial enzymes is produced in Europe. The two large producer countries of enzymes are Denmark and Holland. At least 75% of all industrial enzymes are hydrolytic in action and proteases remain the dominant type because of their huge use in the dairy (coagulants) and detergent industries and collectively these account for approximately 40% of all enzyme sales. Carbohydrases used in baking, brewing, distilling, starch and textile represent the second largest group. In 1983, there were some 30 different enzyme types in common commercial use. This has already more than doubled and is being added to at an increasing rate.[2,3]

In contrast to extensive uses of microbial enzymes in the processing of food and other purposes in the world, the status of these in India is in its infancy. The use of microbial enzymes in food industry is confined to few cases in India but the details of these applications is not known precisely due to the trade secrets. Bacterial α-amylase and fungal glucoamylase are used in Indian starch processing industry for the production of dextrose, liquid glucose and potable alcohol. These two enzymes have been used in fermentation of potable alcohol from tapioca flour, sorghum, corn, wheat, rice, etc. by employing simultaneous saccharification and fermentation of the prehydrolysed slurry. The use of starchy biomass for the production of potable alcohol is gaining significance in the present day scenario due to rise in molassses costs as a result of Government's decontrol policy as well as due to deteriorating quality of molasses due to technological advancements for sugar recovery in Indian sugar industry. This stresses upon the need to develop a process for inhouse production of amylases so as to lower the cost of starch hydrolysis.

Amylases have traditionally been obtained from submerged cultures of bacteria, yeasts and molds because of ease of handling and greater control of environmental factors such as temperature and pH. Solid state fermentation (SSF) constitutes an interesting alternative and is gaining significance since the metabolites so produced are concentrated and purification procedures are less cumbersome and expensive. Some of the advantages of SSF over submerged cultures especially for work involving fungi are simplicity of equipment and low moisture content which prevents bacterial contamination.[4,5,6] In addition less solvent is needed to extract the enzymes which greatly reduces the energy requirements and pollution problems.[7] Over the years there have been several reports on the production of amylases by solid state fermentation.[8,9,10,11]

The present study has been undertaken to obtain a bacterial and a fungal isolate for the production of thermostable liquefying and saccharifying amylases respectively by solid state fermentation technique using simple and low cost production media and evaluation of these enzyme preparations for the hydrolysis of starchy biomass for alcohol production.

MATERIALS AND METHODS

Microorganisms

The bacterial strain *Bacillus* sp. SN-1, producing α-amylase and fungal strain, *Aspergillus oryzae* HS-3, producing glucoamylase, in appreciable titres, were isolated from local soil.

Amylase Production by Solid State Fermentation

Different sets of 250 ml Erlenmeyer flasks, each containing 5g rice bran or wheat bran moistened with 1.5 parts of distilled water were autoclaved and used as basal media for bacterial and fungal solid state fermentations. The basal medium for bacterial α-amylase production was supplemented with 0.5% (w/v) glucose, 0.5% (w/w) soyabean meal, 10mM $CaCl_2$ and inoculated with 2 ml of (A_{600}.0.8) 12 h old shake flask culture *of Bacillus* sp. SN-1 prepared in the nutrient broth, and incubated at 37°C for 96 h. On the other hand the basal medium for fungal glucoamylase production was supplemented with 1% lactose, 1% soyabean meal, 10mM $CaCl_2$ and inoculated with 2 ml of spore suspension (2×10^s spores/ml) of *Aspergillus oryzae* HS-3 grown on potato dextrose agar and incubated at 37°C for 192 h.

The α-amylase and glucoamylase were extracted twice separately, after regular intervals of 24 h from the bacterial and fungal solid cultures by mixing with 25 ml of 0.1 M Phosphate buffer (pH 7.0), and 0.1M citrate-PO_4 buffer (pH 5.0) respectively, and keeping the flasks on a rotary shaker (200 rpm) for 30 min at room temperature. The two pooled extracts from the respective flasks were centrifuged (10,000 rpm, 20 min, 4°C) and the cell free clear supernatants were used for determining the α-amylase and glucoamylase activities respectively which were expressed in terms of Units/g of original substrate used in the basal medium.

Amylase Assay

The α-amylase and glucoamylase activities were determined from the corresponding bacterial and fungal extracts in terms of liquefying[12] and saccharifying activities[13] respectively. One Unit of the liquefying activity is equivalent to the enzyme causing 10% reduction in the starch-iodine colour in the substrate solution in 10 min. On the other hand, one unit of saccharifying activity is equivalent to the enzyme that releases one micromole of glucose from the substrate solution in 1 min.

Characterization of Amylase Preparations

The α-amylase and glucoamylase preparations from *Bacillus* sp. SN-1 and

Aspergillus oryzae.HS-3 concentrated, separately, and partially purified by precipitating with 2.5 volumes of, previously chilled (-20°C), acetone, at 4°C with continuous stirring and the precipitates were dissolved in 0.1M Phosphate buffer (pH 6.5) in case of α-amylase and 0.1M Citrate-phosphate buffer (pH 5.0) in case of glucoamylase, respectively. These were then characterised in terms of the effect of temperature and pH on their activity and stability. [1] The temperature activity profiles for α-amylase and glucoamylase activity were determined by assaying the liquefying and saccharifying activities at different temperatures (30°, 40°, 50°, 60°, 70°, 80°, 90° and 100°C) while the pH activity profiles were determined by assaying the liquefying and saccharifying activities at 50°C using substrate solutions made in different buffer systems with variable pH values ranging from 3.0–8.0. [2] The thermostability profiles for α-amylase and glucoamylase preparations were determined by incubating at various temperatures of 60 and 50°C respectively for 7h without and with 10mM $CaCl_2$.

Hydrolysis of 15% Corn Starch Gels with Amylase Preparations

One set of 100 mL Erlenmeyer flasks, containing 10 ml of 15% gelatinized corn starch gel in 0.1 M Phosphate buffer (pH 7.0), was incubated with 1,00,000 U of α-amylase preparation from *Bacillus* sp. SN-1. Another set of flasks containing equal volume of 15% starch gel made in 0.1M Citrate phosphate buffer (pH 5.0) was incubated with 2000 U of glucoamylase preparation from *Aspergillus oryzae* HS-3. The flasks were kept in constant temperature water baths at 60° and 50°C respectively and the extent of liquefaction and saccharification was determined at regular time intervals for 7 h, by measuring a decrease in residual starch and an increase in reducing sugar content. The starch content was estimated by iodine method[12] while the reducing sugars were assayed in terms of glucose by DNSA method.[13]

Application of Amylase Preparations from *Bacillus* sp. SN-1 and *Aspergillus oryzae* HS-3 in the Hydrolysis of Wheat Mash for Alcohol Production

The basic procedure employed for the hydrolysis of wheat biomass involved the grinding of 100 g wheat grains to 20-25 mesh particle size, and dispersing, with constant stirring into 400 ml tap water, supplemented with 10mM $CaCl_2.2H_2O$. This was followed by the addition of 1,00,000 Units of bacterial α-amylase and maintaining the temperature at 75°– 80°C in a water bath for 30 min. The mash was then gelatinised by cooking under pressure in an autoclave at 105°C for 30 min and thereafter another dose of 1,00,000 Units of α-amylase, was added after maintaining the temperature at 65–70°C for 30 min. This was followed by the addition of 2000 Units of fungal glucoamylase preparation and 1 g of active dry yeast culture of *Saccharomyces cerevisiae* (viable cell count of 1×10^{10}/g) after cooling the liquefied mash for simultaneous saccharification and fermentation. Uninoculated mashes were used for determining the total sugars liberated after the saccharification of the mashes. The mashes were then incubated at 32°C for 72 h and the residual starch, sugars and alcohol contents were determined by the standard procedures.[12,13,14]

Conversion Efficiency in Mashing was Calculated as

$$\frac{\text{Reducing sugars produced in wheat mash}}{\text{Reducing sugars obtainable from starch in wheat}} \times 100$$

Sugars obtainable from starch hydrolysis as determined from stoichiometrical calculations are 1.11 g/g.

Fermentation efficiency was calculated as

$$\frac{\text{Alcohol produced in fermented mash}}{\text{Alcohol obtainable from fermentable sugars in mash}} \times 100$$

Alcohol obtainable from sugars according to Gay-Lussac equation is 51.1% which comes out to be 0.64/g of glucose after employing the specific gravity factor.

RESULTS AND DISCUSSION

Due to potential usefulness of the bacterial α-amylase and fungal glucoamylase in the hydrolysis of starchy biomass, the development of method for cheaper production of enzyme is very important. One alternative low cost production method is the use of Solid State Fermentation (SSF). In this study SSF has been found to be a cheap way of producing high levels of α-amylase by laboratory isolates of *Bacillus* sp. and glucoamylase by *Aspergillus* sp. where the productivity levels have been found to be quite high . The selected strains *of Bacillus* sp. SN-1 and *Aspergillus oryzae* HS-3 colonised well on the solid media composed of rice bran and wheat bran, respectively. The visual observations after 24 h revealed spreading of white cellular mass on the entire surface of the solid media. The enzyme synthesis boosted up with the growth, exhibiting the maximum levels of α-amylase (75000 U/g) and glucoamylase (3460 U/g) respectively, after 48 and 96 h of incubation (Fig. la,b). The yields of both the enzymes obtained in the present study were very high in comparison to the published reports on amylase production in solid state cultures.[8,10,11,15]

The nature of solid substrate is the most important factor in SSF. This not only supplies the nutrients to the culture but also serves as an anchorage for the microbial cells. Therefore, the particle size and the chemical composition of substrate are of critical importance.[16] An ideal solid substrate provides all necessary nutrients to the microorganism for optimum function. Various substrates used by different workers for amylases include wheat bran[17&18], soyabean meal, potato meal and alfalfa meal[19], maize bran[18], corn meal base, defatted ground nut cake, defatted castor seed cake, rice bran, lucerne leaf powder, lucerne stalks, rice husk, sweet potato, maize gluten (free from starch) and soya residue obtained after extraction of milk.[20] Rice bran and wheat bran proved to be most suitable substrates for the colonisation of the bacterial and fungal strains, in the present study, as indicated by

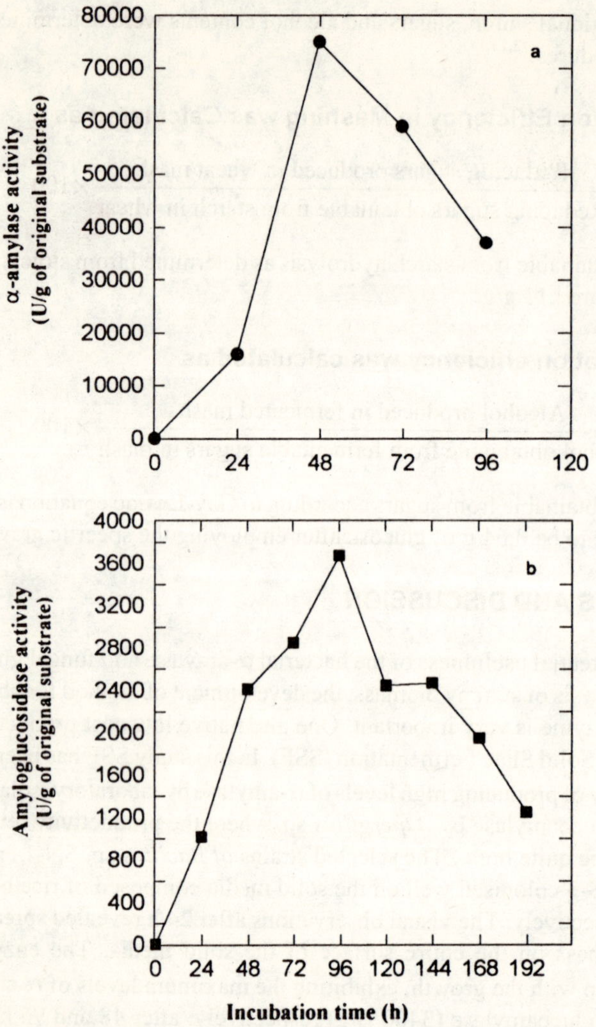

Fig. 1. Effect of incubation period on (a) α-amylase production by *Bacillus* sp. SN-1 and (b) amyloglucosidase production by *Aspergillus oryzae* HS-3 in solid state fermentation.

the maximum visible growth on the surface of substrate and the high enzyme yields which is possibly due to the presence of various suitable nutrients in rice and wheat brans and/or due to its most suitable particle size and consistency required for anchorage, colonisation and enzyme excretion by the *Bacillus* and *Aspergillus* strains.

Characterization of Amylase(s) Preparation(s)

Since the efficiency of starch liquefaction and saccharification improves at temperatures above 50°C, the thermostable amylases are of great significance. In this regard the α-amylase from *Bacillus* sp. SN-1 which was optimally active at 60°C and displayed 87, 78 and 50% of activity at 40, 70 and 80 °C indicating its thermostable nature. (Fig. 2a). The enzyme was very stable at 60°C and revealed 94% residual activity after 7 h incubation (Fig. 4a) and thus could be a good candidate for the efficient liquefaction of starch. On the other hand glucoamylase activity showed a maxima at 50°C. The activities 78% and 76% were recorded at 60° and 40°C respectively (Fig. 2b). The half life of glucoamylase was recorded after 6 h at 50°C. The addition of l0mM $CaCl_2$ significantly improved the thermostability of the enzyme with a half life of more than 7 h (Fig. 4b). The requirement of Ca^{2+} by α-amylases for their stability at higher temperatures has also been reported in *Bacillus* spp.[21,22]. Since the glucoamylase was quite active even at the room tem-

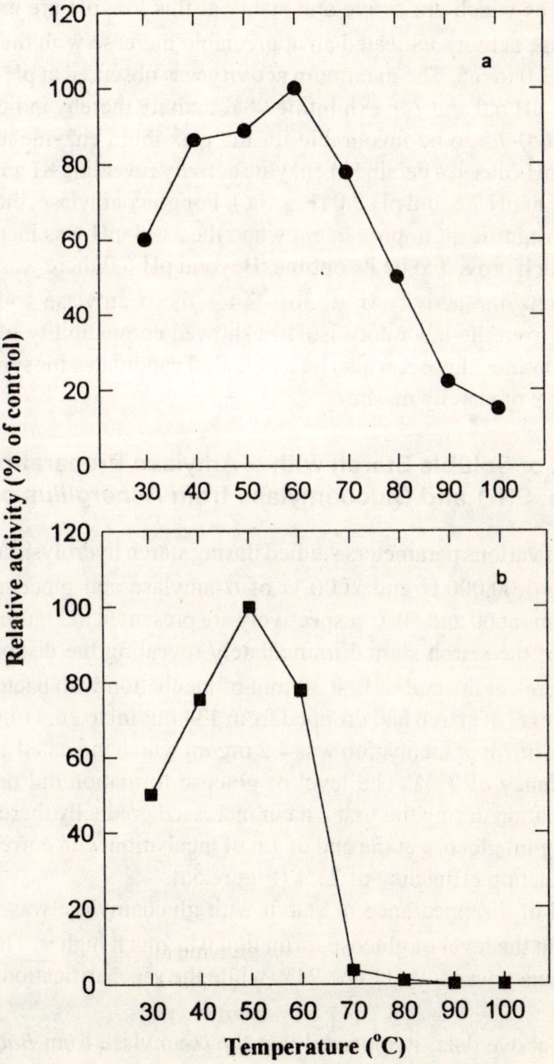

Fig. 2. Effect of temperature on activity of (a) *Bacillus* sp. SN-1 α-amylase and (b) *Aspergillus oryzae* HS-3 amyloglucosidase.

perature (40°C), this can also be a potential candidate for use in combination with the culture of *Saccharomyces cerevisiae,* in the conventional processes, taking place in the alcohol industries, for simultaneous saccharification and alcohol fermentation of starchy mashes where the temperature of the fermenters is in the range of 30-40°C.

As the conventional process of starch hydrolysis involves the action of α-amylase and the glucoamylase, several workers have demonstrated the synergistic effect of these two enzymes on starch degradation.[11] To exploit the opportunity of synergism, the temperature and pH optima of the α-amylases and AMGs should be compatible. However, the pH optima of the most AMGs is at or near 4.0. However, the α-amylases which are active and stable at this low pH are extremely rare.[23] The α-amylase activity indicated an appreciable increase with the increase in assay pH from 3.0 to 6.5. The maximum activity was observed at pH 6.5 and was almost same at pH 6.0 and 7.0, exhibiting 98% activity thereby indicating the range between pH 6.0-7.5 to be favourable for the maximum enzyme activity. Further rise in assay pH caused a decline in enzyme activity revealing 81 and 76% of maximum activity at pH 7.5 and pH 8.0 (Fig. 3a). For glucoamylase, the enzyme activity showed a significant improvement when the assay pH was increased from 3.0 to pH 6.0 which proved to be its optima. Beyond pH 6.0 there was a considerable decrease in enzyme activity (Fig. 3b). Since the α-amylase and glucoamylase preparations from the laboratory isolates showed compatibility in their temperature and pH optima, these can also be a potential candidates for synergistic use for the hydrolyisis of starchy mashes.

Hydrolysis of Soluble Starch with α-Amylase Preparations from *Bacillus* sp. SN-1 and Glucoamylase from *Aspergillus oryzae* HS-3

The results of various parameters studied during starch hydrolysis and glucose formation using 1,00,000 U and 2000 U of α-amylase and glucoamylase in 15% starch solutions at 60 and 50°C respectively are presented in Figures 5 a and b. The breakdown of the starch started immediately revealing the disappearance of almost 82% starch at the end of first 30 min of incubation with bacterial α-amylase wherein the level of starch had dropped from 134 mg/ml to 26.4 mg/ml. The starch left at the end of 7 h of incubation was 4.2 mg/ml which indicated an overall liquefaction efficiency of 97%. The level of glucose formation did not pick up with starch degradation during the first 2 h but increased gradually thereafter yielding a total of 33 mg/ml glucose at the end of 7 h of incubation with corresponding overall saccharification efficiency of 23% (Figure 5a).

The trend of disappearance of starch with glucoamylase was similar as with α-amylase but the level of glucose formation was much higher. The overall liquefaction efficiency was found to be 94% while the saccharification efficiency was 73%.

From the above data, it appeared that the α-amylase from *Bacillus* sp. SN-1, having a liquefying efficiency of 97% and glucoamylase from *Aspergillus oryzae* HS-3, having a saccharifying efficiency of 73% in 15% starch solution, can be em-

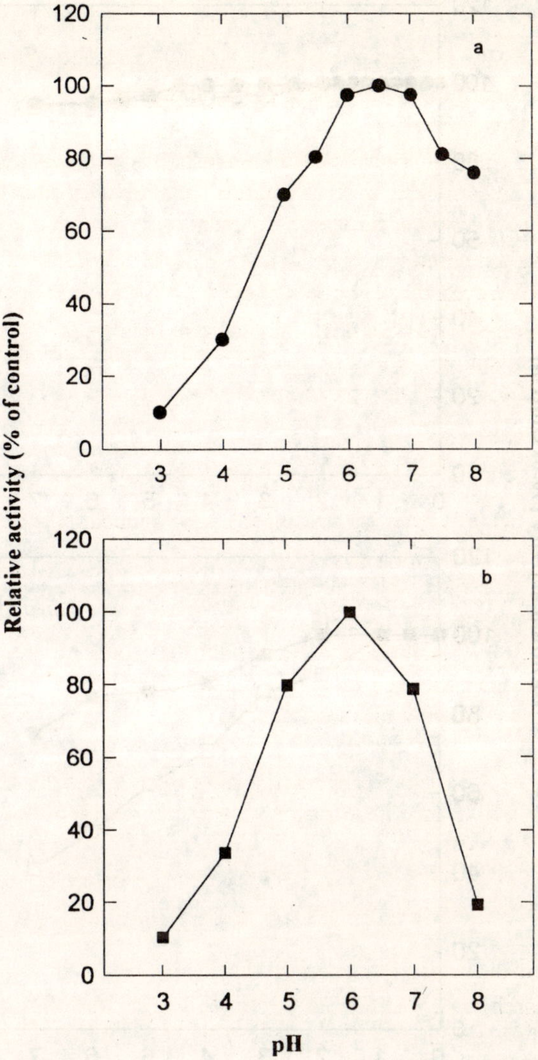

Fig. 3. Effect of pH on activity of (a) α-amylase and (b) glucoamylase .

ployed for the hydrolysis of wheat starch for various biotechnological applications.

Application of α-Amylase Preparation from *Bacillus* sp. SN-1 and Glucoamylase Preparation from *Aspergillus oryzae* HS-3 in the Hydrolysis of Wheat Starch for Alcohol Production

The two enzyme system, bacterial α-amylase and fungal glucoamylase worked very well in the hydrolysis of wheat flour for alcohol fermentation with *Saccharomyces cerevisiae.* The total sugars liberated after 72 h were 11.9 mg/ml

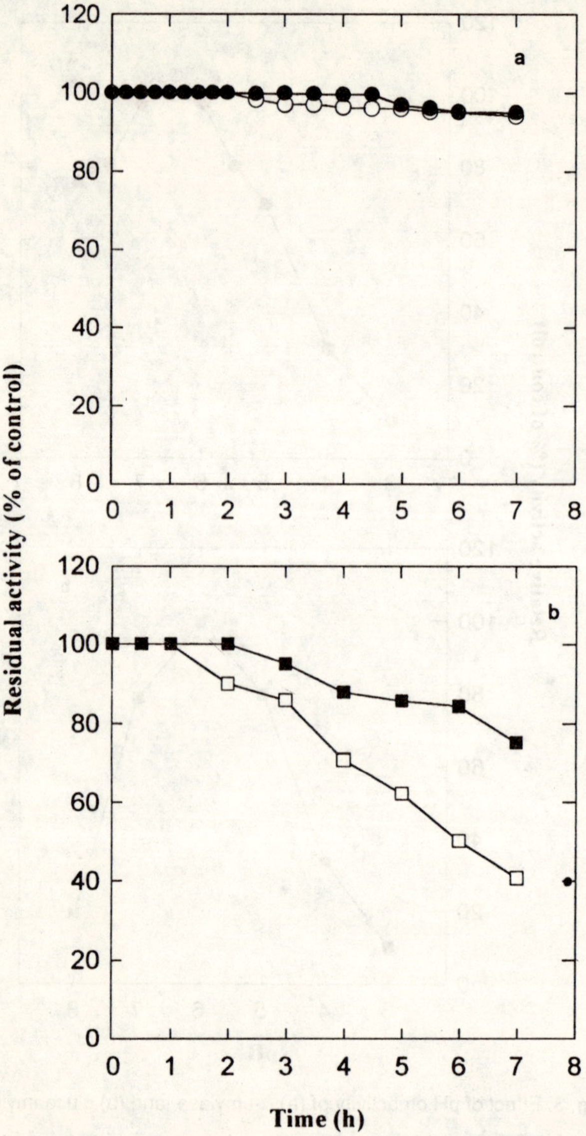

Fig. 4. Thermal stability of (a) *Bacillus* sp. SN-1 a-amylase at 60°C and (b) *Aspergillus oryzae* HS-3 glucoamylase at 50°C in the absence and presence of 10 mM $CaCl_2$. Hollow symbols represent the data without $CaCl_2$ while the solid symbols indicate the supplementation of $CaCl_2$.

indicating the overall mashing efficiency of 87.8%. The alcohol content formed after the fermentation of sugars (formed after the mashing) was 6.8% (v/v) revealing the fermentation efficiency of 89.9% (Table 1).

The attainment of significant mashing, fermentation and overall efficiencies along with alcohol yields indicates the potential of the enzyme preparations from

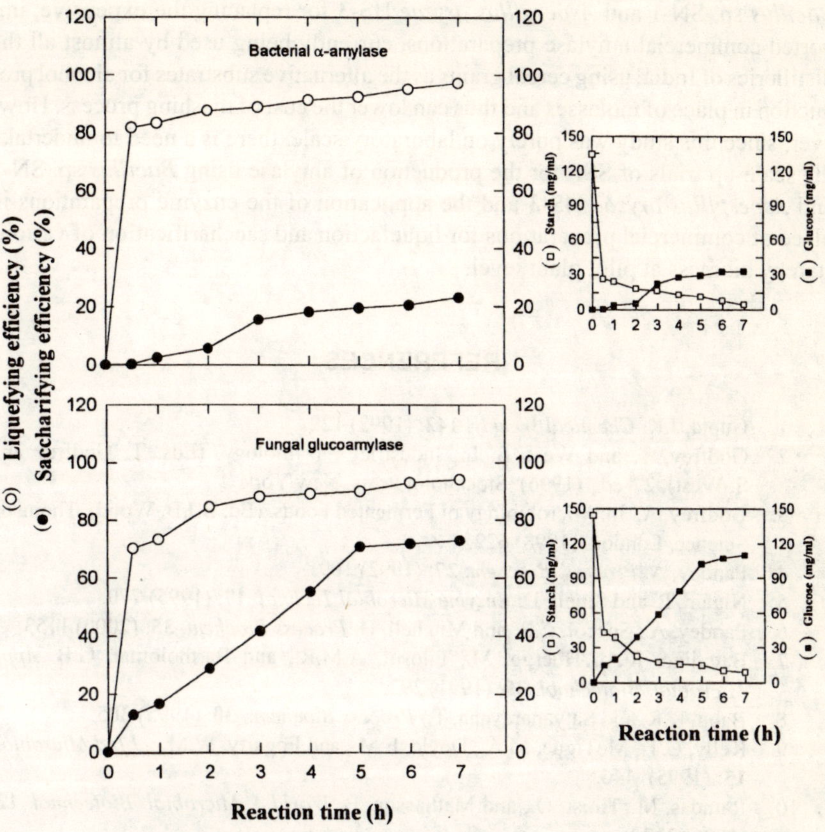

Fig. 5. Profiles of liquefaction, saccharification, residual starch and reducing sugars in 15% starch gel digestion by *Bacillus* sp. SN-1 (α-amylase) and *Aspergillus oryzae* HS-3 (amyloglucosidase) at 60°C and 50°C respectively.

Table 1: Various mashing and fermentation parameters attained in wheat grains involving liquefaction with laboratory amylase preparations

Wt. of Wheat used (g)	100
Volume of mash made (ml)	500
Total Sugar in mash after hydrolysis (%)	11.90
Initial specific gravity	1.054
Initial °Brix	13.90
Initial pH	5.52
Final pH	4.52
Final specific gravity	1.002
Final °Brix	0.60
Alcohol in fermented mash (% ,v/v)	6.83
Mashing efficiency (%)	87.87
Fermentation efficiency (%)	89.90
Over all efficiency (%)	78.90
Alcohol Yield (ml/100g)	34.19

Bacillus sp. SN-1 and *Aspergillus oryzae* HS-3 for replacing the expensive, imported commercial amylase preparations, currently being used by almost all the distilleries of India, using cereal grains as the alternative substrates for alcohol production in place of molasses and thus can lower the cost of mashing process. However, since this study was purely on laboratory scale, there is a need to undertake the scale-up trials of SSF for the production of amylase using *Bacillus* sp. SN-1 and *Aspergillus oryzae* HS-3 and the application of the enzyme preparations in place of commercial preparations for liquefaction and saccharification of various starchy biomass at pilot plant level.

REFERENCES

1. Gupta, J.K. *Chemical Weekly.* **142**: (1992) 129.
2. Godfrey, T. and West, S. In: Industrial Enzymology, (Eds. T. Godfrey and S. West), 2nd ed. (1996). Stockholm Press, New York. 1.
3. Godfrey, A. In: Microbiology of Fermented Foods. (Ed. B.J.B. Wood), Thomson Science, London, (1998) 622.
4. Pandey, A. *Process. Biochem.* **27**: (1992) 109.
5. Nigam, P. and Singh, D. *Enzyme Microbial Technol.* **17**: (1995) 770.
6. Pandey, A., Soccol, C.R. and Mitchell, D. *Process Biochem.* **35**: (2000) 1153.
7. Battalino, R.A., Huergo, M., Pilosuf, A.M.R., and Bartholomai, G.B. *Appl. Microbiol. Biotechnol.* **35**: (1991) 292.
8. Babu, K.R. and Satyanarayana, T. *Process Biochem.,* **30**: (1995) 305.
9. Kelly, C.T., McTigue, M.A., Doyle, E.M., and Fogarty, W.M. *J Ind Microbiol.* **15**: (1995) 446.
10. Ramdas, M., Hoist, O., and Mattiasson, B. *World J. Microbiol. Biotechnol.* **12**: (1996) 267.
11. Mamo, G. and Gessesse, A. *J Ind Microbiol Biotechnol.* **22**: (1999) 622.
12. Fuwa. H. *J. Biochem.* **41**: (1954) 583.
13. Miller, G.L. *Anal. Chem.* **32**: (1959) 426.
14. AOAC, (1975). Official Methods of Analysis, 16th edition, (Ed. P. Cunniff), Benjamin Franklin Station, Washington.
15. Ramesh, M.V. and Lonsane, B.K. *Biotechnol. Lett.* **9**: (1987) 501.
16. Lonsane, B.K., Ghildyal, N.P., Budiatman, S., and Ramakrishna, S.V. *Enzyme Microb. Technol.* **7**: (1985) 258.
17. Tobey, J.F. and Yousten, A.A. *Dev. Ind. Microbiol.* **18**: (1976) 499.
18. Qadeer, M.A., Anjum, J.I. and Akhtar, R. *Pakistan J. Sci. Ind. Res.* **23**: (1980) 25.
19. Beckord, L.D., Kneen, E., and Lewis, K.H. *Ind. Eng. Chem.* **37**: (1945) 692.
20. Lulla, B.S. and Subrahmanyan, V. *J. Sci. Ind. Res.* Sect. B **13**: (1954) 410.
21. Krishnan, T. and Chandra, A.K. *Appl. Environ. Microbiol.* **46**: (1983) 430.
22. Kochhar, S. and Dua, R.D. *Biotechnol. Lett.* **12**: (1990) 393.
23. Crabb, W.D. and Mitchinson, C. *Trends. Biotechnol.* **15**: (1997) 349.
24. Sani, A., Awe, F.A., and Akinyanju, J.A. *J. Ind. Microbiol.* **10**: (1992) 55.

Polygalacturonase from *Aspergillus niger*: Optimization of Production Conditions and Its Application

R. GUPTA, A. KUMAR, N. KUMAR *and* T. C. BHALLA

Department of Biotechnology
Himachal Pradesh University
Summerhill, Shimla 171005 (H.P.)

ABSTRACT

Aspergillus niger van tieghem isolated from banana produced extracellular pectinases when grown in liquid medium containing pectin as sole carbon source. The production conditions for polygalacturonase (PG), which showed maximum activity out of various pectinolytic enzymes were optimized. The production medium with citrus pectin as the carbon source, casein hydrolysate as the nitrogen source and pH 3.5 yielded an enhanced extracellular PG activity, as high as 60.35 IU. The enzyme preparation was used in apple juice extraction and clarification. A temperature of 45°C was found to be optimum for both juice clarification and extraction while holding time of 1 h and 3 h was found to be optimum for clarification and extraction respectively.

INTRODUCTION

Pectic enzymes have considerable commercial significance in disintegration of plant tissues, particularly in fruit and vegetable processing and manufacturing industries. The applications of these enzymes include increased yield of juice and solids from plant material, reduction in viscosity of concentrates and modification and solubilization of pectic substances to affect sedimentation and clarification of juices .[1] The commercial preparations of pectic enzymes used in food industry often come from fungal sources and normally contain a mixture of pectinolytic enzymes associated with hemicellulases and/or cellulases. But there are some cases where only one type of pectolytic enzyme is required. The cloud in orange juice can be stabilized by use of high levels of polygalacturonase activity.[2] The range of enzymes used for juice clarification is being extended through new recombinant and non-recombinant fungal strains.[3]

This study was undertaken to isolate microorganisms which produce pectin degrading enzymes and to optimize production conditions to augment the yield of the enzyme. The temperature and the holding time of polygalacturonase for apple juice extraction and clarification were optimized.

MATERIALS AND METHODS

Microorganism

The organism used in the present study was isolated from the rotten banana fruit and identified as *Aspergillus niger* Van Tieghem at the Institute of Microbial Technology, Chandigarh. This organism grew optimally at 30°C and pH 4.5. It was maintained on potato dextrose agar (PDA) slants and stored at 4°C.

Production of Pectolytic Enzymes

The medium used for the production of PG was Rivieręʼs broth with the following composition: $NaNO_3$, 0.1g; $Na_2HPO_4.7H_2O$, 0.118g; KH_2PO_4, 0.09 g; $MgSO_4.7H_2O$, 0.05g; KCl, 0.05g; yeast extract, 0.05g; casein hydrolysate, 0.05g; pectin 1 g in 100 ml distilled water. The medium (50 ml) in 250 ml Erlenmeyer flasks was seeded with 2×10^5 spores . The flasks were incubated for 72 h at 30°C in an incubator rotary shaker at 150 rpm. The broth was harvested by filtration using Whatman filter paper disks. Filtrate was used as crude enzyme for determining pectolytic activities.

Enzyme Assay

Polygalacturonase (PG) activity was determined by colorimetric method using polygalacturonic acid as substrate. The reducing sugars released were measured following the method of Miller.[3] The assay mixture contained 0.5% polygalacturonic acid in 0.05 M sodium citrate buffer, pH 4.8. The mixture was incubated at 50°C for 15 min. The reaction was stopped by adding 3.0 ml of dinitrosalicylic acid reagent. Colour development after 10 min of boiling was measured at 540 nm. One unit of activity was defined as the amount of enzyme that catalysed the release of 1 micromole of galacturonic acid under the assay conditions.

Effect of Carbon and Nitrogen Sources and pH on Enzyme Production

The effect of various carbon sources, fructose, lactose, maltose, galactose, apple pectin and citrus pectin was studied on the production of enzyme. Similarly, the effect of different nitrogen sources, ammonium sulphate, ammonium nitrate, ammonium chloride, ferrous ammonium sulphate, calcium nitrate and casein hydrolysate was studied. The pH of the production medium was varied from pH 2.2 to 4.2.

Optimization of Conditions for Apple Juice Extraction

Eight sets of 100 g apples were mashed and 100 units of enzyme preparation in a volume of 5 ml was added to each. The mashes were incubated at 30°, 35°,40° and 45°C along with the controls for 3 h. Three separate mashed and enzyme treated samples were incubated at 35° C for 2, 3 and 4 h respectively. The juices were extracted by pressing the mash after incubation. The volume of the juice extracted and its clarity were determined.

Optimization of Conditions for Apple Juice Clarification

Two hundred grams of apples procured locally were washed and mashed in a waring blender. The juice obtained was extracted and filtered through cheese cloth. The volume of the juice obtained and the residual pulp weight after pressing were determined. The reaction mixture containing 2 ml of apple juice and 25 units of enzyme in a volume of 200 µl was incubated at 30°, 35°, 40° and 45°C for a period of 30 min to determine the optimum temperature for clarification of juice. Similar reaction mixtures were set up and incubated at 35°C for 30, 60, 90 and 120 min to determine optimum clarification time. The clarity of juice was determined by measuring its transmittance at 650 nm after centrifuging it at 1500 rpm for 4 min.[5, 6] In each experiment as a control, the juice was incubated without enzyme under identical conditions.

RESULTS AND DISCUSSION

Effect of Carbon and Nitrogen Sources and pH on PG Production

Production of PG from *Aspergillus niger* was significantly affected by the type of carbon and nitrogen source used. Citrus pectin emerged as the most suitable carbon source for enzyme production (Table 1). Earlier reports show high production

Table 1: Effect of various carbon sources on PG production by *Aspergillus niger*

Carbon Source (1%)	*Enzyme Activity (IU)*
Fructose	15.6
Lactose	4.15
Citrus Pectin	32.16
Galactose	4.42
Apple Pectin	23.31
Maltose	2.84

of pectinases by fungi such as *Fusarium oxysporum* on pectin containing medium.[7] The presence of pectin in the medium as a carbon source has been found to be neccessary to induce pectinase production in *Aspergillus niger*, *Botryodiplodia theobromae*, *Geotrichum candidum* and *Corticium rolfsii*.[8] However, cultures of yeast producing pectinases have been shown to grow on glucose.[9,10] Among nitrogen sources, casein hydrolysate acted as an efficient nitrogen source to produce polygalacturonase by this organism (Table 2). It has been reported earlier that nitrogen limitation decreases the enzyme production[11], which agrees with our results showing higher enzyme production with readily available nitrogen source such as casein hydrolysate.

Table 2: Effect of various nitrogen sources on PG production by Aspergillus niger.

Nitrogen Source (0.1 %)	*Enzyme Activity (IU)*
Ammonium Sulphate	37.11
Ammonium Chloride	23.05
Casein Hydrolysate	60.35
Calcium Nitrate	18.72
Ammonium Nitrate	24.39
Ferrous Ammonium Sulphate	0.213

The effect of pH of the medium on enzyme production is shown in Fig 1. The

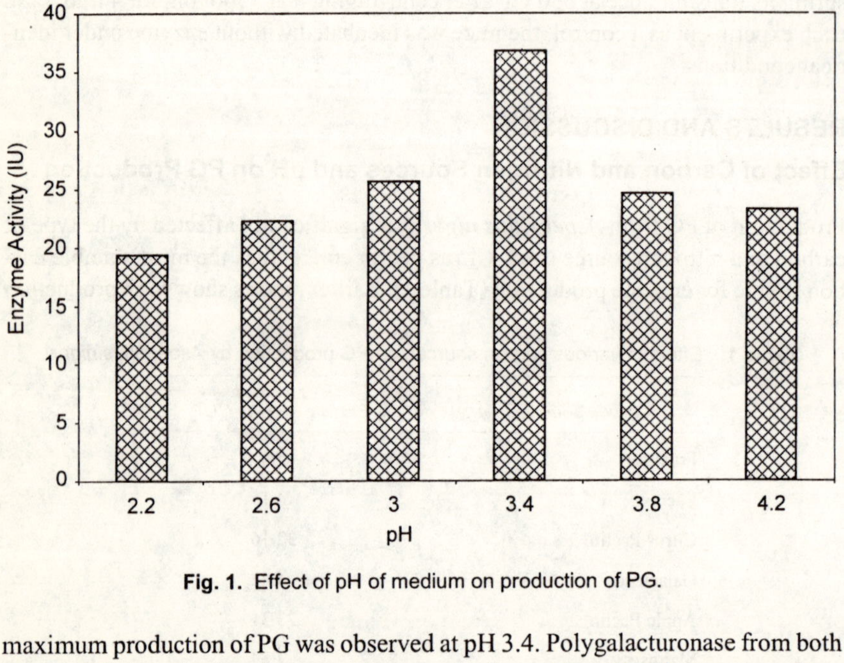

Fig. 1. Effect of pH of medium on production of PG.

maximum production of PG was observed at pH 3.4. Polygalacturonase from both fungi and bacteria has been produced only in acidic pH. The enzyme from *Saccharomyces fragilis* has been produced at a pH of 3.5.[12] The same has been produced from *Fusarium oxysporum*[7] and *Clostridium multifermentans*[13] at pH 5.5. However, lyases have been produced at alkaline pH in contrast to polygalacturonases.

Optimum Conditions for Apple Juice Extraction

When treated with 100 units of enzyme at 45°C for 3 h, the pulp was degraded to an extent of 33%. The clarity of the juice obtained was 50%. The filterability of the juice increased correspondingly with temperature and holding time (Table 3). There was no significant increase in the juice yield if held for more than 3 h. A de-

Table 3: Effect of temperature and holding time on apple juice extraction by PG of *Aspergillus niger*

Temperature (°C)	*Volume (ml)*	*Clarity (%T)*	*Time (h)*	*Volume (ml)*	*Clarity (% T)*
30	45	40.2	2	48	40.8
35	58	45.4	3	55	42.2
40	61	48.2	4	57	41.8
45	67	50.0	5	57	42.3

tailed investigation on the efficacy of the improvement in the juice yield and quality from Stanley plums using five commercial preparations of pectinolytic enzymes has been carried out.[6] The enzyme treatment to the fruit mash before pressing and extraction of the juice resulted in increased yields with higher clarity and better filterability. In a study conducted earlier[5], pectinase treatment of grape mash resulted in increased clarity of the juice extracted. The solid residues left after juice extraction were also less. In yet another study[14], treatment at 45°C for 120 min of palmyrah palm (Borassus) fruit juice by pectinase has shown highest juice yield.

Optimum Conditions for Apple Juice Clarification

The results on the effect of temperature and holding time on apple juice clarification are presented in Table 4. A clarification of 99.2% was obtained at a temperature of 45°C. However at temperatures of 40° and 45°C the untreated juice showed a high transmittance suggesting self clarifying action to some extent. Fruits such as apple[15] and papaya[16] have a high acting pectinmethylesterase. Pectinmethylesterases have been implicated in the cloud loss in citric juices[14]. Thus, the self clarifying activity might be attributed to the activation of some inherent pectinolytic enzymes such as pectin esterase and lyase in the fruit juices. With holding

Table 4: Effect of temperature and holding time on apple juice clarification by PG of *Aspergillus niger.*

Temperature (°C)	*% Transmittance*		*Time(h)*	*% Transmittance*	
	*Control **	*Test***		*Control**	*Test***
30	20.3	89.66	0.5	21.2	89.66
35	17.4	97.0	1.0	22.4	97.0
40	84.0	98.6	1.5	24.5	97.2
45	62.9	99.2	2.0	27.3	97.2

*Juice incubated without enzyme.

**Juice incubated with enzyme.

time of 1 h, a clarification of 97% was observed . A holding time of more than one hour did not produce any significant increase in clarity of the juice.

CONCLUSION

Polygalacturonase from *Aspergillus niger* gave a high activity when produced using cheap nitrogen source casein hydrolysate. A temperature of 45°C and a holding time of 3 h was optimum for juice extraction. A temperature of 45°C and a holding time of 1 h was found to be optimum for juice clarification. In general, the yield and quality of juice obtained is affected by treatment temperature and maceration time. The PG discussed above is promising one for application in fruit juice extraction and clarification at large scale.

REFERENCES

1. Whitaker, J.R., Microbial enzymes and biotechnology, Elsevier applied Science Publishers, London. 1990.
2. Baker, R.A. and Bruemmer, J.H., *J. Agric. Food Chem.*, **20** (1972) 1169.
3. Lang, C. and Dornenberg, H., *Appl. Microbiol Biotechnol.*, **53** (2000) 366.
4. Miller, G. L., *Anal Chem*, **31** (1959) 426.
5. Sreenath, H.K. and Santhanam, K., *J. Ferment. Bioeng.*, **73** (1992) 241.
6. Chang, T.S., Siddiq, M., Sinha, N.K., and Cash, J.N., *J. Food Process Preserv.*, **19** (1995) 89.
7. Guevara, M.A., Gonzalez-Jaen, M.T., and Estevez, P., *Can J Microbiol* , **43** (1997) 245.
8. Ugwuanyi, J.O. and Obeta, J.A.N. , *J Sci Food Agric* , **73** (1997) 432.
9. Barnby, F.M., Morpeth, F.F. and Pyle, D.L., *Enzyme Microb. Technol.*, **12** (1990) 891.
10. Gomez-Ruiz, L., Garcia-Garibay M and Barzana, E., *J. Food Sci.,* **53** (1988) 1236.
11. Hgouvieux-Cotte-Pattat, N., Condemine, G., Nasser, W. and Reverchon, S., *Ann. Rev. Microbiol* , **50** (1996) 231.
12. Luh, B.S. and Phaff, H.J., *Arch. Biochem. Biophys.*, **33** (1951) 212.
13. Macmillan, J.E. and Vaughn, R.H., *Biochem.*, **3** (1964) 564.
14. Agbo N'zi, G. and Simard, R.E. *Plant Foods Hum. Nutr.*, **42** (1992) 55.
15. Macdonald, H. M and Evans, R. *J. Sci. Food Agric.*, **70** (1996) 321.
16. Fayyaz, A., Asbi, B.A., Ghazali, H.M., Che Man, Y.B., and Jinap, S., *Food Chem.*, **47** (1993)183.

Development of Fermentation Process for High Value Beverage from *Vitis vinifera* and *Allium sativum* L.

A. CHAWLA, H.K. TEWARI, H.K. SAHOTA, K. SHVETA

and

*T.S. JULKA

*Deptt. Of Microbiology & * University Health Centre*
Punjab Agricultural University
Ludhiana 141004 (Punjab)

ABSTRACT

A high value beverage has been developed from *Vitis vinifera* var. Beauty seedless (Red grapes) and *Allium sativum* L. (Multicloved garlic). Natural antioxidants in grapes and garlic lower Low Density Lipoproteins (LDL). Allicin is produced when alliinase, an enzyme confined to cell vacuoles of garlic, on crushing, comes in contact with alliin present in cytoplasm. Ajoene, produced from allicin prevents aggregation of blood platelets responsible for clotting of blood and arterial blocking. The low-alcoholic beverage contains phenolic compounds, vitamins, minerals and other health enhancing natural constituents. It contains health protective factors with added food value. Platelet aggregation inhibitory factor present in the beverage is heat stable. Moderate consumption of the high value beverage at meal time may improve health of the public. Fermentation process for the production of the beverage will be discussed.

INTRODUCTION

Allium sativum L. (Multicloved garlic) is a perennial herb belonging to family Alliaceae. It is used as a valuable condiment for foods and a popular remedy for various ailments and physiological disorders. The selective enzymatic processing of odourless precursor alliin, present in cytoplasm, by enzyme alliinase, confined to cell vacuole, yields allicin, which is responsible for the characteristic odour of crushed garlic cloves[1]. Hypolipidaemic[2] and hypocholesterolaemic[3] effects of garlic are due to allicin. Allicin is the specific inhibitor of acetyl CoA synthetase enzyme necessary for fatty acid biosynthesis[4]. Ajoene formed by molecular rearrangement of odourous allicin, is a potent antiplatelet agent[5]. Allicin has been reported to be the most potent antimicrobial agent against pathogenic strains of gram positive bacteria, when compared with ampicillin and kanamycin[6].

Vitis vinifera (Grape) is an important and economic proposition for farmers among the horticultural crops grown in India. Grape is fairly a good source of minerals like calcium, phosphorus, iron and vitamins[7]. Grape berries, particularly the skin contains resveratrol and catechin which are involved in preventing cardiovascular diseases and cancer[8]. Procyanidins present in grapes stabilize the collagen in the arterial wall, inhibit localized histamine formation in the arteries and accelerate the removal of cholesterol[9]. Catechin, epicatechin and gallic acid in grapes are active as antioxidants in inhibiting LDL oxidation[10].

Fermentation process has been developed for the production of high value beverage from beauty seedless variety of red grapes and multicloved garlic (*A.sativum*).

MATERIALS AND METHODS

Substrate

The samples of *A.sativum* were collected from the local market, Ludhiana, India and red grapes of variety beauty seedless were collected from the vineyards located at Bathinda, India.

Microorganism

Saccharomyces cerevisiae var.ellipsoideus strain montrachet for fermentation of beverage was obtained from the culture collection of Department of Microbiology, Punjab Agricultural University, Ludhiana, Punjab. The *S cerevisiae* was maintained on glucose yeast extract (GYE) medium[11].

Effect of Boiling Time of Garlic on Fermentation

Garlic was boiled for 3, 5 and 10 min to inactivate the enzyme alliinase, crushed and 5 g added to 100 ml grape juice. The mash was fermented at 28 ± 2ºC using 10% inoculum and stirring as and when required. The fermentation was studied in terms of fall in TSS of mash.

Effect of Garlic Concentration on Fermentation

To study the effect of garlic concentration, raw garlic @ 1.0, & 2.0 g and alliinase inactive garlic @ 5.0 & 10.0g were added to 100 ml of grape juice and inoculated with 10.0 % *S.cerevisiae*. The fermentation was carried out at 28 ± 2ºC with stirring as and when required. The fermentation was studied in terms of fall in TSS of mash.

RESULTS AND DISCUSSION

Effect of Boiling Time of Garlic on Fermentation

The Fermentation was inhibited when garlic boiled for 3 min was used @ 5g/100ml grape juice and fermented at 28 ± 2°C using as inoculum 10.0 % (v/v)

S. cerevisiae as inoculum (Table 1). This indicates that enzyme is not completely inactivated within 3 min. boiling time and some allicin is formed which inhibits yeast and thus fermentation. Boiling of garlic for 5 min. is sufficient for inactivating alliinase as boiling for higher duration did not show any change in fermentation. So whole garlic bulb was boiled for 5 min. to inactivate the enzyme alliinase.

Table 1: Effect of boiling period of garlic (*A. sativum*) on fermentation

Fermentation period (h)	*Garlic boiling period (min.)*			
	0	*3*	*5*	*10*
	TSS (°B)			
0	21.8	21.8	21.8	21.8
12	21.8	21.8	16.6	16.6
16	21.8	21.6	13.8	13.8
20	21.8	21.0	10.8	10.6
24	21.6	16.2	8.2	8.0
36	2.16	13.4	6.8	6.6
40	21.6	13.4	6.4	6.4
44	21.6	13.4	6.4	6.4
Substrate Utilization	*Fermentation Period (h)*			
≥ 50%	—	—	20	20
≥ 75%	—	—	—	—
Maximum	24 (0.92%)	36 (38%)	40(71%)	40 (71%)

Fermentation conditions: Garlic concentration : 5g/100 ml grape juice; Temperature : 28 ± 2°C; Inoculum : 10.0 % (v/v)

It has been reported that garlic extract when exposed to heat treatment of 80-90°C for 5 min, the antimicrobial activity of the extract was completely destroyed.

Effect of Garlic Concentration on Fermentation

The maximum substrate utilization has been more with raw garlic @1% as compared to 0 and 2% of garlic in grape juice (Table 2) Results indicate that 1% garlic is stimulatory for fermentation rather than being inhibitory. The concentration of garlic higher than 1% is inhibitory to yeast as maximum substrate utilization has been 11% with 2% garlic as compared to 76% substrate utilization recorded with 1% garlic. Thus grape juice fermented efficiently with 1% (w/v) raw garlic at 28 ±

Table 2 : Effect of concentration of raw garlic (*A. sativum*) on fermentation.

Fermentation period (h)	*Concentration of raw garlic in grape juice (g/100 ml)*		
	0	*1*	*2*
		TSS (°B)	
0	21.8	21.8	21.8
12	16.6	17.8	19.4
16	13.8	15.4	19.4
20	11.0	12.6	19.4
24	9.4	8.4	19.4
36	7.6	5.4	19.4
40	5.6	5.2	19.4
44	5.6	5.2	19.4
Substrate Utilization		*Fermentation Period (h)*	
≥ 50%	20	20	—
≥ 75%	—	36	—
Maximum	40 (74%)	40 (76%)	12 (11%)

Fermentation conditions: Temperature : 28 ± 2°C; pH : 5.0; Inoculum : 10.0% (v/v)

2°C with 10% (v/v) inoculum. The inhibitory effect of garlic extract on yeasts and molds has been demonstrated.[13]

The effect of concentration of alliinase inactive garlic on fermentation of grape juice is shown in Table 3.

The results in Table 3 indicate that boiled garlic is far less inhibitory to fermentation as compared to raw garlic. The 70% substrate was utilized even with 10% (w/v) boiled garlic while only 11% substrate was utilized with 2% (w/v) raw garlic. It indicates that heating inactivates the antimicrobial principle present in garlic. But slight inhibition of fermentation was still observed with alliinase inactive garlic. The antimicrobial activity of garlic extract at elevated temperature of 55°C and 60°C has been demonstrated.[14] The possible existence of other antimicrobial agents either present in garlic or developed during heat treatment has also been suggested.

To summarize, boiling inactivates the enzyme alliinase which converts alliin to allicin, the antimicrobial principle in garlic. The efficient fermentation of grape juice supplemented with 1.0% (w/v) raw garlic and 10.0% (w/v) alliinase inactive garlic (boiled for 5 min) was efficiently carried out at 28± 2°C using 10% (v/v) inoculum of *S. cerevisiae*.

Table 3: Effect of concentration of alliinase inactive garlic (*A. sativum*) on fermentation

Fermentation period (h)	*Concentration of alliinase inactive garlic in grape juice (g/100 ml)*		
	0	*5*	*10*
		TSS (°B)	
0	21.8	22.0	22.2
12	16.6	16.4	16.8
16	13.8	13.6	13.8
20	11.0	10.4	10.6
24	9.4	8.0	8.2
36	7.6	6.4	6.6
40	5.6	6.4	6.6
44	5.6	6.4	6.6
Substrate Utilization	*Fermentation Period (h)*		
≥ 50%	20	20	20
≥ 75%	—	—	—
Maximum	40 (74%)	36 (70%)	36 (70%)

Boiling period (garlic) : 5 min.

Fermentation Conditions: Temperature : 28 ± 2°C; pH : 5.0; Inoculum : 10.0% (v/v)

REFERENCES

1. Cavallito, C.J. and Bailey, J.H., *J. Am. Chem. Soc.*, **66** (1944) 1950.
2. Augusti, K.T. and Mathew, P.T., *Indian J. Exp. Biol.*, **11** (1973) 239.
3. Augusti, K.T. *Indian J. Exp. Biol.,* **15** (1977) 489.
4. Focke, M., Feld, A. and Lichtenthaler, K., *FEBS Letters,* **261** (1990) 106.
5. Jain, M.K. and Apitz–Castro, R., *Current Science*, **65** (1993) 148.
6. Ahsan M and Islam, S.N., *Fitoterapia,* **67** (1996) 374.
7. Amerine, M.A., Berg, H.W. and Cruess, W.V. The Technology of wine making, The AVI Publishing Co. Inc, Westport, 1967.
8. Meishiang, J., Lining, C., George, O.U., Kalra, V.S., Cathyl, F., Christopher, W.W.B., Mehta, R.G., Richard, C.M. and John, M.P. *Science*, **275** (1997) 218.
9. Tibor, F. and Jorge, M.R.S., *J. Agric. Food Chem.*, **45** (1997) 1156.
10. Meyer, A.S., Yi, O.S., Pearson, D.A., Waterhouse, A.L. and Frankel, E.N., *J. Agric. Food Chem.*, **45** (1997) 1638.
11. Tuite, J., Plant Pathological methods—Fungi and bacteria, Burges Publishing Co., Minerapolis Minn., USA, 1969.
12. Sleem, Z.M. and Al-Delaimy, K.S. *J. Fd. Prot.* **45** (1982) 1007.
13. Sreenivasa Murthy, V., Shashikanth, K.R. and Johar, D.S., *J. Scient. Ind. Res.*, **19** (1960) 61.
14. Shashikanth, K.N., Basappa, S.C. and Sreenivasa Murthy, V.J. *Fd. Sci. Tech.*, **18** (1981) 44.

Fungal Biotechnology for Nutritional Upgradation of Crop Residues

M. ARORA, V.K. SEHGAL *and* V.K. THAPAR

Department of Processing & Agricultural Structures
Punjab Agricultural University
Ludhiana 141004. Punjab

ABSTRACT

For efficient fermentation of crop residues, a simple static batch type solid state fermenter, comprising of five aluminium horizontal trays (76 x 36 cm), with 0.3 mm dia perforations and with temperature and humidity control was developed. Two non-toxic fungi: *Rhizopus oryzae* and *Coprinus cinereus* were used to improve the nutritive value of potato processing waste (peels, trimmings etc.) and paddy straw, respectively, in the above developed fermenter. Enzymes produced by *Rhizopus oryzae* degraded the highly polymerised carbohydrates of potato waste by its optimal growth in 48 h resulting in almost two-fold increase in the protein content. Similarly *Coprinus cinereus* under non-sterile conditions and without any pre-treatment, saccharified the complex carbohydrates of paddy straw in three days, resulting in least dry matter loss, 17% increase in protein content with enhancement of nylon bag dry matter digistibility (*in vivo*) from 40 to 62% and the efficiency of fungus or nutrient availability being 27.4%.

INTRODUCTION

Fungal biotechnology has played a major role in large scale production of biomass including edible proteins, enzymes, biochemicals, vitamins and many other useful products. Fungi have the ability to secrete a variety of exocellular enzymes, prominent being hydrolases, polysaccharases and oxidoreductases. They have a natural role in decomposition of lignocellulose (cellulase, xylanasse, ligninase, pectinase).[1,2] The other exocellular enzymes that are commercially produced are amylases[3], lipases and proteases[4]. These extra cellular enzymes can be used for efficient bioconversion of various crop residues for animal feed purposes. In India, there is a shortage of livestock feed and very few attempts are made to exploit agricultural wastes for the same. Hence an attempt has been made for nutritional upgradation of two types of agricultural wastes viz. paddy straw and potato processing waste (peels, trimmings etc.) with suitable non-toxic fungi in a solid state fermenter.

MATERIALS AND METHODS

Solid state fermenter

A solid state fermenter was developed for fermenting 7.5-10 kg of agricultural wastes in one batch. The fermentation equipment comprised of five horizontal, aluminium perforated trays (76 × 30 cm) placed one above the other with 6 cm gap between each tray. The size of the perforation was 0.3 mm in diameter. These specifications were selected keeping in view the type of substrate to be used and to provide adequate aeration, for the growth of fungi during the course of fermentation. The trays stacked on the metal stand were slid into an aerated, humidified enclosure maintained at 30-37°C.

Fermentation of paddy straw with *Coprinus cinereus*

Cyprinus cinereus, a non-toxic lignocellulolytic fungus belonging to class basidiomycetes was procured from Institute of Microbial Technology, Chandigarh. Paddy straw was fermented in the solid state with 10% grain inoculum of *Coprinus cinereus* in aluminium perforated trays (1.5 kg of wet straw in each) for 12 d at 37°C under non-sterile conditions.

Fermentation of potato processing waste with *Rhizopus oryzae*

Rhizopus oryzae, an amylolytic fungus was procured from Institute of Microbial Technology, Chandigarh. Potato processing waste was obtained from a local potato chips processing industry which comprised mainly of potato peels, slices and trimmings. Dried potato waste was moistened with three times water, autoclaved, cooled and inoculated with 10% inoculum of *Rhizopus oryzae.* About 2.5 kg of wet wt. of inoculated potato waste was spread on each tray of the fermenter and fermented at 30°C for 48 h.

Analytical determinations

Crude and true protein (alkali soluble) were determined following the procedures of earlier workers.[5,6] Cellulose content, total ash, crude fibre and soluble/insoluble phosphorous were estimated using the standard methods.[7,8] The standard method[9] was followed to study the Nylon bag dry matter digestibility of the samples. The proportion of the dry matter digested was calculated from the amount of sample incubated and that which was left in the bag after incubation in the rumen of a buffalo calf. The digestible dry matter recovery and the efficiency of the fungus was determined by the following equations:

$$\left.\begin{array}{c}\text{Digestible Dry Matter}\\ \text{Recovery (DDMR)}\end{array}\right\} = \frac{\left\{\begin{array}{c}100 - \text{Dry Matter Loss } \times\\ \text{Nylon Bag Dry Matter Digestibility (NBDMD)}\end{array}\right\}}{100}$$

$$\left.\begin{array}{r}\text{Efficiency of the Fungus or}\\ \text{Nutrient Availability}\end{array}\right\} = \frac{\text{DDMR} - \text{NBDMD of Untreated Straw} \times 100}{\text{NBDMD of Untreated Straw}}$$

RESULTS AND DISCUSSION

Nutritional upgradation of paddy straw by *Coprinus cinereus*

Paddy straw was fermented under farm conditions with 10% grain inoculum of *Coprinus cinereus* in fermenter for 0–12 d. The samples were drawn at 24 hrs interval for determination of crude protein, dry matter loss and nylon bag dry matter digestibility. Maximum nutritional upgradation was achieved within three days incubation because of least dry matter loss (15.26%) and maximum digestibility (62.26%) as compared to that of untreated straw (40.48%). Increase in protein was 1.68 fold and the digestible dry matter recovery of the fermented straw was 51.06% over 47.21% of unfermented straw, while the efficiency of the fungus (nutrient availability) was 27.4% (Fig. 1). Fungi which adqueately, synthesize ligno-cellulolytic enzymes increase the nutrient availability from the fermented product [10,11].

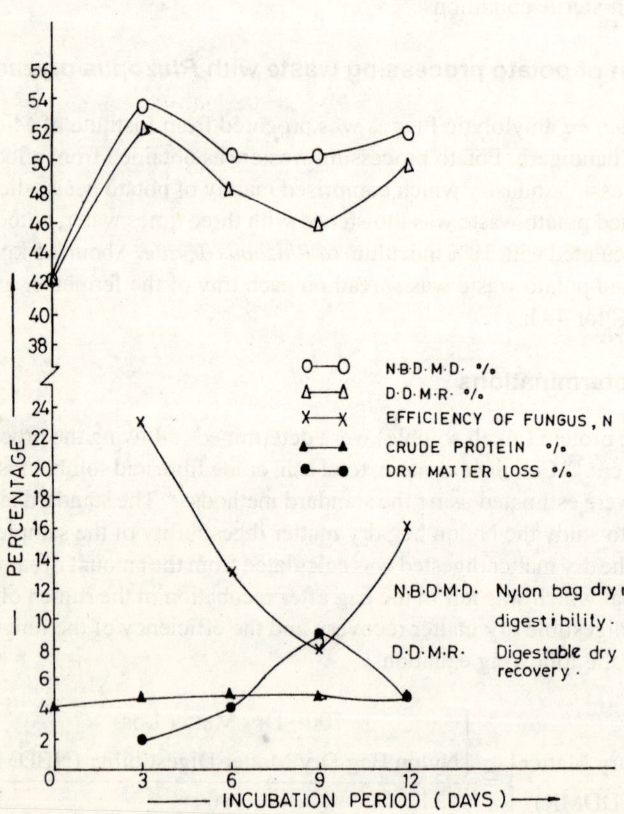

Fig. 1: Fermentation of Paddy Straw with Coprinus Cinerius in Metal Perforated Trays. (Unsterilised Conditions).

Nutritional upgradation of potato processing waste by *Rhizopus oryzae*

Potato processing waste was fermented in the solid state under sterile conditions by a non-toxic amylolytic fungus *Rhizopus oryzae* within 48 h, sporulation of the fungus was observed. It was inferred from the results depicted in Fig. 2 that there was an improvement in the nutritional value of the fermented potato waste, as observed by the enhancement of protein content. A consistent relationship was recorded between the per cent loss of carbohydrates and the increase of the culture protein with time. In 48 h, almost two-fold increase of the protein content in final product of *Rhizopus oryzae* fermented potato waste was obtained in a batch type, static fermenter. Similar results were obtained, i.e. two-fold enhancement of protein content in final product in 48 h by solid state fermentation of sugarbeet pulp by two thermophilic fungi, namely *Thermoascus aurantiacus* and *Sporotrichum*

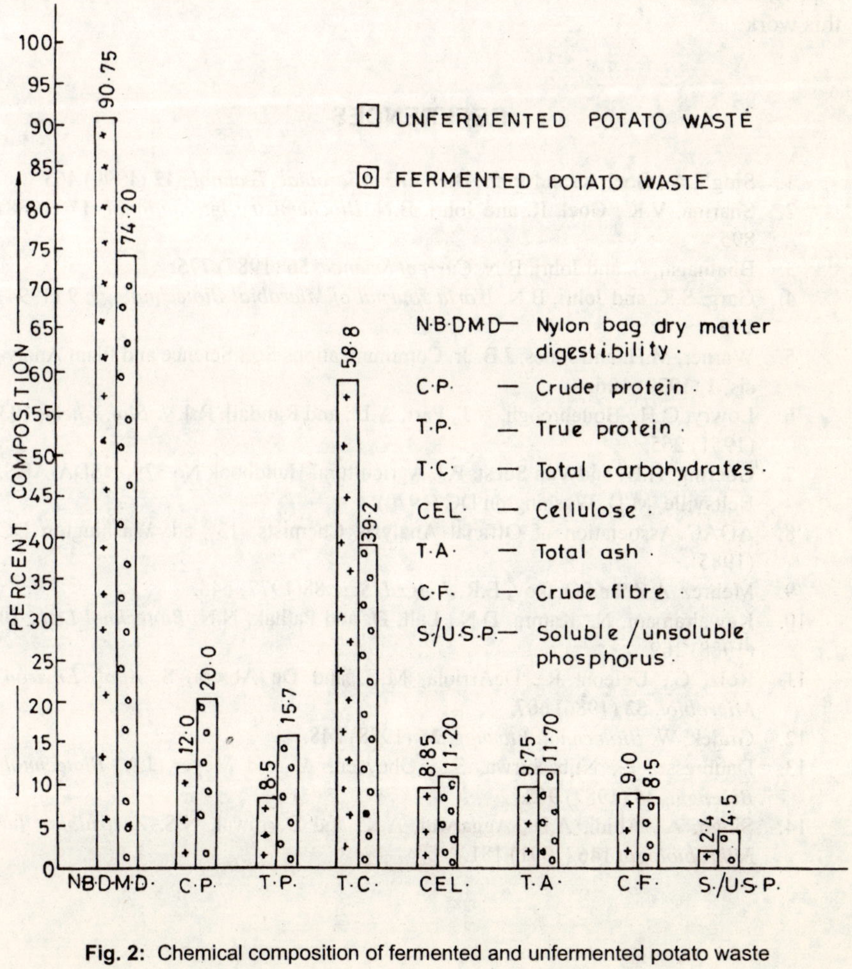

Fig. 2: Chemical composition of fermented and unfermented potato waste

thermophile by other workers[12]. A similar fermentation equipment was also used[13] to enrich the protein of cassava from 1% to 10.7% by *Rhizopus oryzae* in 65 h. The final fermented product at 30°C had higher mineral content. The nylon bag dry matter rumen digestibility decreased due to high dry matter loss. However, considerably high dry matter digestibility (74.2%) coupled with high crude protein (20%) suggests the possible use of fermented potato processing waste as a ruminnt feed supplement. Elsewhere 77% dry matter digestibility and 30.4% protein in maize cobs fermented by *Aspergillus niger* in six days has been reported[14] for usage as animal feed component. Hence it can be concluded that the fungal biotechnology was suitable for efficient fermentation of the two type of agricultural wastes used for nutritional upgradation as animal feed component.

ACKNOWLEDGEMENTS

The authors are grateful to Institute of Microbial Technology, Chandigarh for the supply of cultures and the financial assistance provided by ICAR for completion of this work.

REFERENCES

1. Singh, A., Goel, R. and Johri, B.N. *Enz. Microbial. Technol.,* **12** (1990) 463.
2. Sharma, V.K., Goel, R. and Johri, B.N. *Biochemistry International,* **17** (1988) 895.
3. Bhatnagar, S. and Johri, B.N. *Current Science,* **56** (1987) 775.
4. Garg, S.K. and Johri, B.N. *World Journal of Microbial Biotechnology,* **9** (1994) 139.
5. Warner, M.H. and Jones, J.B. Jr. Communications Soil Science and Plant Analysis, **1** (1970) 106.
6. Lowry, O.H., Bouebrough, N.J., Farr, A.L., and Randall, R.J. *J. Biol. Chem.* **193** (1951) 265.
7. Goering, H.K. and Van Soest, P.J. Agricultural Handbook No.379, USDA-ARS, Beltsville, M.D. Washington DC (1970).
8. AOAC Association of Official Analysis Chemists. 13th ed. Washington DC (1985).
9. Mehrez, A.Z. and Orskov, E.R. *J. Agril. Sci.,* **88**(1977) 645.
10. Kewalramani, N., Kamra, D.N., Lall, D. and Pathak, N.N. *Biotechnol Lett.,* **10** (1988) 369.
11. Rolz, C., Deleon, R., DeArriola, M.C. and De Abrera, S. *Appl. Environ. Microbiol.* **52** (1986) 607.
12. Grajek, W. *Biotechnol. Bioengg.* **25** (1988) 248.
13. Daubresse, P., Ntibashirwa, S., Gheysen, A. and Meyer, J.A. *Biotechnol. Bioengg.,* **24** (1987) 962.
14. Singh, A., Abidi, A.B., Aggarwal, A.K. and Darniwal, N.S. *Zentralblatt fur Mikrobiologie* **146** (1991) 181.

Effect of Process Variables on the Quality Characteristics of Wheat Flour-Milk Protein Co-Extrudates

S. SINGH *and* U. BAJWA

Department of Food Science & Technology
Punjab Agricultural University
Ludhiana 141004 (Punjab)

ABSTRACT

The effect of blending 20, 40 and 60% milk protein in wheat flour and texturizing in a extruder at feed moisture of 21, 24 and 27%, temperature of 110°, 130° and 150°C and screw speed of 75, 100 and 125 rpm on the physical parameters and sensory characteristics of co-extrudates was studied. The yield of co-extrudates increased with increase in milk protein content and decreased with increase in temperature of extrusion. The yield decreased with increase in feed moistures and increase in screw speed. The expansion ratio of co-extrudates was found to be maximum at milk protein content of 60%, feed moisture of 21% and temperature of 130°C. The expansion ratio increased with increase in screw speed. The density of co-extrudates was found to be lowest in the blend containing 40% milk protein and extruded at 130°C. The density was lowered by augmenting feed moisture and screw speed. The moisture content of co-extrudates varied significantly with milk protein content. It decreased with temperature and increased with feed moisture and screw speed. The overall acceptability scores improved with increase in the level of milk protein and temperature of extrusion.

INTRODUCTION

Milk is a high quality food and a very important source of nutrients for most of the vegetarian population of India. More than 54% of the milk produced in India is consumed in the form of different milk products. Products like cheese, paneer, concentrated milk etc. are invaluable sources of proteins to the consumers. The milk proteins exhibit unique properties of structure formation and form a basis of the commercial production of structured protein products.

Extrusion brings about changes in the physico-chemical properties of proteins and starches and, therefore, it is said to have the greatest meaning in the modification of structure of proteins and starches.[1,2] As a result of these changes, quality of final product is considerably improved.

The initial thrust for research into the production of textured protein material was provided by USA and in particular its soybean industry. In the recent past, textured soy protein products have been studied quite extensively and now the research workers have shown an equally increasing interest in other protein sources like meat by-products, oilseed proteins, gluten, single cell proteins, etc.

Extensive studies have been carried out on texturizing milk protein by extrusion, intended mainly for inclusion in meat products. Extrusion cooking is capable of converting soluble rubbery (when coagulated) milk protein into material with crumbly and chewy texture. However, most processes used for manufacturing textured milk protein products are patented and lack details to reproduce them. Hence the information available on texturizing milk proteins is scanty. The present investigation was, therefore, proposed with the objective to study the effect of varying milk protein, moisture, temperature and screw speed on the quality characteristics of flour-milk protein co-extrudates.

MATERIALS AND METHODS

Wheat flour, citric acid, salt and oil were purchased from local market. The buffalo skim milk was procured from the milk processing plant of Department of Food Science & Technology. Punjab Agricultural University (PAU), Ludhiana.

Preparation of the Co-extrudates

Coagulation of Milk Proteins

Skim milk was heated to a temperature of 85°C and coagulated with 2% citric acid solution. The whey was drained and the coagulum was washed three times with water. The coagulum thus obtained was pressed to drain out the excess water. It was then disintegrated with hand and dried to a moisture content of 35-40% at room temperature. Thereafter, it was passed through a manual grinder to further reduce the size of protein particles and was dried at 40-45 C for 3-4 h in an air drier to achieve a moisture content of approximately 15%.

Preparation of Blend

Weighed quantity of milk protein and flour were conditioned separately to moisture contents of 21, 24 and 27% and stored overnight in a refrigerator (5°C). The protein and flour were blended in the ratio of 20:80, 40:60 and 60:40, one hour before extrusion.

Extrusion Processing

C.W. Brabender Laboratory Extruder (Model 2003 Hackensack, NJ, USA) with 19 mm diameter barrel having 20:1 ratio of barrel length to diameter with three heating zones was used for the extrusion processing of different blends. Blends were extruded at an exit zone (third zone) temperatures of 110°, 130° and 150°C by keeping the temperature of the first and second zones at 60° and 70°C, respectively, in each case. The blends were extruded at extruder screw-speeds of 75, 100 and 125 rpm (revolutions per minute). The speed of feeding screw was kept constant to have a regular flow. The co-extrudates were prepared as per following scheme:

Scheme of Process Variables

Variables		*Levels*	
Milk protein (%)	20	40	60
Temperature (°C)	110	130	150
Moisture (%)	21	24	27
Screw speed (rpm)	75	100	125

Determination of Physical Characteristics

Yield

The yield of the co-extrudates was determined by recording the weight of extrudate coming out from the extruder die in one minute. The average of three replications was taken and expressed in g/min.

Expansion Ratio

This was determined by taking the average diameter of 7-10 extruded cut pieces and dividing by the diameter of the die used in extrusion, i.e. 4 mm. The value was expressed as the ratio of diameter of the extruded product (mm) to the diameter of the die (mm):

$$\text{Expansion Ratio} = \frac{\text{Diameter of Extruded Product}}{\text{Diameter of Die}}$$

Density

This was determined by taking the average weight and volume of 3 pieces from each sample. To arrive at an estimate of volume, first the average length of three pieces of the product was obtained (h). The volume was then calculated from the formula $\pi r^2 h$, where r is the radius of the extruded piece and h is the length. The mass of co-extrudate was calculated at 14% moisture content and density expressed as mass per unit volume (g/cm^3).

Moisture Content

The moisture content of milk protein was determined using AOAC method.[3] A weighed (2g) sample was dried at 100°C for 2 h. Moisture content of flour was determined by AACC method[4] by drying 2 g flour at 130°C for one hour in a hot-air oven. The loss in weight was expressed as moisture percent.

Sensory Analysis

Sensory analysis of fried (180°C, 1 min) texturized milk protein product prepared by the addition of 2% salt, was carried out by a semi-trained panel consisting of 11 members. The co-extrudates were evaluated on a 9-point hedonic scale for sensory attributes of colour, appearance, texture and flavour. Overall acceptability was calculated by taking average of all these attributes.

Statistical Analysis

The data were analysed using factorial experiments in complete randomized design (CRD)[5]. The mean values obtained for each parameter were compared on the basis of critical difference (CD).

RESULTS AND DISCUSSION

Proximate Composition

The proximate composition of milk proteins and wheat flour is presented in Table. 1. Milk protein contained 83% total solids. The protein, fat and ash contents of the milk protein were 80.83, 0.74 and 8.06%, respectively. The protein, fat and ash contents of the wheat flour were 12.44, 1.36, and 0.74%, respectively (dry weight basis).

Table 1: Proximate composition* of milk protein and wheat flour

Characteristics	*Milk protein*	*Wheat flour*
Protein (%)	80.83	12.44
Fat (%)	0.74	1.36
Ash (%)	8.06	0.74
Carbohydrate (%) (By difference)	10.37	86.38

* Values presented are on dry weight basis

Effect of Process Variables on the Yield of Co-extrudates

Effect of Milk Protein Content

The milk protein content in the blend had a significant effect on the yield of co-extrudates (Table 2).There was a significant increase in the yield with the increase in milk protein content from 20 to 40%. However, further increase in milk protein content did not have any significant effect on the yield of co-extrudates. Yield at 60% protein was comparable to that of 40% protein level.

Effect of Feed Moisture

Yield of the co-extrudates decreased significantly with the increase in the blend moisture from 21 to 27% (Table 2). It has been observed that as the moisture content of durum semolina was increased from 18 to 27 per cent, the yield decreased.[6]

The increase in yield was reported with similar die areas and extruder power requirement, by lubrication of the material and this effect could also be achieved by adding water to the material at the extruder head inlet[7]. However, this was not observed in the present study because of the varied blend composition.

Table 2: Effect of various process variables on the yield (g/min) of co-extrudates

Variables	*Yield (g/min)*	*C.D. (0.05)*
Milk Protein (%)		
20	82.78[a]	
40	92.94[b]	
60	92.63[b]	1.44
Feed Moisture (%)		
21	95.10[C]	
24	87.83[b]	
27	85.42[a]	1.44
Temperature (^{0}C)		
110	90.72[b]	
130	89.46[ab]	
150	88.17[a]	1.44
Screw speed (rpm)		
75	71.29[a]	
100	90.33[b]	
125	106.68[C]	1.44

Values with different superscripts differ significantly ($P < 0.05$)

Effect of Temperature

The yield of co-extrudates was affected significantly ($P < 0.01$) with increase in temperature (Table 2). The yield declined as the extrusion temperature was raised from 110° to 150°C.

Effect of Screw Speed

The increase of screw speed had a significant ($P < 0.01$) effect on the yield of co-extrudate (Table 2). An increase of 49.64% was observed with the increase in screw speed from 75 to 125 rpm. The increase in yield with increase in screw speed might be attributed to the increased displacement of material at higher speed.

Effect of Process Variables on the Expansion Ratio of Co-extrudates

Effect of milk protein content

The significantly different expansion ratio of co-extrudates ($P < 0.01$) with increase in protein content from 20 to 60% (Table 3) was observed. The expansion ratio was not affected significantly with the increase in the level of milk protein content in the blend from 20 to 40%. However, with further increase in protein content to 60% the mean expansion value increased to 1.238 showing a significant increase over that of blend containing 40% protein.

Table 3: Effect of various process variables on the expansion ratio of co-extrudates

Variables	*Expansion ratio*	*C.D. (0.05)*
Milk Protein (%)		
20	1.220[3]	
40	1.216[a]	
60	1.238[b]	0.006
Feed Moisture (%)		
22	1.247°	
24	1.207[a]	
27	1.220[b]	0.006
Temperature (°C)		
110	1.179[a]	
130	1.261[C]	
150	1.234[b]	0.006
Screw Speed (rpm)		
75	1.181[a]	
100	1.234[b]	
125	1.258[c]	0.006

Values with different superscripts differ significantly ($P < 0.05$)

A negative effect on the product expansion of the sample extruded at 170° C has been reported with the addition of 5% milk protein in defatted corn germ flour.[8] The share of milk proteins in the formation of structure seems to have influence on the formation of protein-starch complexes or their closure in the structure of gelatinized starch. The expansion of extrudates reaching a maximum at a higher milk protein content of about 36% of total protein has also been observed. On co-extrusion of milk protein raffinates and non-fat dried milk powers with corn flour, the decrease in extrudates expansion ratio at levels higher than 5% non-fat dry milk or milk protein raffinate has been reported[9].

A variety of changes in protein and starch lead to the formation of new structures during extrusion. The share of milk protein in the formation of structure seems to have an influence on the formation of protein starch complexes[8]. This may be possible cause of variable expansion ratio of co-extudates at varied protein concentration. Apart from this, heating and pressure, mechanical forces also play an important role in forming the structure during extrusion.[10]

Effect of Feed Moisture

The moisture content of the feed had a significant ($P < 0.01$) effect on the expansion ratio. Maximum expansion (1.247) of co-extrudates occurred when the moisture content of blend was 21% (Table 3). The expansion ratio decreased as the feed moisture was increased from 21 to 24%, but with further increase in feed moisture to 27%, the expansion value recovered, but, not to that level of 21% moisture level.

The increase in moisture content upto 33% of milk protein preparation resulting in increased index of the extruded protein product has been reported.[11] These workers concluded that humidity of raw material influenced its rheological properties in decisive manner. When the water content was higher than 33%, the viscosity of protein mass was too low as to expect that the forces of intra-molecular friction could sufficiently raise the product temperature, thus causing a structural modification of proteins[10]. The increase in moisture content has been shown to decrease expansion ratio of product obtained by extruding wheat, rice and potato blends[12].

Effect of Temperature

The expansion ratio of co-extrudates differed significantly ($P<0.01$) with temperature of extrusion (Table 3). It recorded a significant increase of 6.96% with the temperature was raised from 100° to 130°C. Thereafter, there was fall in expansion ratio with further rise in temperature to 150°C.

In the extrusion of buckwheat flour and its mixture with milk protein the maximum expansion of extrudate has been reported to occurr at 130°C. Evaporation of moisture and changes in protein and starch, leading to formation of new structure were most intensive. However, while extruding sound and sprouted wheat a decrease in the expansion ratio beyond 160°C has been reported[12].

Effect of Screw Speed

It could be inferred from Table 3 that increase in the screw speed from 75 to 125 rpm resulted in a steady and significant increase in expansion ratio of co-extrudates. While extruding defatted glandless cotton seed meal, it has been observed by earlier workers, that as the screw speed increased, expansion ratio also increased. This could be attributed to the fact that due to increase in screw speed there was a greater amount of mechanical work resulting in functional heat generation, thereby, causing an increase in temperature, which caused the material to be fully cooked and become more plastic. Therefore, as the material emerges from extruder die the flashing moisture and plastic nature of the material causes the product to expand more and cause increase in its expansion ratio.

Effect of Process Variables on the Density of Co-extrudates

Effect of Milk Protein Content

Increase in levels of milk protein content in the blend did not have any significant effect on the density of co-extrudates (Table 4). The addition of milk protein to buckwheat flour before the extrusion has been reported to affect its density adversely.[13] Product with the addition of 25% milk protein revealed a density higher in the average by 0.2g/cc, than a product without the addition of milk protein.

Table 4: Effect of various process variables on the density (g/cm) of co-extrudates

Variables	*Density (g/cm³)*	*C.D. (0.05)*
Mil protein (%)		
20	0.981	
40	0.968	
60	0.985	N.S.
Feed Moisture (%)		
21	1.070^{c}	
24	0.977^{b}	
27	0.886^{3}	0.015
Temperature (°C)		
110	0.962^{b}	
130	0.945^{b}	
150	1.027^{b}	0.015
Screw Speed (rpm)		
75	0.993^{b}	
100	0.986^{b}	
125	0.954^{a}	0.015

Values with different superscripts differ significantly ($P < 0.05$)

Effect of Feed Moisture

It is observed from the data given in Table 4 that with increase in the moisture content of blend from 21 to 27% density decreased significantly ($P < 0.01$) and showed a minimum value of 0.886 g/cm^3 at 27% moisture content.

The earlier researchers[15] reported that increased dough moisture resulted in lower bulk density. Increasing moisture content could increase the mobility of the protein molecules thereby allowing reactive sites to come into proximity and cross link adjacent molecules. This could produce an elastic and expanded matrix. This condition might cause a decreased density. Added moisture could also affect density by reducing the viscosity of dough at the die allowing it to expand to a greater extent at a given discharge temperature. The good quality products extruded from a mixture of cereal with soya, containing 28 to 30% moisture were obtained by earlier workers.[16,17]

Effect of Temperature

Temperature had a significant ($P < 0.01$) effect on the density of co-extrudates (Table 4). As the temperature increased from 110° to 130°C a fall in density was observed. However, with further increase in temperature to 150°C density increased sharply. The lowest density (0.945) was obtained at 130°C on extrusion of buckwheat flour and 25% milk protein at 100°C has been reported.[13] However, increase in temperature of the extrusion process to 130°C made it possible to obtain extrudates with a high porosity and greater fragility, with numerous air bubbles surrounded with thin walls, in the cross section.

A decrease in the density with increase in the temperature of extrusion has been reported.[11] The extruded milk protein preparation had the lowest density (0.72 kg/dcm^3) at 140°C. While extruding cotton seed meal a decrease in density with increased temperature of extrusion has also reported[14]. A similar relationship between the density and the temperature for extruded soy products produced using a Brabender laboratory extruder was also observed[18]. As the product temperature is increased, the materials are fully cooked and become more plastic. Therefore, as the cooked material emerges from extruder die the flashing moisture and plastic nature of materials causes the product to expand more and thus reduce its density[18].

Effect of Screw Speed

The changes in the screw speed had a significant ($P<0.01$) effect on the density of co-extrudates (Table 4). The density of co-extrudates decreased with increase in screw speed from 75 to 125 rpm. There was a decrease of 3.25% in density when screw speed was increased from 100 to 125 rpm.

A decrease in density of extrudate with an increase in screw speed during extrusion of buckwheat flour was recorded[13]. Earlier, apart from heating and pressure, mechanical forces were found to be important factors forming the structure during extrusion[10]. The mechanical energy supplied to the screw was used for displacement of the material along the heads of the extruder and it was partially changed into heat. About 30% energy fed to the screw has been reported to be transformed into heat which lead to an increase of the temperature of the process[15]. The increasing screw speed has been reported to result in greater shear which tends to enhance cross-linking between protein molecules during extrusion of plant protein[15].

Effect of Process Variables on the Moisture Content of Co-extrudates

Effect of Milk Protein Content

The level of milk protein had a significant ($P < 0.01$) effect on the moisture content of co-extrudates (Table 5). The moisture content increased appreciably when the milk protein content in the blend was raised to 40% from an initial 20%. However, the moisture content dropped when the level of milk protein in the blend was further increased to 60%.

Table 5: Effect of various process variables on the moisture content (%) of co-extrudates

Variables	*Moisture (%)*	*C.D. (0.05)*
Milk Protein (%)		
20	14.99[b]	
40	16.31[C]	
60	14.02[a]	0.16
Feed Moisture (%)		
21	13.61[a]	
24	15.19[b]	
27	16.51[c]	0.16
Temperature (°C)		
110	16.45[C]	
130	15.32[b]	
150	13.55[a]	0.16
Screw Speed (rpm)		
75	14.99[a]	
100	14.93[a]	
125	15.40[b]	0.16

Values with different superscripts differ significantly ($P < 0.05$)

Effect of Feed Moisture

The moisture content of the blend had a significant ($P < 0.01$) effect on the moisture content of co-extrudates (Table 5). The moisture of co-extrudates increased significantly with the increase in the moisture content of feed. A linear effect of feed moisture on the final moisture content of the extrudates has been reported[19].

Effect of Temperature

The effect of variation in temperature of extrusion on the final moisture content of co-extrudates was found to be significant ($P < 0.01$) [Table 5]. As the temperature of extrusion was raised from 110° to 130°C and subsequently to 150°C. The final moisture of co-extrudates lowered significantly from 16.45 to 15.32% and then to 13.55%, showing a decrease of 6.87 and 11.55%, respectively.

Similar results have been reported for extrusion cooking of triticale, com starch and full fat soybean[19,20]. The importance of temperature can be readily gauged by the increased extruder air temperature differential, leading to greater moisture losses as the product leaves the die[19].

Effect of Screw Speed

The screw speed had a significant (P<0.01) effect on the moisture content of the co-extrudates. The co-extrudates obtained at screw speed of 75 and 125 rpm did not differ significantly in moisture content (Table 5). However, extrudates produced at 100 rpm had significantly higher moisture content as compared to those obtained at 75 and 125 rpm. It has been found that screw speed determines dwell time which in turn influences the product moisture[19].

Effect of Process Variables on the Overall Acceptability Scores of Co-Extrudates

Effect of milk protein content

Incorporation of increasing levels of milk protein in blend, improved the overall acceptability scores of co-extrudates although the effect was found to be statistically non-significant (Table 6).

Table 6: Effect of process variables on the overall acceptability scores of co-extrudates

Variables	*Overall Acceptability Score*	*C.D. (0.05)*
Milk Protein (%)		
20	6.60	
40	6.71	
60	6.81	N.S.
Temperature (°C)		
110	6.29^{a}	
130	6.83^{b}	
150	7.00^{b}	0.36

The organoleptic qualities of extrudates has been reported to improve by the small addition of milk protein in the defatted corn germ flour during extrusion but higher milk protein concentrations caused a decrease of organoleptic acceptability[8].

Effect of temperature

It could be inferred from result presented in Table 6 that temperature of extrusion improved the overall acceptability scores of co-extrudates. This effect of temperature was found to be statistically significant (P < 0.05).

REFERENCES

1. Bass, P.A. and Clay, H.J. Seventh World Cereal and Bread Congress *Abstr.* (1982) 94.
2. Menser, F., Van Lengerich, B. and Kohler, F. International symposium Chemistry and Technology of Starch. Krakow (1982).
3. AOAC. 1983. Official Methods of Analysis. 13th ed., Association of Official Analytical Chemists, Chicago.
4. AACC. 1990. Approved Laboratory Methods. 8th ed. American Association of Cereal chemists, St. Paul, Minnesota, USA
5. Snedecor, G.W. and Cochran, W.G. Satistical Methods. Sixth Edition, Oxford and IBH Publish. Co., New Delhi, (1968) 339.
6. Sekhon, K.S., Singh, B. and Nagi, H.P.S. Third Agricultural Science Congress held at Punjab Agricultural University, Ludhiana, (1997) 334.
7. DeMuelenaere, H.J.H. and Buzzard, J.L. *Fd. Tech.* **23**: (1969) 71.
8. Peri, C., Barbieri, R. and Casiraghi, E.M. *J.Fd. Tech.* **18**: (1983) 43.
9. Singh, R.K., Nielsen, S.S., Chambers, J.V., Martinez-Serna, M. and Villota, R. *J.Fd. Process. Preser.* **15**: (1991) 285.
10. Aguilera, J.M., Kosikowski, F.V. and Hood, L.F. *J.Fd. Sci.* **41:** (1976) 1209.
11. Szpendowski, J., Smietana, Z. and Zuraw, J. *Milchwissenschaft*. **38**: (1983) 577.
12. Singh, N., Sekhon, K.S. and Nagi, H.P.S. *J.Fd. Sci. Technol.* **36**: (1994) 233.
13. Smietana, Z., Fornal, L., Szpendowski, J., Soral-Semietana, M. and Fornal, J. *Acta Almen. Polonica.* 11 : 275 (FSTA. **18**:11M53, 1986).
14. Taranto, M.V., Meinke, W.W., Cater, C.M. and Mattil, K.F. *J.Fd. Sci.* **40**: (1975) 1264.
15. Holay, S.H. and Harper, J.M. *J. Fd. Sci.* **47**: (1982) 1869.
16. Harper, J.M. *Fd. Tech.* **32**: (1978) 67.
17. Nagucki, A., Kugimugs, W., Hagane, Z. and Saio, K. Seventh World Cereal and Bread Congress. *Abstr.* **91** (1982).
18. Gumming, D.B., Stanley, D.W., and deMan, J.M. Can. *Inst. Fd. Sci. Technol.* J. **5**: (1972) 125.
19. Owusu-Ansah, J., Voort van de, F.R. and Stanley, D.W. Can. *Inst. Fd. Sci. Technol.* J. **15**: (1982) 257.
20. Howath, E. and Czukor, B. *Acta Alimen. Polonica* **22**: (1993) 151.

Maintenance of Glutathione in Vegetables by Modified Atmosphere Packaging

D.R. RAI
Scientist
Central Institute of Post Harvest Engineering & Technology
PAU Campus, Ludhiana 141 004 (Punjab)

ABSTRACT

QUALITY AND QUANTITY OF THE fresh produce changes rapidly after harvest through respiration and loss of moisture, leading to physico-chemical changes resulting in the losses which have been reported to the order of 5-25% in developed countries and 20-50% in developing countries. Adequate and appropriate packaging of the vegetables may protect them from microbial and chemical contamination, oxygen, light and water vapour. Glutathione is a tri-peptide composed of three amino acids and it retards the aging process in human body system by decreasing the frequency of DNA breaks and aiding in their repair, wherever necessary. It is a powerful antioxidant and is vital for optimum healing and survival. However, glutathione levels decrease with old age and they need to be supplemented through external sources especially, food intake. It is present in most of the vegetables and mushrooms and deteriorates under normal atmospheric conditions. Therefore, it requires to be preserved or maintained through the use of proper packaging materials.

This paper discusses the efforts made to preserve glutathione in some vegetables by packing them in different type of packaging materials and keeping them under long term storage in cold rooms. Qualitative analysis of the stored vegetables was carried out throughout the storage and optimisation of packaging materials vis a vis stored vegetables was done in order to maintain/preserve glutathione.

INTRODUCTION

Fresh produce changes in quantity and quality between harvest and consumption. These losses have been observed to the order of 5 to 25% in developed countries, while they are of the order of 20-50% in developing countries, depending upon the commodity[1]. Fruits, vegetables and flowers are living tissue and are subjected to continuous changes after harvest leading to loss in their quantity and quality. Moreover, fresh commodities being diverse in morphological structure (roots, stems, leaves, flowers, fruits), composition and physiology, present other problems. Respiration of the produce does not cease after harvest, therefore stored organic materials such as carbohydrates, protein and fat etc are broken into simple end products with the release of energy. The released energy affects the refrigeration and ventilation requirements of the produce. Production of ethylene by the plants affect many aspects of growth, development and senescence. Other changes that take place during development and maturation of the fresh commodity include loss of chlorophyll, development of carotenoids and anthocyanins, water loss, pathological break down etc.

These changes cannot be stopped altogether, but they can be delayed by proper handling, packaging, transportation and storage of the produce. Although low temperature storage helps in prolonging the shelf life of the stored produce, but it is not a sufficient technique as some crops suffer chilling injuries and tissue break down, and can become move attractive for pathogens. Proper packaging can protect the produce from contamination, spoilage, light, oxygen and water vapour and can lead to increase in shelf life along with maintenance of the physico chemical constituents of the stored produce. Controlled atmosphere and modified atmosphere techniques are the standard techniques being followed for creation and maintenance of a suitable gaseous environment different from the normal atmosphere around the produce. Application of modified atmosphere packaging (MAP) leads to an atmospheric condition inside the package depending upon the permeability characteristics of the packaging film to O_2, CO_2, N_2, ethylene and water vapour etc. This technique is being followed mostly for whole sale or retail packs.

Glutathione is a tripeptide composed of amino acids namely cystine, glycine and glutamic acid. In human body system it retards the process of aging by decreasing the frequency of DNA breaks and aiding in their repair, wherever necessary. Glutathione is a powerful antioxidant and is readily available from pine bark extract, bilberry extract and turmeric etc. However, glutathione levels decrease with old age and they need to be supplemented through external sources especially by food intake. It is present in most of the vegetables and mushrooms and deteriorates during storage under normal atmosphere. Therefore, its needs to be protected through the application of suitable packaging materials. This paper discusses the efforts made to preserve glutathione in some vegetables by packing them in different type of packaging materials and keeping them under long term storage in cold rooms.

MATERIALS AND METHODS

Preliminary experiments were carried out to determine the glutathione content of various vegetables and mushrooms in order to select the suitable ones for the study. For this purpose, fresh vegetables were procured from a local packaging house and were immediately transferred to cold room maintained at 15°C. Vegetables such as asparagus and oyster mushroom (*Pleurotus ostreatus*) which were found to be considerably rich in glutathione were selected for the study. Unpackaged vegetable/mushroom was purchased from a local packaging house and were immediately transferred to cold rooms maintained at 15°C and 75% R.H. Packaging was carried out in the cold rooms itself and the packages were kept for storage after packing in micro-perforated OPP films having gas permeability characteristics as shown in Table 1 and in normal OPP film packages used for commercial packing. Gas concentration inside the film bags were noted down during the storage period using a gas chromatograph (Shimadzu GC-8A, Tokyo, Japan) and the samples were analysed for gulutathione concentrations. Multiple experiments were carried out using different loads of asparagus and oyster mushroom and optimisation of in pack gas concentrations vis a vis glutathione concentration was carried out. Samples were prepared by homogenising the respective vegetable for 1 min in 96.4 mM sulphosalicyclic acid and 1.205 mM Na_2 EDTA solution . The extracted samples were centrifuged at 7000 rpm and 3°C for 10 min (centrifuge GS-15 R, Beckman, USA). The supernatant was filtered through a membrane filter and 0.5 ml of the supernatant was mixed with 1 ml of 0.5 M phosphate buffer (pH 8.25) and was derivatised by forming mixed bisulfides with 50 ml of 10 mM of DTNB in 25 mM phosphate buffer (pH 7.0) and was allowed to stand for 2 min. The reaction mixture was neutralised by adding 0.25 ml of 0.5 M H_3PO_4 and was analysed by using a HPLC[2] (Jasco Corporation, Tokyo, Japan). After a series of experiments, the search for suitable micro-perforated OPP film was narrowed down to MA-C and its suitability was compared with an anti microbial film, commercially known as sakisui for asparagus. Similarly for hiratake mushroom, two micro perforated films known as MA-D and MA-E, were finally compared with commercially used OPP film packages.

Table 1: Gas Permeability Coefficients for Micro-perforated OPP films at 15°C.

Film	*Thickness (μm)*	*Gas permeability coefficients* ($mmol.m^{-1}.h^{-1}.kPa^{-1}$)		
		O_2	CO_2	N_2
MA-A	35	6.77×10^{-7}	2.02×10^{-6}	1.91×10^{-7}
MA-B	35	5.37×10^{-6}	5.57×10^{-6}	5.04×10^{-6}
MA-C	35	1.57×10^{-5}	1.65×10^{-5}	1.56×10^{-5}
MA-D	35	4.46×10^{-5}	4.54×10^{-5}	4.30×10^{-5}
MA-E	35	5.71×10^{-5}	5.99×10^{-5}	5.61×10^{-5}

RESULTS AND DISCUSSION

Results for gaseous conditions inside the film packages for asparagus are shown in Fig. 1. As evident from the data gaseous conditions inside MA-C film packages for O_2 and CO_2 stabilised after first day of storage and thereafter no change was observed in these film packages (reached to 6% and 14% for the storage period for O_2 and CO_2, respectively). On the other hand, as can be seen from Fig. 1, no equilibrium could be achieved for asparagus packed in MA-E film packages. Among micro perforated OPP film packages, under modified atmosphere packaging, under these conditions in MA-C film, glutathione content reached to 62% of its initial content in bud portion, while it became almost double in stem portion. In anti-microbial film, the gaseous conditions reached to 1% and 9% of O_2 and CO_2, respec-

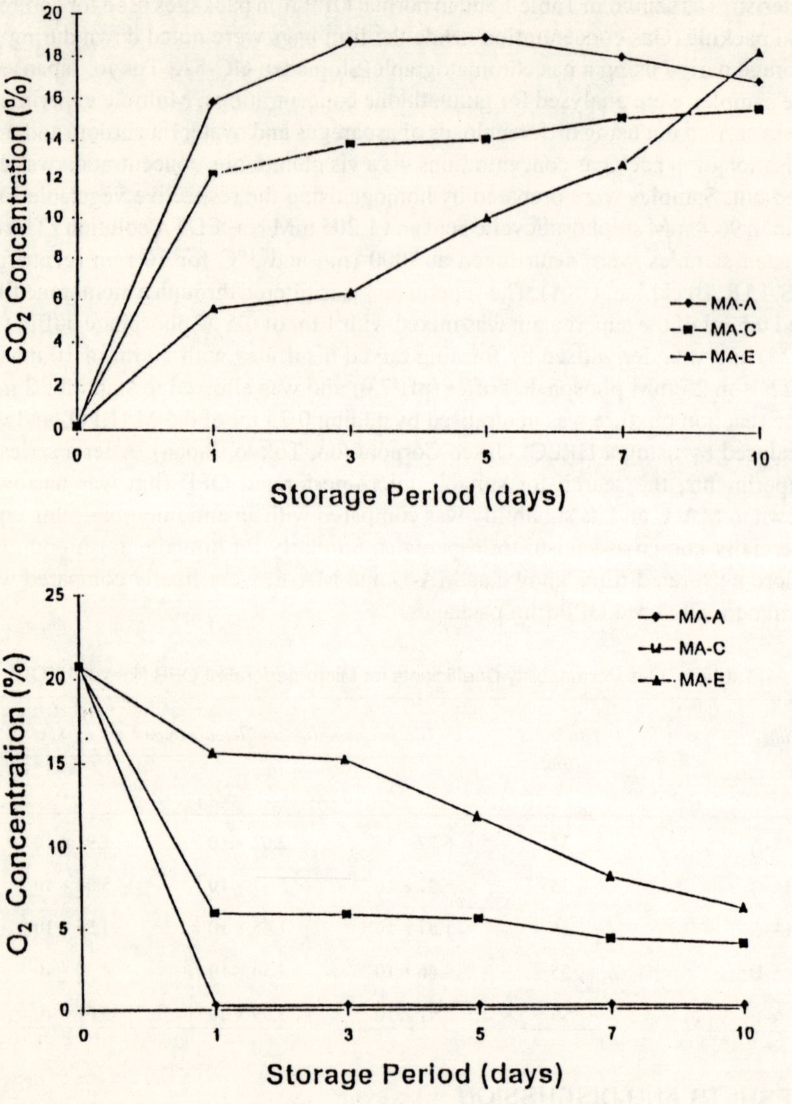

Fig. 1

tively, at the end of storage leading to anaerobic respiration which resulted in a decrease in glutathione levels in bud and stem portions (Table 2). On the other hand, control samples which were open to atmosphere lost 84% and 20% of the glutathione in bud and stem portion, respectively after nine days of storage, and also they were unfit for human consumption. Therefore, during these experiments, after packing 140 g of asparagus, was packed in a bag having dimensions of 275 × 165 mm—gas conditions ideal for glutathione maintenance were achieved.

Table 3 shows the results for hiratake mushroom under modified atmosphere packaging. As evident from Table 3, in MA-D and MA-E film packages,

Table 2: Changes in glutathione content among different type of film packages for asparagus spears after 9 days of storage at 15°C and 75% R.H.

Package Type	*Glutathione (mg/100g fw)*			
	Bud		*Stem*	
	Initial	*After 9 days*	*Initial*	*After 9 days*
MA-C	83.66	52.00	40.29	79.49
Anti-microbial	83.66	21.00	40.29	26.42
Control	83.66	13.79	40.29	32.00

Table 3: Changes in glutathione content among different type of film packages for hiratake mushroom after 9 days of storage at 15°C.

Package Type	*First Experiment*				*Second Experiment*			
	Glutathione (mg/100g fw)				*Glutathione (mg/100g fw)*			
	Pileus		*Stipe*		*Pileus*		*Stipe*	
MA-D	38.0a	37.8[b]	12.3[a]	15.5[b]	14.7[a]	19.8[b]	11.9[a]	21.5[b]
MA-E	38.0	34.6	12.3	20.0	14.7	19.1	11.9	17.1
OPP film	—	—	—	—	14.7	16.0	11.9	9.5

[a] Initial [b] After nine days [c] Not used

glutathione was more or less stable throughout the storage period in pileus as well as in stipes of the mushroom. Whereas, in case of commercially used OPP film packages, till nine days although not much decrease in glutathione content was observed, but these samples were unfit for human consumption after third day of storage due to nauseating odour emanating from them. These results can be attributed to the gaseous conditions, prevailing in these film packages, as shown in Fig. 2, where micro perforated OPP (MA-D and MA-E) film packages had an atmosphere to the range of 12-14% O_2 and 7-9% CO_2, and OPP film packages had anaerobic gas conditions, for a hiratake sample of 50 g having bag size of 275 × 220 mm.

Taking into account the data obtained from different experiments, it was concluded that using MA-C film package having dimensions of 275 × 165 mm and asparagus weight of 140 g, asparagus can be stored for nine days maintaining glutathione content whereas, by packing 50 g of hiratake mushroom in MA-D/MA-E film packages having bag dimensions of 275 × 220 mm, it can be stored safely for 9 d, as compared to commercial OPP film packages.

Variation of In Pack Atmosphere for Hiratake Mushroom under MAP in MA-D Film Packages

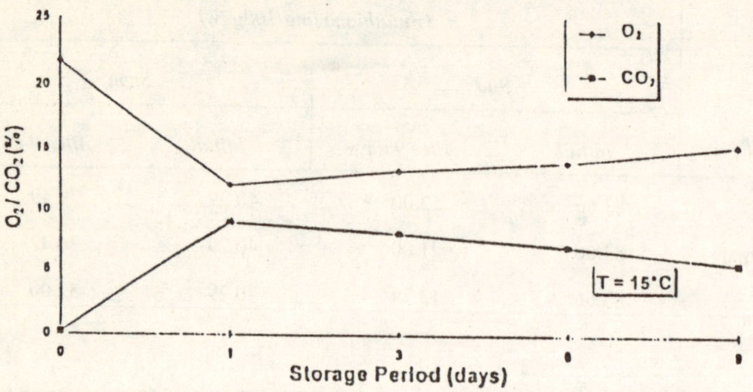

Variation of In Pack Atmosphere for Hiratake Mushroom under MAP in MA-E Film Packages

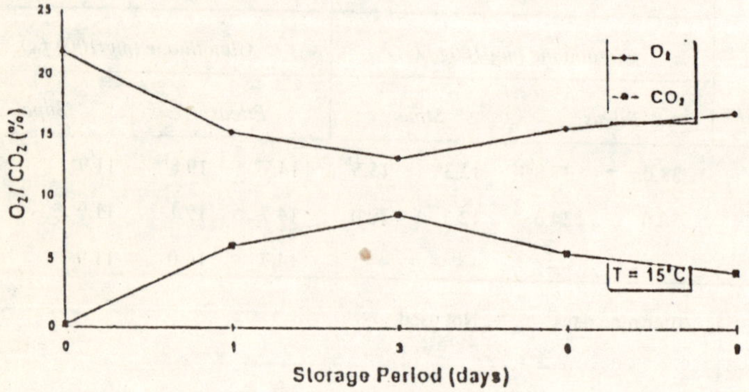

Variation of In Pack Atmosphere for Hiratake under MAP in Commercial OPP Film Packages

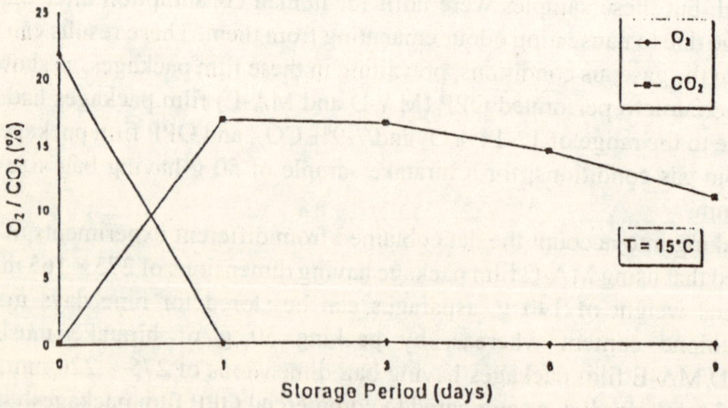

Fig. 2.

REFERENCES

1. Kader, A.A. Post harvest Technology of Horticultural Crops. Publication 3311, University of California, Division of Agriculture and Natural Resources, Oakland, California. (1992).
2. Hirata, T., Nakatani, A., Ishikawa, Y., Yamada, C. and Katsuura, S., *Nippon Shokuhin Kogyo Gakkaishi*, **42**: (1995) 996.

Development of Value Added Products from Tissue Cultured Banana

K.K. Surendranathan, N.K. Ramaswamy
and
R.K. Mitra

Bioprocess Development and Biochemical Applications Section
Nuclear Agriculture and Biotechnology Division
Bhabha Atomic Research Centre
Mumbai 400 085

ABSTRACT

Banana is the largest grown and consumed fruit crop in India. The emergence of tissue culture technique facilitated the easy availability of a large number of plantlets to farmers and this revolutionised banana cultivation. This resulted in several fold increase in its production. However, because of the highly perishable nature this nutritive fruit has a very short post-harvest shelf life. The increased production and the perishable nature could eventually result a market glut and price crash. This could adversely affect banana cultivation. Development of non-conventional value-added products from banana could increase the demand for this crop, thus sustain and encourage it's cultivation. The major basic products developed are banana juice, ripe and raw banana powder. Unlike other fruits, banana cannot be extracted by squeezing or crushing. A technique was developed to activate the *in vivo* pectolytic enzymes and inactivate pectin-forming enzymes. The yield was 600-700 ml juice/kg. Generally ripened banana forms a sticky mass when dried or powdered. Appropriate processing of ripe banana and further dehydration and grinding could develop a very fine powder having natural banana flavour. Raw banana powder though not a novel product was useful for preparation of some commercially important products such as biscuits, cake and baby food. Product evaluation by members from food industry and related professional bodies as well as by the public gave very good ratings for all the products.

INTRODUCTION

Banana is the largest produced, maximum consumed amongst the fruits cultivated in India. It is affordable to all sections of the society and available through out the year. It is highly nutritive and very delicious. India ranks first among the banana cultivating countries of the world with an annual production share of 14.2% of the total harvest[1]. The other major producers are African and Latin American countries. Philippines, China and Indonesia are other major Asian producers of this fruit crop.

In India 445 thousand hectors are used for the cultivation of banana. Kerala, Karnataka, Tamilnadu and western Maharashtra are the major banana cultivating states. The annual crop production is over 10461 thousand tons with a monitory turnover of Rs. 40 Crores. Banana export generates a foreign exchange of Rs. 12.67 crores. The Gulf countries and the east European nations are the major countries to which banana is exported.

During the past two decades the large-scale cultivation of this fruit crop has undergone considerable changes. Banana is perhaps one of the major crops which has accepted tissue culture as a mode of propagation especially in India. Tissue culture helps in rapid multiplication of plants from certified and disease free plant germplasm. It also helps in continuous production of the plants all through out the year and takes less time for large-scale production of the plants. Besides it provides uniformity in flowering period and shorter harvesting time with an enhanced yield of fruits. All these factors considerably popularised and enhanced the cultivation of this fruit crop with respect to the area of cultivation and its production.

The changing scenario due to the excessive production of this fruit now compels to look into several other factors. Banana when ripened is a soft and delicate fruit having a post harvest shelf life of 5-10 days. The softness makes the fruit injury prone during transportation. Further ripening related triggering of endogenous ethylene level during bulk storage makes all the fruit ripen within a day or two and very short shelf life period makes the fruit rot by the time it reaches its destination. Hence it was always considered a 'problem fruit' with respect to transportation. Because of these reasons the enhanced production of this fruit resulted in a local market glut and a resultant price crash. This generated considerable disinterest amongst the farmers cultivating it on a large scale.

It is most important to overcome this problem and generate increased demand for this fruit. One of the ways is by probing the possibilities of converting this into a cash crop by developing products of commercial interest. A number of conventional products prepared from banana are already available in the market. They are banana puree, wafers and jam. Amongst these, jam or wafers can be made from few varieties only. Further these products meet the market demand and hence no scopes for further production. Hence development of some non-conventional products, which could have multiple market potentials may be able to salvage this crop. Two of the basic products that could find wide applications in food industry are banana juice and banana powder.

Fruit juices are the most common and demanding product made out of most of the fruits. They also rank first amongst all products developed from fruits. Juices are extracted by crushing the fruits. However, some fruits like banana, though have over 80% moisture content, is not used for juice production, as it is not possible to extract juice by simple means. The major factor responsible for this is the fruit consistency due to the high pectin and other oligo-saccharide contents. Hence development of a process for the production of banana juice could have high commercial potentials. Further because of its abundance throughout the year, the impact the banana juice could make in the industry could be substantial.

Another product that could be developed is banana powder. Raw banana powder is a well-known product being used for ages especially as a healthy infant food. The technology involved is quite simple; peel the raw fruit, cut the pulp into thin discs and grind it into fine powder after drying. However preparation of ripe banana powder is not that simple. The ripe fruits when dried after all these process turns into a sticky mass due to high sugar content, oligosaccharides and pectin. Development of a successful process to make ripe banana powder, thus could be quite novel. Further, if proper applications having enormous demand can be identified for both the raw and ripe powder, banana will be one of the fruits in high demand always.

This paper presents the development of processes for the production of banana juice without adding any external ingredients such as enzymes or water; and banana powder from ripe banana. It also examines the scope of development of commercially important products from these products.

MATERIALS AND METHODS

Processing of Banana

Tissue cultured bananas harvested from the field were separated into hand cleaned and the latex flow was allowed to stop. An antifungal paste made out of a commercial antifungal agent, Bavistin, was applied on the surface of the cut area. The fruits were stored in perforated polythene bags (700 gauge) for ripening at 25°C. [Varieties of banana grown by the Plant Cell Culture Technology Section, NA & BT Division from the tissue cultures of banana was used for these studies]. 'Basrai 10 Grey' a variety obtained on irradiation of multiple shoots of 'Basrai' banana was the major variety used for these studies. Other varieties were Williams, Grandnine, Panamala, Trikoni, and Lalkela. The ripened fruits were peeled and the pulp was thoroughly homogenised in a Waring Blender. To the pulp, 2% (w/w) of Acetone Powder prepared from the peel was added (Banana peel acetone powder was prepared by homogenising the banana peel in chilled acetone (–80°C) and immediate filtering off and drying the powder at 0°C) and shaken for two hours. The pulp was then centrifuged at 5000 rpm for 30 min or filtered through a filter press. The juice obtained was bottled and autoclaved at 15lb for 10 min and stored.

Preparation of Raw and Ripe Banana Powder

Raw banana powder was prepared, using the conventional procedure. The freshly harvested fruit was peeled off and the pulp was cut into thin slices and dried. The dried pulp was then finely powdered, packed and stored.

Ripe banana powder was prepared after partial de-sugaring of the homogenised pulp. The de-sugared pulp was then spread into a thin layer and dried. This dried pulp was powdered, packed and stored under desiccation.

Development of Products from Banana Powder

The major products developed by using banana powder which have good commercial scope were banana biscuits, banana cakes and banana baby food. Normal flavouring ingredients required for the production of biscuits or cake were added to the banana pulp, which replaced the major constituent, wheat flour (Maida). Other ingredients were according to the standard procedure though the flavouring ingredients can be modified depending on the taste requirements.

The banana juice and the products developed from banana powder were analysed for sugar content, solid content, pH, ash content etc. The food sample/extract was passed through Dowex-1 and Dowex-50 columns for separation of the organic and amino acids. The reducing sugar concentration was determined by the Arsenomolybdate method and non-reducing sugar was assayed by Thiobarbiturate[6] method using the respective standard graphs. The total solid content was measured by dehydrating the sample at 60°C and weighing after cooling. The drying process was continued until no change in mass was observed during consecutive weighing. The fibre content and ash content were determined as detailed in 'Nutritive Value of Indian Foods.

RESULTS AND DISCUSSION

Banana juice harvested and stored in perforated polythene bags at room temperature ripened uniformly. The fully ripened fruit was processed for the production of the juice.

Banana Juice: The juice obtained by the method detailed earlier could extract out 75 to 85 % of the total moisture content. The juice yield was 650 to 700 ml per kg pulp. A delicious, golden yellow viscous juice having an excellent keeping quality was obtained (Fig. 1). The yield varied from variety to variety. The maximum yield (700ml/kg pulp) was obtained from 'Basrai-10 Grey' variety of fruits. Relatively lesser juice was obtained from 'Lalkela' variety (500-550 ml /kg pulp). From fruits like pineapple, oranges or mangoes, juice is extracted out by simple squeezing or crushing. These fruits also yielded only a maximum of 85% of total moisture content. The process acquired further significance due to the fact that no substance other than banana was added to facilitate extraction of the juice.

Fig 1. Taste Panel Studies of Banana juice.

The juice obtained was having an excellent shelf life of 6 to 8 months at 2 to 4°C under pasteurised condition. Depending on the variety, the specific gravity of the juice varied between 1.08 and 1.12; the pH, between 4.4 to 4.7; solid content, between 24 to 27%and total sugar between 16 to 28% and ash content from 0.4 to 0.6%. Taste Panel studies conducted amongst panellists (30 members) and general public (more than 120 members) adjudged the juice an excellent drink.

The de-sugared ripe banana powder contained 8 to 12% soluble sugar. However, probably due to a high concentration of oligo-saccharide, a sticky feeling was observed on direct consumption of the powder. Mixing of this powder with milk or other beverages gave a tasty drink.

Development of products using ripe or raw banana powder or a combination of both was also attempted. The most successful products were biscuits, cake and infant food.

Biscuits: Biscuits made using ripe or raw banana powder as the major ingredient appeared identical to the control samples in which wheat flour (Maida) was used. The textural quality was comparable to the commercial samples when tested on an 'Instron' unit. These studies also revealed that even biscuits made using raw banana of plantain variety, which has a large fruit bulk ('Manthan') was found to be of excellent quality. Addition of ripe banana powder gave more banana flavour to the product.

The taste-panel studies conducted employing expert panellists (30 members) and general public (more than 120 members) adjudged the banana biscuits novel, tasty and appealing (Fig. 2). Further it provided 15-fold higher fibre content when compared to the control sample prepared using 'Maida' only.

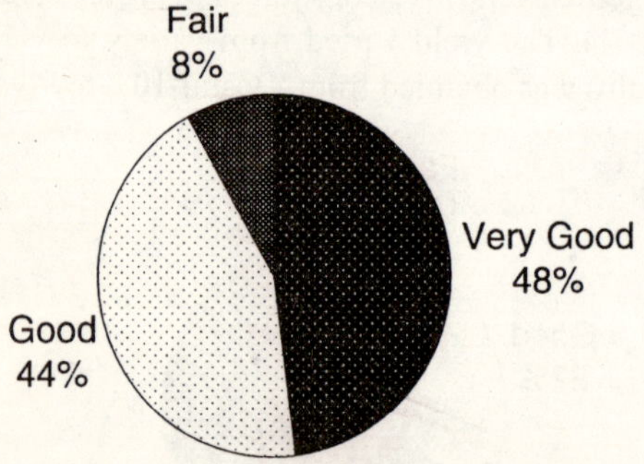

Fig. 2. Taste panel assessment of banana biscuits

Cake: Banana cake also showed excellent product quality as assessed by the taste-panel members (Fig. 3). The softness, physical consistency and appearance of this product were comparable to commercial products. Further, it was also shown that the natural banana flavour or modified choice flavour could be obtained from this product.

Baby food: Another major product developed using banana powder was baby food. Traditionally, infants of 2 months to two years of age are fed with a porridge

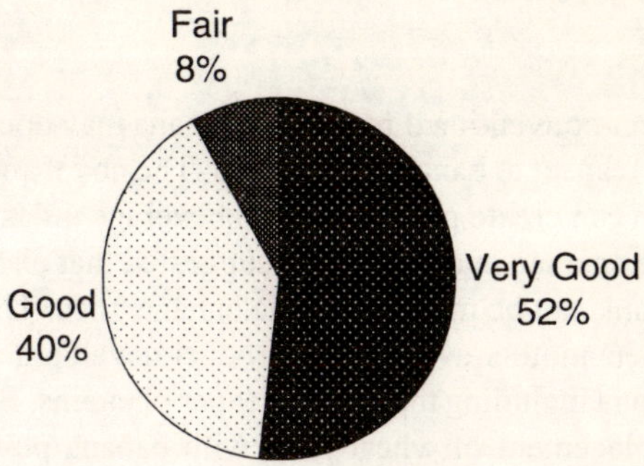

Fig. 3. Taste Panel Studies of Banana Cake

made using raw banana powder especially in southern India. Taking cue from this, using raw banana powder (Safed Velchi) an infant food has been developed during the course of present study. Besides raw banana powder other ingredients were, germinated wheat flour and a bioactive preparation made by lyophilising raw banana. The banana varieties tested were 'Safed Velchi' (Kunnan), 'Grandnine' 'Williams' and 'Panamala'. The bioactive compound served as an active enzyme source, ensured the digestibility of the baby food. The porridge was prepared by making a paste of the baby food with warm water (~60°C). For maximum hydrolytic activity, it was kept for 10 min before feeding it to the baby. This product was nutritionally very much comparable to some of the commercially available baby foods.. Proximate analysis[8] suggested that this preparation contained 10% protein, 1.5% fat, 80% carbohydrate, 2.3% ash and a calorific value of 370 Kcal. The major minerals were Manganese, Zinc, Iron, Sodium, Potassium, and Magnesium. Taste-panel evaluations suggested that 'Safed Velchi' was the most preferred variety and the acceptability was as good as the commercial product (Fig. 4). Though the cost structure has to be calculated taking into account the marketing strategies, from the cost of raw materials and processing required it can be assumed that it should be several fold cheaper than the cheapest baby food available in the market. Besides, being very cheap, reaching this infant food to the lowest income group could help to a great extent in solving the malnutrition problem amongst the infants of this class.

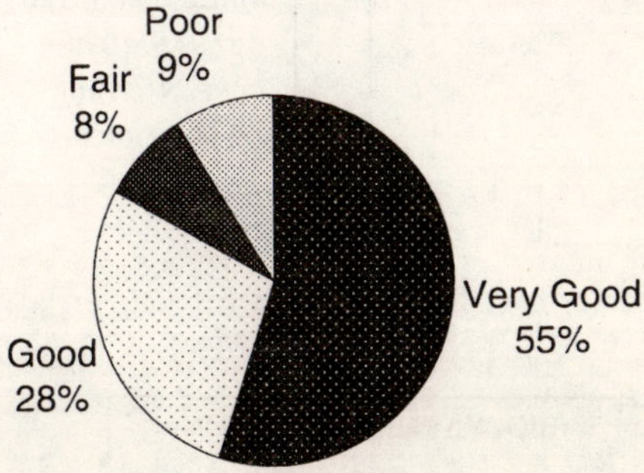

Fig. 4. Taste Panel Studies of Banana Baby Food

CONCLUSION

Development of non-conventional food from banana may open the floodgate for a new thinking with respect to banana cultivation. The novel and simple technology of juice production can create great impact to beverage industry. The studies carried out by Surendranathan and coworkers suggested that excellent wine and carbonated banana drink can be made using the banana juice. The production of biscuits and cakes open another avenues for product development and utilization of all varieties of banana including the much cheaper plantains. Besides being a nutritive substitute, replacement of wheat flour with banana powder for biscuit and cake production especially in places where wheat has to be transported further enhances its commercial significance. Development of banana baby food highlights the scope to eradicate malnutrition amongst poor children. Thus development of non-conventional products from banana besides sustaining the interest of farmers in cultivating this most nutritive fruit crops, also make it an industrially important and viable commodity. Figure 5 depicts the scope for product development from banana.

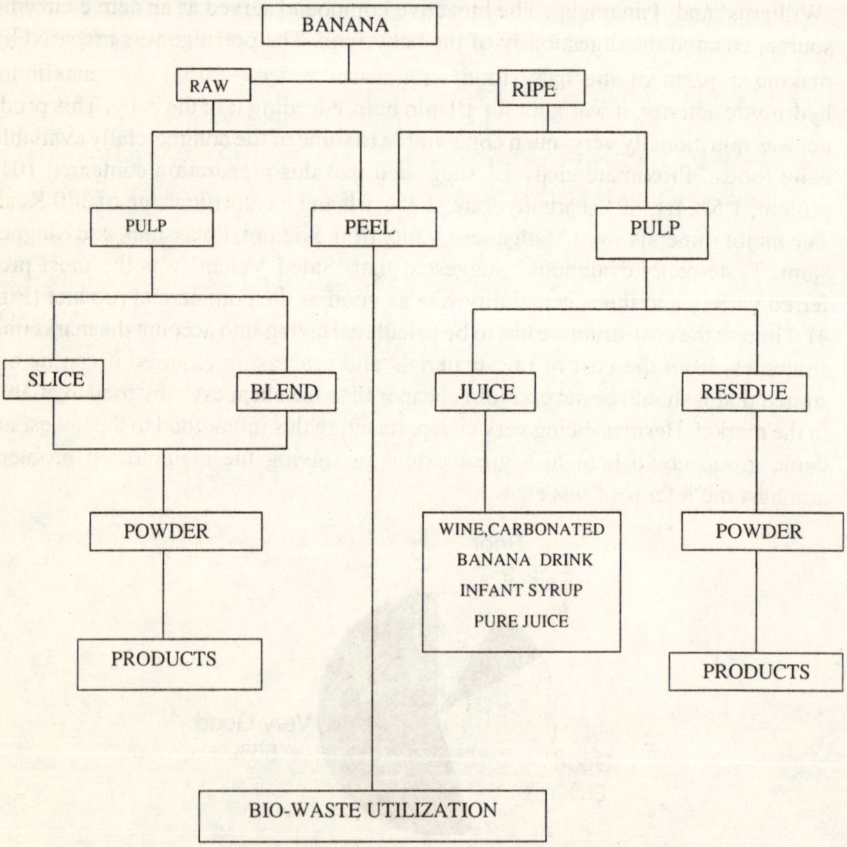

Fig. 5. Flow-chart on development of commercially important products from banana

Acknowledgement: The authors sincerely thank Ms. Cendry Cardoz, Tulsi Jayachandran, (Summer Trainees) and Jagdish Nair (DAE Fellow) for their help at all stages of this projects. Our sincere thanks are also due to Dr. V. A. Bapat, Head, Plant Cell Culture Technology Section, and his sectional colleagues especially, Dr. T. R. Ganapati, Mr. V. A. Kulkarni, and Mr. L. Srinivas.

REFERENCES

1. FAO Report, 1998.
2. Surendranathan, K.K. and Nair, P.M., *Phytochemistry*, **11** (1972) 119 .
3. Surendranathan, K.K. and Nair, P.M., *Plant Sci. Lett.*, **12** (1978) 169.
4. Surendranathan, K.K. and Nair, P.M., *Phytochemistry*, **12** (1976) 241 .
5. Surendranathan, K.K. and Nair, P.M., *Phytochemistry*, **15** (1973) 371 .
6. AOAC; Official Methods of Analysis, Washington DC 55, (1990).
7. Gopalan, O., Rama Sastri, B.V. and Balasubramanyam, S.G. In: Nutritive value of Indian Foods (Revised), NIN, ICMR Hyderbad, (1988).
8. Thantry, S.S. and Kumar, S.C. Elemental Analysis in Biological Systems. *Analytical Chem. Div.*, BARC, (1998).

Temperature Management in Relation to Post-composting Supplementation during *Agaricus bisporus* Cultivation

S. GARCHA *and* P.K. KHANNA

Department of Microbiology
Punjab Agricultural University
Ludhiana 141 004 (Punjab)

ABSTRACT

White button mushroom *Agaricus bisporus* (Lange) Sing is the most widely cultivated variety throughout the world. In India, it is mainly grown by seasonal growers in indoor environmental conditions where the yields are low. Hence, it is imperative to utilize supplementation as an important tool to enhance yields. Supplementation in composts increases temperature which can prove to be detrimental to the growth and spread of the *A. bisporus* mycelium. Therefore, supplementation calls for judicious use and greater management to counter this increase in temperature.

INTRODUCTION

White button mushroom *Agaricus bisporus* (Lange) Sing is the most widely cultivated variety throughout the world which constitutes about 37.6% of the total world production of 4.9 million tonnes.[4] In India, more than 95% of total mushroom production estimated to be 50,000 tonnes is of this variety. In India, *A. bisporus* is mainly cultivated by seasonal growers under highly impoverished indoor environmental conditions where yields are low due to narrow adaptability of the strains used and their poor performance under stress conditions. Different approaches have been used to improve the productivity of strains and post-composting supplementation with nitrogen rich material is one such practice. However, over-heating of the compost normally occurs upon supplementation which is sometimes detrimental to the spawn growth as nitrogen increases the rate of metabolism. Adequate provision to control the compost bed temperature after supplementation and during cropping, therefore, has to be made as supplementation imposes additional demands on management.

MATERIAL AND METHODS

Agaricus bisporus strain U-3 was used for the cultivation which was carried out in natural environmental conditions at Mushroom Research Complex, PAU, during the seasons 1997-98 and 1998-99. Wheat straw based compost was prepared by the long method of composting spread over 24 d by following the composting schedule of –2, 0, 4, 8, 12, 15, 18, 21, 24d. The standard wheat straw based compost with 1.5% N was prepared with the following formula:

Compost Ingredients	*Dry weight (Kg)*	*%N*	*Total N (Kg)*
Wheat straw	300	0.4	1.20
Wheat bran	15	5.5	0.82
Calcium Ammonium Nitrate (CAN)	9	23.0	2.07
Urea	3	46.67	1.40
Super-phosphate	3	—	—
Muriate of potash	3	—	—
Molasses	5	—	—
Gypsum	30	—	—
Lindane (20%)	0.15	—	—
γ-BHC	60 ml	—	—
	368.15 g		5.49

Per cent nitrogen in the compost is calculated as:

$$\%N = \frac{\text{Total N}}{\text{Dry weight}} \times 100$$

The amount of urea was adjusted to obtain other compost mixtures with initial nitrogen level of 1.25 and 1.75 per cent. The total weight of all ingredients was maintained by adjusting the amount of gypsum added.

Agaricus bisporus strain U3 was used for cultivation using the two supplements namely – cottonseed meal and soyabean meal added at the time of spawning and at the time of casing. Mushroom cultivation was carried out in polythene bags (18″ × 24″) containing 7 kg of fresh compost per bag filled upto a depth of 10″–12″ which was spawned thoroughly @ 0.6-0.8% on fresh weight basis of compost. Five replicate bags were used for each treatment. The bags were arranged in a room with sufficient light and cross ventilation facilities. Maximum humidity in the room was maintained by spraying water on the floors and the walls. After complete impregnation of compost in about two weeks, casing was done with formaldehyde disinfected casing soil mixture of FYM: sandy soil (4:1, v/v) added @ 2–2.5 kg per bag to a uniform thick layer of 1″–1.5″. Spraying of water was than continued directly on the cased bags till the end of cropping.

Two supplements namely, cottonseed meal (CSM) and soyabean meal (SBM) were added @ 1.5% at the time of spawning and at the time of casing by thorough mixing. Yield data was recorded after the pinheads started appearing and number of fruit bodies obtained were also counted. A temperature profile i.e. temperature in the bags was recorded immediately after supplementation for a period of 7 d till it dropped to the level of control, i.e., unsupplemented bags.

RESULTS AND DISCUSSION

In this investigation temperature was the control parameter. The temperature for a period of 7 d was recorded immediately after supplementation. Following supplementation with cottonseed meal an increase in temperature to a maximum of 6°C was observed irrespective of the time of addition in all the three composts i.e. 1.75, 1.5 and 1.25% N. In case of soyabean meal it rose to a maximum of 7.5°C in 1.75%N compost when supplementation was carried out at the time of casing. In other two composts it gave a temperature of 7°C (Figs. 1 and 2). The temperature rise is not conducive for the growth and spread of *A. bisporus* mycelium. After casing it is essential that the temperature remains low as the mycelium changes from vegetative state to fruiting sporophytic stage. The increase in temperature beyond a certain level proves detrimental to the mycelium's growth which is also reflected in the yield data.

The experiment was repeated using two supplements on compost with 1.5% N level as it gave the best and consistent performance. Again rise in temperature upon supplementation was recorded and it was observed that the rise in temperature with cottonseed meal and soyabean meal supplementation was upto a maximum of 6°C and 7°C above the temperature recorded in the unsupplemented bags, respectively (Fig. 3). As observed earlier the increase in temperature peaked on the fourth day and then decreased to the level of unsupplemented bags. This increase was more with soyabean meal and less with cottonseed meal. The corresponding yield performance is given in Table 1.

Table 1: Effect of incorporation of cottonseed meal and soyabean meal @ 1.5% at the time of spawning and casing on the yield of *A. bisporus* (U3)

Supplement(s)	*At Spawning*		*At Casing*	
	Total Fruit body		*Total Fruit Body*	
	Number	*Yield (g)*	*Number*	*Yield (g)*
Cotton seed meal	700	9455	730	10275
Soyabean meal	477	7080	474	6470
Control	413	6155	—	—

Crop season : Nov.- Jan.; Yield data – 50 days

The experiments were conducted under natural environmental conditions when the prevailing room temperature was around 25-27°C. The negative effect of addition of supplements could be attributed to temperature increase which proved detrimental to the growth of *A. bisporus*. This was corroborated by the observation that the increase in bed temperature was lesser with the addition of cottonseed meal that gave the maximum yield.

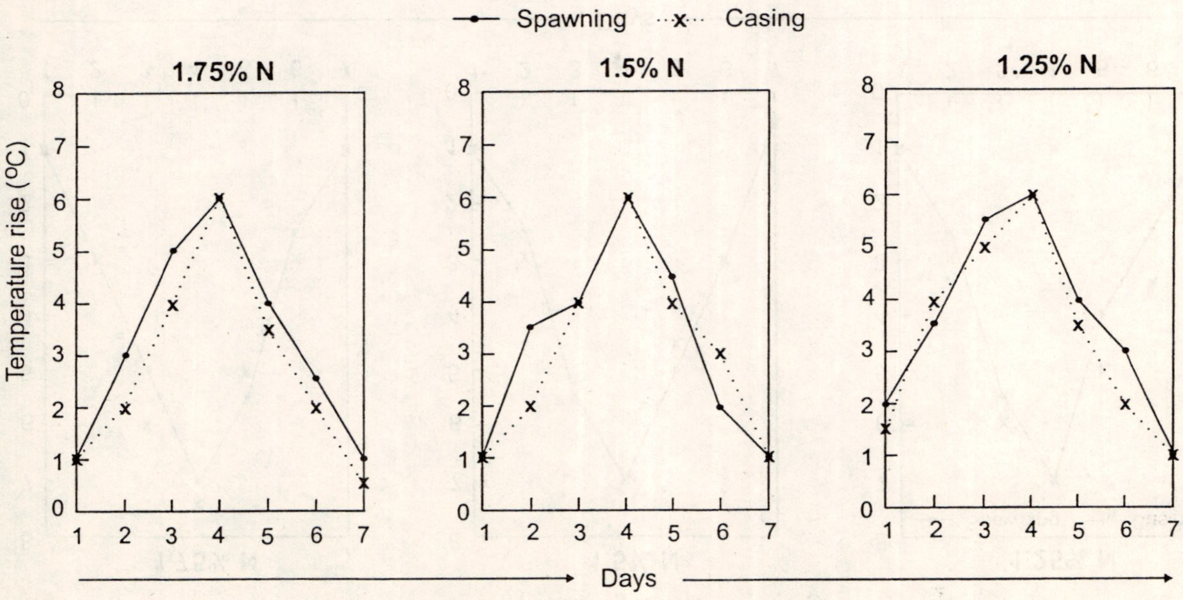

Fig. 1. Temperature rise in compost with different nitrogen levels after supplementation with cotton seed meal.

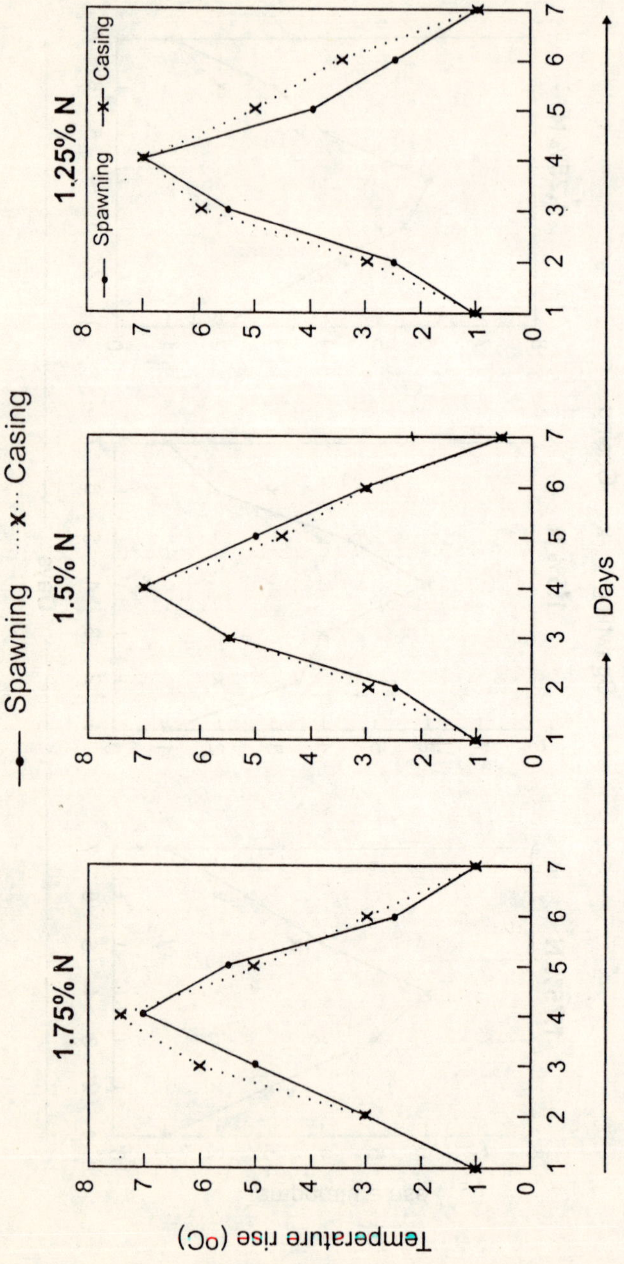

Fig. 2. Temperature rise in compost with different nitrogen levels after supplementation with soyabean meal.

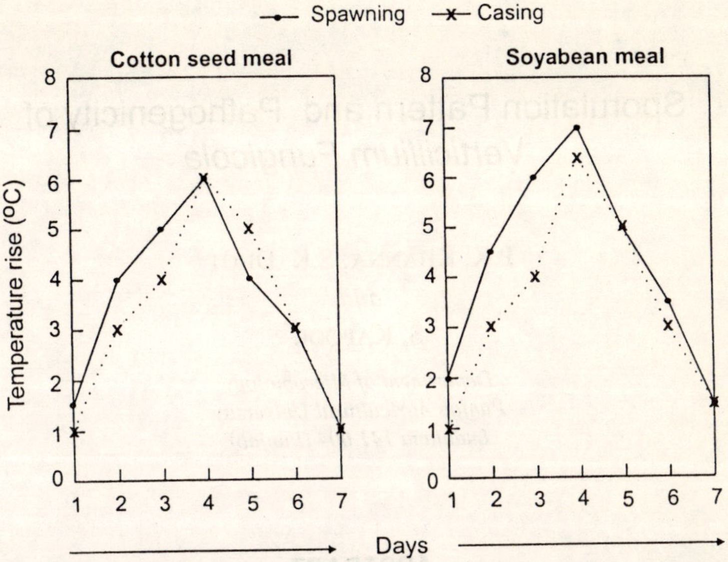

Fig. 3. Temperature rise in the bags after supplementation.

It was earlier observed that overheating of the compost occurred upon supplementation as nitrogen increased the rate of metabolism.[2] It was further emphasized that adequate provision to control compost temperature after supplementation and during cropping must be made as supplementation imposes additional demands on management.[3] Supplementation rates should be determined in a way that one is in a position to respond quickly to temperature rises and further that upper limit for supplement use is the level at which temperature surges can be controlled and is justified by higher returns in mushroom production.[1]

REFERENCES

1. Dahlberg, K.R., *Mushroom News*, **7** (1993) 4.
2. Randle, P.E., Fermor, T.R., Flegg, P.B., Matcham, S.E., Smith, J.F. and Wood, D.A., *Mushroom J.*, **129** (1983) 345.
3. Randle, P.E., *Mushroom J.*, **152** (1985) 263.
4. Sharma, S.R. In: Advances in Mushroom Biology and Production (Eds. R.D. Rai, B.L. Dhar and R.V. Verma), Nirmal Vijay Printers, New Delhi, (1997) 193–203.

Sporulation Pattern and Pathogenicity of *Verticillium Fungicola*

P.K. KHANNA, S.K. DHOT
and
S. KAPOOR

Department of Microbiology
Punjab Agricultural University
Ludhiana 141 004 (Punjab)

ABSTRACT

Verticillium fungicola causes dry bubble diesease and hence considerable loss to the crop of white button mushroom *Agaricus bisporus*. Studies revealed spore production and germination to be maximum at 20°C on a medium adjusted to a pH of 6.5. These spores could withstand a temperature treatment of 50°C for 1 h. in an aqueous state and 110°C for 1 h. in a dry state. Interaction studies revealed both *V. fungicola* and *A. bisporus* growing side by side without inhibiting the growth of each other. Pathogenicity studies, however, showed that a spore load of 10^6-10^7 spores/cap produced the most severe symptoms on mushroom caps after 48 h. This study can help to design a strategy for the control of dry bubble disease of white button mushroom caused by *V. fungicola* infection.

INTRODUCTION

The white button mushroom (*Agaricus bisporus*) is the most abundantly cultivated variety of mushroom and constitutes about 37% of the total world production of 5.0 MT of mushrooms[1]. In India this mushroom is cultivated by most of the growers on compost prepared by long method in improvised mushroom growing houses without effective environmental controls. This increases the chances of a number of microbial diseases and competitors on the beds as compared to the highly mechanized and controlled growing conditions in the advanced countries[2]. Among the various factors responsible for low productivity of button mushroom in India, fungal diseases play a major role[3,4]. The predominant fungal flora isolated from the diseased mushroom beds are *Trichoderma* spp, *Verticillium fungicola*, *Populospora byssina*, *Geotrichum* sp., *Penicillium* spp. and *Aspergillus* spp.[4,5].

The occurrence of *Verticillium* on the mushroom beds has been a common observation and has been reported by several workers. This first report of this mould from India[6] showed that it caused fungal spots on mushroom caps leading to a considerable loss of the mushroom crop. Two major groups of symptoms are associated with the infection of *V. fungicola* and the disease is named in accordance with the nature of the symptoms. The first major group of symptoms is superficial infection of mature mushroom caps causing *Verticillium* spots of cinnamon-brown colour (necrotic lesions). These spots enlarge and colease leading to complete browning of the cap. The second group of symptoms is characterized by a puff ball like mass and is called bubble or dry bubble disease. If further stipe splitting takes place it is called stipe blow out[7]. Temperature and humidity seem to be the primary factors leading to high inoculum concentration build up in the growing rooms, once the primary infection of the beds has occurred. The present study was envisaged to study the sporulation pattern and pathogenicity in *V. fungicola* in order to devise some effective control measures.

MATERIALS AND METHODS

Culture and Maintenance

The culture of *Verticillium fungicola* was procured from the Station de la Recherche sur les Champignons INRA, Bordeaux, France and was maintained on Malt Extract agar medium (Malt extract 20g, Dextrose 20g, Peptone 1g, Agar agar 20g, Distilled water 1000 ml).

Spore Production

The spore production of *V. fungicola* was observed at four temperatures (15, 20, 25 and 30°C) on malt extract medium and the spore count was taken at the end,of 19 days by taking five bits per petri plate from the sporulating area of the culture using a cork borer. This represented a total area of 1.41 cm^2. The bits were transferred to 2 ml of sterilized water blank and shaken to get a uniform suspension. The spore count was taken on a haemocytometer and expressed as spores/cm^2.

Spore Germination

The spores of *V. fungicola* were allowed to germinate on Malt Extract agar at four temperatures (15, 20, 25 and 30°C). A 0.5 ml suspension of spores with a standardized count of 30 spores/ml was spread plated and the plates incubated at the desired temperatures. The effect of pH (4.0 to 8.0) on the germination of the spores was also monitored by altering the pH of the Malt Extract medium. The effects of the temperature and pH was expressed as percentage germination of spores after 48h and 72h of incubation.

Lethal Temperature for Spores

Lethal temperatures for the spores of *V. fungicola* were determined both in aqueous suspension and in dry conditions at different temperatures. Five different time intervals (periods) ranging from 30 min to 24h for each temperature were evaluated. For the aqueous suspension, 1 ml of the suspension was given heat treatment at temperatures from 30 to 50°C (intervals of 5°C) for 30 min, 1, 2, 4, and 24h. The suspension was spread plated on Malt Extract medium and incubated at 25°C.

Spores in the dessicated state were skimmed using Whatman No. 1 discs (10 mm dia). These discs were then heat treated at 70 to 130°C (intervals of 10°C) for 30 min, 1, 2, 3 and 4h in an oven. The spore suspensions were made and spread plated as above. The results were expressed in terms of growth and percentage germination.

Interaction Studies between V. *fungicola* and *Agaricus bisporus*

The interaction between the two organisms was studied by growing them alongside in the same petri dish. On one side 6 mm bit of *V. fungicola* was placed and on the other side a bit of *A. bisporus*. The inoculated plates were incubated at 25°C for 14 d and observations recorded for growth behaviour of both the fungi.

Pathogenicity Studies of *V. fungicola*

V. fungicola was tested for its ability to produce disease symptoms on the mushroom caps by inoculating distilled water suspension of its spores or macerated mycelial extract grown in malt extract broth for 7 d. The experiment was conducted by the method of other workers[8]. The effect of spore density (spore count/cap) on the appearance of disease symptoms was also observed using this method for 72h.

RESULTS AND DISCUSSION

Spore Production and Germination

The spore production in *V. fungicola* at four different temperatures showed that maximum spore count of $106.08 \times 10^4/cm^2$ was obtained at 20°C and closly followed by the spore count at 15°C. No growth of this fungus could be obtained at 30°C (Fig. 1). It has been earlier been reported that an isolate of *Verticillium* was able to produce more than 105 spores/cm growth[9]. The intensity of spore production of two pathogens, *Mycogone pemiciosa* and *V. fungicola* and showed that *Verticillium* could form as many as 1.2×10^6 spores/cm^2 which was approximately 10 times more than the spore count of *Mycogone.*[10]

The spores produced by this organism were also tested for their germination at different temperatures. It was observed that germination started after about 24 h (Fig 1) and about 73.33% spores germinated after 48 h at 25°C. After 72 h, 100% germination was observed at 20°C. Therefore the present studies indicate that spore production and germination is best at 20°C and if lower temperature is maintained during fruiting it would be less favourable for the growth of the pathogen. The effect of the pH on the germination of spores showed that 86.5 to 100% germination took place after 72 h in a pH range of 5.5 to 7.5 (Fig. 2). The most optimum pH was 6.5 with 100% germination. The earlier workers[11] had reported an optimum pH of 6.0 for the germination of *Mycogone perniciosa.*

Lethal temperature for spores

The heat treatment of the spores of *Verticillium* showed that they could withstand a

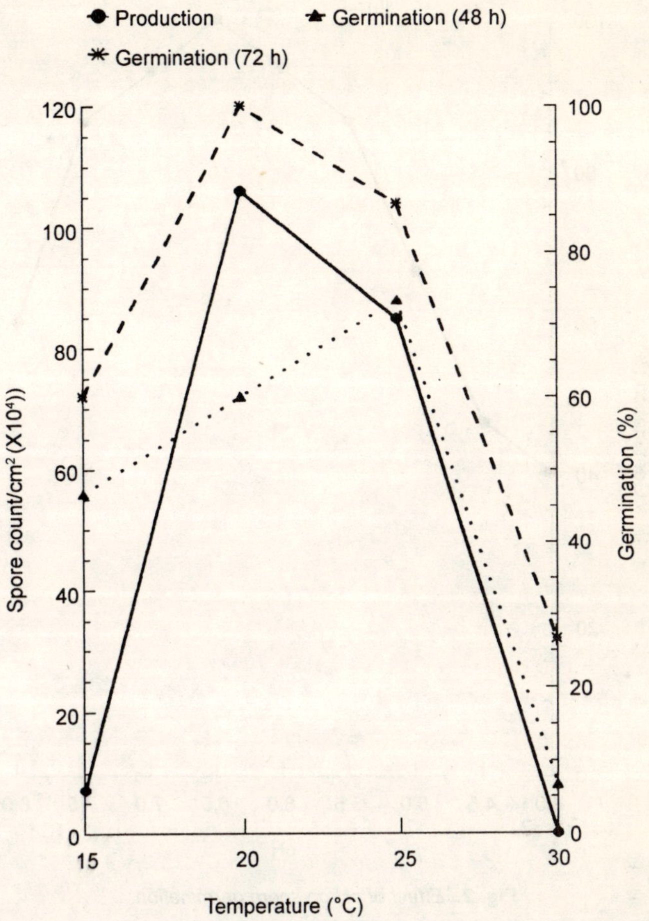

Fig. 1. *Spore production and germination of V. fungicola at different temperatures.*

temperature of 40-50°C for 1h with a residual germination efficiency of 6.6% when in the aqueous state. However, in the dry state the spores could withstand a treatment of 110°C for 30 min whereas they could withstand a temperature of 110°C for 1h in the dessicated state (Figs. 3 & 4). The comparison of lethal temperatures for *Verticillium* (44°C for 10 min and 38°C for 1 h in aqueous state and 125°C for 1h in dry state) and *Mycogone* and showed that *Verticillium* spores can withstand higher temperatures[12].

Interaction studies between *V. fungicola* and *A. bisporus* and Pathogencity studies

Both *V. fungicola* and *A. bisporus* could grow simultaneously on Malt Extract agar at 25°C. There was no inhibition of growth even after 14 d of incubation when a diameter of 56 mm and 39 mm was recorded for *V. fungicola* and *A. bisporus*, respectively (Table 1).

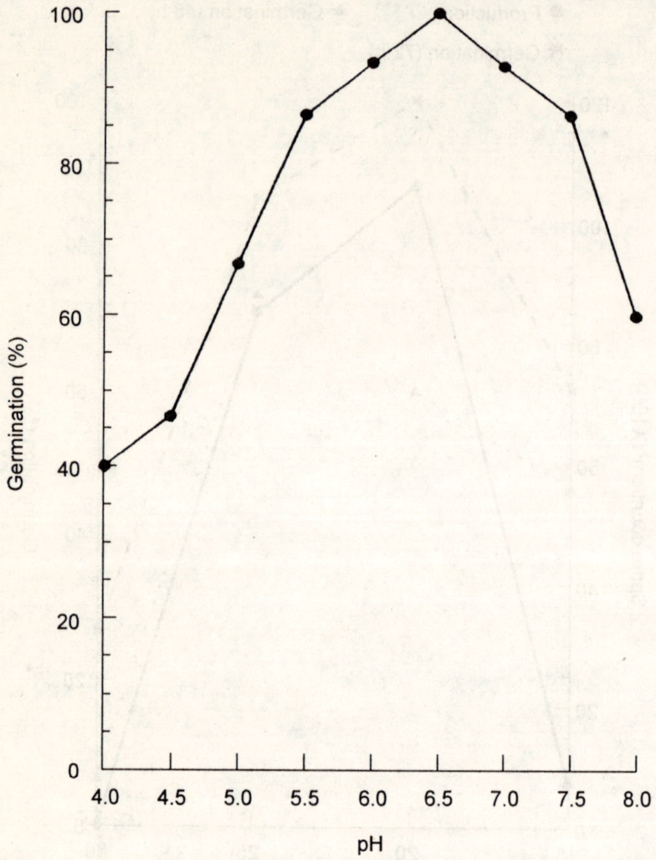

Fig. 2. *Effect of pH on spore germination*

Table 1: Study of interaction between *V. fungicola* and *A. bisporus*

Incubation period (h)	*Colony diameter (mm)*	
	V. fungicola	*A. bisporus*
0	6.0	6.0
4	15.0	9.0
7	34.0	20.0
10	43.0	29.0
14	56.0	39.0

Medium: Malt Extract agar pH: 5.0 Temperature: 25°C

The inoculation of spore and mycelial extract on the mushroom caps of *A, bisporus* resulted in development of disease symptoms. It was observed that brown spots were produced when spore suspension was used and the symptoms started after 24h and intensified with further passage of time. A spore load of 106 or

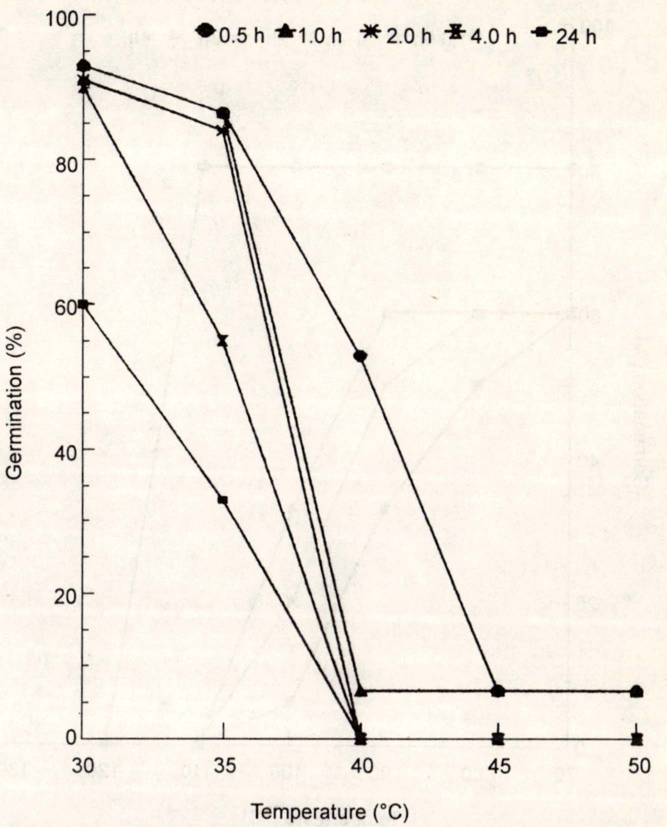

Fig. 3. *Evaluation of lethal temperature for spore suspension of V. fungicola.*

Table 2: Pathogenicity of *V. fungicola* on mushroom caps

Suspension used	*Incubation time (h)*	*Symptoms*
Mycelial extract	24	E
	48	+
Spore suspension	24	+
	48	+++
Inoculated filterate	24	—
	48	—
Uninoculated filterate	24	—
	48	—
Water blank	24	—
	48	—

— : No change in colour E : Light yellowish
+ : Yellowish brown +++ : Intense brown.

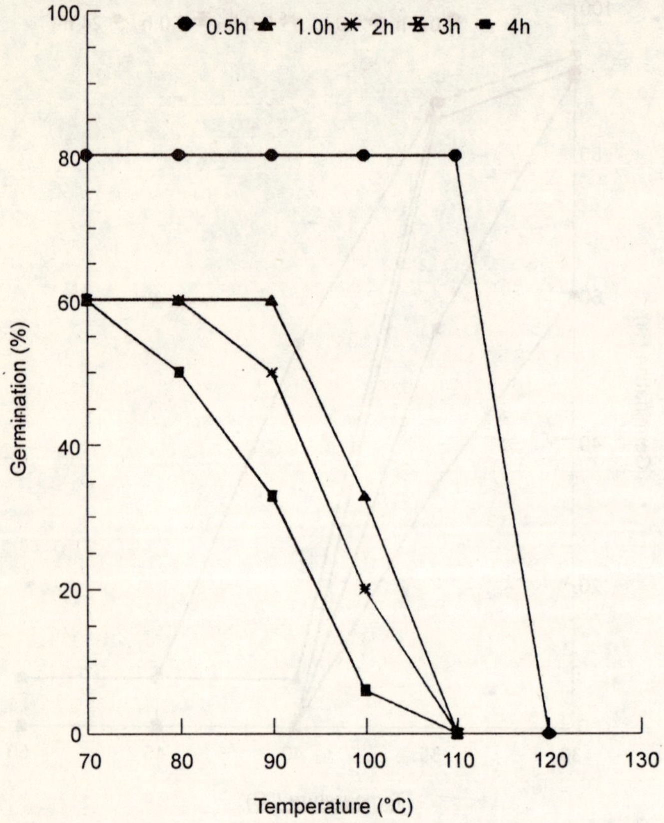

Fig. 4. *Evaluation of lethal temperature for dry spores of V. fungicoal.*

more/cap was responsible for appearance of dark brown necrotic spots after 48h. The mycelial extract produced yellowish brown spots after 48h of incubation which remained unchanged on prolonged incubation (Table 2). A similar technique for studying the appearance of disease symtoms of bacterial blotch caused by *Pseudomonas tolasii* has been reported[8]. The infection of cut fruit bodies of *A. bisporus* with *Veiticilum* causing brown necrotic spots after 48-72h of incubation has been reported[13]. The use of cap inoculation test for carrying out pathogenicity test of *Verticillium* isolates has been reported[14]. A concentration of 2×10^7 spores/ml was observed to cause the development of brown dry spots.

CONCLUSIONS

The present studies can help to design a strategy for the control of dry bubble disease of white button mushroom caused by *V. fungicola* infection as the environmental conditions in the growing rooms will affect the rate of disease development. As the Verticillium fungus likes warm and moist conditions for growth and reproduction, higher temperature and humidity are responsible for faster development and spread of the fungus. With a more favourable environment the life cycle is faster resulting in more spore production in shorter time. A manipulation of the crop environment may reduce the amount of disease development e.g. reduction of the air temperature from 200°C to 14°C and relative humidity from 90 to 80% would decrease the incidence of the disease as the conditions would be less favourable for the growth of the pathogen.

REFERENCES

1. Verma, R.N. National and international scenario of mushroom production and trade: Conference on Mushroom industry in India—a decade of achievements and future perspective, (Mushroom Society of India) (1999).
2. Bhatt, N., and Singh, R.P. *Mush. Sci.*, XV (2000) 587.
3. Sohi, H.S. *Indian Phytopathol.*, **45** (1992) 147.
4. Sharma, S.R. Management of mushroom disease: Advances in Horticulture Mushrooms, Malhotra Publishing House, New Delhi (1995).
5. Sharma, S.R. and Vijay, B. *Mushroom Res.*, **5** (1996) 13.
6. Seth, P.K., Kumar, S., and Shandilya, T.R. *Indian Horticulure*, (1973) 17.
7. Flegg, P. *The Mush J*, **519** (1993) 22.
8. Khanna, P.K. and Olivier, J.M. *Indian Phytopathol.*, **9** (1939) 147.
9. Lambert, D.H. and Wuest, P.J. *Phytopathology*, **66** (1976) 1144.
10. Bech, K. and Kovacs, G. *Mush. Sci.*, IX (1981) 381.
11. Hsu, H.K. and Han, Y.H. *Mush. Sci.*, XI (1981) 403.
12. Bech K., Jacobsen B.D. and Kovacs G. *Mush. Sci.*, XI (1989) 739.
13. Thapa, C.D. and Jandaik, C.L. *Indian J. Mush.*, **10-11** (1984-85) 54.
14. Wong, W.C. and Preece, T.F. *Plant Pathol.*, **36** (1987) 577.

Whey Utilization for the Preparation of RTS Beverages

P.S. PANESAR, H.K. SHARMA

and

M. KAUR

Department of Food Technology
Sant Longowal Institute of Engineering & Technology
Longowal, Distt. Sangrur 148 106 (Punjab)

ABSTRACT

Whey, a byproduct of dairy industry contains many valuable constituents of milk such as lactose, protein and other nutrients. Whey is produced in large amounts by the manufacture of casein, channa, cheese and paneer. The utilization of whey in any form of the product would serve the dual purpose i.e. reduction in environmental pollution and preparation of valuable products. In the present investigation, an attempt has been made to utilize this dairy byproduct for the preparation of whey based beverages. The different proportion of lemon and ginger were mixed with whey and TSS was maintained using sugar. The various samples were graded using sensory evaluation method. Among the various samples of this beverage, the drink having TSS 16°, pH 4.6 and acidity 0.59 was found to be of the best quality. The prepared drink was bottled and pasteurized. Finally, shelf life study of this drink was also carried out under refrigerated and room temperature conditions.

INTRODUCTION

Whey may be defined broadly as the serum or watery part of milk remaining after separation of the curd that results from the coagulation of milk by acid or proteolytic enzymes. Its composition will vary substantially, depending on the variety of cheese produced or the method of casein manufacture employed. On average, whey contains about 65g/kg of solids, comprising about 50 g lactose, 6 g protein, 6 g ash, 2 g non-protein nitrogen and 0.5 g fat. The protein fraction contains about 50% β-lactoglobulin, 25% α-lactoalbumin and 25% other protein fractions including immunoglobulins[1] . However, there will be wide variations in composition depending upon milk supply and the process involved in production of whey.

Wheys can be conveniently classified[2] into the following groups:

Sweet whey—Titratable acidity 0.10–0.20%

Medium acid wheys—Titratable acidity 0.20–0.40%

Acid wheys—Titratable acidity 0.40%

In general, wheys produced from rennet-coagulated cheeses develop low levels of acidity, while the production of fresh acid cheeses such as ricotta or cottage cheese yields medium acid or acid wheys. Whey from caseins produced by acid addition is categorised as high-acid whey whereas whey from rennet casein is sweet whey. The dairy industry in India on an average has been reported to generate ample amount of whey per litre of milk processed, depending upon the process employed, product manufactured and house keeping exercised. The utilization of whey in any form of the product would serve the dual purpose[3] i.e.

- reduction in environmental pollution
- preparation of valuable products.

Utilization of whey for the production of alcohol, citric acid, biogas, lactic acid has been reported[4-6]. The preparation of beverages from the combination of whey and fruit juices has been described by many research workers. There is still enough scope for the standardization of the technology of such products. In the present investigation, an attempt has been made to prepare a beverage from the combination of whey, lemon and ginger.

MATERIALS AND METHODS

Procurement of Materials

Milk was procured from the Verka Milk Plant, Sangrur. Fruits for juices and sugar were purchased from the local market of Longowal. Citric acid was obtained from M/s Loba Chemie. Whey, used in the study was prepared in the laboratory.

Preparation of Whey

Milk, containing 3.5% fat and 9.0% solids not fat (SNF) was subjected to Cream Separator (Modern Dairy Appliances Company, Delhi) thrice to minimize the fat content in the whey. Separated milk was coagulated with the addition of citric acid (0.2%). Protein was coagulated at its isoelectric pH. Thereafter, the whey was collected and pasteurized at 63-64°C for 30 min. Pasteurized whey was subjected to filtration.

Deproteinization of Whey

The removal of protein was carried out following the procedure of earlier workers[7] with little modifications as follows:

The whey obtained after filtration and separation was adjusted to pH 6.4 ± 0.2 and held at 90°C in a temperature controlled water bath for 15 min and subsequently filtered.

Addition of Juice and Sugar

Addition of juice to whey was carried out as given in Table 2. Sugar was added on the basis of required brix (Table 3).

Addition of Preservative

The preservative (sodium benzoate) was @ 150 ppm, (after dissolving it in minimum amount of water) was added.

Packaging, Processing and Storage

The prepared drink was then filled in cleaned and sterilized dark brown glass bottles of the capacity of 250 ml. The crown coking of the bottles was carried out under hot conditions. Thereafter, filled bottles were sterilized for 20 min at 15 psi followed by cooling at room temperature and then finally stored at refrigeration temperature for their shelf life study.

Analysis

Fat and Ash content of the whey were determined by standard methods[8]. Lactose content was estimated by the standard procedure of earlier workers[9]. TSS and pH of the whey and finished products (whey beverages) were measured by Hand Refractometer (Erma, Japan) and digital pH meter respectively. Acidity (%citric acid) of beverages was measured as per the method given by Rangana[9].

Sensory Evaluation Method

Sensory analysis was carried out by a panel of judges using 10 points scale for each of the attribute viz. color, taste, and aroma[9]. The points allowed for overall acceptability were as follows:

Excellent-9, Very Good-8, Good-7, Fair-6, Undecided-5, Poor-4, Slightly poor-3, Very poor-2 & Extremely poor-1.

RESULTS AND DISCUSSION

Whey having 94-95% moisture, 0.1% fat, 4.0% lactose and 0.46% ash (Table l) was used in different formulations (Table 2) adopted for the preparation of beverages. Whey, Water, Lemon and Ginger juice were taken in different proportion and the TSS was maintained between 15-20° through sugar. Beverages prepared were carried out for their sensory evaluation. The formulation, having 50 parts of whey, 35 parts of water, 9 water of lemon juice and 6 parts of ginger juice and maintained 16° brix, evaluated to be the best among all the samples and subjected for the further study. The beverages obtained were studied for its shelf life. The quality in terms of percent acidity, °brix and pH was checked after weekly intervals, (Table 3). Hardly any change was observed in the percent acidity, °brix and pH of the beverages even after storage of five weeks. Sensory analysis was also

Table 1: Physico-chemical characteristics of whey

Parameters	*Composition*
TSS (°Brix)	5-6
Moisture (%)	94-95
pH	4.5-4.6
Fat (%)	0.1
Lactose (%)	4.0
Total Ash (%)	0.46

Results average of three individual experiments.

Table 2: Formulation* adopted in the preparation of whey based beverages

Constituents	*1st*	*2nd*	*3ra*	*4th*
Whey	50	50	57	55
Water	35	30	28	30
Lemon juice	09	12	09	10
Ginger juice	06	08	06	05

* Represent in % (Results average of three individual experiments.)

Table 3: Changes in Acidity of a RTS Beverage prepared from Lemon and Ginger during storage

Storage Period	*% Acidity*	*°Brix*	*pH*
Control	0.59	16	4.60
3rd day	0.60	16	4.60
2nd week	0.61	16	4.54
3rd week	0.60	16	4.50
4th week	0.64	16	4.46
5th week	0.62	15-16	4.40

Results average of three individual experiments.

carried out after different time intervals (Table 4). Samples stored for different times intervals were also found to be very good in its color, taste and aroma and liked by the panel for its consumption.

Finished product having TSS 16°, pH 4.6 and acidity 0.59 was found to be best in quality. During storage for 35 d, there was hardly any change in the quality of the beverage. A further study with respect to standardization of its shelf life has to be undertaken before it commercial exploitation.

Table 4: Sensory Evaluation of the RTS Beverage prepared from Lemon and Ginger during storage

Storage Period	*Sensory Score*
Control	8.5
3rd day	8.0
2nd week	8.0
3rd week	8.0
4th week	8.0
5th week	8.0

Results average of three individual experiments.

REFERENCES

1. Matthews, M.E., *New Zealand J. Dairy Sci. Technology*, **13** (1978) 149.
2. Robinson, R.K. Modern Dairy Technology, Advances in Milk Processing, Elsevier Applied Sci. Pub. Essex, **1** (1986) 273.
3. Jones, R.H. Pollution Control in Dairy Industry, Noyes Data Corp. Pub. Park Ridge, New Jersey, USA, (1974) 154.
4. Hossain, M., Brooks, J.D. and Maddox, I.S., *New Zealand J. Dairy Sci. Technology*, **18** (1983) 161.
5. Lang, F., *Milk Industry*, **82** (1980) 30.
6. Stieber, R.W. and Gerhardt, P. *J. Dairy Sci.*, **62** (1979) 1558.
7. Irvine, D.M., Hill, A. and Bullock, D.H. *Modern Dairy*, **63** (1984) 18.
8. ISI, Handbook of Food Analysis, Part XI, Dairy Products, Indian Standard Institution, New Delhi, (1981).
9. Ranganna, S, Handbook of analysis and quality control for fruits and vegetables products, Tata McGraw-Hill Publishing Company, New Delhi, (1986).

Formulation of Whey-Mango Concentrate

S. SINGH*, A.K. SINGH*
and
D.N. GANDHI**

*Dairy Technology Division**
*Dairy Microbiology Division***
National Dairy Research Institute
Karnal 132 001 (Haryana)

ABSTRACT

Mango pulp was preserved by heating at 90°C for 2 min to inactivate the native enzymes. Paneer whey was concentrated (36-38%, total solids) in the manufacture of whey-mango concentrate. Different acidulants such as 1 N HCl, 25% solution of citric acid and 25% solution of glucono-D-lactone (GDL) were added to lower the pH of concentrated whey from 5.2 to 3.6. Heat treatment of whey concentrate to 90°C for 1 min resulted in whey protein precipitation. Mango pulp (25% total solids) at 10, 15 and 20% and sugar 6-8% level were added to whey mango concentrate and pH was adjusted to 3.8 to 4.5. The concentrate was subjected to in-bottle heat treatment at 90°C for 1 min and cooled immediately at room temperature. Whey-mango-concentrate was subjected to organoleptic evaluation after adjustment to 14% total solids. The concentrate at 15 : 77 ratio of mango pulp : paneer whey concentrate received maximum sensory score for flavour, consistency, colour and appearance and overall acceptability score at pH 4.2. Addition of carboxy methyl cellulose (CMC) at 0.05% level markedly improved the consistency of the ready-to-serve (RTS) beverage. Acceptable whey-mango-concentrate had 55% total solids, 48% total soluble solids, acidity 1.36% as lactic acid and viscosity 28 cP.

INTRODUCTION

India is emerged as largest milk producer in the world with milk production of 71 million tonnes per annum in 1998[1]. Now dairy industry is focussing towards value added products such as cheese and indigenous coagulated milk products such as paneer, chhana, casein and shrikhand due to more demand. This leads to generation of large quantity of whey as byproduct. It has been reported that about 3 million tonnes of whey is annually generated in India, which contains about 2 lakh tonnes of valuable nutrients[2]. Whey, a component of milk contains 20% of the protein found in milk. Whey is proven to be an extraordinary nutritional material with all of the essential and nonessential amino acids. Whey proteins also has the most naturally occurring biological value of any known protein, having 68% more than egg white and has the highest naturally occurring branched chain amino acid content[3].

India is also top fruit producing country in the world with maximum production of mango and banana fruits. Fruit pulp and juices are rich source of many vitamins, minerals, digestible carbohydrates and organic acids that stimulate appetite and help in proper digestion of foods. However, many mango varieties are not suitable for table purposes due to sour taste and absence of typical mango flavour. The farmers are forced to dispose of sour varieties of mango at much lower price. Nutritive attributes of whey and mango can be combined to develop delicious beverages. This will provide a healthful blend of fruit based calcium and whey proteins[4]. The formulation of whey based mango beverage would serve as an alternative to carbonated soft drinks.

Currently, whey based beverages are available in the form of single strength juice and concentration of whey is important for wide spread availability, less packaging and transportation cost. The present investigation is undertaken out to develop the formulation of paneer whey mango concentrate that can be reconstituted into delicious ready-to-serve (RTS) whey-mango beverage.

MATERIALS AND METHODS

Paneer whey: Fresh paneer whey was collected from experimental dairy of the Institute at the time of draining of whey. Paneer whey contained 5.6–5.8% total solids, 0.1% fat, 0.45–0.48% protein, 0.58–0.60% ash and at pH 5.2.

Clarification and concentration of whey: Paneer whey was clarified in tri-process to remove the residual fat and was concentrated in single vacuum pan at 25" Hg and at 54-56°C. Paneer whey was concentrated to 36-38% total solids.

Acidulants: Different acidulants such as 1N HCl, 25% citric acid (w/v), 25% glucono-D-lactone (w/v), were added separately to lower the pH of paneer whey concentrate from 5.2 to 5.0, 4.5, 4.0 and 3.5. The adjusted whey concentrate was subjected to heat treatment at 90°C for 1 min after the addition of sugar to study the whey-protein stability.

Carboxy methyl cellulose: Carboxy methyl cellulose (CMC) was added at varying levels (0.05–0.15%, w/v) in mango-whey concentrate to improve the consistency and mouthfeel in RTS whey-mango beverage.

Formulation of whey-mango concentrate: Concentrated whey was sweetened with cane sugar, filtered through muslin cloth and was mixed with CMC and mango pulp. The pH was adjusted with 25% glucose-D-lactose (GDL, w/v). The whey-mango concentrate was filled in 200 ml milk bottles and pasteurized at 90°C for 1 min (Fig. 1).

Clarified whey

↓

Whey concentration in vacuum pan (36 – 38% total solids)

↓

Addition of sugar

↓

Addition of CMC

↓

Addition of mango pulp

↓

pH adjustment

↓

In – bottle pasteurization (93°C / min)

Fig. 1: Flow diagram for the manufacture of whey-mango concentrate

Total solids in RTS beverage: Whey-mango concentrate was reconstituted to varying total solids from 12 to 18% with addition of water.

Adjustment of pH in whey-mango concentrate: The varying pH levels such as 3.4, 3.6, 3.8 and 4.0 was adjusted with 25% glucose-D-lactose (w/v).

Sensory evaluation: Sensory evaluation of RTS beverage was carried out on 9-point hedonic scale by a trained panel of 8 judges. The data were subjected to analysis of variance to compare their sensory rating [5].

Chemical analysis: The total solids of whey, mango pulp and whey-mango concentrate were determined as per the method of BIS[6]. The protein content in whey was estimated by semi microkjeldahl method[7]. The total soluble solids in whey-mango concentrate were measured with the help of 'Erma' hand refractometer. The viscosity of whey-mango concentrate was determined with Rheomat RM 108 E/R at 20 °C. Shear stress and shear rate relationship was determined.

RESULTS AND DISCUSSION

Formulation of whey-mango concentrate: Whey-mango concentrate having 15% mango pulp (26% total solids) and 77% whey concentrate was found to be most acceptable to judges in RTS beverage. Judges observed RTS bevarages containing 10% of mango pulp less consistant, whereas the bevarage with 20% pulp were more consistant (Table 1). Earlier findings also reported that acceptable quality of whey based mango beverage can be developed with 15% mango pulp and 78% paneer whey[8].

Table 1: Formulation of whey – mango concentrate

Ingredients	*I*	*II*	*III*
Mango pulp (%)	10	15	20
Whey concentrate (%)	82	77	72
Sugar (%)	8	8	8
pH	4.2	4.2	4.2

Acidulants on whey protein stability: Citric acid and 1 N HCl as acidulants could not stabilize the whey proteins at pH levels from 5.2¯3.5 followed by heat treatment at 90 °C (Table 2). Sugar also could not provide stability to whey proteins. Moreover, GDL as acidulant also could not stabilize the whey proteins in pH range between 5.2–4.0 after addition of sugar and subsequent heat treatment to 90 °C. However, whey proteins were stabilized with GDL at pH range 3.5–4.0 after addition of sugar and heat treatment to 90 °C (Table 2). Disaccharide sugars were more effective than monosaccharides in increasing the thermal stability of whey proteins, particularly β¯lactoglobulin[9]. The quantity of 1 N HCl was minimum (0.25 gm) while the quantity of GDL was maximum (1.8 gm) to lower the pH of paneer whey concentrate from 5.2–3.5 in 100 gm whey concentrate (Fig. 2).

Table 2: Effect of acidulant on whey protein stability

Acidulant	*pH/whey protein precipitation*		
	5.2–4.5	*4.5–4.0*	*4.0–3.5*
1 N HCl	+	+	+
Citric acid (25%, w/v)	+	+	+
Glucono–D–Lactone (25%, w/v)	+	+	–

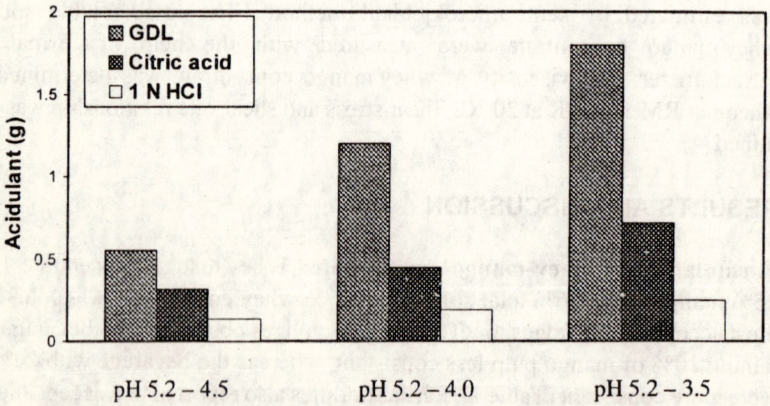

Fig. 2: Amount of acidulant in lowering the whey concentrate pH

Carboxy methyl cellulose (CMC) on overall acceptability score of whey-mango concentrate: RTS beverage with 0.05% CMC scored maximum for overall acceptability score with desirable consistency (Table 3). However, as the CMC level was increased from 0.05%, the concentrate became too viscous/soggy and hence scored less. The higher flavour score obtained by the sample containing 0.05% CMC than that of samples containing higher levels of CMC may be due to the flavour-fixation ability of CMC[10].

Table 3: Carboxy methyl cellulose on consistency of whey–mango concentrate*

Carboxy methyl cellulose (%)	*Overall acceptability score*	*Consistency level*
0.05	8.0	Desirable consistency
0.10	7.0	Highly viscous
0.15	6.5	Highly viscous

* Average of three values

Total solids level in RTS beverage: RTS beverage at 14% total solids was preferred most for sensory score. The attributes for flavour, consistency, colour and appearance and overall acceptability were significantly higher ($P < 0.1$) than other levels of total solids (Table 4). RTS beverage at 12% total solids was least accepted due to thin consistency. However, judges observed more sweet taste in RTS mango beverage with 16% and 18% total solids. Whey based mango beverage at 15% total solids received maximum flavour score (7.8) on 9-point hedonic scale[8].

Table 4: Effect of total solids in whey – mango concentrate on sensory score of RTS beverage *

Attributes	*12% total solids*	*14% total solids*	*16% total solids*	*18% total solids*
Flavour	6	8	7	6.5
Consistency	6	8	7	7
Colour and appearance	7	8	7.5	7.5
Overall acceptability	6	8	7	7

*Average of three values

Optimum pH level in whey-mango concentrate: Whey-mango concentrate at pH level 4.2 was found to be most acceptable in terms of mean sensory score for flavour and overall acceptability score (Table 5). Adjusting the pH level to 3.8, 4.0 and 4.5 significantly lowered ($P < 0.01$) the flavour and overall acceptability score. The intensity of typical mango taste decreased with increase in pH of whey mango concentrate to 4.5. Subsequently, the judges disliked the whey -mango concentrate adjusted at pH 3.8 and 4.0 due to sour taste in RTS beverage. Whey based RTS mango beverage received maximum sensory score at pH 4.5 (Table 5). RTS mango beverage was preferred most by the judges at pH 4.5[8]. However, whey-kinnow beverage was most acceptable at pH 4.25[11].

Table 5: Effect of pH levels in whey-mango concentrate on sensory score of RTS beverage*

Attributes	*I* 3.8 (4.1)**	*II* 4.0 (4.3)**	*III* 4.2 (4.5)**	*IV* 4.5 (4.8)**
Flavour	6	7	8	5.5
Consistency	8	8	8	8
Colour and appearance	8	8	8	8
Overall acceptability	6	7	8	7

* Average of three values
** pH value in bracket reflects the pH of RTS beverage

Sensory score of whey-mango concentrate: Whey-mango concentrate with 15 : 77 ratio of mango pulp: whey concentrate received maximum sensory score for flavour, consistency, colour and appearance and overall acceptability scores (Table 6). Whey-mango concentrate obtained from 15% mango pulp was not acceptable to judges due to more consistency. However, 10% mango pulp in whey-mango concentrate was not preferred due to flat flavour and thin consistency.

Table 6: Mean sensory score of whey mango concentrate

Attributes	*I* Mango pulp : Whey concentrate (10 : 82)	*II* Mango pulp : Whey concentrate (15 : 77)	*III* Mango pulp : Whey concentrate (20 : 72)
Flavour	7.8	8.5	6.5
Consistency	7	8	7
Colour and Appearance	7	8	8
Overall acceptability	7.3	8.2	7
pH	4.2	4.2	4.2

Physico-chemical properties of whey-mango concentrate: The acceptable whey-mango concentrate (15:77, mango pulp: whey concentrate) was found to contain 55% total solids, 48% total soluble solids, acidity (% lactic acid) 1.36, viscosity (cP) 28 (Table 7). The relationship between shear rate and shear stress realized that consistency increased with increase in the levels of mango pulp. The whey-mango concentrate reflected more towards non-Newtonian fluid with increase in the level of mango pulp in the concentrate.

Table 7: Physico-chemical properties of whey-mango concentrate*

Constituents	*I* *Mango pulp : Whey concentrate (10 : 82)*	*II* *Mango pulp : Whey concentrate (15 : 77)*	*III* *Mango pulp : Whey concentrate (20 : 72)*
Total solids (%)	48	55	58
Total soluble solids (%)	40	48	53
Acidity (%)	1.31	1.36	1.39
Viscosity (cP)	20	28	38
Shear stress–shear rate	$0.12\ D^{0.4054}$	$0.17\ D^{0.5478}$	$0.20\ D^{0.6104}$

* Average of three values

REFERENCES

1. Kurup, M.P.G., *Indian Dairyman*, **52**: (2000) 25.
2. Khamuri, K. and Rajorhia, G.S., *Indian Dairyman*, **50** (1998) 13.
3. WWW.musclextreme. Com/products/proteins. Html (website).
4. Kravcenko, E.F. *IDF Bulletin*, **233** (1988) 61.
5. Snedecor, G.W. and Cochran, W.G., In: Two-way classification. Statistical methods. 6 th Edition, **141** (1967) 299.
6. BIS: 4799–Part II 1961, Bureau of Indian Standards, Manak Bhawan, New Delhi.
7. Maneffe, G.S. and Overman, O.R., *J. Dairy Sci.*, **23** (1940) 1177.
8. Singh, S., Ladkani, B.G., Kumar, A. and Mathur, B.N., *Ind. J. Dairy Sci.* **47** (1994) 586.
9. Ibrahim, F.S., Madkor, S.A. and Hewedy, M.M., *Egyptian J. Dairy Sci.* **23** (1995) 177.
10. Somagyi, L. P., In: Processing fruits: science and Technology, vol. 1, Principles and application, Technomic Publishing Co., Pennsylvania, USA, (1996) 293.

Effect of Incorporation of Sweet Cream Butter Milk Solids on Physical and Sensory Characteristics of Cookies

H. KAUR *and* U. BAJWA*

Department of Food Science & Technology
Punjab Agricultural University
Ludhiana 141 004 (Punjab)

ABSTRACT

The investigation was undertaken to explore the possibility of utilization of butter milk in cookie-making. The effect of 25 to 100% replacement of water with butter milk in cookie dough, and the effect physical and sensory characteristics of cookies, was studied. The spread ratio of cookies increased significantly with the increase in levels of butter milk solids. The browning of cookies was enhanced and crispness improved by augmenting the butter milk solids in cookie-dough.

INTRODUCTION

Butter milk is an important dairy by-product obtained by churning cream for butter manufacture. It is a good source of proteins, carbohydrates, minerals, vitamin-A, α-tocopherol and cholesterol[1]. Production of butter milk in India is estimated at 275 million kg per annum. It is generally utilized in the production of flavoured beverages, standardization of market milk or admixing with skim milk for the manufacture of non-fat dried milk. Reports are available which indicate its utilization in the manufacture of paneer and cheese[2].

Butter milk powder has wide use as an ingredient for batters, cake mixes and chocolate preparations, but the conversion of buttermilk to buttermilk powder is an expensive process. The objective of present study was to utilize butter milk in cookie making and to evaluate the effect of 25, 50, 75 and 100% replacement of water with buttermilk on physical and sensory characteristics of cookies.

MATERIAL AND METHODS

Flour and other dry ingredients

Flour used in the present study was purchased from local market. It contained 10.4% protein, 0.3% fat and 0.27% ash. Sugar, shortening, sodium bi-carbonate, glucose and salt were also procured from the local market.

Butter Milk

Cream required for the study was procured from the milk processing plant of the Department of Food Science & Technology, Punjab Agricultural University, Ludhiana. It was pasteurized at 71°C for 20 min, cooled immediately and stored overnight in refrigerator. Cream was then churned in a manually operated butter churn to obtain butter milk, which was stored in a refrigerator for further use. Butter milk was standardized to 10.0% total solids before its use in cookie-making. It contained 3.98, 0.67 and 0.32% protein, fat and ash respectively.

Cookie-Manufacture

AACC (1990) method[3] was followed for baking cookies. It included the following ingredients:

Ingredients	*Quantity (g)*
Flour	100
Sugar	57.8
Sodium bicarbonate	1.1
D-glucose	0.878
Salt	0.933
Fat	28.44
Water	Optimum (ml)

The control dough was mixed using optimum level of water. Water was replaced at 25 to 100% levels with butter milk in test samples. The dough was sheeted and cut into circular cookies of 4.5 cm diameter and baked for 12 min at 380°F.

Cookie Quality

Cookie width (W) and thickness (T) was measured according to standard AACC (1990) procedures[3]. Spread ratio was calculated by taking the ratio of average width to thickness for four cookies at a time.

Sensory Evaluation

Samples of cookies were evaluated by a panel of six semi-trained judges on a 9-point hedonic scale for sensory attributes of top grain texture, flavour and mouthfeel and overall acceptability was calculated by taking average of all these attributes.

Statistical Analysis

The effect of water replacement with butter milk was tested using analysis of variance[6]. The means for various physical and sensory characteristics were compared using Duncan's multiple range test[5].

RESULTS AND DISCUSSION

Butter milk obtained directly from churning of cream was utilized to replace 25, 50, 75 and 100 per cent water in cookie-dough. It contained 10.0% total solids and 3.58% protein.

Effect on Spread Ratio of Cookies

As inferred from Table 1, width of the cookies increased significantly (P< 0.01) with enhanced levels of butter milk, thereby significantly improving the spread ratio of cookies. Spread ratio was 5.66 in control and increased to 6.10 in 100% butter milk cookies causing an increase of 7.77%. Browning of cookies was enhanced with the increased amount of butter milk probably due to augmented mallard reaction between the additional protein and lactose present, in recipe, although this effect was not reflected in sensory scores for colours (Table 2).

Table 1: Effect of water replacement with buttermilk on organoleptic scores of cookies

Replacement (%)	*Appearance*	*Colour*	*Texture*	*Flavour*	*Mouthfeel*	*Overall acceptability*
0	6.83	7.16	6.83	7.17	6.83	7.11
25	7.00	7.50	7.661	7.50	7.50	7.37
50	7.33	7.50	7.83	7.83	7.83	7.71
75	7.50	8.16	8.16	7.50	7.83	7.85
100	7.30	7.66	8.00	7.66	8.00	7.71

ANOVA

Source of variation	*d.f.*	*MSS*					
		Appearance	*Colour*	*Texture*	*Flavour*	*Mouthfeel*	*Overall accept ability*
Levels	4	0.450	0.800	1.616	0.367	1.300	·0.755
Error	25	1.027	0.480	0.473	0.480	0.72	0.386

Values with different superscripts in a column (1, 2, 3 4) differ significantly (P < .05).
Values presented are means of six replications.

Effect on Sensory Characteristics

Effect of various levels of butter milk on sensory attributes of appearance, colour, flavour, mouthfeel and overall acceptability of cookies was found to be non-significant except for texture, which was affected significantly (P<0.05). Crispness of cookies improved with increase in levels of butter milk, thereby, resulting in an improvement of sensory scores for the attribute of texture.

Table 2: Effect of water replacement with buttermilk on spread ratio of cookies

Replacement (%)	*Width (cm)*	*Thickness (cm)*	*Spread ratio*
0	6.011	1.06	5.66
25	6.092	1.06	5.74
50	6.243	1.05	5.94
75	6.223	1.03	6.04
100	6.253	1.02	6.10

ANOVA

Source of variation	*d.f.*	*MSS*		
		Width	*Thickness*	*Spread Ratio*
Levels	4	0.033	0.0007	0.0942
Error	10	0.0012	0.0014	0.0129

Values with different superscripts in a column (1, 2, 3, 4) differ significantly (P<0.01)
Values presented are means of three replications.

The increase in width of cookies with incorporation of butter milk might be due to the additional phospholipid content acting as emulsifier. This was in accordance with earlier finding[4] which reported that as the level of the milk proteins increased, the mouthfeel tended to become finer in characater and the colour became brown due to the maillard browning reactions between the protein and lactose present.

CONCLUSION

Incorporation of butter milk improved both spread ratio and crispness of cookies. This also improved the nutritive value of cookies and partially solved the problem of butter milk disposal.

REFERENCES

1. Tylkin, V.B., Grishina, V.E. and Svidovskii, A.N. *Tovarovedenie,* **8**: (1975) 42.
2. Aarora, K.L. and Mathur, B.N. *Indian Dairyman*, **32**: (1980) 707.
3. AACC 1990. Approved Laboratory Methods. 8th ed. American Association of Cereal Chemists. St. Paul, Minnesota, U.S.A.
4. Cooper, H.R., Patten, J.D. and Fletcher, R.H. *J. Fd. Sci.*, **49**: (1984) 377.
5. Duncan, D.B. *Biometrics*, **11**: (1955) 1.
6. Snedecor, G.W. and Cochran, G.C. Statistical methods. Oxford Univ Press. TBH Publ. Co. New Delhi, (1968) 339.

Use of Electrochemical Methods for Testing of the Purity of Agro-Products: Development of a Model

J. DUTT[1], T. SINGH[2]
and
R.P. SINGH[2]

1. Deptt. of Chemical Engineering & Technology
Punjab University
Chandigarh 160014 (Punjab)
2. Deptt. of Mathematics & Statistics
Punjab Agricultural University
Ludhiana-141004 (Punjab)

ABSTRACT

The fast growth of agro-process technology including food processing has enhanced the scope for development of various techniques for testing the purity of agro-products especially those to be consumed by human beings. Electrochemical techniques offer a simple mean to detect and remove traces of impurities associated with the products. With the advent of fast computing mechanics, complex chemical processes can be easily investigated using electrochemical techniques. In the present paper, a mathematical model is developed for a process in which the impurities undergoing a chemical reaction at the electrode surface can be detected. The model is developed for linear sweep voltammetry under hydrodynamic condition for tubular electrodes and the effects of experimentally controllable parameters on the concentration profile of the electrolyte have been studied theoretically.

INTRODUCTION

To provide adequate nourishment for the growing population, especially in developing countries is becoming a difficult affair. Where as, on the one hand, extensive form of agriculture, expansion of cultivated land and possibility of breeding and utilizing of modem genetic and biotechnological methods are not fully exhausted in these lands, on the other hand, a problem of preserving produced food stuff from material loss as well as loss of their nutritive value, needs to be addressed. In this respect, the chemisation of the agriculture and food industry played a significant role. The use of industrial fertilizers, polymers for improvement of physical or physicochemical properties of soil, pesticides which include fungicides, herbicides, rhodenticides, acaricides, nematocides are important factors for such chemisation.

Almost all pesticides are physiologically active substances, which are more or less toxic in accordance with their chemical properties. Because these materials can occur as contaminants either in an initial form—their degradable products, it is necessary, therefore, to detect and evaluate the presence of these materials from the hygienic -toxicological aspects. In today's complex industrial environment, the possibility of occurrence of trace contaminants of food and beverages as well as number of possible sources of contamination has significantly increased. Concern about low level of exposure to contaminants, coupled with the improved analytical capability to detect smaller amount of both organic compounds as well as inorganic elements, has generated a desire for surveillance of food supply by both regulatory authority as well as manufacturing and processing industries.

With advancement of analytical technology, a number of analytical techniques, such as mass spectrometry[1], X-ray fluorescence spectrometry[2], neutron activation[3], and electrochemical techniques[4] have been developed having excellent sensitivity and high speed. The electrochemical techniques include AC and DC current polarography, anodic and cathodic stripping and voltametry. The polarographic techniques are suitable for dilute solution with sensitivities of 10^{-5} -10^{-6} moles per litre. These electrochemical techniques have been reviewed by earlier workers[5]. The use of polarographic methods as detection means for chlorinated pesticides has received some attention. Pesticide such as DOT[6], Methoxiclor[7], hexachlorocyclohexane[7] have been successfully determined by this technique. Polarography is used for determination of N-trichloromethylthio fungicide[8]. A study dealing with polarogaphic determination of twenty organophosphorus pesticides has been presented[9]. The use of polarography for determination of parathion, Malathion and Quthion are well documented[10].

In our recent work we have observed that linear sweep voltammetry under hydrodynamic conditions is a more sensitive technique than polarography. The theory of this technique has been developed in this Iab[11–15] and the theoretically derived results have been verified experimentally. In the present investigations attempts have been made to use linear sweep voltammetry under hydrodynamic conditions and develop a mathematical model for detection and removal of impurities associated with food products.

DEVELOPMENT OF MODEL

For an electrochemical process in which an irreversible homogeneous chemical reaction is interspersed between two reversible electrode charge reaction is represented as:

$$\begin{aligned} A + n_1 e^- &\leftrightarrow B \\ B &\xrightarrow{k_f} C \\ C + n_2 e^- &\leftrightarrow D \end{aligned} \qquad (1)$$

and is known as ECE process, n_1 and n_2 are the number of electrons involved in first and second electrode charge reactions respectively and k_f is the chemical reaction rate. All these reactants and products are in liquid state and assumed to flow, laminarly, through a tubular electrode. Since convection along the axial direction of the electrode and diffusion along the radial direction will occur, the mathermatical model incorporating these phenomena is

$$\frac{\partial C_A}{\partial t}+V_a(r)\frac{\partial C_A}{\partial z}=D_A\left[\frac{\partial^2 C_A}{\partial r^2}+\frac{1}{r}\frac{\partial C_A}{\partial r}\right] \quad (2)$$

$$\frac{\partial C_B}{\partial t}+V_a(r)\frac{\partial C_B}{\partial z}=D_B\left[\frac{\partial^2 C_B}{\partial r^2}+\frac{1}{r}\frac{\partial C_B}{\partial r}\right]-k_f C_B \quad (3)$$

$$\frac{\partial C_C}{\partial t}+V_a(r)\frac{\partial C_C}{\partial z}=D_C\left[\frac{\partial^2 C_C}{\partial r^2}+\frac{1}{r}\frac{\partial C_C}{\partial r}\right]+k_f C_B \quad (4)$$

$$\frac{\partial C_D}{\partial t}+V_a(r)\frac{\partial C_D}{\partial z}=D_D\left[\frac{\partial^2 C_D}{\partial r^2}+\frac{1}{r}\frac{\partial C_D}{\partial r}\right] \quad (5)$$

where C_A, C_B, C_C, C_D, are molar concentration of species A, B, C, D respectively. D_A, D_B, D_C, D_D are their respective diffusion co-efficients, v_a is the axial flow rate of electrochemical species flowing laminarly through the tubular electrode with radius a, and length l and r, z are their radial and axial directions, respectively.

This system of transient convective-diffusion equations are subjected to following initial and boundary conditions:

$$t=0, 0\le r\le a, 0\le z\le 1; C_A=C_A^*, C_B=C_C=C_D(\approx 0) \quad (6)$$

$$t>0, r=0, 0\le z\le 1; C_A\to C_A^*, C_B=C_C=C_D\to 0 \quad (7)$$

$$t>0, r=a, 0\le z\le 1; D_A\frac{\partial C_A}{\partial r}=-D_B\frac{\partial C_B}{\partial r}=-f_A(t)$$

$$D_C\frac{\partial C_C}{\partial r}=-D_D\frac{\partial C_D}{\partial r}=-f_C(t) \quad (8)$$

$$t>0, 0\le z\le 1, r=a; \frac{C_A}{C_B}=\exp\left[\frac{n_1 F}{RT}(E(t)-E_{AB}^{\circ})\right]$$

$$D_C\frac{\partial C_C}{\partial r}=-D_D\frac{\partial C_D}{\partial r}=-k'C_C=-f_C(t) \quad (9)$$

$$\frac{C_C}{C_D}=\exp\left[\frac{n_2 F}{RT}(E(t)-E_{CD}^{\circ})\right]$$

where $i_1(t), E_{AB}^{\circ}$, and $i_2(t), E_{CD}^{\circ}$ are current densities and standard electrode potential at the electrode surface for first and second charge transfer reaction respectively. C_A^* is the initial bulk concentration of electroactive specie A; $E(t)$ is the potential sweep function. For linear sweep voltammetry potential varies linearly with time i.e.;

$$E = E_i - vt \tag{10}$$

where E_i is initial electrode potential and v is potential scan rate.

SOLUTION OF THE MODEL

For further analysis of the above model, it is assumed that diffusion co-efficients of electroactive species are quite close to each other (i.e. $D_A = D_B = D_C = D_D = D$(say)) which leads to a well known cosequence that

$$C_A + C_B + C_C + C_D = C_A^* \tag{11}$$

The above system of equations 2-5 is solved using following transformations

$$\Psi = \frac{C_B + C_C}{C_A^*} \tag{12}$$

$$\Phi = \frac{C_B}{C_A^*} \exp(k_f t) \tag{13}$$

and other non-dimensional parameters and mathematical techniques as already reported in our earlier work[11, 12, 13, 14, 15]. The expressions for C_A, C_B, C_C, C_D are respectively.

$$C_A = C_A^* - \frac{1}{\Lambda k_f^{1/2} D^{1/2}} \int_0^{t'} f_A(\tau) g(\zeta, t'-\tau) d\tau \tag{14}$$

$$C_B = \frac{1}{\Lambda k_f^{1/2} D^{1/2}} \int_0^{t'} f_A(\tau) g(\zeta, t'-\tau) e^{-\Lambda^{-2}(t'-\tau)} d\tau \tag{15}$$

$$C_C = \frac{1}{\Lambda k_f^{1/2} D^{1/2}} \int_0^{t'} f_A(\tau)(1 - e^{-\Lambda^{-2}(t'-\tau)} - f_C(\tau)] g(\zeta, t'-\tau) d\tau \tag{16}$$

$$C_D = \frac{1}{\Lambda k_f^{1/2} D^{1/2}} \int_0^{t'} f_C(\tau) g(\zeta, t'-\tau) d\tau \tag{17}$$

Thus, a system of differential equations (2-4) is transformed to a system of integral equations (14-17) in which the unknown current functions to be evaluated are $f_A(t')$ and $f_C(t')$. Substituting the expression of C_A, C_B, C_C and C_D in equation (9), we get

$$e^{(u_1 - y)} \int_0^y \chi(z) K_2(y-z) dz = 1 - \int_0^y \chi(z) K_1(y-z) dz \tag{18}$$

and

$$\int_0^y Q(z) K_1(y-z) dz + e^{(u2 - y\gamma)} \int_0^y Q(z) K_1(y-z) dz = \int_0^y \chi(z)(K_1(y-z) - K_2(y-z) dz \tag{19}$$

where

$$\sigma = \frac{nF\nu}{RT};\ y = \sigma t';\ z = \sigma\tau \tag{20}$$

$$\delta = \frac{k_f RT}{n_1 F\nu} \tag{21}$$

$$u_1 = \frac{n_1 F}{RT}(E_i - E^{\circ}_{AB}) \tag{22}$$

$$u_2 = \frac{n_2 F}{RT}(E_i - E^{\circ}_{CD}) \tag{23}$$

$$\Delta E^{\circ} = E^{\circ}_{AB} - E^{\circ}_{CD} \tag{24}$$

$$K_1(y-z) = \sum_{n=1}^{\infty} \frac{1}{\lambda_n} \exp\left[-\frac{\lambda_n}{\sigma}(y-z)\right] \tag{25}$$

$$K_2(y-z) = \sum_{n=1}^{\infty} \frac{1}{\lambda_n} \exp\left[-\left(\frac{\lambda_n}{\sigma} + \delta\right)(y-z)\right] \tag{26}$$

$$\chi(z) = \frac{f_A(z)}{C_A^* \sigma} \tag{27}$$

$$Q(z) = \frac{f_C(z)}{C_A^* \sigma} \tag{28}$$

These integral equations are numerically, solved to give the values of $\chi(z)$ and $Q(z)$. Since current and potential functions are related to $\chi(z)$, $Q(z)$ and $\sigma t'$ as

$$i_1(t') = n_1 FA\Lambda\sqrt{k_f D\sigma}\, C_A^* \chi(\sigma t') \tag{29}$$

$$i_2(t') = n_2 FA\Lambda\sqrt{k_f D\sigma}\, C_A^* \chi(\sigma t') \tag{30}$$

$$\text{Total current } = i(t') = i_1(t') + i_2(t') \tag{31}$$

$$E(t') - E^{\circ}_{AB} = \frac{RT}{n_1 F}(u_1 - \sigma t') \tag{32}$$

$$E(t') - E^{\circ}_{CD} = \frac{RT}{n_2 F}(u_2 - \sigma t') \tag{33}$$

Thus, a relationship between $i(t')$ and $E(t')$ is theoretically, established and theoretical current potential voltammograms can be obtained, which can help to give the values of parameters like peak current for two reactions, steady state currents and peak potential for the system which give the measurement of concentration of the electroactive species produced in the process.

EXPERIMENTAL VERIFICATION

To confirm the theoretically derived results, experiments were conducted for the reduction of *p*-nitrosophenol at mercury-plated electrode. The reaction scheme for ECE process is represented as

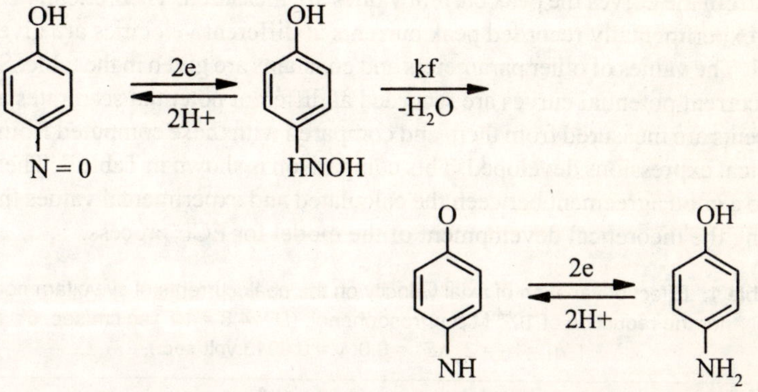

p-nitrosophenol was prepared and purified by the procedure described by earlier worker[16]. The supporting electrolyte consisted of 0.1 M acetic acid, 0.1 M potassium acetate, 0.1 M potassium nitrate and 20% by volume of alcohol. The pH of the solution was 4.9. A stock solution of 10^{-2}M of nitrosophenol was prepared. The solution of required concentration (i.e. 10^{-4} M) was prepared by dilution with the solution of supporting electrolyte. Current potential curves were recorded on an X-Y recorder attached to the potentiostat, with X-axis of the recorder serving as a voltage axis. The axial velocity of the solution flowing through the electrode varied from 0.5315 to 2.2282 cm/sec and potential scan rate varied from 0.0100 to 0.1000 volts/sec. The value of diffusion coefficient D and rate constant 'k_f' were taken as 4.8×10^{-6} cm^2 sec^{-1} and 0.60 sec^{-1} respectively[16].

RESULTS AND DISCUSSION

The theoretical expression for variation of total current measured at the electrolyte surface, which, intrinsically measured the variation of concentration with the linear sweep variation of potential are

$$i(t') = n_1 FA\Lambda\sqrt{k_f D\sigma}\, C_A^* \left[\chi(\sigma t') + \frac{n_2}{n_1} Q(\sigma t') \right] \qquad (34)$$

$$E(t') - E_{AB}^{\circ} = \frac{RT}{n_1 F}(u_1 - \sigma t') \qquad (35)$$

It is found that variation of u_1 affects the current potential voltammogram by shifting it along the potential axis only and there is no effect on the shape of the voltammogram as well as on peak or steady-state current values. So for further investigation, the value of u_1 is fixed at 8. The effects of two experimentally controllable parameters viz; axial velocity of the flow of electrolyte through the tubular electrode and potential scan rate are investigated on the peak currents theoretically and verified experimentally.

The current potential curves were recorded for the reduction of 10^{-4} M p-nitrosophenol at different axial velocities ranging from 0.5305 to 2.2282cm/sec and from the curves the peak current values are measured. Theoretically computed and experimentally recorded peak currents at different velocities are given in Table 1. The values of other parameters and constants are given in the table. Similarly the current potential curves are recorded at different potential scan rates and peak currents are measured from them and compared with those computed from the theoretical expressions developed. This comparison is shown in Table 2. There seems to be a good agreement between the calculated and experimental values thus, validating the theoretical development of the model for ECE process.

Table 1: Effect of variation of axial velocity on the peak currents of cv voltammogramms for the reduction of 10^{-4} M *p*-nrtrosophenol. ($D = 4.8 \times 10^{-6}$ sq cm/sec. $\delta = 0$, $n_1 = n_2 = 2$, $\Delta E° = 0.0$, $v = 0.0013$ volt/sec.).

Axial velocity va, cm/sec.	*Peak Current ip*	
	Theoretical	*Experimental*
0.5305	22.0477	22.0
1.0000	21.4230	21.5
1.0610	21.3801	21.4
1.2500	21.2753	21.3
1.5915	21.1936	21.2
1.6960	21.1809	21.2
2.0000	21.1649	21.2
2.2282	21.1674	21.2

Table 2: Effect of variation of scan rate on the peak currents of cv voltammogramms for the reduction of 10^{-4} M *p*-nitrosophenol. ($D = 4.8 \times 10^{-6}$ sq cm/sec. $k_f = 0.6$, $n_1 = n_2 = 2$, $\Delta E° = 0.0$, $v_a = 0.5305$ cm/sec.)

Axial velocity va, cm/sec.	*Peak Current ip*	
	Theoretical	*Experimental*
0.0100	22.0477	22.0
0.0200	21.4230	21.5
0.0500	21.3801	21.4
0.0750	21.2753	21.3
0.0900	21.1936	21.2
0.1000	21.1809	21.2

REFERENCES

1. Evans, C.A. Jr. and Morrison, G.H., *Anal. Chem.*, **40** (1968) 869.

2. Kabta, J. and Lazar, V.A., 'Instrumental methods of analysis of soil and plant tissues' (Ed. L.M. Walsh) *Soil Sci. Soc. America*. Madison, WS (1971) 67.
3. Huskin, L.A. and Ziege, K.E., 'Instrumental methods of analysis of soil and plant tissues' (Ed. L.M. Walsh) *Soil Sci. Soc. America*. Madison, WS (1971) 185.
4. Siros, J.C., *Analyst*, **87** (1962) 900.
5. Neeb, R., Methodicum chimicum., vol. 1: Analytical Methods, (Ed. F. Korte) Acad. Press. N.Y. (1974) 799.
6. Fater, F. and Monieu, H., *Z. Anal. Chem*., **204** (1964) 119.
7. Jonzyk, H.B. and Roezinki Panstwowego, *Zahlain Hig*., **16** (1965) 223.
8. Nangniot, P. and Dardenne, G. *Bull. Inst. Agron Stak. Resch, Gembloux*; **37** (1963) 116.
9. Nangniot, P. *Anal. Chem Acta*, **31** (1964) 166.
10. Ott, D.E. and Gunther, F.A. *Analyst*, **87** (1962) 70.
11. Singh, T., Dutt, J. and Kaur, J. S. *Electroanal. Chem*. **304** (1991) 17.
12. Singh, T., Singh, R.P., and Dutt, J., Proceedings of Conference 'Mathematics & its Applications in Engg. and Industry', (India) Dec. (1996) 529.
13. Singh, T., Singh, R. P. and Dutt, J., J. *Mathematical. Chem*. **17** (1995) 335.
14. Singh, T., Singh, R.P. and Dutt, J., *J. Mathematical. Chem*. **23** (1998) 297.
15. Singh, T., Singh, R.P. and Dutt, J., *Indian J. Pure & Applied Math*. **31**(4) (2000) 363.
16. Nicholson, R.S. and Shain, I. *Anal. Chem*. **37** (1965) 190.

2. [illegible] and [illegible], "The present state of methods of analysis of soil and plant [illegible]" (Ed. L.M. Walsh), Soil Sci. Soc. Am., [illegible] (1971) 67.
3. [illegible] and Zinga [illegible], [illegible] instrumental methods of analysis of soil and plant tissue (Ed. L.M. Walsh), Soil Sci. Soc. Am., Madison, WS, 19, [illegible] 1982.
4. [illegible], Analyst, 87 (1962) 990.
5. [illegible], Method [illegible] Analytical Methods (Ed. [illegible]) Series, [illegible] (1978) 104.
6. [illegible], Z. Anal. Chem., 204 (1964) 149.
7. [illegible], H.E. and [illegible], Punjab [illegible], 16 (1965) 223.
8. [illegible] and [illegible], Rev. Roum. Chim., [illegible] (196[illegible]) [illegible]
9. [illegible], Indian J. Chem., 11 (196[illegible]) 66.
10. [illegible], D.E. and Hamilton, P.A., [illegible]
11. [illegible], [illegible], Dutt, J. and [illegible], S. [illegible], 104 (1990) 27.
12. [illegible] and Dutt, J., Proceedings of conference [illegible] spectroscopy in science and industry, [illegible] (India) Dec. (1990) 349.
13. [illegible], Singh, R.P. and Dutt, J., [illegible] (1985) 33.
14. Singh, R., Singh, R.P. and Dutt, J., [illegible] Chem. [illegible]
15. Singh, [illegible], Singh, R.P. and Dutt, J., [illegible] & Applied [illegible], 34(4) (2000) [illegible]
16. Nicholson, R.S. and Shain, I., Anal. Chem., 37 (1965) [illegible]